NANOTECHNOLOGY SCIENCE AND TECHNOLOGY

# NANOFLUIDS

# RESEARCH, DEVELOPMENT AND APPLICATIONS

# NANOTECHNOLOGY SCIENCE AND TECHNOLOGY

Additional books in this series can be found on Nova's website under the Series tab.

Additional e-books in this series can be found on Nova's website under the e-book tab.

NANOTECHNOLOGY SCIENCE AND TECHNOLOGY

# NANOFLUIDS

## RESEARCH, DEVELOPMENT AND APPLICATIONS

YUWEN ZHANG
EDITOR

nova publishers
*New York*

**Library of Congress Cataloging-in-Publication Data**

Nanofluids (Nova Science Publishers)
Nanofluids : research, development and applications / [edited by] Yuwen Zhang (Department of Mechanical and Aerospace Engineering, University of Missouri, Columbia, MO, USA).
pages cm
Includes bibliographical references and index.
ISBN 978-1-62618-165-6 (hardcover)
1. Nanofluids. I. Zhang, Yuwen, 1965- editor of compilation. II. Title.
TJ853.4.M53N36 2013
620.1'06--dc23

2013005761

*Published by Nova Science Publishers, Inc. † New York*

# CONTENTS


| | | |
|---|---|---|
| **Preface** | | **vii** |
| **Chapter 1** | Critical Issues in Nanofluids Research and Application Potentials<br>*Milivoje M. Kostic* | **1** |
| **Chapter 2** | Nanofluids Heat Transfer: Preparation, Characterization and Theoretical Aspects<br>*Agus P. Sasmito, Saif A. Khan and Arun S. Mujumdar* | **55** |
| **Chapter 3** | Formulation and Thermophysical Properties of Nanofluids<br>*S. Suresh and M. Chandrasekar* | **91** |
| **Chapter 4** | Thermal Conductivity of Nanofluids and Heat Propagation Velocities<br>*Kenneth D. Kihm and Chan Hee Chon* | **119** |
| **Chapter 5** | Magnetic Nanofluids: Synthesis, Properties and Applications<br>*G. P. Singh and S. Ram* | **137** |
| **Chapter 6** | Application of Oxide-Water Nanofluids in Forced Convection Loops: Performance Evaluation<br>*Shijo Thomas and C. B. Sobhan* | **199** |
| **Chapter 7** | Heat Transfer Enhancement of Jet Impingement Cooling with Nanofluids<br>*Yimin Xuan, Qiang Li, Peng Tie and Feng Yu* | **243** |
| **Chapter 8** | Nanofluid Bubble Formation and Boiling Heat Transfer<br>*Saeid Vafaei* | **273** |
| **Chapter 9** | Nanofluid Oscillating Heat Pipe<br>*Hongbin Ma, Corey Wilson, Il Yoon and Yuwen Zhang* | **329** |
| **Index** | | **367** |

# PREFACE

This book presents current research related to the synthesis, characterization, and heat transfer of nanofluids. Nanofluids are stable colloidal suspensions of solid nanomaterials in base fluids. While nanoparticles were first added to base fluids to obtain nanofluids; other nanomaterials, like nanorods, nanotubes, nanowires, nanofibers, nanosheets, or other nanocomposites, are used to synthesize the nanofluids. The types of base fluids cover a wide range of liquids that include water, oil, ethylene-glycol (automotive antifreeze), refrigerants, polymer solutions, or even bio-fluids. The special properties of nanomaterials and their interactions with base fluids lead to substantially different properties of nanofluids compared with that of base fluids. Significant physical insights into complex physical phenomena in nanofluids are gained via the utilization of advanced theoretical tools and state-of-the-art experimental measurement techniques.

Chapter 1 – The objective of this chapter is to present critical research issues and application potentials of selected nanofluids, as related to the research accomplished and a number of hypotheses posed by different investigators – the goal is to shed some light on the subject as opposed to merely "generating heat." There are significant discrepancies among the experimental data available, as well as between the experimental findings and the theoretical model predictions, related to nanofluids thermo-physical characteristics. Oddly, there are more hypothetical theories proposed than reliable experimental results to verify those theoretical models. The nanofluids were hyped-up in the past, but it would be a mistake to hype-down nanofluids now and make premature judgments based on inconsistent and incomplete research to-date.

Regardless of ever-increasing number of research studies in this area, the basic research of convenience (repetition of similar experimentation with unstable nanofluids) remains in the initial stage; the promising nanofluids are still to be developed and results are still to be experimentally re-confirmed and established.

Development of many industrial and new technologies is limited by existing thermal management, and need for enhanced heat transfer and high-performance cooling. Nanofluids, stable colloidal mixtures of nanoparticles (including nanorods, nanofibers, nanotubes, nanosheets, and functional nanocomposites, even nano-droplets and nano-bubbles) in common fluids, have a potential to meet these and many other challenges. Colloidal nano-mixtures with functionally-stable and active-like nanostructures that may self-adjust to the process conditions, require systematic surface-chemistry study and enhancements (coatings with functional layers, surfactants, etc.), in addition to investigation of thermo-physical characteristics and phenomena.

A comprehensive, systematic and interdisciplinary experimental research program is necessary to study, understand and resolve critical issues in nanofluids research to date. The research must focus on both, synthesis of nanofluids and a careful exploration and optimization of thermo-physical characteristics. Development of new-hybrid, drag-reducing nanofluids may lead to enhanced flow and heat transfer characteristics. The nanoparticles in these fluids yield increased heat-transfer while the long-chain polymers are expected to enhance flow properties, including active and functional interactions with nanoparticles, thus providing potential for many applications yet to be developed and optimized.

Chapter 2 – Over the past decade, research in heat transfer enhancement using nanofluids – suspensions of nanometer-sized solid particles in a base liquid – has received considerable attention all over the world. However, the value of thermophysical properties of suspensions and heat transfer characteristics reported in literature by several authors are in disagreement. Several theoretical works to predict the mechanism of heat transfer enhancement, the effective thermophysical properties of the suspension as well as convective heat transfer, range from a homogeneous model to complex two-phase flow models have been proposed. Validation of theoretical studies with good experimental data is essential for proper understanding of the phenomena. This chapter provides insight into the recent development of nanofluids from preparation, characterization to theoretical aspects and ways to improve nanofluid stability so as to be able to commercialize and implement potential applications.

Chapter 3 – In this chapter, the formulation of nanofluids and experimental investigations on thermophysical properties together with the mechanisms proposed for the alteration in their values or characteristics due to the addition of nanoparticles are discussed. This chapter describes the two kinds of methods which have been frequently employed in producing nanofluids namely single-step method and the two-step method. Important thermophysical properties of nanofluids like thermal conductivity, viscosity, density, specific heat capacity and surface are also explained in this chapter.

Chapter 4 – The empirical correlation shows that the base fluid temperature plays an important role for effective thermal conductivity enhancement of nanofluids with decreasing nanoparticle sizes. Furthermore, it shows that the Reynolds number, function of temperature, has a more dominant role than other factors such as base fluid thermal conductivity, particle size, and diffusivity. Despite that the Brownian velocity of nanoparticles has been traditionally used for the Reynolds number in modeling, it is too slow to properly describe the effective thermal conductivy enhancement of nanofluids for many cases. The more comprehensive model based on a faster heat transfer velocity, in the same order as the speed of sound, agrees better with wide range of experimental data.

Chapter 5 – Innovative designs of different kinds of stable ferrofluids (magnetic or ferroelectric nanoparticles (NPs) dispersed in a suitable carrier liquid) are described in this chapter with well-known chemical synthesis routes, and basic physics underlying behind tailoring functional properties for medicals, biosensors and other applications. Common nanostructured flocculates described in this chapter include magnetic oxides such as $\gamma$-$Fe_2O_3$, $Fe_3O_4$ and spinel ferrites $AB_2O_4$ or perovskites $ABO_3$, A = a divalent cation and B = a tri or tetravalent cation in general. Surface modified NPs dispersed and immobilized in different polymer and ionic fluids give a road map of understanding the formation and properties of different ferrofluids with stable properties at elevated temperatures for applications. The most fascinating properties of a practically useful ferrofluid are its stability and rheology with self-adjustable magneto-viscosity and magnetic/ferroelectric features upon perturbations at moderate magnetic or electric fields. Applications of ferrofluids primarily depend on the basic properties such as ability to manipulate the parent features subject to an applied perturbation field (such as magnetic, electric, optical, etc.), rheology, absorption capacity of electromagnetic energy at selective frequencies and heating-up, thermal conductivity, magnetocalory, biological activities, ferroelectrics, light-energy transfer, magnetic switching, and other perturbation sensitive features. Those cover diversified fields such as chemical and biochemical separations, magnetic target drug delivery, magnetic resonance imaging contrast agents, hyperthermia therapy, disease diagnosis, chromatography, electrochemical biosensors, optical detection and luminescence detection, heat sinker and dampers, multicolored pigments, magnetic switches and valves, phosphors, and many more in the time to come in the near future.

Chapter 6 – Nanofluids, as alternate heat transfer fluids, have gained enormous attention from research laboratories and industries in recent times. The thermophysical properties and the overall convective heat transfer performance of nanofluids have been investigated extensively. This chapter aims to review the methodologies adopted for quantification of the forced convection behavior of oxide-based nanofluids. The factors that lead to the enhancement of heat transfer in nanofluids are also discussed. The influence of the two major thermophysical properties, the thermal conductivity and the dynamic viscosity are discussed. The effects of parameters like inertia, Brownian diffusion, thermophoresis, diffusiophoresis, Magnus effect, fluid drainage and gravity are also brought out. Apart from the literature review, results from experiments at the authors' laboratory on forced convection experiments at high pressure loops are also presented, which show a moderate increase in heat transfer coefficient at pressures as high as 5 bar.

Chapter 7 – Jet impingement cooling is a vital technique for thermal management of electronic devices of high heat flux by impinging fluid on a heater surface due to its high local heat transfer rates. One may find plenty of publications on the heat transfer characteristics of jet impingement cooling technique such as the distance between the jet outlet and the heated surface, the magnitude of the jet velocities, turbulence intensity, the angle of impingement, the array of jets and the roughness of the heater surface. But little attention is paid to the effect of the thermal properties of the impingement fluid, which may be another way to further improve the heat transfer performance of the jet impingement. With development of enhanced heat transfer techniques in conventional systems, the conventional working fluid with low thermal conductivity turns into a bottleneck factor for promoting the performance effectiveness of thermal systems. Improvement of the thermal properties of impinging fluids may become a promising trick of augmenting heat transfer process in the jet impingement cooling system.

Recent advances in the nanotechnology have permitted the creation of a new and special class of fluids, named 'nanofluids'. The thermal conductivity and heat transfer performance of nanofluids are superior to those of the original pure carrier fluids because the suspended nanoparticles remarkably improve energy transport capability of the suspensions. Nanofluid offers the exciting possibility of increased heat transferability of jet impingement.

In this current study, experiments were conducted to investigate the heat transfer enhancement of the nanofluid in the impinging jet cooling system. Two types of nanofluids were used as the working fluid. Cu-water nanofluid was studied in submerged single jet impingement cooling while Cu-ethylene glycol/water nanofluid was introduced into free surface jet array impingement cooling, respectively. In submerged single jet impingement, nanofluids with volume fraction ranged from 1.5% to 3.0% were studied for different jet-to-target spans. Experiments have revealed that the suspended nanoparticles increase the heat transfer performance of the base liquid in the jet impingement cooling system. It had been found that an enhancement of 18.5% compared with the base fluid at the Reynolds number of 8000 when nanofluids are used. In addition, the experiments have shown that the suspended nanoparticles almost result in no extra addition of pressure drop in both submerged single jet and free surface jet array impingement.

Chapter 8 – The purpose of this chapter is to demonstrate the effects of nanoparticles on boiling heat transfer coefficient and heat flux. Effects of nanoparticles on force balance at the triple line, behavior of triple line, wettability, roughness, active nucleation site density, initiation of nucleation, dynamics of bubble growth, bubble departure, advancing and receding contact angles, wetting and rewetting, boiling heat transfer coefficient and critical heat flux will be discussed in this chapter.

Nanoparticles have two main roles on boiling heat transfer phenomenon, a) suspended nanoparticles modify the physical properties of nanofluids; b) deposited nanoparticles modify the characteristics of heated substrate. Behavior of the triple line might be changed by deposited nanoparticles as well as the suspended nanoparticles inside nanofluids. This chapter attempts to explore and discuss the effects of both mechanisms on boiling heat transfer phenomenon in details.

Droplet and bubble formation methods are developed to investigate the effect of nanoparticles on force balance and behavior of triple. In case of droplet method, gravity is in favor of the expansion of triple line and reduction of contact angle which will reduce the accuracy of evaluation of effects of nanoparticles on behavior of triple line. Droplet contact angle changes with volume, so droplet method can not be a unique method to study the effects of nanoparticles on behavior of triple line including surface wettability, thus asymptotic contact angle will be discussed to investigate the behavior of triple line, regardless of gravity.

In case of bubble growth method, gravity affects as buoyancy force and lifts the bubble upward, so triple line can move more freely and effect of nanoparticles on triple line could be noted easier. Formation of bubbles inside nanofluids has a great capacity to express effects of nanoparticles on behavior of triple line, bubble growth, waiting time, bubble formation time, bubble frequency, departure process, receding and advancing dynamic contact angles, wetting and rewetting. Indeed, effects of nanoparticles depend on concentration, and characteristics of nanoparticles such as shape, size, coating and material.

Chapter 9 – An Oscillating Heat Pipe (OHP) is a heat transfer device that can be utilized to transfer a large amount of heat from heating to cooling sections. A typical OHP is fabricated from a long capillary tube bent into many turns with an evaporator and a condenser located at two opposite sides, and an adiabatic region between the evaporator and condenser. In order to form a train of vapor bubbles in an OHP, the liquid and vapor must not be able to stratify in the channel. When nanoparticles are added in the base fluid in an OHP, the OHP effectively utilizes 1) the oscillating motion in the OHP to suspend nanoparticles; 2) the higher heat capacity of nanofluid; 3) the higher thermal conductivity; 4) the nanoparticle effect on the fluid field; and 5) the nanoparticle effect on the surface wetting conditions. The unique feature of the oscillating motion fully utilizes the nanofluid features to significantly enhance the heat transport capability in an OHP.

In the fabrication process of OHPs, preparation, charging the OHP with nanofluids and sealing it are inevitable and requiring careful attention for successful testing. Nanofluid was first used as the working fluid in a 12-turn OHP by the Center for Thermal Management at the University of Missouri in 2006. The results of using nanofluids showed a significant improvement over results which used the traditional base fluid of water as the working fluid. The improvement was especially significant at low power; however, even at high power the improvement was measurable. The parameters affecting heat transfer performance were systematically investigated by researchers around the world. It was reported that the thermal resistance for the $Al_2O_3$-water nanofluid was lower than that of pure water, while the thermal performance of $SiO_2$-water nanofluid was worse than pure water. There is an optimal nanofluid concentration for the OHP. It is also worth noting that the effect of nanoparticle deposition on thin-film evaporation has not been investigated and might play a significant role in the thermal performance of the nanofluid OHP.

In: Nanofluids
Editor: Yuwen Zhang
ISBN: 978-1-62618-165-6


*Chapter 1*

# CRITICAL ISSUES IN NANOFLUIDS RESEARCH AND APPLICATION POTENTIALS

*Milivoje M. Kostic*[*]
Department of Mechanical Engineering, Northern Illinois University
DeKalb, IL, US

## ABSTRACT

The objective of this chapter is to present critical research issues and application potentials of selected nanofluids, as related to the research accomplished and a number of hypotheses posed by different investigators – the goal is to shed some light on the subject as opposed to merely "generating heat." There are significant discrepancies among the experimental data available, as well as between the experimental findings and the theoretical model predictions, related to nanofluids thermo-physical characteristics. Oddly, there are more hypothetical theories proposed than reliable experimental results to verify those theoretical models. The nanofluids were hyped-up in the past, but it would be a mistake to hype-down nanofluids now and make premature judgments based on inconsistent and incomplete research to-date.

Regardless of ever-increasing number of research studies in this area, the basic research of convenience (repetition of similar experimentation with unstable nanofluids) remains in the initial stage; the promising nanofluids are still to be developed and results are still to be experimentally re-confirmed and established.

Development of many industrial and new technologies is limited by existing thermal management, and need for enhanced heat transfer and high-performance cooling. Nanofluids, stable colloidal mixtures of nanoparticles (including nanorods, nanofibers, nanotubes, nanosheets, and functional nanocomposites, even nano-droplets and nano-bubbles) in common fluids, have a potential to meet these and many other challenges. Colloidal nano-mixtures with functionally-stable and active-like nanostructures that may self-adjust to the process conditions, require systematic surface-chemistry study and enhancements (coatings with functional layers, surfactants, etc.), in addition to investigation of thermo-physical characteristics and phenomena.

[*] Email: kostic@niu.edu.

A comprehensive, systematic and interdisciplinary experimental research program is necessary to study, understand and resolve critical issues in nanofluids research to date. The research must focus on both, synthesis of nanofluids and a careful exploration and optimization of thermo-physical characteristics. Development of new-hybrid, drag-reducing nanofluids may lead to enhanced flow and heat transfer characteristics. The nanoparticles in these fluids yield increased heat-transfer while the long-chain polymers are expected to enhance flow properties, including active and functional interactions with nanoparticles, thus providing potential for many applications yet to be developed and optimized.

# 1. Introduction and Critical Issues

Regardless of ever expending nanofluid research effort, comprising many inconsistent experimental results and unproven hypothetical theories, followed by growing number of related publications, including a number of review articles, the state of the nanofluid research is still inconsistent and often conflicting, thus incomplete and inconclusive. Many challenges are to be resolved and unforeseen opportunities are to be pursuit in the future. The nanofluids were hyped-up in the past, but it will be a mistake to hype-down nanofluids now and make premature judgments based on limited research and inconclusive results.

It is not objective here to review the nanofluids' related literature, since it has been done in a number of publications. Many reviews on nanofluid research are provided [1-6], and more recently [7-9]. Therefore, the objective here is to present this author's views on selected nanofluids' critical research issues and application potentials, as related to the research accomplished and a number of hypothesis posed by different investigators – the goal is to shed some light on the subject as opposed to merely 'generating heat'.

Development of nanofluids, nano-technology based heat-transfer and other fluids, i.e., suspensions of different nano-materials in common and novel, base fluids (with rather complex structures and interactions), is a new challenge but also unforceen opportunity. It may open the road for development of many, complex nanofluids (including organic nanofluids) with diverse additives (including known surfactants, interfacial surface enhancers and other polymers), with many and unprecedented applications in existing critical areas as well as emerging and novel applications. The trend in further development of nanomaterials is to make them multifunctional and controllable by external means or by local environment, thus essentially turning them into useful nano-devices.

Nanofluids are stable colloidal suspensions of nano-materials (nanoparticles, nanorods, nanotubes, nanowires, nanofibers, nanosheets, other nanocomposites, or even nano-droplets and nano-bubbles) in common, base fluids, such as water, oil, ethylene-glycol mixtures (antifreeze), refrigerants, heat transfer fluids, polymer solutions, bio-fluids, and others. Nanoparticles are very small, nanometer-sized particles with their smallest dimension usually less than 100 nm (nanometers). The smallest nanoparticles, only a few nanometers in diameter, may contain a few thousand atoms. These nanoparticles can possess properties that are substantially different from their parent materials, and they may interact quite differently within their dynamic molecular structure with the base fluids, than the corresponding microparticles, and respond differently within different force-flux processes accompanied with mass-energy transfers. Similarly, nanofluids may have properties that are substantially different from their base fluids, like much higher thermal conductivity, and other flow and heat transfer characteristics.

Since Choi coined the term "nanofluids" [10] for carbon and metal-based nanoparticles in common heat-transfer fluids, research has intensified, due to the substantially increased thermal conductivity of those nanofluids, and the tremendous potential for many applications. Biologists Turner et al. [11] and physicists Pozar and Gubbins [12] have used the term nanofluids to describe bio-nanoparticles, like DNA and other protein molecules in aqueous solutions, or for fluids confined in slit nano-pores or other nano-meter sized enclosures.

Argonne National Laboratory is recognized for pioneering scientific activities in nanofluids research, including innovative production methods, thermal characterization, and theoretical studies that correlate enhanced thermal conductivity with static and dynamic mechanisms between nanoparticles and base-fluid molecular interlayers. At many other institutions around the world, including some industrial companies and collaborative associations, is underway [6, 13]. Number of publication rate increased considerably in recent years, from several per year rate in 1990's to over hundred per year recently, with several hundreds of archived publications so far [in *Science Citation Index* journals].

Regardless of accumulated research outcomes, many results are incomplete, some conflicting and without conclusive results. The state of the art in nanofluid research is still in initial phases, in part due to rather complex nature of nanoparticle materials and even more complex nanoparticle-base fluid interactions, often involving diverse surfactants and stabilization additives to prevent inherited tendency of nanoparticle conglomeration and clustering in the unstable colloidal mixture. The inherited complexity and inconsistency make it almost impossible to establish well-defined and reliable experimental results to verify existing hypotheses and to improve and develop new ones. There are many other issues to be addressed and resolved. It is hypothesized by this author that nanofluid thermal conductivity may be a function of temperature gradient (thus heat flux), in addition to temperature level dependence, the way non-Newtonian fluid viscosity is dependent on shearing rate, i.e., velocity gradient (thus flow rate).

Even a so-called "benchmark study" [14] is based on very limited and unfortunate choice of types of nanofluids, thus with rather limiting results; therefore, it may produce a disservice to the future research in this important area with unprecedented potentials. It may mislead (since "benchmark study" title) that the classical mixture theory predicts thermal conductivity of (most) nanofluids. It would be surprising that simple static mixture-theory (that includes only volume fraction ratio) will predict complex dynamic interactions of diverse constitutions, size and shapes of nanoparticles in base fluids. Mostly alumina nanoparticles (conveniently obtained) were tested and not those known to demonstrate anomalous TC enhancements, like metallic and CNT nanoparticles with relevant concentrations. Why such nanofluids were not reproduced by the participating institutions in the benchmark study?

Regardless of many publications of similar research, in essence, the nanofluids research is still in initial phase, when the present limited research (repetition of convenience) should substantially expend in type and scope, not to mention many other and yet to be discovered/engineered functional nanoparticles and innovative additives to enhance and optimize nanofluids properties and flow characteristics. The issue, in this author's opinion, is to discover or reproduce the nanofluids which show enhanced properties. Researching different nanoparticles with different additives in different base fluids using difference synthesis methods is a challenge, but (exactly because of it) also opportunity with many application potentials! That what the future research should be focusing on.

There are many important variables and issues related to making and using nanofluids, which is confirmed by significant discrepancy in the reported experimental data. Type of nanoparticles, their size, shape and distribution are important but not easily measured and usually not well-defined nor properly reported in the publications. Type of base fluids used, method of nanofluid production, use of surfactants and stabilization additives, including pH adjusters, etc. Conglomeration and clustering of nanoparticles in the nanofluid mixture is taking place before and after the nanofluid is made and during its use, and depends on many factors, especially on additives used. Two nanofluid samples with all of the parameters being the same but different type and amount of surfactants and/or pH adjusters used may result in quite different thermo-physical properties and flow and heat-transfer characteristics of apparently the same nanofluids. These and other unknown factors may explain anomalous and controversial results obtained by different researches. Furthermore, ultrasonic vibration is commonly utilized to enhance dispersion and breaking the clusters of nanoparticles. It is obvious that the duration and the intensity of the ultrasonication will affect the dispersion characteristics, however, the clusters will form again and their size will increase in time after ultrasonication [15]. Therefore, by using apparently the same samples, different results could be obtained just by varying the time between the ultrasonication and measurement of nanofluid characteristics.

In the advanced electronics and new emerging industries there is a great need for efficient thermal management and cooling. In many other industries such as energy production and utilization, manufacturing, transportation and commercial and residential buildings, thermal management is critical and nanofluids could yield significant benefits.

## 2. Production of Nanofluids and Challenges for Commercialization: Improved One-Step Method

There is a number of methods used to manufacture nano-materials, referred often as nanoparticles, and for production of nanofluids. Nanoparticles of various materials and forms have been produced by physical or chemical synthesis techniques. Typical physical methods include the mechanical grinding method and the inert-gas condensation technique [16]. Chemical methods for producing nanoparticles include chemical precipitation, chemical vapor deposition, micro-emulsions, spray pyrolysis, and thermal spraying. The specific processes for making metal nanoparticles include mechanical milling, inert-gas-condensation technique, chemical precipitation, spray pyrolysis, and thermal spraying. The production methods are reviewed by several researchers, including [6] and more recently by [9, 17].

Strong temperature dependence of thermal conductivity of nanofluids is very important and has potential to expend the possible application areas of nanofluids. Additional important issues during application of nanofluids are flow erosion and settling. Before commercialization of nanofluids, possible problems associated with these issues should be investigated and resolved. It should also be noted that, customary increase in viscosity of nanofluids over the base fluids is an important drawback due to the associated increase in pumping power. Therefore, further experimental research is required in that area in order to improve flow properties and to determine the feasibility of nanofluids.

Nanofluids of various qualities have been produced mainly in small volumes and for research purposes, but large-scale production at low cost of well-dispersed, stable nanofluids is required for commercial applications. The lack of large-scale production is a major barrier to testing and use of nanofluids in the transportation, industry and other applications.

Most of the nanofluids used in research so far are produced by a two-step process. First, nanoparticles are produced as a dry powder, typically by inert gas condensation [16]. The second step involves dispersion of dry nanoparticle powder into a base fluid, like water, oil or ethylene-glycol. An advantage of the two-step process is that the inert-gas condensation technique has been scaled up to commercial nano-powder production [18]. A deficiency of this method is the tendency of nano-powders to agglomerate during storage and dispersion in the base fluids, particularly with heavier metallic nanoparticles. Surfactants and other surface-stabilization additives can be used to achieve more homogeneous and more stable suspensions, however they may influence the nanofluids properties. In addition to mechanical mixing, ultra-sonic mixers are used to break up agglomerates and give more uniform dispersions. In general, although the process works well for some oxides, it has not been able to yield metallic nanofluids with substantially enhanced thermal conductivity.

By contrast, the one-step physical and chemical methods have potential to produce better quality nanofluids. For example, a physical one-step method consists of a direct-evaporation process, involving nanoparticle source evaporation and direct condensation and dispersion onto the base fluid in a single step. This method has been developed by Yatsuya et al. [19] and improved by Wagner et al. [20]. The one-step method has been employed by Choi and Eastman [21] in Argonne National Laboratory (ANL) and successfully used to produce nanofluids with very small copper nanoparticles (about 10 nm) and exceptionally high thermal conductivity [22]. However, the one-step process is intrinsically more difficult to reproduce, since particles cannot be characterized and pre-sorted before addition to the fluid. Consequently, the ANL method, although an excellent idea, needed to be substantially improved in order to yield improved control of nanoparticle sizes and sustained nanofluid production. The improvement of the Argonne one-step method for nanofluid production has been realized and a U.S. patent has been issued recently (Kostic et al. [23]). Details of the improvement will be presented below.

## 2.1. Improvement of Physical One-Step Production Method

For nanofluids with high-conductivity metals such as copper, a single-step technique is preferable to the two-step process to enhance dispersion and prevent oxidation of the particles. Argonne National Laboratory developed a one-step physical method for creating nanofluids, by nanoparticles being formed and dispersed in a fluid in a single process. This patented single step method involves direct evaporation and has been used to produce non-agglomerating copper nanoparticles that remain uniformly dispersed and stably suspended in ethylene glycol [21]. The technique consists of condensing nanoparticles from the vapor phase directly into a flowing low-vapor-pressure ethylene glycol or oil in a vacuum chamber. The well-dispersed nanofluids of Cu in ethylene glycol enhance the thermal conductivity of the base fluid by up to 40% at the particle volume concentration of 0.3 %, significantly larger than the prediction of effective medium theory [22]. Although the one-step physical method has produced nanofluids in small quantities for research purposes, it has a number of deficiencies, and needed to be improved, since lack of consistent nanofluid production limits the progress of the future research in this and related areas.

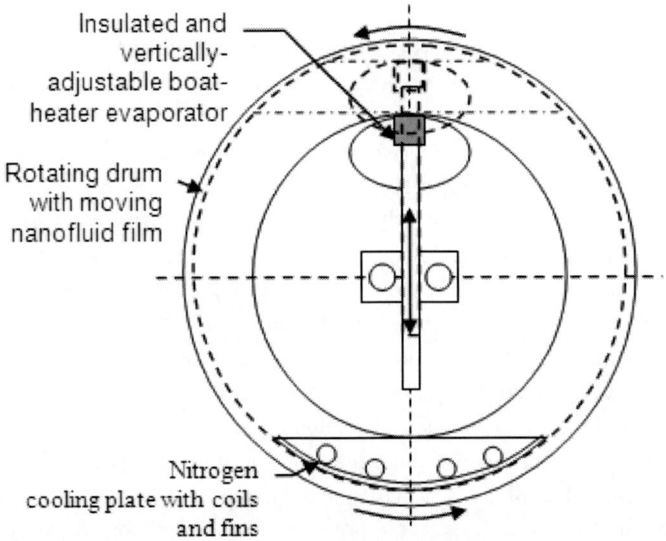


Figure 1. Improved new-design for the one-step, direct-evaporation nanofluid production apparatus (U.S. Patent No. 7,718,033) [23].

An improved "One-step method for the production of nanofluids" has been developed and a U.S. patent has been issued recently [23], see also Figure 1. The improved method and system is provided for producing nanofluids by a so called one-step direct metal evaporation and its deposition on a moving liquid film and further dispersion and mixing of nanoparticles within the fluid.

The improved method and system is achieved by the following: better positioning (longitudinal instead of crosswise) and variable-adjustable distance between the metal evaporation source and liquid film, which achieves smaller particle deposition path and smaller fluid-film exposure over the heated source and thus smaller nanoparticle size, but at the same time it provides larger liquid film area and thus larger deposition rate. Furthermore, better heater evaporation source insulation reduces the heating power which is possible to be balanced by an improved new heat-exchanger design of nitrogen cooler and increased drum rotation, thus providing lower liquid temperature and pressure which contributes to smaller nanoparticle size, and provides for continuous, steady-state production of nanofluids with desired nanoparticle size distribution. Additional improvement features include a liquid feed-in, inert gas flashing, visual observation, and better process heating control all of which further contribute to continuous, steady-state operation and control of temperature and pressure for production and optimization of desired nanofluid qualities and quantities.

The improved one-step process and system for production of nanofluids includes placing a base fluid, such as ethylene glycol or oil, in a rotating cylindrical drum situated in a vacuum chamber. The rotating axis of the drum is preferably horizontal, and the drum with solid backendplate contains annular front-endplate to prevent the liquid from running out but providing entrance for metal evaporator heater and nitrogen cooler elements. An electric motor rotates the drum at a designated rotational speed. As the drum rotates, it drags liquid filled in the bottom part of the drum along its inside cylindrical surface, forming a thin film of liquid on the inner surface above the pool of liquid in the bottom of the drum. An evaporation boat heater is positioned in close proximity to the upper inside surface of the cylindrical drum. The metal evaporates at a given rate and the gaseous molecules rise outwards and condense onto the liquid film on the revolving drum. The liquid is cooled by inserting a metallic heat exchanger in the liquid and passing liquid nitrogen through the exchanger. In a steady-state process, the cooling capacity of the heat exchanger balances the heat input from the evaporator and heat gains from the surroundings.

In a most preferred form of the invention, the method and system includes: (a) strategic positioning of the boat-heater that evaporates the metal close to the moving liquid film in the axial drum direction, instead of perpendicular to the drum axis, and with adjustable spacing of the evaporator relative to the liquid film; (b) a boat/heater that is thermally well insulated with foam, foil and radiation shields; (c) better liquid cooling by substantially increasing drum rotational speed, improving the design of the heat exchanger that cools the base liquid with liquid nitrogen flow, including optimized cooler-to-drum gap and fin spacing, and addition of cooling fins to the rotating drum if needed; and (d) well controlled, adjustable evaporator temperature and balanced cooling to provide for desired nanoparticle size distribution in a steady-state process, for controlled and continuous nanoparticle production.

This improved process and system allows for much better process control with the resulting improvement in the quality and quantity of the nanofluids produced by the process. The system improvements, such as placement of the boat/heater evaporator closer to the top of the rotating drum, increasing the drum rotation speed, and use of fins to increase surface area for cooling give rise to numerous advantages. These and other improvements are described in much greater detail elsewhere [23].

## 3. Nanofluids Stability and Influence of Additives: Friction-Drag Reduction and Issue of Heat-Transfer Reduction (Development of Polymer-Nanofluids)

As already stated, the conglomeration and clustering of nanoparticles in the nanofluid mixture is taking place before and after the nanofluid is made and during its use, and depends on many factors, especially on additives used. Two nanofluid samples with all of the parameters being the same but different type and amount of surfactants and/or pH adjusters used, may result in quite different thermo-physical properties and flow and heat-transfer characteristics of apparently the same nanofluids. These and other unknown factors may explain anomalous and controversial results obtained by different researches.

The particle size distribution of nanoparticles is another important factor and reporting the average particle size is not sufficient to characterize a nanofluid. It is also known that cylindrical and rod-shaped particles offer higher thermal enhancement when compared to spherical particles. More consistent research should be made for the investigation of the properties and thermal performance of well-characterized 'static' and dynamic constitution of nanofluids, since there are more hypothetical theories proposed than reliable results to verify those theoretical models.

In addition to investigating the effects of basic parameters such as particle size, base fluid, and particle volume fraction, the researchers should well control and report all important parameters, including additives used, details of ultrasonic treatment, and others. No wonder that effects of many parameters on the thermal conductivity and viscosity of nanofluids has not been fully understood yet, which are important prerequisites for better understanding, development and optimization of nanofluid flow and thermal performance.

Another important, but overlooked issue with use of additives in nanofluids is a possibility to drastically reduce friction drug in turbulent flow (desired effect; used in Trans-Alaska oil pipeline system), which is usually accompanied with commensurate heat-transfer reduction, the letter being an undesirable, adverse effect. These drag and heat-transfer reduction phenomena, described next, are well-known by some researchers but are not widely known and thus overlooked by nanofluid researchers.

### 3.1. Friction-Drag Reduction and Issue of Heat-Transfer Reduction -- Development of Polymer-Nanofluids

A class of fluids, known as "drag-reduction" fluids, has intrigued many investigators, ever since Toms' discovery [24] that the friction drag of common fluids with minute concentrations of certain polymer additives, under turbulent flow conditions, is considerably smaller (several times) than the expected values. Drag-reduction fluids are also nanofluids, i.e. solutions of minute concentrations of certain nano-size (sometimes micro-size) additives, like high-polymers, soap and surfactant aggregates, or fibers, in common fluids, like water or oil. The pressure drop and heat-transfer in turbulent pipe with drag-reducing fluids (classified as non-Newtonian fluids if with higher concentration of additives) is several times lower than for the corresponding Newtonian fluids as discussed in an extensive review by Metzner [25]. These phenomena can be characterized by the so-called Virk's minimum asymptotic friction value [26]. Unfortunately, heat-transfer is also reduced. Hartnett and Kostic [27] have studded the drag-reducing fluids in laminar non-circular duct flows and discovered enhanced heat transfer. They reviewed the flow and heat transfer phenomena of Newtonian and non-Newtonian fluids in rectangular ducts [27], and Kostic [28] presented an overview on turbulent drag and heat transfer reduction and laminar heat transfer enhancement of certain non-Newtonian fluids in non-circular duct flows.

Complex fluids, like polymer solutions have functional, process-adjusting, active molecular-structures, and exhibit well-known shear-rate dependent viscosity, see Figure 2. Randomly-oriented, long-chain macro-molecules increase substantially the zero-shear-rate (zero velocity) solution viscosity, but under shearing stresses, they self-align with the flow and viscosity is substantially reduced with the shear-rate increase. This is more dramatically demonstrated in turbulent flow, where a minute concentration (50 *ppm*) of long-chain macromolecules may reduce turbulent friction fivefold by suppressing transverse turbulence fluctuations, and in turn substantially reduce turbulence dissipation and over-all friction, see Figure 3 [28]. Regrettably, the turbulent heat transfer is commensurately reduced. Similarly, the nanoparticles could re-align and process-adjust during flow and heat-transfer processes. There is a need to selectively choose or develop new nanoparticles and drag-reduction additives to optimize flow and heat transfer characteristics of hybrid nanoparticle+additive based nanofluids.

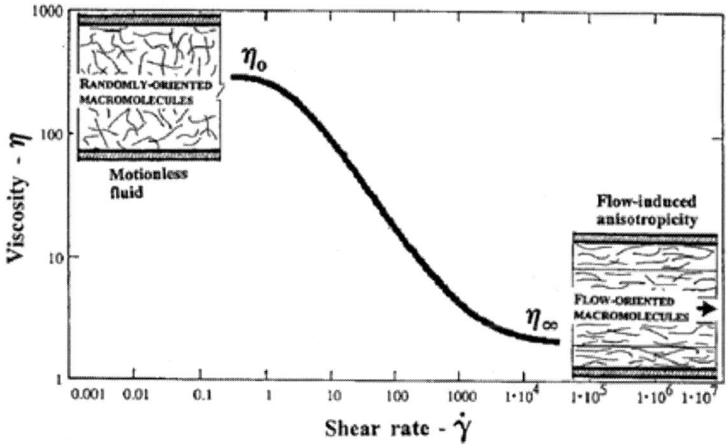


Figure 2. Shear-rate dependent viscosity due to process-adjusting active structure of an aqueous polymer solution [28].

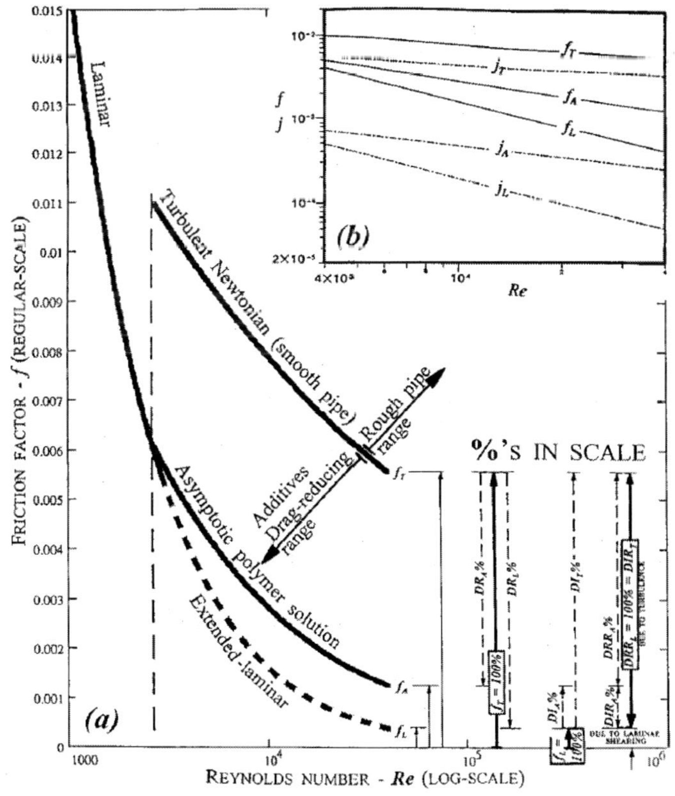


Figure 3. Friction-factor vs. Reynolds-number curves for laminar, turbulent, and polymer drag-reducing asymptotic flows (in semi-log scale as opposed to usual log-log scale) [28].

The polymer-nanofluids, could be developed, and are expected to be even more complex and active fluid mixtures, and thus have more degrees-of-freedom to 'self-adjust' under different process- and/or field-conditions. To develop, study, understand, and optimize polymer-nanofluid functionalities, by reducing instabilities while promoting those fluid structural activities that enhance flow and heat transfer characteristics as well as other characteristics, will be a research challenge and potential for many applications.

Another research challenge could be to combine the selected nanoparticles and drag-reducing additives, and thus develop and optimize a Drag-Reducing nanofluid (dubbed here DR-nanofluid). The expectation is to obtain a hybrid nanofluid with improved flow and heat transfer characteristics. The drag-reducing fluids may contain water, different co-solvents, lubricants, co-polymer emulsions, dispersants and surfactants, rheological agents, surface energy controlled agents, and pH and ionic strength adjusting agents. The formulation principle and techniques for drag-reducing fluids and nanofluids should be similar to those formulations of water-based organic-inorganic hybrid emulsions. Different techniques could be developed and employed to synthesize and formulate the nanofluids with polymer additives.

Furthermore, developing nanofluids with polymer additives or "POLY-nanofluids" may have other applications, beyond the advanced heat-transfer fluids, for creating more functional nanostructures, since polymer molecules may provide an enhanced web-like structure for nanoparticles in base fluids. Development of nanofluids, with enhanced or entirely different properties from their base fluids, is a new challenge and opportunity, and may have unprecedented application potentials, not only in thermal management and efficient cooling, but also in industry, buildings, environmental and bio-medical applications, as well as emerging critical applications.

## 4. Thermal Conductivity Measurements: Possible Issues with Nanofluids

Thermal conductivity measurements of suspensions of nanoparticles in common, base fluids conducted in Argonne national Laboratory almost two decade ago, discovered "anomalous" enhancement of thermal conductivity, compared to the suspensions of the same type and concentration but larger-size particles, and as "expected" values, predicted with the heterogeneous mixture theory. The "*nanofluid*" name was introduced to represent such suspension of nano-particles in a common fluid. Exceptional enhancement of nanofluids thermal conductivity and potential for development of substantially improved heat transfer fluids, have propelled research and publications, almost exponentially. A number of review articles have been published [1-6], and more recent extensive experimental investigation [29].

Most of the measured results, reviewed in the above references, show significant enhancement of nanofluids thermal conductivity (TC), but some with large discrepancy in the values, while some others without significant enhancement over predictions of classical mixture theory [30]. Usually, three measurement methods have been used: steady-state method; transient oscillation method, and in most cases the transient Hot-Wire Thermal Conductivity (HWTC) method. In addition to the enhancement of nanofluids TC with nanoparticle concentration, a strong dependence of TC on temperature is observed. Since the measured nanofluids TC enhancement has not been understood and justified yet, there is a need to investigate influence of measurement methods on the obtained results, since all apparatus calibration have been performed with classical fluids with known TC. However, there may be some unknown issues with measurements of complex and dynamic (usually unstable) nanofluid structures. For example, influence of electromagnetic fields on nano-particle fluid molecule interactions, possible nano-convection effects related to measurement methods, or influence of temperature gradients on TC results.

Therefore, in this section, only a couple of selected issues will be covered, namely, a proposed improvement of a Hot-Wire Thermal Conductivity (HWTC) apparatus, and a design and comparative measurement using steady-state Parallel-Plate Thermal Conductivity (PPTC) apparatus. Both are described in details to facilitate their replication and possible future improvements.

A new and improved HWTC apparatus for thermal conductivity measurements of common fluids and nanofluids has been recently developed, designed and fabricated [31, 32]. A platinum (Pt) wire of 50.8 $\mu m$ diameter with a Teflon insulation coating of 25.4 $\mu m$ thickness was used as the hot-wire heater and temperature sensor for the present application. The new apparatus employs innovative solutions for easy calibration of uniform Pt-wire tension and thus minimizing the strain influence on temperature measurement (i.e., minimizing the well-known and unwanted "strain-gage effect" on Pt-wire electrical resistivity); measurement of Pt-wire voltage drop independently from power wiring (four wires); and an effective off-centered mechanical design to minimize the test fluid sample size (about 35 *mL*), but at the same time providing additional space for wiring, including three inside thermocouples for fluid temperature uniformity verification. Data acquisition hardware and LabVIEW® application software are optimized to minimize signal noise and enhance acquisition and processing of useful data.

In addition, a steady-state, parallel-plate thermal conductivity (PPTC) apparatus has been developed and used for comparative measurements of complex POLY-nanofluids [33, 34], in order to compare results with the corresponding measurements using the transient, hot-wire thermal conductivity (HWTC) apparatus. The related measurements in the literature, mostly with HWTC method, have been inconsistent and with measured thermal conductivities far beyond prediction using the well-known mixture theory [30]. The objective was to check out if existing and well-established HWTC method might have some unknown issues while measuring TC of complex nano-mixture suspensions, like electro-magnetic phenomena, undetectable hot-wire vibrations, and others.

These initial and limited measurements have shown considerable difference between the two methods, where the TC enhancements measured with PPTC apparatus were about three times smaller than with HWTC apparatus, the former data being much closer to the mixture theory prediction. However, the influence of measurement method is not conclusive since it has been observed that the complex nano-mixture suspensions were very unstable during the lengthy steady-state measurements as compared to rather quick transient HWTC method. The nanofluid suspension instability might be the main reason for very inconsistent results in the literature. It is necessary to expend investigation with more stable nano-mixture suspensions.

### 4.1. Improved, Transient Hot-Wire Thermal Conductivity (HWTC) Apparatus for Nanofluids

A new and improved, transient hot-wire thermal conductivity apparatus has been developed to measure the thermal conductivity of fluids, polymer solution, nanofluids and poly-nanofluids (a mixture of nano particles, polymers and conventional heat transfer fluids) [31, 32]. The apparatus is a part of a research program at Northern Illinois University with an objective to resolve some critical issues in nanofluids research and to develop and optimize new hybrid, drag-reducing polymer-nanofluids with enhanced thermo-physical characteristics [36, 37]. Thermal conductivity, a measure of material's ability to conduct heat, is a very important property for thermal analysis.

The mathematical model for the hot-wire method is based on an ideal, infinitely long and thin continuous line source dissipating heat, of heat flux $q$ per unit length, applied at time t = 0, in an infinite and incompressible medium. The general assumption is that heat transfer to the infinite medium, of thermal conductivity $k_f$ and thermal diffusivity $\alpha_f = k_f / \rho_f C_f$, is by conduction alone and thus increases the both temperatures in time, of the heat-source and test-medium. It is also assumed that the line heat-source has uniform instant temperature everywhere, but transient in time (virtually achieved with small diameter and long wire with large thermal conductivity and/or small heat capacity). The governing equation is derived from the Fourier's equation for one-dimensional (1-D) transient heat conduction in cylindrical coordinates,

$$\frac{1}{\alpha_f}\frac{\partial T}{\partial t} = \frac{1}{r}\frac{\partial}{\partial r}\left(r\frac{\partial T}{\partial r}\right) \tag{1}$$

where, $T = T_0 + \Delta T$ is the temperature of the medium at any time, $t$ and arbitrary radial distance, $r$ ; $T_0$ is the initial temperature of the source and medium, and $\Delta T$ is the temperature difference between the medium and initial temperature. The Eq. (1) is the subject of the following boundary conditions:

$$\lim_{r\to 0}\left\{r\left(\frac{\partial T}{\partial r}\right)\right\} = -\frac{q}{2\pi k_f} \quad \text{at } t \geq 0 \text{ and } r = 0, \tag{2}$$

$$\lim_{r\to\infty}\{\Delta T(r,t)\} = 0 \quad \text{at } t \geq 0 \text{ and } r = \infty, \tag{3}$$

where, $\rho_f$ and $C_f$ are density and specific heat capacity of the test medium, respectively. The infinite series solution of this problem is outlined by Carslaw and Jaeger [38]. After initial, short transient period (i.e., $t >> r^2/4\alpha_f$ ; note, this transiency is much longer for finite wire diameter, including insolation if any), except for the first term containing time $t$, the higher order terms could be neglected, resulting in a very good approximation as,

$$\Delta T = T(r,t) - T_0 = \frac{q}{4\pi k_f}\left\{-\gamma + \ln\left(\frac{4\alpha_f t}{r^2}\right)\right\} \tag{4}$$

where $\gamma$ =0.5772 is the Euler's constant.

For constant fluid medium properties and a fixed and arbitrary radius *r*, after differentiation of Eq. (4), the radius is eliminated from the equation, and the following relation is obtained,

$$k_f = \frac{q}{4\pi} \cdot \frac{1}{d(\Delta T)/d\ln(t)} \tag{5}$$

Therefore, if temperature of the medium is measured as function of time at any fixed radial position, including at the contact with the line source (i.e. the temperature of the 'thin' line source), the thermal conductivity of the test medium, $k_f$, is proportional to the source heat flux and inversely proportional to the temperature (or temperature difference) gradient with regard to the natural logarithm of time, see Eq. (5).

The advantage of the hot-wire method is its simplicity and consequently low cost of construction. Furthermore, the wire itself acts as both the heating source and temperature sensor for measurement. Another advantage is that convection heat transfer effects can be minimized and identified when present as deviation of the linearity in the plot of $\Delta T$ as a function of *ln(t)*.

#### 4.1.1. Application of Hot-Wire Method for Nanofluids

A bare metal wire centered in a fluid medium is generally used for thermal conductivity measurement of fluids by the transient hot-wire method. Nanofluids containing metal particles are electrically conductive, so application of a bare wire could lead to ambiguous results in the measurements. Some of the problems identified by Nagasaka and Nagashima [39] in the application of the ordinary transient hot-wire method to electrically conducting liquids are: (a) possible current flow through the liquid, resulting in ambiguous measurement of heat generated in the wire; (b) polarization of the wire surface; (c) distortion of output voltage signal due to influence of the conducting liquid cell.

In order to overcome these errors, it is recommended that the bare metal wire should be coated using electrically insulating material. The effect on temperature distribution due to thin insulation coating has been analyzed [39] and outlined by Yamasue et al. [40]. The temperature rise $\Delta T$ of the hot-wire is given as,

$$\Delta T = \frac{q}{4\pi k_f}\left[\ln t + A_o + \frac{1}{t}(B_o \ln t + C_o)\right] \tag{6}$$

The terms $A_o$, $B_o$ and $C_o$ are defined as follows:

$$A_o = \ln\left(\frac{4\alpha_f}{r_o^2\gamma}\right) + \frac{2k_f}{k_i}\ln\frac{r_o}{r_w} + \frac{k_f}{2k_w} + \ldots \tag{7}$$

$$B_o = \frac{1}{2k_f}\left\{r_w^2\left(\frac{k_i}{\alpha_i} - \frac{k_w}{\alpha_w}\right) + r_o^2\left(\frac{k_f}{\alpha_f} - \frac{k_i}{\alpha_i}\right)\right\} \tag{8}$$

$$\begin{aligned} C_o = {} & \frac{r_w^2}{8}\left\{\left(\frac{k_f - k_i}{k_w}\right)\left(\frac{1}{\alpha_w} - \frac{1}{\alpha_i}\right) + \frac{4}{\alpha_i} - \frac{2}{\alpha_w}\right\} \\ & + \frac{r_o^2}{2}\left(\frac{1}{\alpha_f} - \frac{1}{\alpha_i}\right) + \frac{r_w^2}{k_i}\left(\frac{k_i}{\alpha_i} - \frac{k_w}{\alpha_w}\right)\ln\left(\frac{r_o}{r_w}\right) \\ & + \frac{1}{2k_f}\left\{r_w^2\left(\frac{k_i}{\alpha_i} - \frac{k_w}{\alpha_w}\right) + r_o^2\left(\frac{k_f}{\alpha_f} - \frac{k_i}{\alpha_i}\right)\right\}\ln\frac{4\alpha_f}{r_o^2\gamma} \end{aligned} \tag{9}$$

where, $r_w$ is the radius of the wire; and $r_o$ is the sum of the radius of the wire $r_w$ and the insulation thickness $\delta_i$. Subscripts *w*, *I*, and *f* represent wire, insulation coating, and liquid, respectively.

Comparison of Eq. (4) and (6) indicates that the term $(1/t)(B_o \ln t + C_o)$ is due to the presence of the insulation layer on the wire. If the term $(1/t)(B_o \ln t + C_o)$ is negligibly small compared to the $(\ln t + A_o)$ term, then the constant $A_o$ shifts (i.e., offsets) the plot of $\Delta T$ against *ln(t)*, without changing the slope. Therefore, the thermal conductivity, $k_f$, is again accurately determined using Eq. (5). Yu and Choi have analyzed the wire temperature rise as a function of time, to determine the influence of insulation coating on the thermal conductivity measurement and have concluded that relative measurement error of thermal conductivity is negligible if the slope of $\Delta T$ against *ln(t)* is measured at later times after start of heating, and that no correction to insulation coating is necessary, even if the insulation coating thickness is comparable to the wire radius.

Electrical insulation coating to bare metal wire has been recommended for electrically conducting fluids. Nagasaka and Nagashima [39] have coated the platinum wire (diameter 40 *μm*) with polyester insulation (thickness 7.5 *μm*) to measure electrically conducting aqueous *NaCl* solution. Whereas, Perkins [41] anodized a tantalum wire (diameter 25.4 *μm*) to form an electrically insulating layer of tantalum peroxide (thickness 70 *nm*), Yu et al. [42] have applied an epoxy insulation coating (estimated thickness 10 *μm*) to a platinum wire (diameter 76.2 *μm*) to measure thermal conductivity of nanofluids. Jwo et al. [43] insulated a Nickel-Chromium alloy wire with Teflon to measure thermal conductivity of CuO nanofluids. More recently, Ma [44] in his thesis has utilized a platinum wire (diameter 25 *μm*) with an *Isonel* insulation coating (thickness 1.5 *μm*) to measure thermal conductivity of various combinations of nano crystalline material and base fluids.

### *4.1.2. Hot-Wire Cell Design*

The main design parameters are: (a) material of hot-wire, (b) radius of hot-wire, (c) insulation coating, (d) length of hot-wire, (e) radius of the test sample outer boundary, and (f) length of the sample.

Platinum has been selected as superior hot wire material. It has higher thermal conductivity (TC) compared to the nichrome and tantalum, also used as hot-wires. Along with the material, hot-wire radius is one of the most important parameters for the cell design. Among commercially available sizes, 25.4 and 50.8 $\mu m$ radius platinum-wires have been selected for the present application, since smaller 12.5 $\mu m$ radii is considered to be too fragile for cleaning and handling of nanofluid samples.

Teflon has been selected as insulating material, as it is highly resistant to chemical reactions, corrosion and stress-cracking at high temperatures. A 50.8 $\mu m$ diameter platinum wire with a Teflon insulation coating of 25.4 $\mu m$ thickness, manufactured by A-M Systems, Inc., has been used as the hot-wire. Care has been taken to avoid any disruption of the coating during hot-wire mounting.

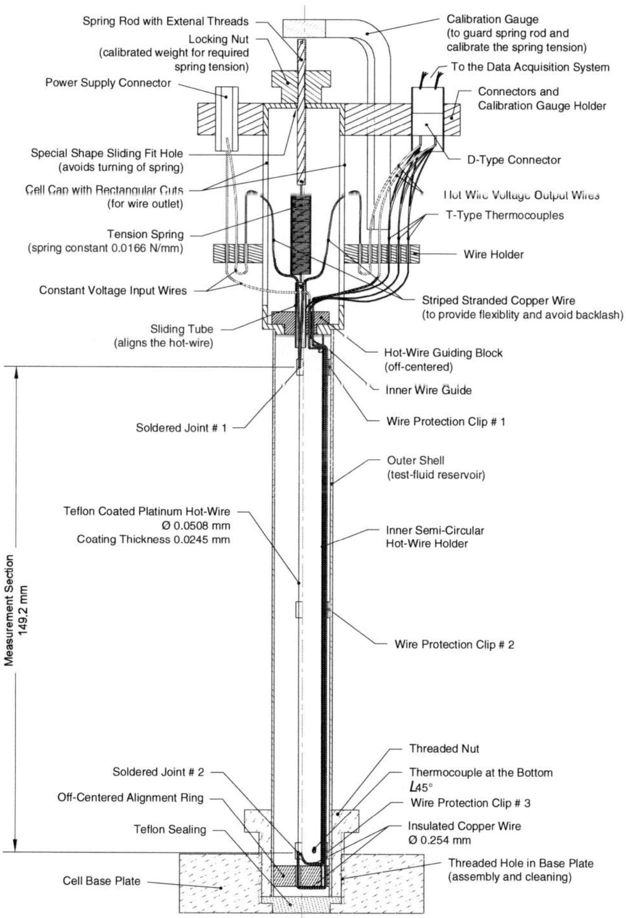


Figure 4. Cross-sectional view of transient hot-wire thermal conductivity apparatus.

In our design [32], the length of the platinum hot-wire was taken as $L_w$ = 0.1484 *m*, based on the 0.139 *m* minimum length of hot-wire, determined according to Kierkus et al. [45] criteria, for our application data. Based on Healy et al. [46] criteria, the minimum hot-wire cell outer boundary radius was determined as 0.0028 *m*, but chosen to be $R_c$ = 0.00718 *m*, and the finite length of the sample to be $L_c$ = 0.170 *m*. The overall sample volume $V_c$ after fabrication is calibrated to be 35 *ml*.

A cross sectional view of the newly designed hot-wire thermal conductivity apparatus with major mechanical components is shown in Figure 4. The major assembly components of the apparatus' cell are: base plate, outer shell, and cell cap with hot-wire. The cell base plate with a threaded hole at the center of the plate (sealed by a Teflon washer) is used for convenient assembling and disassembling the outer shell. The outer shell with 17.4 *mm* inner diameter acts as the sample test fluid reservoir. The cell cap, designed to slide-fit into the outer shell, is hollow inside. The inner semi-circular hot-wire holder with an alignment ring is soldered at the lower end at an offset. A hot-wire guiding block, sliding tube, tension spring, and spring rod are all aligned at an offset inside the cap. The Teflon-coated platinum wire is indirectly connected to the tension spring via copper wires and a sliding rod, which are aligned with the spring mechanism (i.e., sliding tube, tension spring, spring rod and locking nut). A locking nut, fastened to the spring rod, is mounted on the top of the cap. Two symmetric rectangular cuts in the cap provide an opening for routing of electrical and thermocouple wires. A connector and calibration gauge holder and a wire holder, made of Teflon, are mounted on the top and middle section of the cell cap.

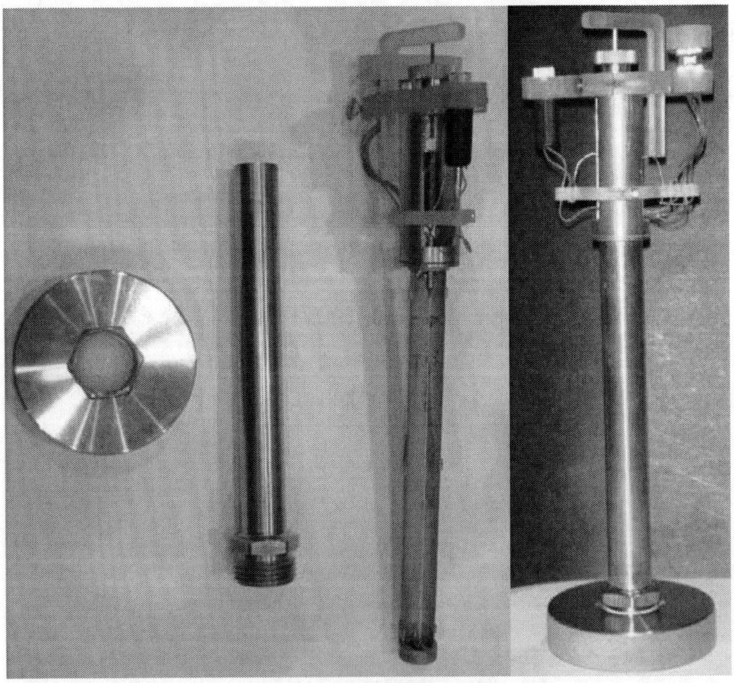

Figure 5. Fabricated transient hot-wire thermal conductivity apparatus.

Three thermocouples, mounted on the outer radius of inner semi-circular hot-wire holder at 15°, 45° and 75° angles, along the length of the sample test section, monitor uniformity of the test fluid temperature. The thermocouple tip is bent towards the platinum hot-wire through the holes on the inner semi-circular hot-wire holder.

Two copper wires at the top soldered-joint of the platinum hot-wire are passed symmetrically through a sliding tube. The inner hollow portion of the sliding tube is filled with epoxy to couple the copper wires with the sliding tube. A clearance between the sliding tube and hot-wire guiding block hole ensure near-frictionless motion of the sliding tube. The spring rod is specially shaped to have external threads. A locking nut and spring rod are screwed together like a nut and bolt. The special shaped sliding fit hole avoids turning of the spring rod when the locking nut is turned for adjusting the tension of the hot-wire. The locking nut has been fabricated to a specific weight that is used for calibration of the platinum hot-wire tension, within 50% of its ultimate tensile strength. An inverted L-shaped gauge has been seated on the holder for calibrating the hot-wire tension and guarding the spring rod movement. One of the two copper wires at another soldered joint of the platinum hot-wire is passed through the off-centered hole of the alignment ring, while the other has been guided through a hole in the inner semi-circular hot-wire holder. The copper wires at the alignment ring act as fixed rigid ends of the hot-wire.

The present cell parameters are $L_c = 0.170\ m$, $2R_c = 0.01437\ m$, $V_c = 35\ ml$, $L_c/2R_c = 11.83$. The parameters of the platinum hot-wire with Teflon coating of thickness 25.4 $\mu m$ are $L_w = 0.1484\ m$, $2r_w = 50.8\ \mu m$, $L_w/2r_w = 2921$, $2R_c/2r_w = 282.9$, and $L_w/2R_c = 10.33$.

The controlled tightness of the thin platinum hot-wire is a very important aspect. An innovative solution, to indirectly connect a tension spring to the platinum hot-wire, while maintaining constant tension, has been incorporated. Also, a unique solution to calibrate the platinum hot-wire tension has been developed. This arrangement minimizes the well known and unwanted strain gage effect on hot-wire electrical resistivity, thus decreasing the strain influence on temperature measurement. An extension spring with a low spring constant is calibrated and used for present application as detailed elsewhere [31].

#### *4.1.3. Instrumentation and Data Acquisition*

A Wheatstone bridge circuit has been employed to measure the resistance change of the hot-wire, with the wire being one of the arms (i.e., resistors) of the bridge. Initially, the bridge is balanced until the voltage output of 10-15 $\mu V$ is achieved. The bridge balancing is performed within a brief period of time, using a constant, low input voltage of 0.1 $V$, to minimize heating of the platinum wire during initial bridge balancing. After the bridge circuit is balanced, a constant, input voltage $V_{in}$, at start-time $t = 0$, is applied to heat the wire, thus resulting in unbalancing of the bridge due to the hot-wire's temperature and thus resistance change. The bridge input $V_{in}$ and output $V_{out}$ voltages are measured using a computerized data acquisition system. A schematic diagram of the Wheatstone bridge circuit used for measurement is shown in Figure 6. The circuit has been fabricated in such a way that the data acquisition system can be easily connected and disconnected for measurement. The schematic of DAQ system is also shown in Figure 6.

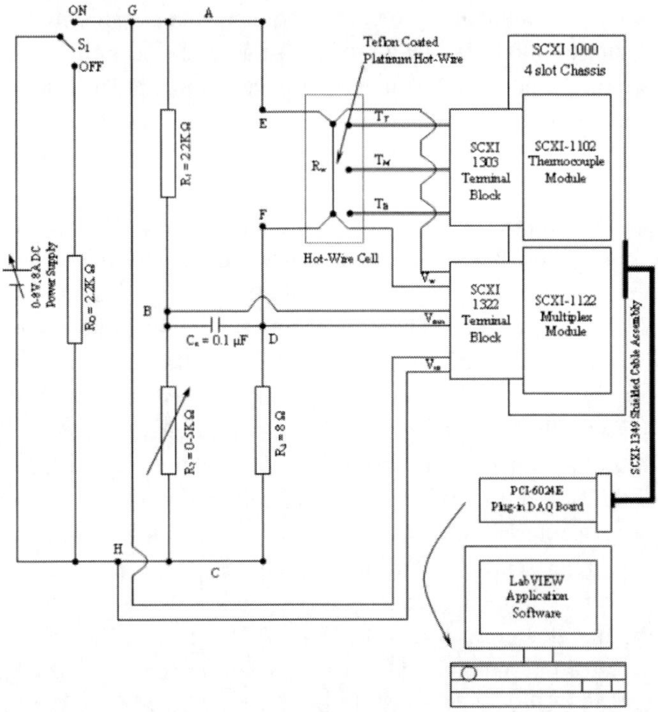


Figure 6. Schematics of electrical circuit with computerized data acquisition system.

Six different signals are measured using National Instruments' <www.ni.com> data acquisition (DAQ) hardware, namely: bridge voltage output; bridge voltage input; hot-wire voltage drop (i.e., voltage drop across platinum wire); and three signals from the thermocouples mounted across the length of the hot-wire cell at the top, middle, and bottom sections. A computer program for acquiring and post-processing the measured data is developed using the LabVIEW® application software.

All measurement parameters are controlled using the LabVIEW® user interface. The input voltage range is configured at 0 – 10 *V*. The bridge voltage output is triggered after the voltage threshold reaches a value of 1 *mV*. The voltage output channel can be configured to a maximum sampling rate is 100 *Hz*, with nominal operation range configured at 0 – 200 *mV*, which provides an overall measurement gain of 100. The bridge voltage output and time are measured and stored simultaneously. Post-processing of the acquired data is then performed in order to calculate the resistance change, temperature change, heat input, and then thermal conductivity of the test fluid.

### *4.1.4. Calibration and Uncertainty Analysis*

Two standard base fluids of well-defined thermal conductivity, ethylene glycol and distilled water, have been used for over-all calibration of the apparatus. The voltage change (i.e., bridge voltage output) for all calibration measurements has been acquired at a sampling rate of 50 Hz. For a typical measurement, the resistances of the Wheatstone bridge circuit are measured as: $R_1$ = 2270.6 Ω, $R_2$ = 2161.1 Ω, and $R_3$ = 7.715 Ω, using a four-wire resistance measurement technique. The reference resistance of the platinum wire is determined as $R_{w0}$ =

8.106 Ω using relevant correlations [1]. The measured length of platinum hot-wire was $L_w$ = 0.1484 *m*. The platinum temperature coefficient of resistance is $\sigma_{TCR} = Z_\sigma / R_w$, where $R_w$ = 8.22 Ω is the resistance measured at 20°C and $Z_\sigma$ = 0.02652 Ω/°C is the slope of the hot-wire resistance with regard to temperature.

Figure 7 shows a typical hot-wire temperature change versus time for ethylene glycol and water. Analysis of the graph indicates that the linearity of the wire temperature change (in semi-log coordinate system) is obtained one second after the experiment starts. The initial deviation from linear temperature change is due to initial transience (where Eq. 5 without truncated higher-order terms is not valid). This initial non-linear response is often overlooked or neglected in the references but may contribute to errors in measurement results. The deviation from linearity at later times, with higher temperature differences, can be attributed to on-set of convection heat transfer and finite boundary effects (Hong and Yang 2005; Hammerschmidt and Sabuga 2000). Under similar testing parameters, the temperature change in time for water is lower compared with that of ethylene glycol, which is attributed to the fact that water, compared to ethylene glycol, has higher thermal conductivity, see Eq. (5).

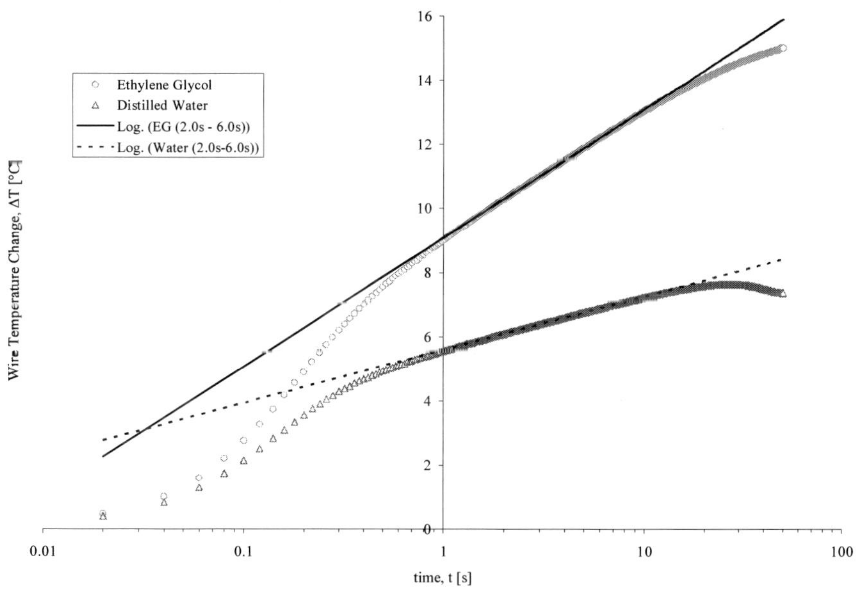


NOTE the required linearity range from 1-10 *s*, reduced from 2 to 6 *s* measurement range for increased accuracy.

Figure 7. Hot-wire temperature change against time (in semi-logarithmic scale) for ethylene glycol and distilled water.

Thermal conductivity measurements of ethylene glycol and water have been repeated 10 times each, and average values reported. The time range from 1 to 10 *s* has been determined to be virtually linear, and measurement sub-range from 2 to 6 *s* is chosen for increased accuracy, see Figure 7. The reference temperature $T_r$, at which the test fluid properties are measured, is evaluated as the process average temperature:

$$T_r = T_0 + \frac{1}{2}\left[\Delta T(t_1) + \Delta T(t_2)\right] \tag{10}$$

where, $T_0$ is the initial temperature of the fluid (determined using thermocouples mounted within the hot-wire cell); and, $\Delta T(t_1)$ and $\Delta T(t_2)$ are measured temperature increases at times $t_1$ and $t_2$, respectively [23]. The standard thermal conductivity values of ethylene glycol and water are obtained from a standard engineering reference.

Ethylene glycol with 99.9 % purity and distilled water, have been used as standard test fluids for over-all calibration of the new HWTC apparatus. The mean reference temperature of ethylene glycol, using Eq. (10), was measured 32.5 *°C*, with the respective reference thermal conductivity of ethylene glycol 0.254 *W/m°C*, while the corresponding values for distilled water were 26.0 *°C* and 0.612 *W/m°C*, respectively.

**Table 1. Measurement uncertainties and repeatability errors of measured thermal conductivity**

| **Fluid** | **Reference [*W/m°C*]** | **Measured [*W/m°C*]** | **Bias Error** | **Precision Error (95 %)** | **Uncertainty in Repeatability** |
|---|---|---|---|---|---|
| Ethylene Glycol (32.5 °C) | 0.254 | 0.253 | - 0.395 % | 2.03 % | 2.06 % |
| Distilled water (~ 26 °C) | 0.612 | 0.619 | 1.2 % | 2.23 % | 2.52 % |

A detailed calibration and the measurement uncertainty analysis have been performed; see Table 1, and elsewhere [31].

## 4.2. Design of a Steady-State, Parallel-Plate Thermal Conductivity Apparatus for Nanofluids and Comparative Measurements with Transient HWTC Apparatus

A steady-state, parallel-plate thermal conductivity (PPTC) apparatus has been developed, fabricated, calibrated and used for comparative measurements of complex POLY-nanofluids [33, 34], in order to compare results with the corresponding measurements using the transient, hot-wire thermal conductivity (HWTC) apparatus [31, 32]. The related available measurements in the literature, mostly with HWTC method, have been inconsistent and with measured thermal conductivities far beyond prediction using the well-known mixture theory [30]. The objective was to use quite different, concomitant method and check out if existing and well-established HWTC method might have some unknown issues while measuring TC of complex nano-mixture suspensions, like electro-magnetic phenomena, undetectable hot-wire vibrations, and others.

### *4.2.1. Parallel-Plate Apparatus Design and Method*

The schematic of the PPTC apparatus with all components labeled and relevant nomenclatures is presented in Figure 8. The objective was to provide controlled one-dimensional heating by conduction through a stationary test fluid specimen and accurate measurements of relevant temperatures in order to simply and accurately measure the fluid thermal conductivity. The main components are described next.

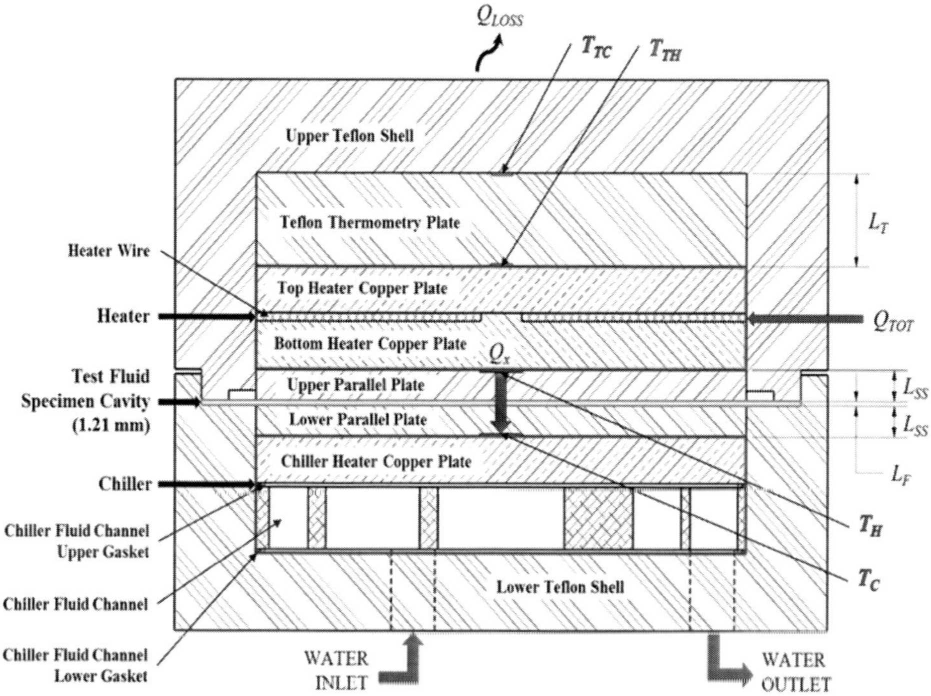


Figure 8. Schematic of the PPTC apparatus with relevant nomenclatures.

**Test Fluid Specimen Cavity**

The test fluid specimen cavity (highlighted yellow on Figure 8) is a critical component of the PPTC apparatus that houses the test fluid sample for measurement of thermal conductivity. The top and bottom of the test fluid cavity are formed by the parallel plates described later. The faces of the parallel plates in contact with the test fluid specimen have a high level of planarity and mirror-polished finish to minimize measurement errors due to surface geometry imperfections of the parallel plates. The side edge of the test fluid specimen cavity is centered by the upper and lower lips located on both the upper and lower Teflon shells, as detailed below. In order to accurately measure the thermal conductivity of the test fluid specimen, a well-defined heat transfer mechanism consisting virtually of conduction only must be provided. This is accomplished by two major design parameters. First, the heater is above and the chiller is below the test fluid (the heat transfer is in the gravity direction) thus suppressing the convection effects due to buoyancy. The second major design parameter is a very small thickness of the test fluid specimen cavity, $L_F$=1.21 *mm*. This rather small thickness, similar to that used in a successful apparatus [47], prevents the convection from developing within the test fluid specimen.

Another important design aspect is the method used to establish a gap with accurately calibrated, very small size cylindrical glass-spacers (1.21±0.01 *mm* thick and 2.0 *mm* diameter) placed between the upper and lower parallel plate assemblies. The glass spacers have been chosen due to their dimensional stability and thermal conductivity being the same order-of-magnitude as that of the measured test specimens'. Three glass spacers were evenly displaced circumferentially close to the outer edge of the bottom parallel plate. Once the test fluid specimen is added to the test fluid specimen cavity, the upper assembly is carefully rested on the spacers.

The final design aspect of the test fluid specimen cavity is the bull's-eye leveling mechanism mounted on an adjustable stand on which the lower assembly rests. It is important for the lower assembly and the upper assembly to be leveled when the test fluid specimen is loaded to provide an even test fluid specimen thickness and minimize convection effect due to gravity.

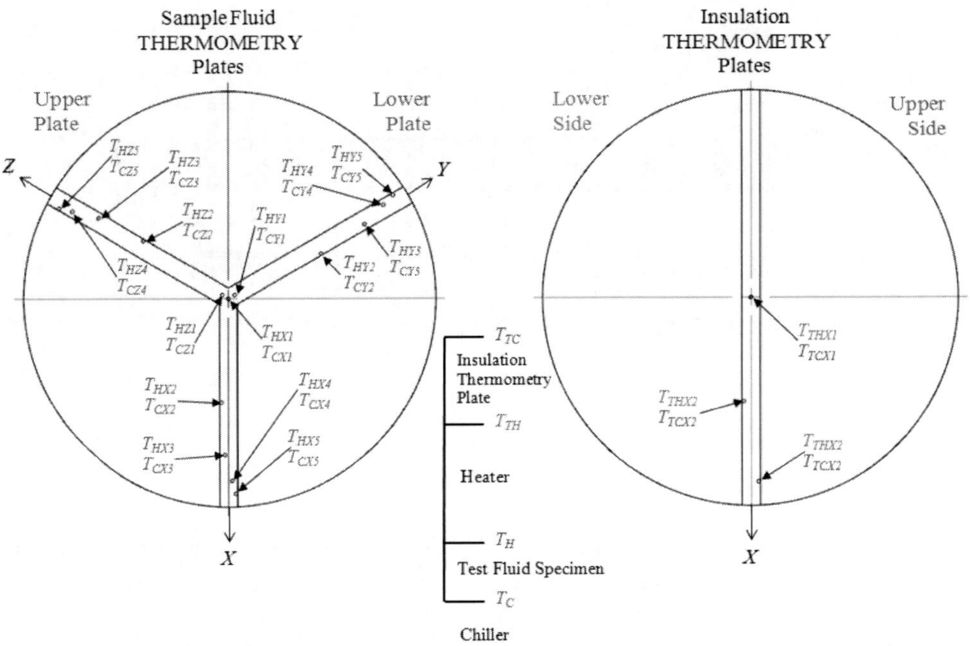


Figure 9. Parallel (Thermometry) Plate and Teflon (Insulation) thermometry plate with thermocouple placement locations and relevant nomencluture.

## Parallel (Thermometry) Plates

The parallel (thermometry) plates provide the plane surfaces that comprise the top and bottom of the test fluid specimen cavity described above. These plates also house the thermocouples used to measure a number of the temperatures at different locations, used to determine the temperature difference across the test fluid specimen, as well as temperature uniformity in the other two directions, see Figure 9. One parallel plate is press fit into the bottom of the upper assembly and one parallel plate is press fit into the top of the lower assembly, resulting in one parallel plate on either side of the test fluid specimen. The parallel plates are made from ANSI 304 stainless steel and are 4.50 *inch* in diameter and have a thickness of 0.25 *inch*. Stainless steel was chosen as the material for the parallel plates to provide corrosion resistance and for easy cleaning of the test fluid specimen cavity. One surface of each of the parallel plates facing the test specimen has a mirror finish. Before the parallel plates are press fit into place, the faces contacting both the heater assembly copper plate and the chiller assembly copper plate are coated with a thin layer of high thermal conductivity paste to reduce the effects of contact thermal resistance. Each face of the parallel plates facing away from the test fluid specimen has radial thermocouple grooves, see Figure 9. These grooves provide locations needed to place the thermocouples and clearance for thermocouple wires. There are three radial thermocouple grooves evenly spaced around the circumference of each parallel plate. These grooves extend from the center to the outside edge of the parallel plates, and are 0.20 *inch* wide and 0.02 *inch* deep. There are fifteen, 30-gauge *T-type* thermocouples mounted on each parallel plate, with each thermocouple groove containing five thermocouples. The thermocouple grooves are then filled with a high thermal conductivity epoxy that isolates and holds the thermocouples in place.

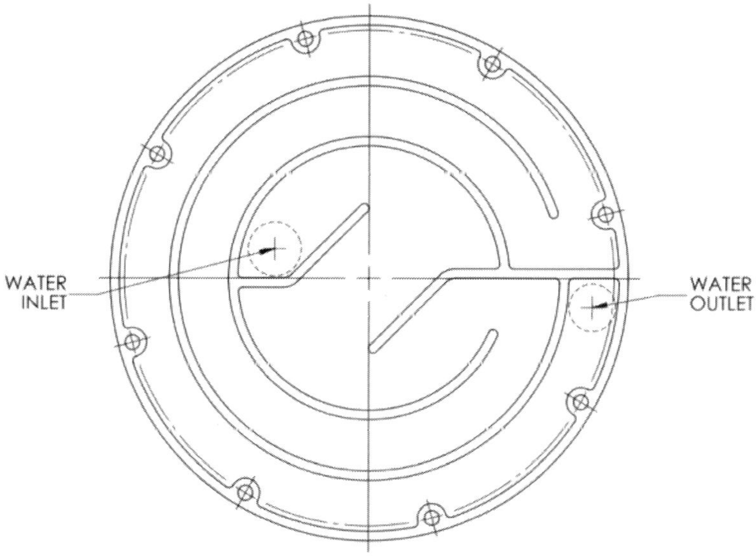


Figure 10. Water Chiller Design.

A large number of thermocouples are utilized in this design for two reasons. Firstly, redundant thermocouples could be used in case of some thermocouple malfunction in the future, since it would be difficult and impractical to disassemble the pressed-fit apparatus in order to repair malfunctioning thermocouples. Secondly, having a large number of evenly spaced thermocouples provides means to verify the radial and circumferential temperature uniformity, needed for evaluation of heat looses, if any, and validation of one-dimensional heat transfer through the thickness of the test specimen, as modeled by the working equation used for evaluation of the thermal conductivity. On Figure 9, the Teflon (Insulation) thermometry plate for evaluation of heat losses from the top of the apparatus is also presented. The Teflon thermometry plate is located above the heater assembly, see Figure 8.

The purpose of the Teflon thermometry plate is two-fold. First, the low thermal conductivity of the Teflon (0.35 *W/m/K*) provides extra insulation to the top of the heater assembly. Second, the thermocouples provide a means to calculate the heat loss through the top of the heater assembly. The Teflon thermometry plate is made of virgin electrical grade Teflon. It is 4.50 *inch* in diameter and has a thickness of 0.75 *inch*. Thermocouples are attached on both sides of the Teflon thermometry plate and are located in thermocouple grooves that are machined into each face of the Teflon thermometry plate.

**Heater Design**

The heater generates the heat flux and thus the temperature difference across the test fluid specimen. The heater is made of a resistance wire that is formed into a spiral and sandwiched between two copper plates to equalize radial and circumferential temperature distribution. The resistance wire of 55% copper and 45% nickel with a diameter of 0.036 *inch* (19 AWG) has Teflon insulation. The resistance is approximately 0.227 *ohms* per foot, resulting in the total of the heater wire resistance of 4.21 *ohms*. The plates made from 145 tellurium copper alloy (of 400 *W/m/K* thermal conductivity) are chosen for easy machining and to ensure that the heat generated by the heater wire will be evenly distributed across the entire surface of the test fluid specimen. The top copper plate is a 0.375 *inch* thick circular disc with a 4.50 *inch* diameter. The bottom plate has the same dimensions, but it includes a 0.80 *inch* high post of 0.50 *inch* diameter. The heater wire is wrapped spirally around the post in the bottom copper plate, providing an evenly distributed heating, and the two plates are screwed together creating a single heater assembly.

The heater assembly and the Teflon thermometry plate are press fit into a Teflon shell, creating the upper assembly of the apparatus. Teflon is used for the heater assembly housing for three reasons. First, the Teflon provides sufficient rigidity for the press fitting of the heater assembly. Second, the low thermal conductivity of Teflon provides insulation, reducing the amount of heat lost to the surroundings. Finally, the Teflon provides for easy cleaning of the apparatus. The upper Teflon shell has an outer diameter of 6.00 *inch* and an overall height of 2.58 *inch*. It also has a raised lip (0.25 *inch* deep and 0.25 *inch* thick) at the bottom surface. This lip helps to center the upper assembly when it is placed on the lower assembly. It also forms the outer side of the test fluid specimen cavity. Finally, the upper assembly is covered with a polystyrene shell. This shell provides a 0.75 *inch* thick layer of insulation around the entire upper assembly. The polystyrene shell has an extremely low thermal conductivity of approximately 0.027 *W/m/K*, and further minimizes the heat loss to the surrounding.

**Chiller Design**

The chiller forms the lower assembly of the apparatus and utilizes cold water as its cooling medium, see Figure 10. A channel made from aluminum is designed to guide the cooling water around the lower assembly and provide even removal of the heat generated by the heater. The basic shape of the fluid channel is a spiral-like, with the water entering the chiller closer to the center and exiting at the outer edge, this enabling more uniform circumferential temperature.

The fluid channel has an outer circumference of 4.50 *inch*, and a depth of 0.50 *inch*. The channel walls are 0.08 *inch* thick, providing sufficient rigidity. The chiller is also comprised of a copper plate that forms the top of the chiller assembly and provides even heat transfer to the chiller fluid. The material and dimensions of the chiller copper plate are identical to those of the upper heater copper plate.

The chiller fluid channel and chiller copper plate are press fit into a Teflon shell, creating the lower assembly of the apparatus, nearly identical to the upper Teflon shell.

**Method and Mathematical Model**

The steady-state, parallel-plate method utilizes the simple mathematical model of one-dimensional heat conduction through a composite three layers of cross-sectional area $A$: the test-fluid specimen of thickness $L_F$ between two identical stainless-steel, parallel thermometry plates of thickness $L_{SS}$ and known thermal conductivity $k_{SS}$, see Figure 8, resulting in the simple equation for calculation of the test fluid thermal conductivity based on known and/or measured quantities (Walleck 2009):

$$k_F = \frac{L_F}{A \cdot \left( \frac{T_H - T_C}{Q_x} - \frac{2L_{SS}}{k_{SS} A} \right)} \quad (11)$$

The heat rate transferred through the test fluid specimen in the axial-gravity direction, $Q_x = Q_{EH} - Q_{top} - Q_{rad}$; where, $Q_{EH} = \frac{V_{heater}^2}{R_{heater}}$ is the heat rate supplied by the electrical resistance-heater, while the other quantities, $Q_{top}$ and $Q_{rad}$ are the top-surface and radial heat losses through the thermal insulation, respectively. The $V_{heater}$ and $R_{heater}$ are measured voltage across and the calibrated resistance of the heater wire. The $T_H$ and $T_C$ are representative temperatures at upper and lower parallel thermometry plates, respectively, see Figures 1 and 4. It is verified, based on extensive temperature measurements, that temperature profile is virtually uniform in radial and circumferential direction, justifying the validity of the above 1-D equation and neglecting the radial heat loss above, while the top-surface heat loss is measured using the Teflon thermometric plate described above, and thus accounted for (usually less than 2%). Typical measured temperatures are presented on Figure 11.

### *4.2.2. Instrumentation and Data Acquisition*

The PPTC apparatus, with all instrumentation and computerized data acquisition, is depicted on Figure 12, while measurement sensors and data acquisition components are presented on Figure 13. Agilent E3644A DC Power Supply with a range of 0-8V, 8A or 0-20V, 4A was used, since the typical heater requires 8.00 *volt*, drawing about 1.875 *amp* electrical current, i.e., about 15 *watt* power.

All thermocouple temperatures (described in previous section) and the supplied heater voltage were measured frequently throughout the testing process at a rate of five times per minute using National Instruments' data acquisition hardware and LabVIEW® application software, described in details elsewhere (Walleck 2009; www.ni.com 2011).

The LabVIEW program automates all measurements and calculations, with very little input from a user; only maximum number of measurements, $N_{MAX}$, recommended 1500 measurements, and the output data file name and location. This condition is used to stop the program if the steady-state conditions, defined within the LabVIEW® program, are not satisfied within the maximum number of measurements. The output file contains all measurements and calculations performed during a single test in text format, convenient for further post-processing if desired. Details are given elsewhere [33, 48].

The instrumentation for all transient thermal conductivity measurements were performed using the Hot-Wire, Thermal Conductivity (HWTC) apparatus developed at Northern Illinois University [31, 32].

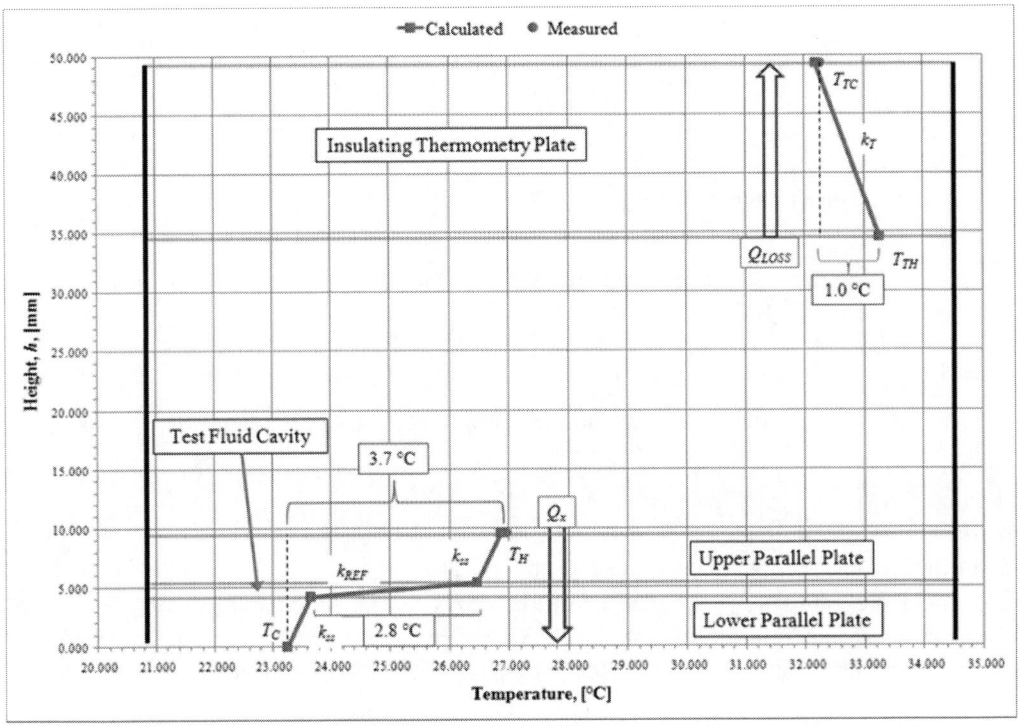


Figure 11. PPTC apparatus typical measured temperature profile (calibration with distilled water).

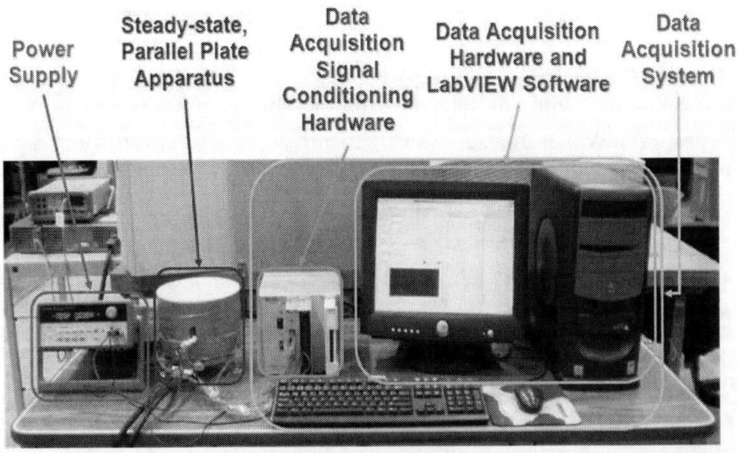


Figure 12. Lab Setup: PPTC apparatus with all instrumentation and computerized data acquisition.

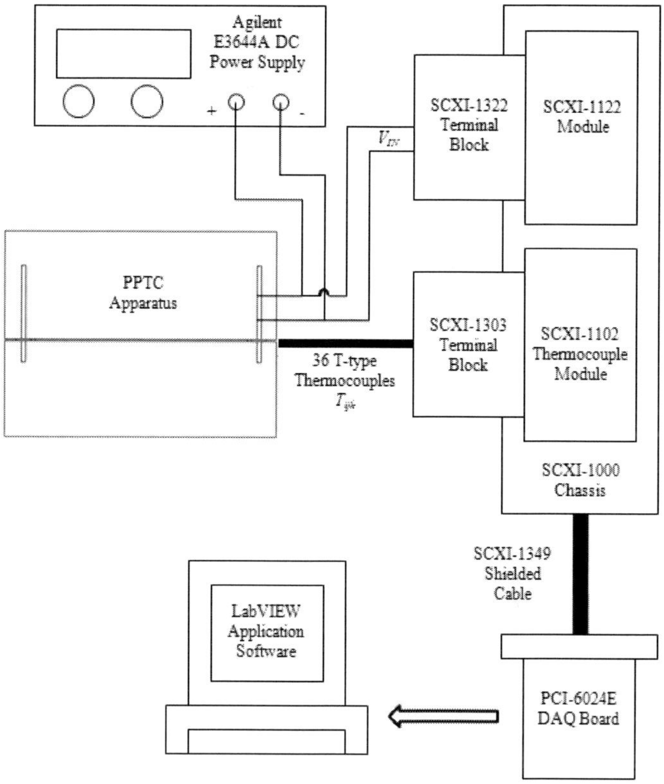


Figure 13. PPTC apparatus instrumentation and data acquisition components.

#### *4.2.3. Calibration and Uncertainty Analysis*

The PPTC apparatus has been thoroughly calibrated in order to insure accurate measurement results. All thermocouples used have been calibrated against a precise RTD standard, reducing thermocouple uncertainty to 0.1°C. A correction factor has been determined in order to minimize errors occurring in fluid thermal conductivity measurements due to heat loss through the apparatus. A conservative and detailed uncertainty error analysis has been performed for the PPTC apparatus using the method of propagation of errors, resulting in a conservative uncertainty within 8% at 95% probability. The unidirectional heat transfer in the apparatus has also been validated through a radial heat conduction analysis based on detailed temperature measurements. Finally, the consistency and over-all accuracy of the PPTC apparatus has been calibrated with repeatability study using distilled water. The PPTC apparatus exhibits a bias error of approximately -4.5% and a precision error of less than 4% with a 95% confidence. The PPTC apparatus exhibits an overall accuracy of approximately 6.5%, however, when bias error is accounted for an accuracy of about 4% is achieved. By repeating measurements on the same sample the accuracy of the mean values of measured quantities and the thermal conductivity could be further improved. The details are presented in [33].

#### *4.2.4. Nanofluid Thermal Conductivity*

A polymer- or POLY-nanofluid consists of a common or standard nanofluid with additional polymer additives [36, 37]. The reason for adding polymers to standard nanofluids is two-fold. First, certain polymer additives in extremely small concentrations, usually on an order of tens weight-parts-per-million (*wppm*), have been shown to substantially reduce the turbulent friction drag [28]. The drag reduction in turbulent flow is of importance because it may increase performance of the POLY-nanofluids in many practical applications. Second, it is expected that some of these polymer additives could increase the stability of the nanofluid suspension by preventing agglomeration of the nanoparticles. If the viscosity of the POLY-nanofluid is too high, it will have increased drag in laminar flow, but still may exhibit drag reduction in turbulent flow [28], the latter being very important since the flows in most thermal systems are turbulent.

Nanofluids containing silica and alumina nanoparticles have been prepared first to satisfy the *zeta*-potential vs. *pH* relationship in order to achieve suspension stability [49, 50]. Then, two different polymers have been added to the 5% by weight silica-nanofluid and the 5% by weight alumina-nanofluids. The first polymer chosen was *polyvinylpyrrolidone*, or PVP, with approximately 9000 *Da* (Daltons) molecular weight, manufactured by BASF under the brand name *Luvitec* K17. This polymer was chosen because it is known to increase the suspension stability of nanofluids [51]. The second polymer chosen is *polyacrylamide*, manufactured by Stockhausen, Inc. under the brand name *Praestol*-2273, since it is known to substantially reduce friction drag in turbulent flows [28].

The polymer concentrations chosen for the POLY-nanofluids are 0.02% and 0.05% PVP by weight and 0.02% and 0.05% *polyacrylamide* by weight (i.e., 100 and 500 *wppm*), for the *silica* POLY-nanofluids; and 0.02% and 0.05% PVP by weight and 0.01% and 0.02% *polyacrylamide* by weight, for the *alumina* POLY-nanofluids. These small concentrations are judged to be sufficient to achieve the desired effects of increased suspension stability and improved frictional drag while maintaining a suitable viscosity. A lower weight concentration of polyacrylamide is necessary for the alumina-nanofluids, as a concentration greater than 0.02% by weight leads to severe agglomeration of the nanoparticles. The thermal conductivity of these POLY-nanofluids was then measured using both, an existing transient, hot-wire, HWTC apparatus [31, 32] and this new steady-state, parallel-plate PPTC apparatus [33]. Comparative measurements have been made using the two quite different methods and apparatus, in order to explore the possible influence of different measurement techniques on the thermal conductivity results of the complex POLY-nanofluids, since the existing data in the literature are very inconsistent and not well justified.

The results are presented on Figure 14 for *silica* POLY-nanofluids and on Figure 15 for *alumina* POLY-nanofluids. The results are expressed as dimensionless thermal conductivity ratio, $k_{PnF}/k_{BF}$, between POLY-nanofluid thermal conductivity, $k_{PnF}$, and the base fluid (water) thermal conductivity, $k_{BF}$, the latter typical value of 0.59 *W/m/K*.

The average thermal conductivity enhancement over the base fluid exhibited by the *silica* POLY-nanofluids is 1.3% when measured using the PPTC apparatus, and 4.4% when measured using the HWTC apparatus. The average thermal conductivity enhancement over the base fluid exhibited by the *alumina* POLY-nanofluids is 3.8% when measured using the PPTC apparatus, and 11.4% when measured using the HWTC apparatus. The effects of the polymer additives PVP and *polyacrylamide* on the silica nanofluids can be classified as statistically insignificant. The thermal conductivity enhancement over the standard alumina nanofluid exhibited by the alumina POLY-nanofluids suggest that a small concentration of PVP can be beneficial to alumina nanofluid thermal conductivity, while a small concentration of polyacrylamide can have a negative effect on thermal conductivity. The viscosity of the POLY-nanofluids was also measured and is presented elsewhere [33]. The POLY-nanofluids containing PVP exhibited a slight increase in viscosity when compared to the base fluid, while the POLY-nanofluids containing polyacrylamide exhibited a larger increase in viscosity when compared to the base fluid.

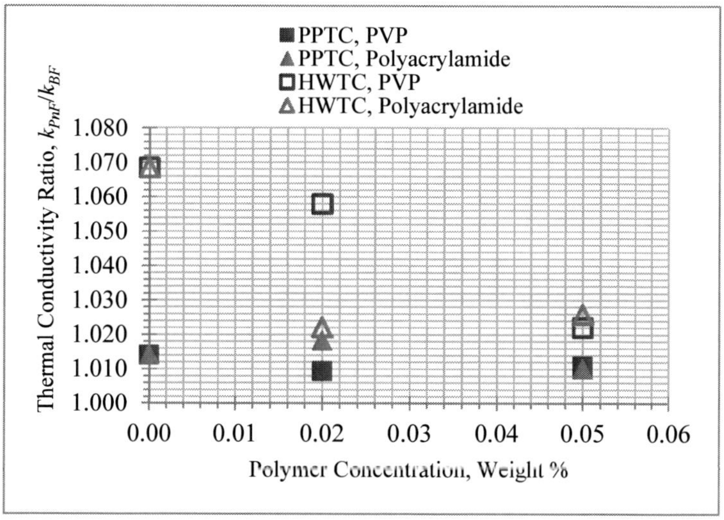


Figure 14. Thermal conductivity ratio versus polymer concentration for 5% by-weight *silica* POLY-nanofluids (solid symbols with PPTC and open symbols with HWTC apparatus).

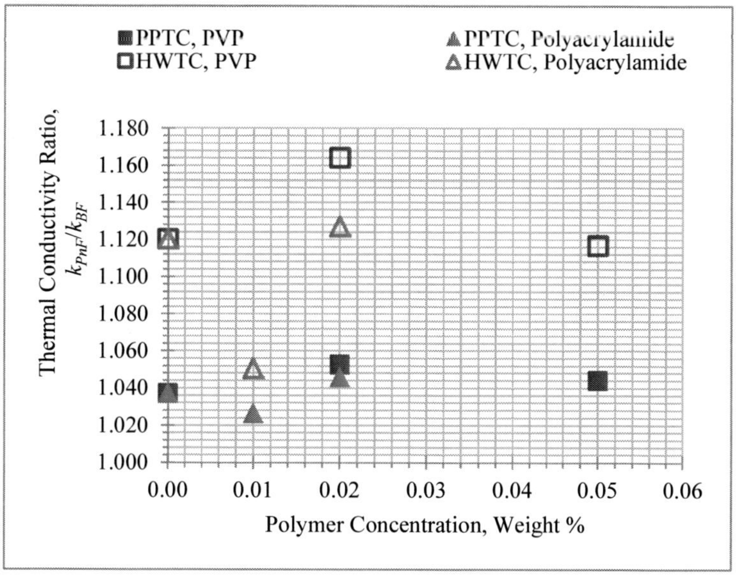


Figure 15. Thermal conductivity ratio versus polymer concentration for 5% by-weight *alumina* POLY-nanofluids (solid symbols with PPTC and open symbols with HWTC apparatus).

Unexpectedly, on average, the differential thermal conductivity enhancements measured by the HWTC apparatus are about three times greater (though with significant scatter) than the corresponding thermal conductivity enhancements measured by the PPTC apparatus. This difference demonstrated that the measurement technique might have a notable impact on the observed thermal conductivity enhancement of both common nanofluids (without, i.e., no/zero polymer additives) and POLY-nanofluids over the base fluid. However, this large difference may be contributed to the suspension instability; namely the POLY-nanofluid suspensions are observed to be degrading during the lengthy steady-state thermal conductivity measurements using the PPTC apparatus (usually takes several hours) as opposed to quick measurements with HWTC apparatus (takes only several seconds). Interestingly, the results obtained using the PPTC apparatus are similar to those predicted by the simple mixture theory [30].

#### *4.2.5. Conclusion*

An apparatus based on steady-state, one-dimensional heat conduction between two parallel plates, has been developed, designed and fabricated with main objective to measure thermal conductivity of fluids, polymer solutions, nanofluids and POLY-nanofluids [28, 33, 36, 37]. The goal was to reduce the overall test sample volume for nanofluids, while maintaining the precision and accuracy of the apparatus. Data acquisition hardware and LabVIEW® application software are optimized to minimize signal noise and enhance acquisition and processing of useful data.

The bias measurement error, based on calibration with distilled water, has been found to be -4.5 %, and precision below 4%. The total uncertainty, after accounting for the bias error in measured thermal conductivity, has been estimated to be within 4 % at 95 % confidence probability.

These initial and limited measurements have shown considerable difference in TC measurements using the two methods, whereby the measured TC increase beyond the base fluid TC by developed PPTC apparatus was about three times smaller than the comparative measurements of apparently the same nano-mixtures using the HWTC apparatus, though with significant scatter, the former data being much closer to the mixture theory prediction [30].

However, the influence of measurement method on the TC results is not conclusive since it has been noticed that the complex nano-mixture suspensions were very unstable during the lengthy steady-state measurements as compared to rather quick transient HWTC method, so the two nano-mixture suspensions were really not the same. The nanofluid suspension instability might be the main reason for very inconsistent results in the literature. It is necessary to expend investigation of the influence of TC measurement methods on the results with more stable nano-mixture suspensions. More testing is necessary to explore the effect of measurement technique on nanofluid thermal conductivity. Also, more testing is necessary to verify the initial POLY-nanofluid thermal conductivity and viscosity results. It is hoped that these unexpected but inconclusive results will initiate constructive criticism and further investigations, related to many open questions.

# 5. Thermal Conductivity Modeling: Possible Issues with Nanofluids

Oddly, there are more hypothetical theories proposed than reliable experimental results to verify theoretical models related to nanofluid thermo-physical characteristics. Observed enhancement of nanofluids thermal conductivity and potential for development of substantially improved heat transfer fluids, have propelled research and publications, almost exponentially. Theoretical work, developing in the absence of a reliable experimental framework, has resulted in an awkward situation of having a larger number of competing theoretical hypotheses than systematic experimental results to prove them.

Nanofluids thermal conductivity depends on many factors, like:

1) nanoparticle type,
2) nanoparticle size,
3) nanoparticle shape,
4) nanoparticle concentration in base fluid,
5) base fluid type,
6) additives and clustering,
7) temperature,
8) possibly temperature gradient, and other unknown factors which influence nanoparticle-base fluid molecule's interactions during the conduction heat transfer.

In addition to general review articles [1-7], including a book [5], and a recent state-of the-art review [8], a number of publications are devoted to theoretical modeling of nanofluids thermal characteristics, mostly thermal conductivity.

Theoretical models could be classified in two broad categories, static and dynamic models. Static models accounts for the different geometrical, static structures of nanoparticle-fluid heterogeneous mixture. From the classical Maxwell mixture theory (for low concentration of uniformly distributed particles in fluid) and several related extensions, including particle geometry and directional clustering/percolation (anisotropicity, including series and parallel limits), matrix-particle layering effects (liquid layering around nanoparticle, etc.). The dynamic models are based on some type of nano-convections phenomena, induced by thermal Brownian motion, thermophoresis, diffusiophoresis, and other electro-magnetic phenomena, including near field radiation, thermal waves, dual-phase lagging, and other special phenomena, like ballistic phonon transport in nanoparticles, etc.

A comprehensive review is given recently by [8], showing significant discrepancies among the experimental data available, and between the experimental findings and the theoretical model predictions. Buongorio [52] analyzed convective transport in nanofluids, considering seven mechanisms that can produce a relative velocity between the nanoparticles and the base fluid:

1) inertia,
2) Brownian diffusion,
3) thermophoresis,
4) diffusiophoresis,

5) Magnus effect,
6) fluid drainage, and
7) gravity

He concluded that, of these seven, only the Brownian diffusion and thermophoresis are important slip mechanisms in nanofluids. These findings for convective transport are also relevant for thermal conductivity due to their interdependency and certain similarities. Some modeling results are contradictive and find both, significant but also little or no influence of certain phenomena on nanofluid thermal conductivity. For example, Aminfar et al. [53] used a Lagrangian-Eulerian approach and found, contrary to some studies, that thermophoretic and Brownian forces do not have remarkable effects on nanofluid conductivity.

Wang and Wei [54] synthesized eight kinds of nanofluids with controllable microstructures by a chemical solution method (CSM) and develop a theory of macroscale heat conduction in nanofluids. Their theory shows that heat conduction in nanofluids is of a dual-phase-lagging type instead of the postulated and commonly used Fourier heat conduction. Due to the coupled conduction of the two phases, thermal waves and possibly resonance may be responsible for the nanofluid conductivity enhancement. Furthermore, they emulsify olive-oil into distilled water to form a new type of "thermal-wave fluids" that can support much stronger thermal waves and resonance than all reported nanofluids, and consequently obtained an extraordinary conductivity enhancement (up to 153.3%). However, Vadasz and Govender [55] used the hyperbolic heat conduction equation to investigate theoretically the heat transfer enhancement observed experimentally in nanofluids suspensions. They ruled-out the possibility that thermal wave effects could explain the improved effective thermal conductivity of the suspension.

Due to complexity of diverse naoparticles-additives-fluids structures and their interfacial dynamic interactions, including inter-coupling of many phenomena, as well as inherited simplification of theoretical modeling, the modeling results are often contradictory and at-best inconclusive. One simple and illustrative example will be presented next, where a "apparently reasonable" (but turned out to be unrealistic) approximation contributed to a huge errors in the modeling results. Namely, it is common practice to approximate temperature distribution and heat flux as unidirectional for heterogeneous mixtures if exposed to "over-all unidirectional" boundary conditions. This approach has been used by Yu and Choi [56] to model and to arrive at an effective (or over-all average) thermal conductivity of heterogeneous mixtures (nanofluids). It is shown by Kostic [57], however, that due to the heterogeneity of system structure and properties, the temperature distribution and heat flow will not be unidirectional (one-dimensional) and the errors due to such unrealistic (physically impossible) approximation may be much higher than anticipated. We could only imagine how uncertain the results are and could be using rather "primitive" modeling of very complex nanofluids structures and their dynamic, multi-coupled interactions.

## 5.1. Effective Thermal Conductivity Errors Due to Assuming Uni-directional Temperature and Heat Flux Distribution within Heterogeneous Mixtures (Nanofluids)

A unidirectional analysis and results of evaluating the effective (*e*) thermal conductivity $k_e$ of nanofluids, a heterogeneous mixture of uniformly distributed spherical particles (*p*) in common fluid (*f*), is provided by Yu and Choi [56] for the cubical arrangement of spherical particles in base fluid, further-on referred to as the "*cubic model*" (without a liquid sublayer, thus corresponding to the Maxwell [30] model, i.e., the effective media theory for dilute solutions where thermal interactions between particles are negligible). The effective thermal conductivity ratio, $k_e^* = k_e / k_f = 1.33$, based on the "cubic model," for volumetric concentration, $C_v = V_p / (V_p + V_f) = 1\%$, of copper particles in water ($k_p^* = k_p / k_f = 401/0.615 = 652$), was substantially higher than the corresponding Maxwell equation value of only 1.03. Since the enhancement difference, for the virtually the same heterogeneous concept, is surprisingly large (33% versus 3% increase), the unidirectional analysis of the cubic model is revisited again and its physical shortcomings are analyzed below, see also Figure 16.

Furthermore, the same problem (the exactly same geometry, material properties and boundary conditions) was solved using a FEM numerical method (thus solving the full, 3-dimensional heat conduction differential equation) and the results agreed with the Maxwell model (within expected numerical errors), which confirms that the unidirectional "cubic model" is unrealistic and thus physically inappropriate.

The erroneous, "cubic model" over-prediction of the effective thermal conductivity (TC) is circumstantial and does not explain experimentally observed increase of TC of nanofluids due to other reasons, like different interfacial particle-fluid interactions, special particle alignments and agglomerations in force-flux fields, particle Brownian motion, and other known and unknown phenomena. The Maxwell and "cubic model" predictions, based on continuum media theory, everything else being the same, should give the same results for the low volumetric particle-concentration, regardless of the spherical particle size. For example, a 1 *cm* copper sphere centered in a cube with a solid material of TC being the same as of water, or 10 *nm* copper nano-sphere centered in "stationary" water cube at the same low volumetric concentration, should have the same effective TC if other phenomena are absent as the two models assume, but their predictions differ substantially.

### *5.1.1. Analysis*

For the so-called uniform distribution of a small concentration of solid particles in base fluid, a uniform cubic arrangement (Yu and Choi [56]) may be assumed and analysis performed for the characteristic cell of a single particle centered in a cube of fluid as presented on Figure 16.

This also corresponds to the Maxwell model, i.e. the random distribution (equally in all directions) of spherical particles which do not interact thermally (far away from each other, i.e. for relatively small particle concentrations). Those continuous (but heterogeneous) media models produce the same results for the intensive properties (per unit of system size) and similar systems, thus the results do not depend on the absolute, but on relative system size.

Therefore, the effective thermal conductivity (TC) for uniform particle distribution will not depend on the particle size but only the concentration in the base fluid for the same properties of the particles and the fluids.

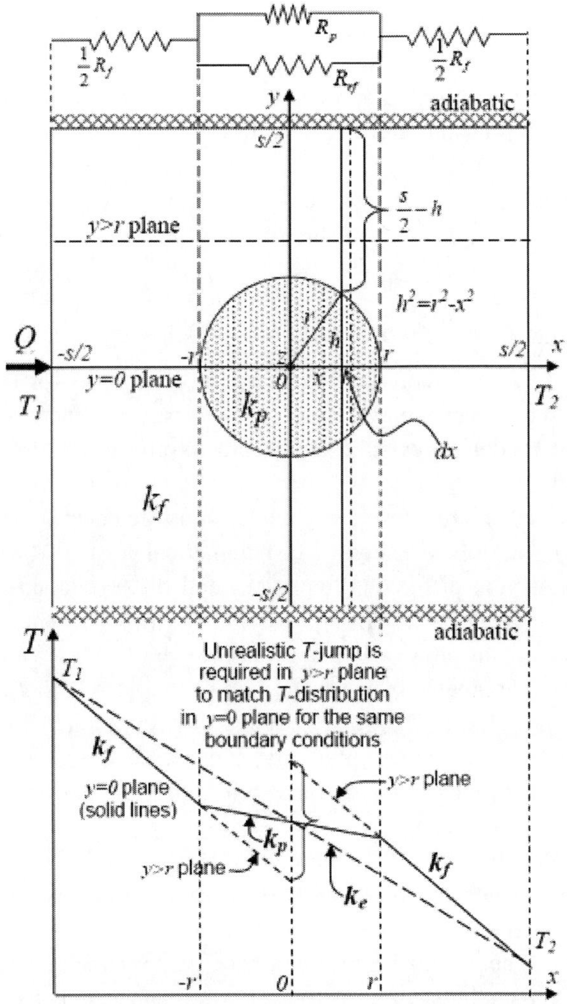


Figure 16. Characteristic heterogeneous cell for "uniform distribution" of spherical particles (*p*) of radius, *r,* centered in a cube of side, *s,* with fluid (*f*) corresponding (thermally) to an effective (*e*) homogeneous mixture.

The objective is to evaluate the so-called "effective thermal conductivity," i.e., to reduce the heterogeneous system or its representative cell to the corresponding homogeneous system with hypothetical effective TC which will provide the same conduction heat transfer under arbitrary boundary conditions (TC being a property). Constant temperature boundary condition (BC) on two opposite cube faces and adiabatic on all other cube faces, are chosen for simplicity, see Figure 16, so that the effective (over-all average) thermal conductivity may be simply determined, using the Fourier law of heat conduction as, $k_e = \dfrac{Q}{A(T_1 - T_2)/s}$; where

$Q$ is the total heat conduction rate through the box of side $s$ from one face at temperature $T_1$ to the opposite face at a lower temperature $T_2$ (in $x$-direction on Figure 16 ); and $A = s \cdot s$ is the boundary face area. Since the boundary heat flux is in the $x$-direction only (the other boundary faces being adiabatic) it is tempting to assume that the heat flux, and thus temperature distribution within the heterogeneous box is unidirectional (on Figure 16 in $x$-direction only), resulting in the following correlations for the steady state and one-dimensional heat conduction, $\frac{\partial()}{\partial t} = \frac{\partial()}{\partial y} = \frac{\partial()}{\partial z} = 0;\ Q(x) = const$, i.e.:

$$Q = A \cdot k_e \frac{\overbrace{T_1 - T_2}^{\Delta T}}{s} = s^2 k_e \frac{\Delta T}{s}$$

$$\Rightarrow \frac{\Delta T}{Q} = \frac{1}{s \cdot k_e} = const \tag{12}$$

$$Q = -A \cdot k \cdot \frac{dT}{dx} \Rightarrow -dT = \frac{Q}{A \cdot k} dx$$

$$\Rightarrow -\int_{T_1}^{T_2} dT = Q \int_{-s/2}^{s/2} \frac{dx}{A(x) \cdot k(x)} \tag{13}$$

$$\frac{\overbrace{T_1 - T_2}^{\Delta T}}{Q} = \underbrace{\int_{-s/2}^{-r} \frac{dx}{(s^2)k_f}}_{R_f/2} + \underbrace{\int_{-r}^{0} \frac{dx}{(\pi h^2)k_p + (s^2 - \pi h^2)k_f}}_{R_{p-ef}/2} + \underbrace{\int_{0}^{r} \frac{dx}{(\pi h^2)k_p + (s^2 - \pi h^2)k_f}}_{R_{p-ef}/2} + \underbrace{\int_{r}^{s/2} \frac{dx}{(s^2)k_f}}_{R_f/2}$$

$$\frac{1}{s \cdot k_e} = 2 \cdot \left[ \int_0^r \frac{dx}{(\pi h^2)k_p + (s^2 - \pi h^2)k_f} + \int_r^{s/2} \frac{dx}{(s^2)k_f} \right], \text{ where } h^2 = r^2 - x^2 \text{ (see Fig.16)} \tag{14}$$

$$\frac{1}{k_e} = 2s \left[ \int_0^r \frac{dx}{(\pi(r^2 - x^2)k_p + (s^2 - \pi \cdot r^2 + \pi x^2)k_f} + \frac{(s/2) - r}{(s^2)k_f} \right] \tag{15}$$

Or, in dimensionless form, by introducing the following substitutions:

$$k_e^* = k_e / k_f;\ k_p^* = k_p / k_f$$

$$x^* = x / r;\ s^* = s / r \tag{16}$$

$$k_e^* = \frac{1}{1 - \frac{2}{s^*} + 2s^* \int_0^1 \frac{dx^*}{s^{*2} + \pi(1 - x^{*2})(k_p^* - 1)}}$$

$$= \frac{1}{1 - \frac{2}{s^*} + 2s^* \int_0^1 \frac{dx^*}{\underbrace{[s^{*2} + \pi(k_p^* - 1)]}_{A^2} - \underbrace{[\pi(k_p^* - 1)]}_{B^2} x^{*2}}}$$

$$= \frac{1}{1 - \frac{2}{s^*} + \frac{s^*}{A \cdot B} \cdot \ln \frac{A+B}{A-B}}, \text{ where,}$$

$$A = \sqrt{s^{*2} + \pi(k_p^* - 1)};$$

$$B = \sqrt{\pi(k_p^* - 1)} \text{ and } s^* = \sqrt[3]{\frac{4\pi}{3C_v}} \tag{17}$$

For a given particle radius $r$ and desired volumetric concentration $C_v$, the required cube size $s$ is obtained from:

$$C_v = \frac{V_p}{V_f + V_p} = \frac{\frac{4\pi}{3} r^3}{s^3} = \frac{4\pi}{3 \cdot s^{*3}}$$

$$\text{or } s^* = \sqrt[3]{\frac{4\pi}{3C_v}} \tag{18}$$

The above correlations are derived using the same modeling as done by Yu and Choi [56] and the result (Eq. 17) is identical as their Eq. (21) [56], also presented here for reference as Eq. (19) with current nomenclature (in original Yu and Choi [56], $r^* = 1/s^*$ ), along with Maxwell [30] correlation, Eq. (20):

$$k_{eYC}^* = \frac{1}{1 - \frac{2}{s^*} + \frac{s^*}{\sqrt{s^{*2} + \pi(k_p^* - 1)} \cdot \sqrt{\pi(k_p^* - 1)}} \cdot \ln \frac{\sqrt{s^{*2} + \pi(k_p^* - 1)} + \sqrt{\pi(k_p^* - 1)}}{\sqrt{s^{*2} + \pi(k_p^* - 1)} - \sqrt{\pi(k_p^* - 1)}}} \tag{19}$$

$$k_{e\,Mxw}^* = 1 + \frac{3C_v}{1 - C_v + \frac{3}{k_p^* - 1}} = \frac{2 + k_p^* + 2C_v(k_p^* - 1)}{2 + k_p^* - C_v(k_p^* - 1)}; \tag{20}$$

where

$$C_v = \frac{4\pi}{3(s^*)^3}$$

As stated above, the effective thermal conductivity ratio, $k_e^* = k_e / k_f = 1.33$, based on the "cubic model," Eq. (17 or 19), for volumetric concentration, $C_v = V_p /(V_p + V_f) = 1\%\ vol$, of copper particles in water ( $k_p^* = k_p / k_f = 401/0.615 = 652$ ), was substantially higher than the corresponding Maxwell equation value of only 1.03 using Eq. (20). The difference of the results, for the virtually the same heterogeneous concept, is surprisingly large (33% versus 3% increase).

The substantial errors of the cubic model results are due to unrealistic assumptions that the heat flux within the heterogeneous particle-fluid cell, see Figure 16, is in $x$-direction only, i.e. that the temperature field is function of $x$ only. As seen on Figure 16, for the given constant-temperature BCs, if the temperature is function of $x$ only as is assumed, the local heat flux must be constant for the steady state conduction heat transfer process through the homogeneous liquid regions, $-s/2<x<-r$ and $r<x<s/2$. However, such constant heat flux will produce different temperature drops along different parallel paths in the non-homogeneous particle-fluid region, $-r<x<r$: for example, through the particle (in $y=0$ plane) and outside of the particle region through the fluid ($|y|>r$ plane), thus "breaking" the temperature continuity from one to the other BCs, which is physically impossible.

This will always be an issue except in three special cases: (1) serial particle distribution so that heterogeneity (change of material properties are in $x$ direction only but not in the other ($y$ & $z$ ) directions so that integration along $x$ will properly account for any variability; (2) parallel particle distribution so that heterogeneity is layered between the same temperature BCs, but the heat flux will be different to accommodate different TC of the parallel layers; and (3) serial distribution of segments with parallel particle distribution but without any contact resistance between the segments to provide for equalization of temperature and redistribution of different heat fluxes in transverse direction as needed for different parallel sub-layers. The last case is unrealistic since there is no ideal thermal contact case, but is widely use in electrical circuitry where different arrangements of electrical resistors are interconnected with connectors of negligible resistance. Similar concepts of thermal resistances in series and/or parallel combinations are used but errors could be much higher than anticipated as is illustrated here.

The effective TC may be derived using the related thermal resistances as depicted at the top on Figure 16, and the effective TC will be the same as the cubic model, Eq. (17), i.e.:

$$R_e = R_f + \underbrace{\left( \frac{1}{\frac{1}{R_p} + \frac{1}{R_{rf}}} \right)}_{R_{p+rf}}; \qquad (21)$$

where

$$R_e = \frac{s}{(s^2)k_e}; \; R_f = \frac{s-2r}{(s^2)k_f};$$

$$R_p = \int_{-r}^{r} \frac{dx}{(\pi h^2)k_p}; \; R_{rf} = \int_{-r}^{r} \frac{dx}{(s^2 - \pi h^2)k_f}$$

$$Note: \; h^2 = r^2 - x^2$$

***5.1.2. The Limiting Cases***

The effective thermal conductivity, $k_{e\_S}$ and $k_{e\_P}$, for the limiting cases of the serial and parallel arrangement of particles in a base fluid, respectively, see Figure 17, may be easily calculated as:

$$R_{e\_S} = R_{p\_S} + R_{f\_S} \; \text{or}$$

$$\frac{s}{(s^2)k_{e\_S}} = \frac{a}{(s^2)k_p} + \frac{s-a}{(s^2)k_f} \tag{22}$$

where

$$a = C_v \cdot s$$

After introducing dimensionless variable (Eq. 16) and simplifying, the above equation for the serial model reduces to:

$$\frac{1}{k^*_{e\_S}} = 1 - C_v + \frac{C_v}{k^*_p} \; \text{or}$$

$$k^*_{e\_S} = \frac{k^*_p}{C_v + (1-C_v)k^*_p} \tag{23}$$

For the parallel particle arrangement in base fluid (see Figure 17 bottom):

$$\frac{1}{R_{e\_P}} = \frac{1}{R_{p\_P}} + \frac{1}{R_{f\_P}} \; \text{or}$$

$$\frac{(s^2)k_{e\_P}}{s} = \frac{(s \cdot a)k_p}{s} + \frac{s(s-a)k_f}{s}; \tag{24}$$

where

$$a = C_v \cdot s$$

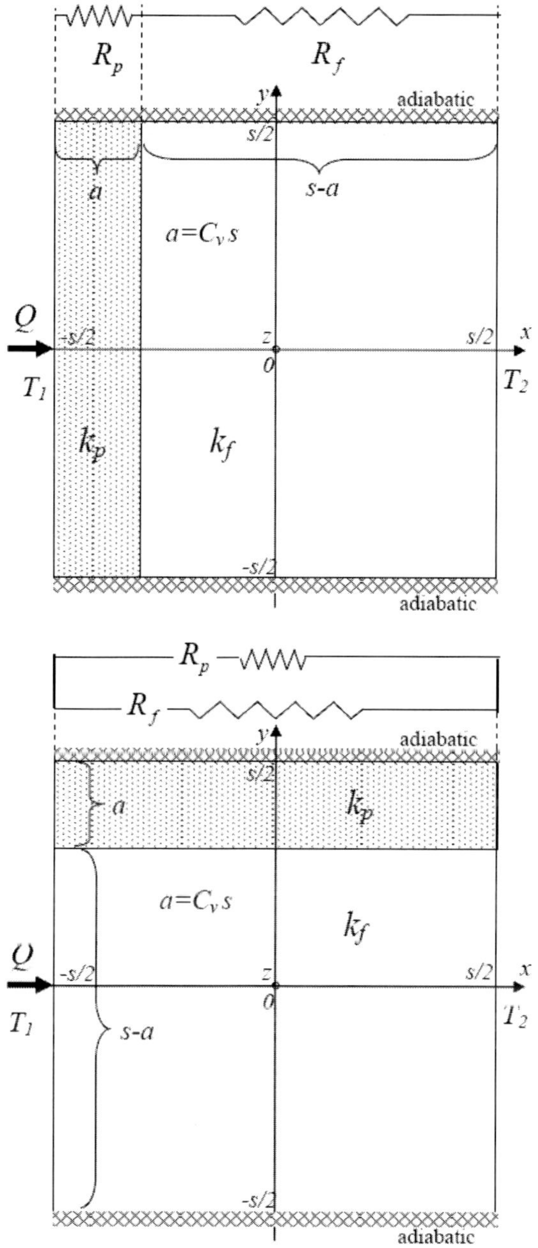


Figure 17. Characteristic heterogeneous cell for limiting case of serial (top) and parallel arrangement (bottom) of particles (*p*) in a cube of side, *s,* with fluid (*f*) corresponding (thermally) to an effective (*e*) homogeneous mixture.

### *5.1.3. A Full 3-D Numerical Solution*

The "cubic model" problem (the same geometry, properties and boundary conditions) was solved using a FEM numerical method [35]. The full 3-Dimensional heat conduction partial differential equation is used (thus equivalent to the Maxwell model) and the results, see Table 2, are in agreement with Maxwell results (within numerical accuracy, e.g., 2.72% ~ 3% increase for 1% *vol* copper particles in water), which is much different from the unidirectional (1-D) cubic model (the latter resulting in 33% effective TC increase for the same 1% *vol* copper particles in water). The obtained temperature distribution, see Figure 18, was changing in all directions around the particle, thus, confirming that the cubic model, 1-Dimensional temperature and heat flow assumptions and results, are not appropriate, i.e., they are erroneous due to unrealistic assumptions for the case it was supposed to model, see also Table 2.

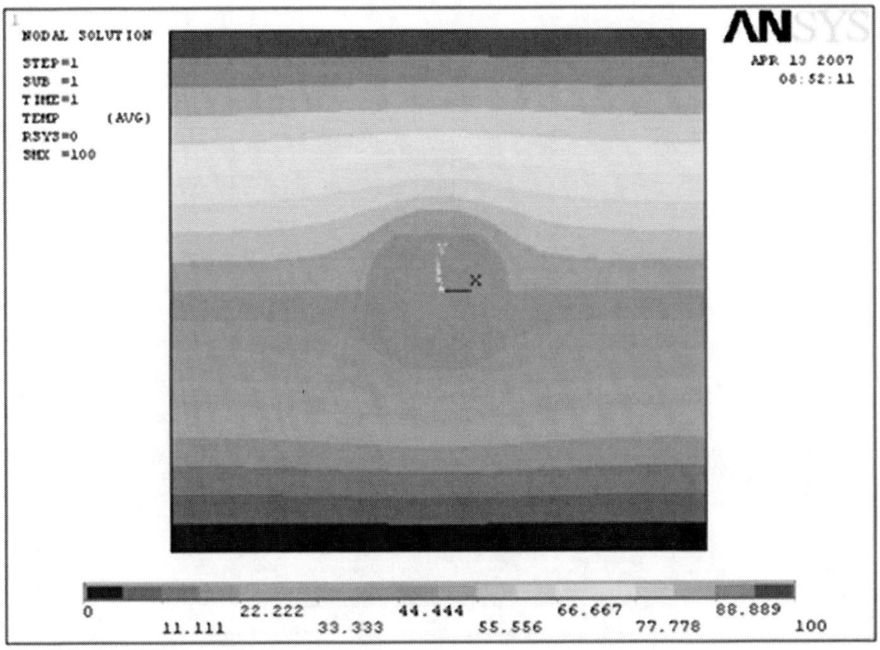


Figure 18. Temperature distribution along the center-plane ($z=0$) for spherical copper particle (1% *vol* concentration) centered in cube of water, determined using a FEM numerical method [solving the full 3-dimensional heat conduction partial differential equation, corresponding to the Maxwell (1873) model].

**Table 2. Effective thermal conductivity ratios $k_e^*$ for various concentrations of copper particles in water ($k_p^* = k_p / k_f = 401/0.615 = 652$)**

| **Particle concen. %$C_v$** | **Cubic model[1)]** $k^*_{e\_YC}$ **Eq. (19)** | **Maxwell model[2)]** $k^*_{e\_Mxw}$ **Eq. (20)** | **FEM Numerical method[3)]** $k^*_{e\_N}$ | **Serial model** $k^*_{e\_S}$ **Eq. (23)** | **Parallel model** $k^*_{e\_P}$ **Eq. (25)** |
|---|---|---|---|---|---|
| 0.1 | 1.109 | 1.003 | 1.003 | 1.001 | 1.650 |
| 0.5 | 1.237 | 1.015 | 1.014 | 1.005 | 4.255 |
| **1** | **1.332** | **1.030** | **1.027** | **1.010** | **7.511** |
| 1.5 | 1.407 | 1.045 | 1.041 | 1.015 | 10.766 |
| 2 | 1.473 | 1.061 | 1.055 | 1.020 | 14.021 |

You and Choi [56]; [2)] Maxwell [30]; [3)] Simham [35].

It should be stated that the continuum media theory accounts for the heterogeneous distribution of properties and geometry only, and the results are as good as the physical assumptions in related modeling. For example, the Maxwell [30] model, Eq. (20), properly accounts for random distribution of small concentration of spherical particles in fluid (thus uniform distribution), but does not account for any other phenomena that may contribute to heat conduction, like special particle alignments and agglomerations in force-flux fields, different interfacial particle-fluid interactions, particle nano-convection due to Brownian motion, and other known and unknown phenomena.

The unidirectional heat flow modeling, like the "cubic model [56]" is similar to the Maxwell model, since it accounts for uniform distribution of spherical particles in a base fluid, but due to unrealistic (actually physically impossible) assumption of unidirectional heat flow and temperature distribution within the heterogeneous system (the representative mixture cell), the results may be very erroneous as demonstrated here. The erroneous, "cubic model" over-prediction of the effective thermal conductivity (TC) [56] is circumstantial and does not physically explain experimentally observed increase of TC of nanofluids due to other phenomena. The Maxwell and "cubic model" predictions, based on continuum media theory, everything else being the same, should give the same results for all similar system scales (i.e., the same small concentrations), regardless of the absolute, spherical particle size.

The serial and parallel particle distribution in base fluid, for a given particle-in-fluid mixture concentration, other phenomena being absent, represent the limiting minimum and maximum enhancement of effective thermal conductivity of the mixture, respectively, due to directional clustering of higher conductive particles suspended in lower conductive fluid. The summary of characteristic results is presented in Table 2 for comparison.

## 6. NANOFLUIDS FLOW AND HEAT TRANSFER CHARACTERISTICS

Thermal conductivity studies have been the focus of the initial nanofluid research, but ultimately, their flow and heat transfer characteristics in real, practical applications will determine their usefulness as advanced flow and heat-transfer fluids. The addition of nanoparticles in a base fluid has been shown to significantly improve the heat transfer capabilities of certain nanofluids. The nanoparticles also adversely impact the flow capabilities of the base fluids, so it is necessary to balance the two properties in order to create the optimal suspension. A review of related literature is provided [9].

An apparatus for exploring friction and heat transfer characteristics of nanofluids in tube flow is presented in next section, including rather peculiar initial results. The apparatus has been recently developed, instrumented and computerized. It has been calibrated using distilled water for which the results are well established and correlated for both laminar and turbulent flow. Initial turbulent friction and heat transfer measurements for silica and carbon nanotube (CNT) nanofluids show peculiar results with substantial friction drag reduction and heat transfer enhancement for lower Reynolds numbers, but heat transfer reduction for higher Reynolds numbers. The apparently anomalous flow and heat transfer behaviors are hypothesized to be due to the conflicting influence of nanoparticle heat transfer enhancement but reduction due to additives used to stabilize inherently unstable nanoparticle suspensions in base fluid. More research is needed to establish conclusive results and development of innovative nanofluids with optimized nanoparticles and additives for improved flow and heat transfer characteristics for existing critical and novel applications.

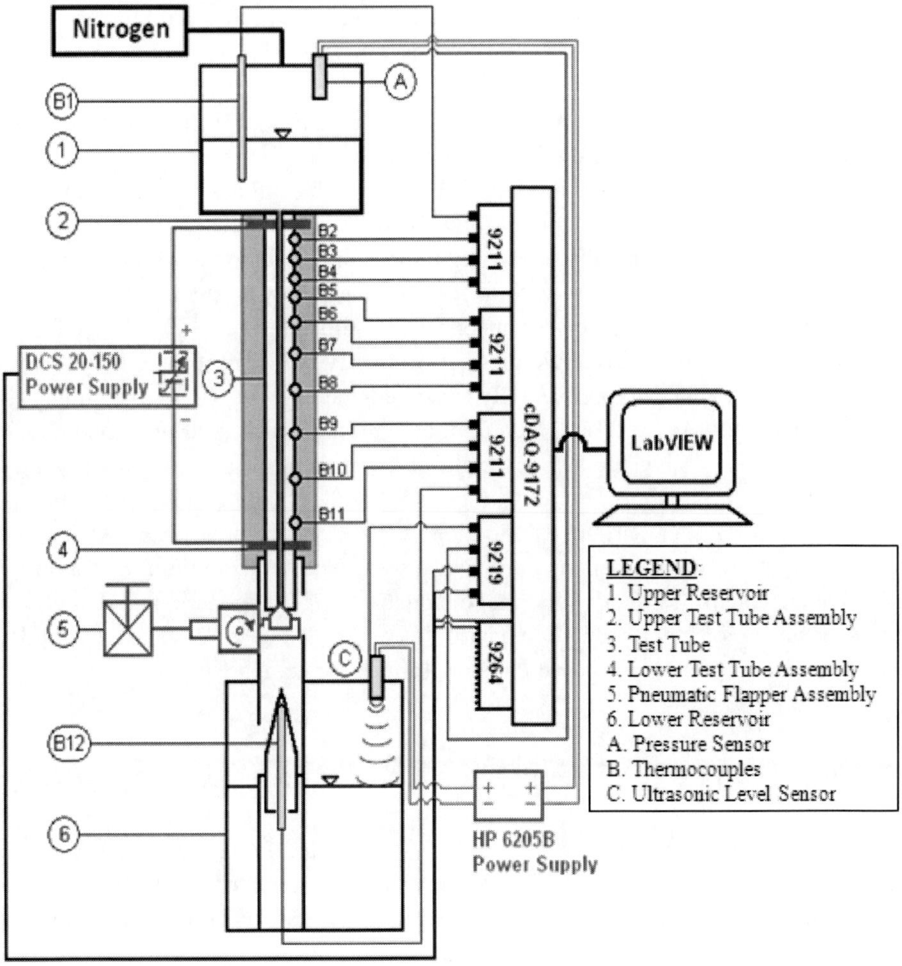


Figure 19. Flow and Heat-Transfer Apparatus Schematics with Instrumentation and Data Acquisition Components.

## 6.1. Flow and Heat-Transfer Apparatus, Instrumentation and Data Acquisition Method

The schematic of the flow and heat-transfer apparatus (with all components labeled) is presented in Figure 19, while details of the apparatus design are described elsewhere [58]. The apparatus has been developed to investigate the flow friction and convective heat transfer characteristics of nanofluids. Instead of a usual closed-loop system where pumps and after-cooling units are required, the developed apparatus utilizes nitrogen pressure-driven flow to test a single batch of fluid. This reduces the complexity of the system while improving its reliability and accuracy.

The upper reservoir, made of heavy-gauge 316 stainless-steel, is designed to accommodate up to 1 liter of a test-fluid under regulated nitrogen pressure up to 1500 psi, and instrumented with a pressure and temperature sensors, see Figure19. It has two heavy-duty stainless-steel flanges for cleaning and maintenance access with appropriate ports for instrumentation, nitrogen line and test-fluid feed line, the latter also used for release of nitrogen pressure when needed.

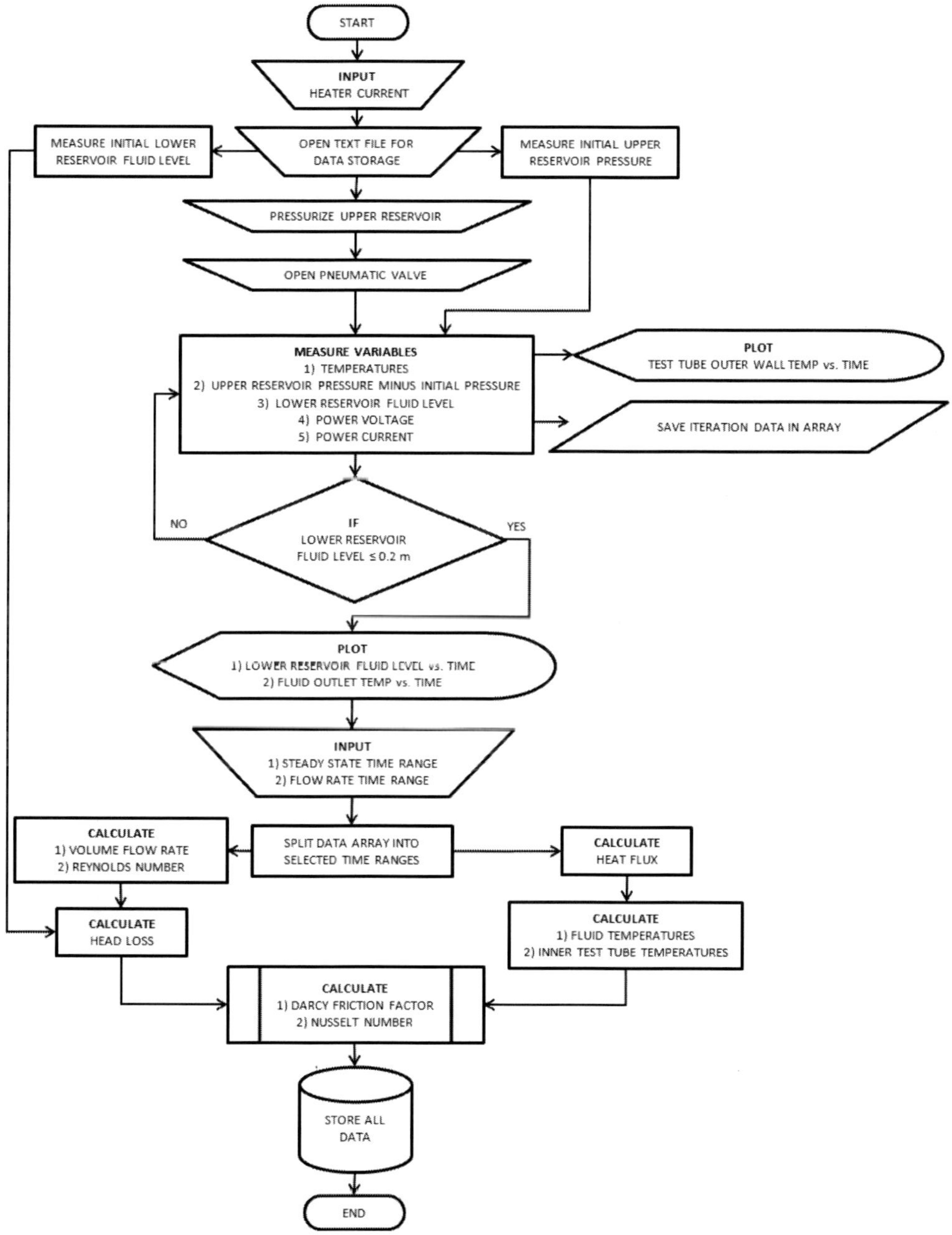


Figure 20. Flow and Heat-Transfer Apparatus Data Acquisition Algorithm.

The 316-stainless-steel test-tube is 36-inch long, with a 0.069-inch inner diameter and 0.125-inch outer diameter, resulting in a dimensionless L/D ratio of 522, long enough to provide fully-developed flow over most of its length. The test tube is mounted in the upper and lower test-tube assemblies by soldering it to the copper connectors wired to a DC low-voltage but high-current power supply with computerized control. Thus the test tube acts as both the flow channel and electrical-resistance heater. Electrical DC current is passed through the test-tube in order to heat up the flowing fluid. To insure minimal heat loss to the surrounding, the test tube is well insulated with thick fiberglass along its entire length, so that virtually all heating power is transferred from the heated stainless-steel test tube (with a thermal conductivity of 13.4 W/mK [58]) to the flowing test fluid inside the tube. A number of Tefon-coated 30 AWG type-T thermocouples are mounted on the outside of the test tube (where radial temperature gradient is negligible) in order to accurately measure the temperature distribution along the tube in the axial direction, and then the corresponding inner wall temperatures are easily calculated. A thermal epoxy with low thermal and high electrical resistivity is used to mount the thermocouples on the test tube, as detailed elsewhere [58].

The lower reservoir, open to the atmosphere, is designed to collect the test fluid, exiting the test-tube, and instrumented with a thermocouple to measure the fluid exit temperature and with an ultrasound level sensor to measure the fluid level and thus its collected volume in time, needed to calculate the flow rate and cross-sectional average fluid-velocity, see Figure 19, with full details in [58].

National Instruments' data acquisition hardware is used (see Figure 19), and LabVIEW® software is developed to acquire and process the test data, see Figures 20 and 21. Details are described in [58].

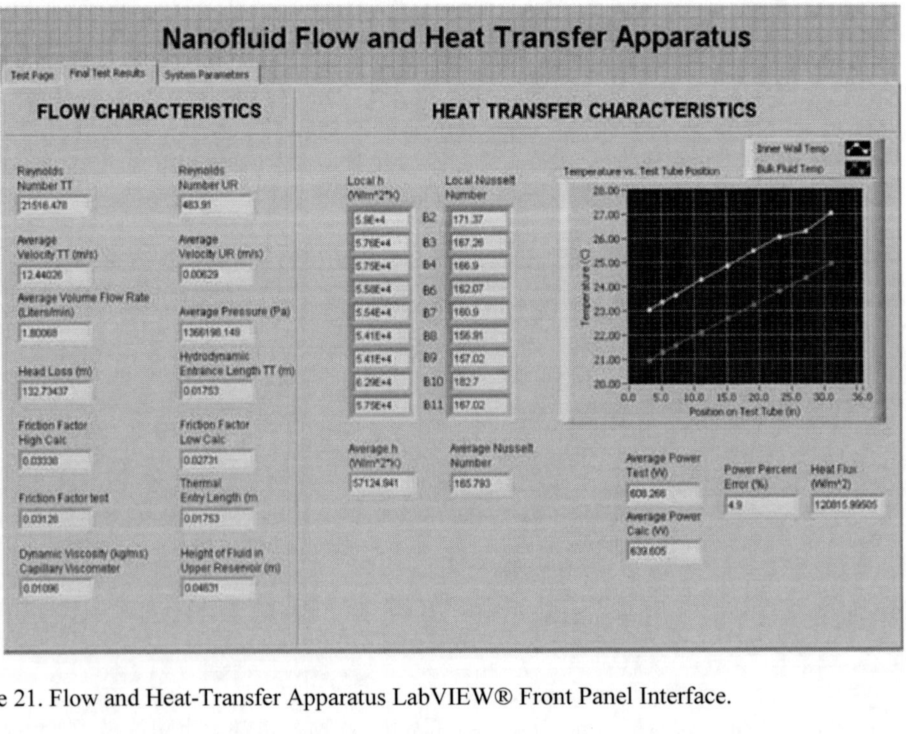


Figure 21. Flow and Heat-Transfer Apparatus LabVIEW® Front Panel Interface.

#### *6.1.1. Experimental and Data Reduction Methods*

An experimental test (or "run") is conducted by filling in the test-fluid in the upper reservoir, and using the developed LabVIEW® program which controls relevant actuators (open the pneumatic valve to start the test-fluid flow, turn on heating, etc.) and acquire measured data in time (DC power supply voltage and current, fluid inlet pressure and temperature in the upper reservoir, test-tube wall temperatures, and collected test-fluid exit temperature and level in the lower reservoir), see Figures 19, 20 and 21. Full details are described in [58]. Prior to execute the LabVIEW® program to monitor the flow data, the upper reservoir pressure is measured while open to the atmosphere and lower reservoir ultrasonic level is recorded, both to be used to zero out the relevant sensors. Then the upper reservoir has to be closed and pressurized to a desired pressure, measured or known fluid properties and desired heating DC current have to be entered via LabVIEW® FrontPanel user-interface, see Figures 20 and 21, and then the program is executed to acquire data at 2 Hz sampling rate (unless otherwise desired) during the flow and heat transfer run, and calculate the results as per algorithm in Figure 20. Selected data are interactively displayed and all data are stored in a file on a local storage media, as described in [58].

During a sample-period the average values of flow rate and heating power are acquired and calculated using the simple correlations for the flow and heat transfer in a steady-state tube flow, since the changes of parameters are negligible during a small sample period (usually 0.5 second). However, the change of flow and heat transfer characteristics in time during the whole batch run is monitored and recorded accordingly, as well as to verify the peace-wise steady-state assumption. The LabVIEW® program automates all measurements and calculations, with very little input from a user.

### 6.2. Calibration and Uncertainty Analysis

In order to validate the operation and accuracy of the apparatus, the system was fully calibrated using distilled water for laminar flow with Re<2000 and turbulent flow with up to Re=40,000. The Darcy friction factor was found to be within 4% of the established values, while the Nusselt number was found to be within 5% of the accepted Gnielinski approximation, both with 95% confidence. In addition, a conservative and detailed uncertainty error analysis has been performed for the measurement results using the method of propagation of errors, resulting in a maximum conservative uncertainty of 8% at 95% probability, for heat transfer coefficient [58]. Most contributing error uncertainties are due to the ultrasonic level sensor (with uncertainty of about 1.5%; calibrated against a lab grade caliper within the operating range of 2 to 14 inches) and thermocouples (with uncertainty of 0.1°C; calibrated against a lab grade RTD sensor). The laminar entry lengths are long and measured water calibration values should be compared with the appropriate entrance region data; however, for turbulent flow data, comparison with the corresponding fully-developed reference data are justified since the entry length is negligible compared to the test tube length of 522 diameters. The turbulent heat transfer results for distilled water are presented on Figure 22 and compared with the well-established Gnielinski correlation. Those results were used for over-all calibration of the apparatus in the range of current interest. The bias over-all error of 4.2% and precision error of about 2% were observed, both within 95% confidence, resulting to over-all uncertainty within 5% at 95% confidence. However, for comparison measurements under similar conditions the bias error will cancel out thus resulting in comparative uncertainty of about 2%, which is much smaller than the conservative uncertainty estimate using the propagation of elemental errors' method. The details are presented in [58].

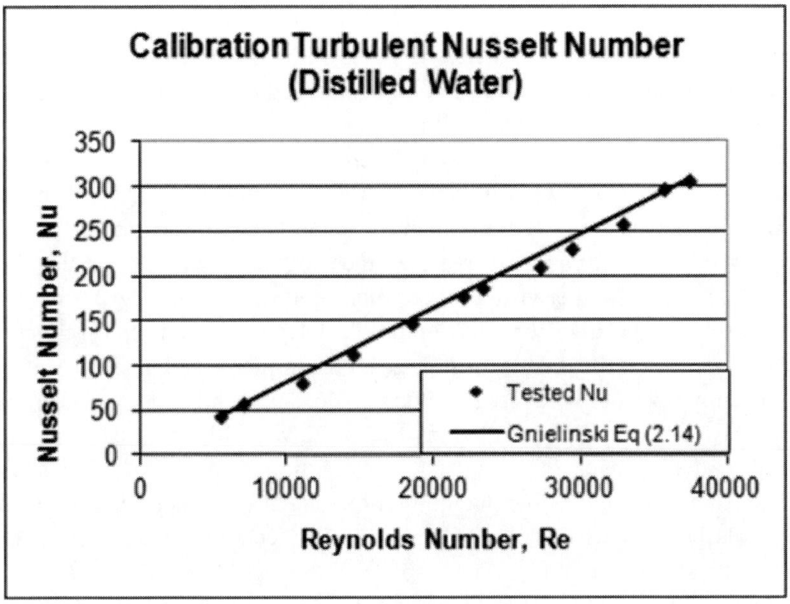


Figure 22. Over-all Calibration of Flow and Heat-Transfer Apparatus with Distilled Water.

## 6.3. Nanofluids Flow and Heat Transfer Results

Two types of nanofluids have been tested and compared with the corresponding distilled water measurements. Several concentrations of silica nanofluids have been prepared by diluting a 40 wt% (by weight) silica dispersion (with effective 50 nm $SiO_2$ nanoparticles of 1.28 W/mK thermal conductivity in water with 9 pH value, manufactured by Allied High Tech Products, Inc.). The NaOH pallets have been added to distilled water until pH value was increased to 9, and then it was used to dilute the original 40 wt% silica dispersion to obtain desired nanofluid concentrations of 1%, 5%, 10%, and 20 % by weight (wt%). After shaking the mixture vigorously, the container was placed in an ultrasonic vibrator for one hour before it is being tested.

First the nanofluids thermal conductivity was measured using HWTC apparatus [32] and dynamic viscosity was measured using a commercial, Brookfield cone-and-plate digital viscometer. The measured thermal conductivities of all silica nanofluids were close to the water value (within the uncertainty of the apparatus), however, the dynamic viscosity has been showing shear-rate dependence at higher silica concentrations, for 20 wt% being more than 20 times higher than water's at 1000 $s^{-1}$ shear rate.

The second type of nanofluids tested were 1 wt% Carbon multi-walled nanotube (CNT or MW-CNT) suspensions in water with unknown additives (sold under the *Nanosolve* name from Zyvex Performance Materials Co.). The first test results (CNT-1) were replicated later

(CNT-2) under similar conditions since the first obtained results have been "unexpected" and somewhat unstable, probably due to unknown additives (trade secret) used to stabilize, otherwise unstable MW-CNT suspensions in water; also possibly not being mixed/ultrasonicated well enough before testing. The thermal conductivity for the first CNT nanofluids (CNT-1) were measured to be close to water's (unexpected result) and the second set (CNT-2, with prolonged ultrasonication) exhibited 23% thermal conductivity enhancement (more reasonable result). The measured dynamic viscosity showed shear rate dependency (non-Newtonian effect; increase from 5.5 to 11 cP while decreasing shear rate from 384 to 77 $s^{-1}$ [58]), probably due to the unknown additives used during manufacturing of the CNT nanofluid.

The limited, flow and heat transfer results for all tested nanofluids are presented on Figures 23 and 24, together with distilled water results for comparison purposes. Surprisingly, all test results showed a reduction of friction factors in turbulent flow for all tested nanofluids, suggesting the additives used for nanoparticles stabilization may have so-called turbulent drag-reducing characteristics [14]. The higher silica concentration, the higher friction factor reduction for the same Reynolds (Re) number, probably due to commensurate higher concentration of additives used, being over 60% of reduction for 10 wt% silica at 40,000 Re number, and even higher for the CNT nanofluids, suggesting that the latter additives used are more effective friction drag reducers than the silica's additives. The CNT nanofluid showed strong reduction of friction factor in turbulent flow, being 75% lower than the distilled water value at 30,000 Re number (see Figure 23). The drag reduction phenomena are not that widely known within the nanofluids' research community and such anomalous phenomena may introduce further confusion in addition to rather unstable nanofluid structure and disagreement among fluid characteristics, between apparently similar nanofluids, where nanoparticle type and concentration are taken into the account, but type and amount of additives used to stabilize the mixture is often unjustifiably overlooked.

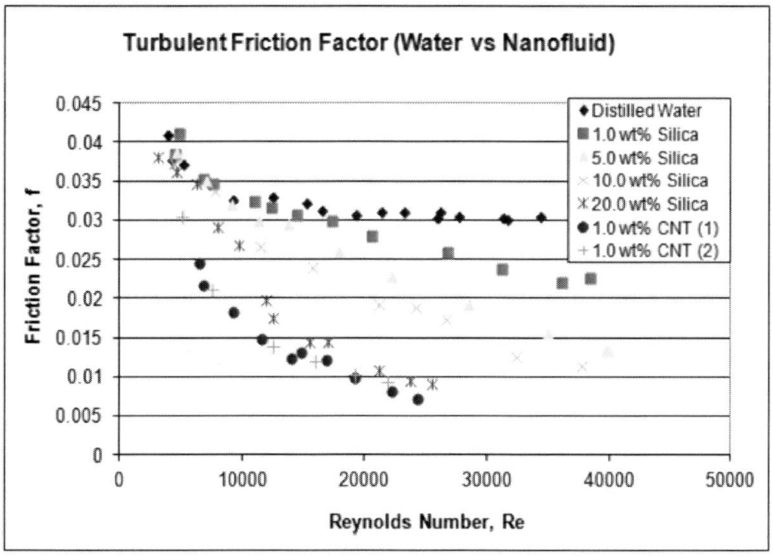


Figure 23. Turbulent Friction Factor Results for Water, Silica and Carbon Nanotube (CNT) Nanofluids.

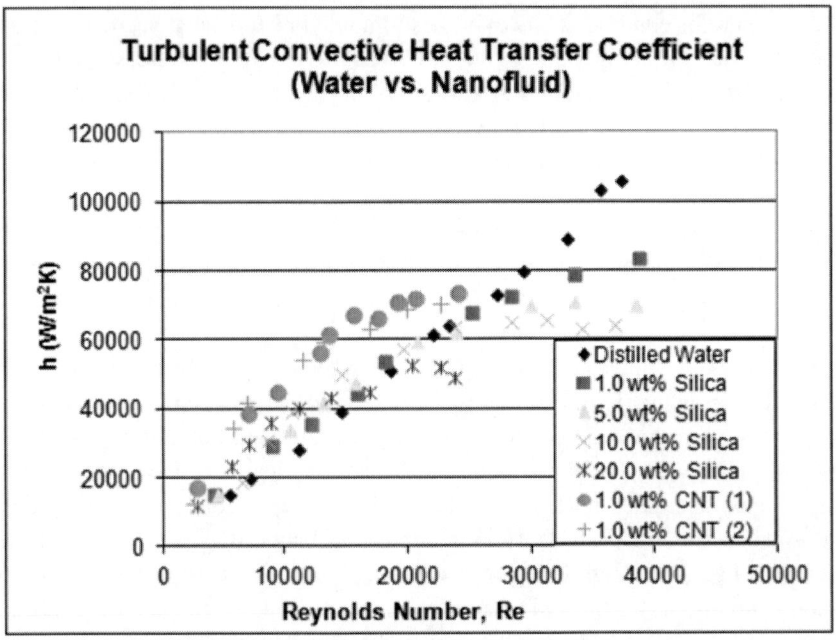


Figure 24. Turbulent Heat-Transfer Results for Water, Silica and Carbon Nanotube (CNT) Nanofluids.

The heat transfer results are presented in Figure 24. Up to about 20,000 Re number, the silica nanofluids have the convective heat transfer enhancement (compared with water results) for all the concentrations (1 to 20 wt%). The heat transfer increase, with the 20 wt% silica nanofluid at about 5,000 Re, was about 155% of the distilled water value. At about 20,000 Re number all the nanofluids begin a trend of reduced thermal performance, whereas at about 30,000 Re number the 20 wt% silica nanofluid had the lowest performance at 68% of the distilled water values. The second type of tested nanofluids was with 1.0 wt% carbon-nanotube (CNT) nanoparticles. The observed enhancement of the convective heat transfer coefficient was up to 100% larger than the distilled water values at 5000 Re number. At 30,000 Re number the trend of diminished convective heat transfer performance was observed as compared to the distilled water values, which is probably due to the turbulent heat transfer reduction due to additives in CNT nanofluid.

## 6.4. Conclusion

The turbulent friction drag reduction was observed for all nanofluids tested. The enhancement of convective heat transfer due to presence of nanoparticles is observed for the smaller values of Re numbers, where turbulent heat transfer reduction due to additives used is not strong enough to neutralize the enhancement. However, for higher values of Re numbers, the turbulent heat transfer reduction is predominant and stronger than the heat transfer enhancement due to nanoparticles, resulting in over-all reduction in convective heat transfer. This hypothetical reasoning based on the two known and opposing phenomena, nanoparticle heat transfer enhancement and additive heat transfer reduction in turbulent flow, may explain many surprising and apparently anomalous behaviors, as well as discrepancies between research results in the literature. However, due to limitation of this initial testing, more measurements are needed with different types of nanoparticles and additives before more conclusive results can be established. It is hoped that these unexpected and inconclusive results will initiate constructive criticism and further investigations, related to many open issues in nanofluids research to date.

## Conclusion

Development of many industrial and new technologies is limited by existing thermal management, and need for enhanced heat transfer and high-performance cooling. Nanofluids, stable colloidal mixtures of nanoparticles (including nanorods, nanofibers, nanotubes, nanoshits, and functional nanocomposites, even nano-droplets and nano-bubbles) in common fluids, have a potential to meet these and many other challenges.

Regardless of extensive research and publications in recent past, the state-of-the-art in nanofluids research is still in initial phases, in part due to rather complex nature of nanoparticle materials and even more complex nanoparticle-base fluid interactions, often involving diverse surfactants and stabilization additives to prevent inherited tendency of nanoparticle conglomeration and clustering in the unstable colloidal mixtures. The inherited complexity and inconsistency make it difficult to establish well-defined and reliable experimental results to verify existing hypotheses and to improve and develop new ones.

Absence of good quality and consistent nanofluids limits the progress of the future research and applications in this and related areas. In this article a number of critical issues are addressed and several selected improvements are described. An improved "One-step method for the production of nanofluids" is presented. It has been developed in Argonne National Laboratory and a U.S. patent has been issued recently.

The related measurements of thermal conductivity (TC) in the literature, mostly with transient hot-wire thermal conductivity (HWTC) apparatus, have been inconsistent and with measured thermal conductivities far beyond prediction using the well-known mixture theory. An improved HWTC apparatus is described. In addition, a steady-state, parallel-plate thermal conductivity (PPTC) apparatus has been developed and used for comparative measurements of complex nanofluids, in order to compare results with the corresponding measurements using the transient HWTC apparatus. Both apparatuses are described in details to facilitate their replication and possible future improvements. The objective was to check out if existing and well-established HWTC method might have some unknown issues while measuring TC of complex nano-mixture suspensions, like electro-magnetic phenomena, undetectable hot-wire vibrations, and others.

The initial and limited measurements have shown considerable difference in TC measurements using the two methods, whereby the measured TC increase beyond the base fluid TC by developed PPTC apparatus was about three times smaller than the comparative measurements of apparently the same nano-mixtures using the HWTC apparatus, though with significant scatter, the former data being much closer to the Maxwell mixture theory prediction. However, the influence of measurement method on the TC results is not conclusive since it has been noticed that the complex nano-mixture suspensions were very unstable during the lengthy steady-state measurements as compared to rather quick transient

HWTC method, so the two nano-mixture suspensions were really not the same. The nanofluid suspension instability might be the main reason for very inconsistent results in the literature.

Another important, but overlooked issue with use of additives in nanofluids is a possibility to drastically reduce friction drug in turbulent flow (desired effect; used in Trans-Alaska oil pipeline system), which is usually accompanied with commensurate heat-transfer reduction, the letter being an undesirable, adverse effect. These "drag reduction" and heat-transfer reduction phenomena are well-known by some researchers but are not widely known and thus overlooked by nanofluid researchers.

There are significant discrepancies among the experimental data available, and also between the experimental findings and the theoretical model predictions, related to nanofluids thermo-physical characteristics. Oddly, there are more hypothetical theories proposed than reliable experimental results to verify those theoretical models.

Due to complexity of diverse nanoparticles-additives-fluids structures and their interfacial dynamic interactions, including inter-coupling of many phenomena, as well as inherited simplification of theoretical modeling, the modeling results are often contradictory and at-best inconclusive. One simple and illustrative example is presented, where a "apparently reasonable" (but turned out to be unrealistic) approximation contributed to a huge errors in the modeling results. We could only imagine how uncertain the results are and could be using rather "primitive" modeling of very complex nanofluids structures and their dynamic, multi-coupled interactions.

Development of new hybrid nanofluids is a new challenge and opportunity. It may open the road for development of diverse, complex nanofluids with polymer additives (including organic nanofluids), with many and unprecedented applications in existing critical areas as well as emerging and novel applications. The trend in further development of nanomaterials is to make them multifunctional and controllable by external means or by local environment thus essentially turning them into useful nano-devices. Colloidal nano-mixtures with functionally-stable and active-like nanostructures that may self-adjust to the process conditions, require systematic surface-chemistry study and enhancements (coatings with functional layers, surfactants, etc.), in addition to investigation of thermo-physical characteristics and phenomena.

A comprehensive, systematic and interdisciplinary experimental research program is necessary to study, understand and resolve critical issues in nanofluids research to date. The research must focus on both synthesis and a careful exploration of surface-chemistry and thermo-physical characteristics. Development of new-hybrid, drag-reduction nanofluids may lead to enhanced both flow and heat transfer characteristics. The nanoparticles in these fluids yield increased heat-transfer while the long-chain polymers are expected to enhance flow properties, including active interactions with nanoparticles, thus providing potential for many applications yet to be developed and optimized. However, frictional drag reduction usually accompanies heat transfer reduction, so the proper selection and optimization of additives may be a daunting endeavor. It is known that some polymer additives are friction drug-reducers in turbulent flow, but also heat-enhancers in laminar flow. Furthermore, nanofluids may be used in micro scale devices, where the flow is usually laminar, but also in large scale devices (common heat exchangers), where both laminar and turbulent flows are encountered.

The ultimate goal is to understand the underlying physical phenomena of diffusion, momentum and energy transport in these novel nanofluids, by correlating and modeling measured nano- and macro-characteristics, thus making possible development and production-optimization of tailor-made nanofluids with significantly enhanced thermo-physical properties, critical for existing and emerging flow and heat transfer applications.

In conclusion, the nanofluids were hyped-up in the past, but it would be a mistake to hype-down nanofluids now and make premature judgments based on inconsistent and incomplete research to-date.

## ACKNOWLEDGEMENT

The author acknowledges prior collaboration with Argonne National Laboratory and support by National Science Foundation (Grant No. CBET-0741078), and additional and continuous support by the College of Engineering and Engineering Technology at Northern Illinois University.

## REFERENCES

[1] Choi, S.U.S., Z.G. Zhang, and P. Keblinski, 2004, Nanofluids, Encyclopedia of Nanoscience and Nanotechnology (H.S. Nalwa, Editor), Vol. 5. Pp.757-773, American Scientific Publisher.

[2] Eastman, J.A., S.R. Phillpot, S.U.S. Choi, P. Keblinski, 2004, "Thermal Transport in Nanofluids," Annu. Rev. Mater. Res., 34, 219–46.

[3] Das, S.K., Choi, S.U.S., and Patel, H.E., 2006, "Heat Transfer in Nanofluids—A Review," Heat Transfer Eng., 27(10), 3–19.

[4] Wang, X.Q. and A.S. Mujumdar, 2007, "Heat Transfer Characteristics of Nanofluids: a Review," *International Journal of Thermal Sciences*, 46(1), 1–19.

[5] Das, S.K., Choi, S.U.S., Yu, W., and Pradeep, T., 2008, *Nanofluids: Science and Technology*, Wiley, Hoboken, NJ.

[6] Yu, W., D.M. France, S.U.S. Choi, and J.L. Routbort, 2007, " Review and Assessment of Nanofluid Technology for Transportation and Other Applications," Tech. Rep. 78, ANL/ESD/07-9, Argonne National Laboratory.

[7] Kakaç, S. and A. Pramuanjaroenkij, 2009, "Review of Convective Heat Transfer Enhancement with Nanofluids," *International Journal of Heat and Mass Transfer*, 52(13-14), 3187–3196.

[8] Özerinç, S., S. Kakaç, and A. G. Yazıcıoğlu, 2010, "Enhanced thermal conductivity of nanofluids: a state-of-the-art review," *Microfluidics and Nanofluidics*, 8(2), 145–170.

[9] Yu, W. and H. Xie, 2012, "A Review on Nanofluids: Preparation, Stability Mechanisms, and Applications," *Journal of Nanomaterials*," 2012, 435873.

[10] Choi, S.U.S., and Eastman, J.A., 1995, "Enhancing Thermal Conductivity of Fluids with Nanoparticles," *Developments and Applications of Non-Newtonian Flows*, D. A. Siginer and H. P. Wang, eds., ASME, New York, FED-231/MD Vol. 66, 99–105.

[11] Turner, S.W., Turner, A.M. Perez, A. Lopez, and H.G. Craighead, 1998, "Monolithic nanofluid sieving structures for DNA manipulation," *J. Vac. Sci. Technol.* B, 16(6), 3835-3840.

[12] Pozar, L.A. and K.E. Gubbins, 1998, "Quasihydrodynamics of Nanofluid Mixtures," *Phys. Rev. E 56*, 5367-5396.

[13] NanoHex, 2012, European collaborative nanotechnology project that aims to develop a revolutionary cooling system for a range of industrial applications.

[14] Buongiorno, J., Venerus, D. C., Prabhat, N., McKrell, T., Townsend, J., Christianson, R., Tolmachev, Y. V., Keblinski, P., Hu, L., Alvarado, J. L., Bang, I. C., Bishnoi, S. W., Bonetti, M., Botz, F., Cecere, A., Chang, Y., Chen, G., Chen, H., Chung, S. J., Chyu, M. K., Das, S. K., Di Paola, R., Ding, Y., Dubois, F., Dzido, G., Eapen, J., Escher, W., Funfschilling, D., Galand, Q., Gao, J., Gharagozloo, P. E., Goodson, K. E., Gutierrez, J. G., Hong, H., Horton, M., Hwang, K. S., Iorio, C. S., Jang, S. P., Jarzebski, A. B., Jiang, Y., Jin, L., Kabelac, S., Kamath, A., Kedzierski, M. A., Kieng, L. G., Kim, C., Kim, J.-H., Kim, S., Lee, S. H., Leong, K. C., Manna, I., Michel, B., Ni, R., Patel, H. E., Philip, J., Poulikakos, D., Reynaud, C., Savino, R., Singh, P. K., Song, P., Sundararajan, T., Timofeeva, E., Tritcak, T., Turanov, A. N., Van Vaerenbergh, S., Wen, D., Witharana, S., Yang, C., Yeh, W.-H., Zhao, X.-Z., Zhou, S.-Q., 2009, "A Benchmark Study on the Thermal Conductivity of Nanofluids," *J. Appl. Phys,* 106, 094312.

[15] Hong, T.K., and Yang, H.S., 2005, "Nanoparticles-Dispersion-Dependent Thermal Conductivity in Nanofluids," *Journal of the Korean Physical Society*, 47, S321-S324.

[16] Granqvist C. G., and Buhrman, R. A., 1976, "Ultrafine Metal Particles," *Journal of Applied Physics,* 47, 22002219.

[17] Li, Y., J. Zhou, S. Tung, E. Schneider, and S. Xi, 2009, "A review on Development of Nanofluid Preparation and Characterization," *Powder Technology*, 196(2). 89–101.

[18] Romano, J.M., Parker, J.C., and Ford, Q.B., 1997, "Application Opportunities for Nanoparticles Made from Condensation of Physical Vapors," *Adv. Powder Methal. Partic. Mater*, 2, 12-13.

[19] Yatsuya, S., Y. Tsukasaki, K. Mihama, and R. Uyeda, 1978, "Preparation of Extremely Fine Particles by Vacuum Evaporation onto a Running Oil Substrate," *J. Crist. Growth*, 45, 490-494.

[20] Wagener, M., Murty, B.S., and Gunther, B., 1997, "Preparation of Metal Nanosuspensions by High Pressure DC-Sputtering on Running Liquids," *Nanocrystalline and Nanocomposite Materials* II (S. Komarnenl, J.C. Parker, and H.J. Wolenberger, Eds.), 149-154, Materials Research Scosiety.

[21] Choi, S.U.S., and Eastman, J.A., 2001, "Enhanced Heat Transfer Using nanofluids," U.S. Patent No. US 6,221,275 B1.

[22] Eastman, J.A., Choi, S.U.S., Li, S., Yu, W., and Thompson, L.J., 2001, "Anomalously Increased Effective Thermal Conductivities Of Ethylene Glycol-based Nanofluids Containing Copper Nanoparticles," *Applied Physics Letters*, 78(6), 718–720.

[23] Kostic, M., Golubovic, M., Hull, J.R., and Choi, S.U.S., 2010, "One-Step Method for the Production of Nanofluids," U.S. Patent No. 7,718,033 B1.

[24] Toms, B.A., 1949, "Some Observations on the Flow of Linear Polymer Solutions through Straight Tubes at Large Reynolds Numbers," *Proc. 1$^{st}$ Int. Cong. Rheol. Vol II*, p.135, North Holland, Amsterdam.

[25] Metzner, A.B., 1965, "Heat Transfer in Non-Newtonian Fluids," *Adv. Heat Transfer, Vol.2*, 357-397.

[26] Virk, P.S., Mickley, H.S., and Smith, K.A., 1970, "The Ultimate Asymptote and Mean Flow Structure in Toms' Phenomenon," ASME, *J. Appl. Mech.,* 37, 488-493.

[27] Hartnett, J.P. and M. Kostic, 1989, "Heat Transfer to Newtonian and Non-Newtonian Fluids in Rectangular Ducts,"*Advances in Heat Transfer*, Vol. 19, 247-356, Academic Press.

[28] Kostic, M., 1994, "On Turbulent Drag and Heat Transfer Reduction Phenomena and Laminar Heat Transfer Enhancement in Non-Circular Duct Flow of Certain Non-Newtonian Fluids," *Int. J. Heat & Mass Transfer*, 37(Suppl. 1), 133-147.

[29] Patel, H.E., Sundararajan, T., and Das, S.K., 2010, "An Experimental Investigation into the Thermal Conductivity Enhancement in Oxide and Metallic Nanofluids," Journal of Nanoparticle Research, 12(3), 1015-1031.

[30] Maxwell, J.C., 1873, *Electricity and Magnetism*, Clarendon Press, Oxford, U.K.

[31] Simham, K.C., 2008, *Development of Computerized Transient Hot-Wire Thermal Conductivity (HWTC) Apparatus for Nanofluids*, M. S. Thesis, Northern Illinois University, DeKalb, IL.

[32] Kostic, M. and Simham, K.C., 2009, "*Computerized, Transient Hot-Wire Thermal Conductivity Apparatus for Nanofluids*," Proceedings The 6th WSEAS International Conference On Heat And Mass Transfer (HMT'09), Ningbo, China, January 10-12, 2009. In *Recent Advances in Heat and Mass Transfer* (Editor: Lifeng Xi), ISBN: 978-960-474-039-0, WSEAS Press, p. 71-78.

[33] Walleck, C.J., 2009, *Development of Steady-State, Parallel-Plate Thermal Conductivity Apparatus for Poly Nanofluids And Comparative Measurements with Transient HWTC Apparatus*, M. S. Thesis, Northern Illinois University, DeKalb, IL.

[34] Kostic, M. and Walleck, C., 2010, "Design of a Steady-State, Parallel-Plate Thermal Conductivity Apparatus for Nanofluids and Comparative Measurements with Transient HWTC Apparatus," Proceedings of the ASME 2010 International Mechanical Engineering Congress & Exposition, IMECE2010-38187 (12 pp), November 12-18, 2010, Vancouver, British Columbia, Canada, ASME, New York.

[35] Simham, K.C., 2007, Numerical Analysis and Evaluation of Effective Thermal Conductivity of Nanofluids with Spherical Nanoparticle in Cubical Arrangement, *MEE 555 Prof. Kostic's Course, Project Report.* Northern Illinois University.

[36] Kostic, M., 2006, "Critical Issues and Applications Potentials in Nanofluids Research," *ASME-MN2006 Multifunctional Nanocomposites International Conference, Honolulu, Hawaii,* ASME Proceedings, MN2006-17036, 2006.

[37] POLY-nanofluids (Polymer-nanofluids) & DR-nanofluids (Drag-Reduction-nanofluids):http://www.kostic.niu.edu/DRnanofluids/; Accessed in December 2011.

[38] Carslaw, H.S. and Jaeger, J.C., 1959, *Conduction of Heat in Solids*, 2nd edition. Oxford: Clarendon Press.

[39] Nagasaka, Y., and Nagashima, A., 1981, "Absolute Measurement of the Thermal Conductivity of Electrically Conducting Liquids by the Transient Hot-Wire Method," *J. Phy. E: Sci. Instrum.*, 14, 435-40.

[40] Yamasue, E., Susa, M., Fukuyama, H., and Nagata, K., Nonstationary, 1999, "Hot-Wire Method with Silica-Coated Probe for Measuring Thermal Conductivities of Molten Metals," *Metallurgical and Materials Transactions* A, 30, 1971-79.

[41] Perkins, R.A., 1994, "Measurement of the Thermal Properties of Electrically Conducting Fluids using Coated Transient Hot Wires," *Presented at the 12th*

*Symposium on Energy Engineering Sciences,* Dept. of Energy, CONF-9404137, 90-97, 1994.

[42] Yu, W., Hull, J.R., and Choi, S.U.S., 2003, "Stable and Highly Conductive Nanofluids – Experimental and Theoretical Studies," *The 6th ASME – JSME Thermal Engineering Conference,* Hawaii Islands, Hawaii.

[43] Jwo, C. S., Teng T. P., Hung C. J., and Guo Y. T., 2005, "Research and Development of Measurement Device for Thermal Conductivity of Nanofluids," *Journal of Physics: Conference Series,* 13(1), 55-58.

[44] Ma, J.J., 2006, *Thermal Conductivity of Fluids Containing Suspensions of Nanometer-Sized Particles*, M.S. Thesis, Massachusetts Institute of Technology.

[45] Kierkus, W. T., Mani, N., and Venart, J. E. S., 1973, "Radial – Axial Transient Heat Conduction in a Region Bounded Internally by a Circular Cylinder of Finite Length and Appreciable Heat Capacity," *Canadian Journal of Physics*, 51, 1182-86.

[46] Healy, J. J., deGroot, J. J., and Kestin, J., 1976, "The Theory of the Transient Hot-Wire Method for Measuring Thermal Conductivity," *Physica C*, 82, 392-408.

[47] Sengers, J. V., 1962, "Experimental Determinations of the Thermal Conductivity of Fluids," *Thermal Conductivity, Vol. 2.* (Ed. Tye, R.P), Academic Press, London.

[48] National Instruments Web site: http://www.ni.com ; Accessed in December 2011.

[49] Singh, B.P., Menchavez, R., Takai, C., Fuji, M., Takahashi, M.T., 2005, "Stability of dispersions of colloidal alumina particles in aqueous suspensions," *Journal of Colloid and Interface Sciences*, 291, 181-186.

[50] Oja, T., 2009, *What is Zeta Potential?* Colloid Measurements Company LLC, Scarsdale, NY.

[51] Wei, X., Zhu, H., Kong, T., Wanga, L., 2009, "Synthesis and thermal conductivity of $Cu_2O$ Nanofluids," *Int. J, Heat and Mass Transfer,* 52, 4371–4374.

[52] Buongiorno, J., 2006, "Convective Transport in Nanofluids," *J. Heat Transfer*, 128(3), 240

[53] Aminfar, H., Motallebzadeh, R., and Farzadi, A., 2010, "The Study of the Effects of Thermophoretic and Brownian Forces on Nanofluid Thermal Conductivity Using Lagrangian and Eulerian Approach," *Nanoscale and Microscale Thermophysical Engineering*, 14, 187–208

[54] Wang L.Q., and Wei X., 2009, "Nanofluids: Synthesis, Heat conduction, and Extension," *ASME J. Heat Transfer,* 131, 033102.

[55] Vadasz, P., 2006, "Heat Conduction in Nanofluid Suspensions," *Journal of Heat Transfer*, 128, 465-477.

[56] Yu, W. and Choi, S.U.S., 2005, "An Effective Thermal Conductivity Model of Nanofluids with a Cubical Arrangement of Spherical Particles," *J. Nanosci. Nanotech,* 5, 580-586.

[57] Kostic, M., 2008, "Effective Thermal Conductivity Errors by Assuming Unidirectional Temperature and Heat Flux Distribution Within Heterogeneous Mixtures (Nanofluids)," Theoretical and Experimental Aspects of Heat and Mass Transfer (Editors: Krope, J., Sohrab, S. H. and Benra, F.-K.), 44-49, WSEAS Press.

[58] Netemeyer, C., 2012, *Development and Calibration of an Apparatus and Investigation of Flow and Heat Transfer Characteristics of Nanofluids*, MS Thesis, Northern Illinois University.

In: Nanofluids
Editor: Yuwen Zhang
ISBN: 978-1-62618-165-6


*Chapter 2*

# NANOFLUIDS HEAT TRANSFER: PREPARATION, CHARACTERIZATION AND THEORETICAL ASPECTS

***Agus P. Sasmito*[1, 2, 3,*], *Saif A. Khan*[1, 2] and *Arun S. Mujumdar*[1, 4]**
[1]Minerals Metals Materials Technology Centre, National University of Singapore, Singapore
[2]Department of Chemical and Bio-Molecular Engineering, National University of Singapore, Singapore
[3]Mechanical Engineering, Masdar Institute of Science and Technology, Abu Dhabi, United Arab Emirates
[4]Department of Mechanical Engineering, National University of Singapore, Singapore

## ABSTRACT

Over the past decade, research in heat transfer enhancement using nanofluids – suspensions of nanometer-sized solid particles in a base liquid – has received considerable attention all over the world. However, the value of thermophysical properties of suspensions and heat transfer characteristics reported in literature by several authors are in disagreement. Several theoretical works to predict the mechanism of heat transfer enhancement, the effective thermophysical properties of the suspension as well as convective heat transfer, range from a homogeneous model to complex two-phase flow models have been proposed. Validation of theoretical studies with good experimental data is essential for proper understanding of the phenomena. This chapter provides insight into the recent development of nanofluids from preparation, characterization to theoretical aspects and ways to improve nanofluid stability so as to be able to commercialize and implement potential applications.

* Corresponding author, Email: ap.sasmito@gmail.com.

# 1. Introduction

Convective heat transfer can be enhanced by active as well as passive methods. While the former usually provide better enhancement, they requires external forces and/or equipment which increase the complexity as well as capital and operating costs of the system. In contrast, passive heat transfer enhancement can be achieved by changing flow geometry or modifying thermo-physical properties of working fluid. Hence, it is generally a more desirable approach compared to an active method. Addition of nanoparticles to base fluid is a passive enhancement technique that is drawing much attention.

It is well known that conventional heat transfer fluids including water, oil and ethylene glycol mixtures yield poor heat transfer rates due to their low thermal conductivity. The thermal conductivity of these fluids can be increased by suspending small (micro or larger size) particles of diverse materials in heat transfer fluids since these solids typically have higher thermal conductivity as compared to the liquids, example of such materials are presented in Table 1. The basic idea of dispersing solid particle is not new. This method was originally proposed by Maxwell [1] about 140 years ago. He developed a theoretical model to predict the effective conductivity of a suspension which applicable for the particle size ranging from milimeter to micrometer. The main drawback of such suspensions is the potential for settling out of the solid particle from the suspension due to their relatively large particle size and density. The lack of stability of such suspensions induces additional flow resistance and possible erosion.

Advancement in nanotechnology provides new opportunities to process and produce particles with average crystallite size below 50 nm. By dispersing such nanoparticles in heat transfer fluids, one can obtain nanofluids, a term introduced by Choi in1995 [2]. A nanofluid is expected to offer higher heat transfer rate; but good stability of the suspension by eliminating possible agglomeration and sedimentation to permit long term application, is a pre-requisite to successful commercialization.

Given its potency in heat transfer application, a vast number of experimental and numerical studies have been conducted in the past decade to investigate and predict various properties and heat transfer performance of numerous nanofluids; refer to Figure 1 for details of increase of the number publication in nanofluids area. Several reviews of nanofluids have also been published, e.g., Wang and Mujumdar [3-5]; Chandrashekar and Suresh [6]; Daungthongsuk and Wongwises [7]; Trisaksri and Wongwises [8]; Kakác and Pramuanjaroenkij [9]; Wang and Fan [10]; Godson et al. [11]; Gadhimi et al. [12]; Murshed et al. [13]; Das et al. [14]; Wen et al. [15]; Li et al. [16]; Yu et al. [17]. While these reviews have covered almost all aspects of the current aspects of experimental and theoretical studies of nanofluids, the progress in nanofluids study need to be re-surveyed since a large number of new studies have been reported, some of which propose new hypotheses and/or new phenomena. In this chapter we present an overview of the literature dealing with recent progress in the study of nanofluids. The preparation of nanofluids is outlined in Section 2; this is followed by a review of characterisation techniques for nanofluids. Section 4 addresses the theoretical aspects – heat transfer enhancement mechanisms and modelling – of nanofluids. Finally concluding remarks are drawn in Section 5.

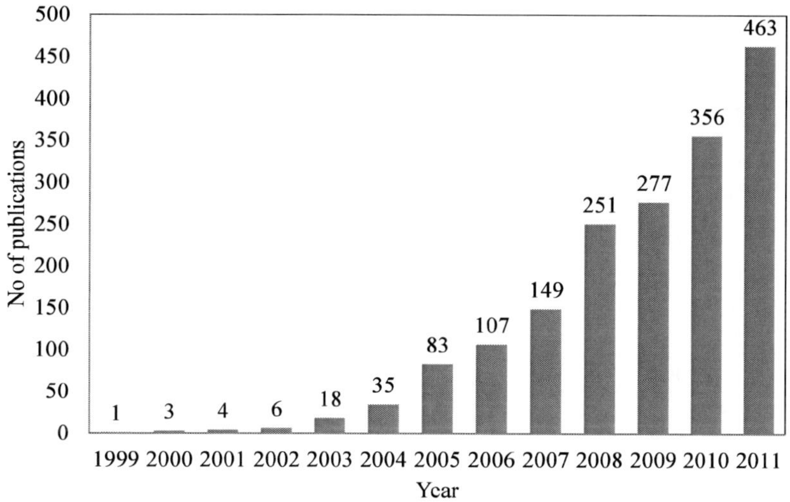


Figure 1. Number of publications on nanofluids in the past decade from ISI Web of Science.

## 2. Preparation Methods

In principle, almost all solid nanoparticles of high thermal conductivity materials can be used as additives to form nanofluids; the most common solid materials used are metallic and non-metallic solids, as summarized in Table 1. The thermal conductivity of solid particles is generally much higher (around two to four orders-of-magnitude higher than that of the base fluid). However, in order to make nanofluids competitive in the market for real applications, the cost of solid materials is an important factor. Sergis and Hardalupas [18] showed, based on their statistical analysis, that most of the published literature deals with $Al_2O_3$ suspension as it is cheaper than others. However, if cost is not an issue and the heat transfer is of paramount importance, carbon nanotubes can be a desired choice.

Preparation of nanofluids is the key first step in the use of nanoparticles to enhance heat transfer, control the thermophysical properties and ensure the stability of the suspension. Broadly, there are two types of preparation methods, i.e., single step and two step method.

### 2.1. Single Step Method

In the single step method, the nanoparticle preparation and nanofluid preparation are carried out simultaneously. The nanoparticles are directly prepared by a physical vapour deposition (PVD) technique or a liquid chemical method (condensing nanophase powders from the vapour phase directly into a flowing low-vapor-pressure fluid, VEROS). In this method, agglomeration of nanopaticles is minimized and the stability of the nanofluid is increased as storage, transportation, drying and dispersion of nanoparticles are avoided. However, this method only applicable for small scale production and, at current stage, it is almost impossible to scale up to industrial scale. Furthermore, this method is only applicable for low vapour pressure basefluids which limit its application. More research is required to address the challenge to scale-up the processes and/or implement it to any basefluids.

**Table 1. Thermal conductivity and indicative prices of various solids and liquids**

| Material | | Thermal Conduct-ivity (W/ m K) | Indicative price ($/100g) |
|---|---|---|---|
| Metallic solids | Copper (Cu) | 401 | 500 |
| | Aluminum (Al) | 237 | 380 |
| | Silver (Ag) | 428 | 400 |
| | Gold (Au) | 318 | 5,500 |
| | Iron (Fe) | 83.5 | - |
| Non metallic solids | Silicon (Si) | 148 | - |
| | Alumina ($Al_2O_3$) | 40 | 70 |
| | Silicon Carbide (SiC) | 270 | - |
| | Carbon nanotubes (CNTs) | 3000-6000 | 930 – 12,500 |
| | Copper Oxide (CuO) | 76.5 | 75 |
| | Titanium Oxide ($TiO_2$) | 13.7 | - |
| Metalic liquids | Sodium (644 K) | 72.3 | - |
| Non metallic liquids | Water ($H_2O$) | 0.613 | - |
| | Ethylene Glycol (EG) | 0.253 | - |
| | Engine Oil (EO) | 0.145 | - |

## 2.2. Two Step Method

In the two step method, the nanoparticle manufacturing and nanofluids preparation is separated. Firstly, dry nanoparticles/nanotubes are produced; after which, they are dispersed in a suitable basefluid. This is simpler than single step method as we can easily buy readily available nanoparticles in the market and then disperse them in the basefluid. However, it is well known that nanoparticles have a high surface energy which, in turn, leads to aggregation and clustering of nanoparticles and after some time, the particles will clog and sediment at the bottom of container. Partial dispersion may occur in the suspension which causes lower heat transfer enhancement compared to single step method, and, hence, high amount of nanoparticles volume fraction (more than 20%) is required. This method works well for oxide particle and carbon nanotubes; however, it less successful for metal nanoparticles [12]. Therefore, more research is required to address the challenge by making a more stable nanofluids using two-step method.

## 2.3. Treatment for Stability Enhancement

In order to make the two step method more reliable, several stability enhancements techniques have been proposed by many researchers, e.g. physical and/or chemical treatment. Basically, to make a high quality suspension, the nanoparticles have to meet two criteria: (*i*) dispersion: nanoparticles are well dispersed into the basefluid and (*ii*) surface charge (zeta potential): the absolute value of zeta potential on the particle surface must be as large as possible, thus engendering a repulsive force between particles in the suspension. Ghadimi et al. [12] classified three most common stability enhancement methods to manage suspension stability and prevent agglomeration and sedimentation of nanoparticles; they are addition of surfactant/activator, modifying the nanoparticle surface chemistry (and thereby surface charge) and providing ultrasonic energy to the agglomerating suspensions. Some researchers apply all methods to achieve better stability, whereas some others implement combination of two methods or only one method satisfactorily. There is no standard guideline on the stability enhancement method and the results may vary with different technique; thus, more experimentation is required and fundamental theoretical background need to be established to provide standard guidelines for nanofluids stability.

### *2.3.1. Addition of Surfactant or Activator*

Addition of surfactant is believed to improve the stability as the hydrophobic/hydrophilic surface of nanoparticles/nanotubes is modified into hydrophilic/hydrophobic counterpart, and vice-versa for non-aqueous liquid. Basically, surfactants can stabilize colloidal suspensions simply through ‘steric’ interactions. The adsorbed surfactants provide a bulky ‘molecular brush’ on the nanoparticle surface that prevents two adjacent nanoparticles from coming too close. These steric interactions are mainly entropic in nature). It should be noted that sufficient additional surfactant is required to ensure coating that can enhance electrostatic or steric interactions and compensate the van der Waals effect [19]. Furthermore, choosing the right surfactant is the most important part - it could be anionic, cationic or non-ionic in nature, depending on the specific nanofluid particulate material at hand. Although a surfactant can improve the stability, it can only effectively be used for low temperature application; at high temperature, i.e., above than 60 C, they can easily desorb from the surfaces, and leads to agglomeration and sedimentation [20].

### *2.3.2. Control of Surface Chemistry*

The chemical nature of the nanoparticle surface is one of the important factors which determine the stability of aqueous nanofluid suspension, as it determines the surface charge and therefore the electrostatic interactions between nanoparticles. At a high surface charge density, strong repulsive force can stabilize a well dispersed solution. For example, Xie et al. [21] showed that acid treatment on CNT suspension improves the suspension stability; they postulated that pH control can convert the hydrophobic to hydrophilic surface due to the generation of a hydroxyl group.

### *2.3.3. Ultrasonic Energy*

Provision of ultrasonic energy into the suspension is effective in breaking down and re-suspending particle agglomerates that eventually form in nanofluids. However, after occasionally passing some sort of the optimized duration of the process, it may cause more serious problem in agglomeration and clogging resulting in fast sedimentation [22].

# 3. CHARACTERIZATION

Addition of even small amount of nanoparticles to a basefluid is known to change the thermophysical properties, viz. thermal conductivity, viscosity, heat capacity and density. Several researchers report appreciably different enhancement values as it is depends on many factors, e.g., preparation, dispersion, nanoparticle type, volume concentration, temperature, base fluid, nanoparticle size, stability treatment, etc.

## 3.1. Thermophysical Properties

### *3.1.1. Thermal Conductivity*

Thermal conductivity is the most important parameter in enhancing the heat transfer performance of a nanofluid. Since thermal conductivity of solid nanoparticles is higher than that of the fluid, it is expected to increase the thermal conductivity and heat transfer performance. There are several methods that are commonly used to measure thermal conductivity of nanofluids, e.g., transient and steady state methods which are summarized in Figure 2. The transient hot wire method is the most popular method among others.

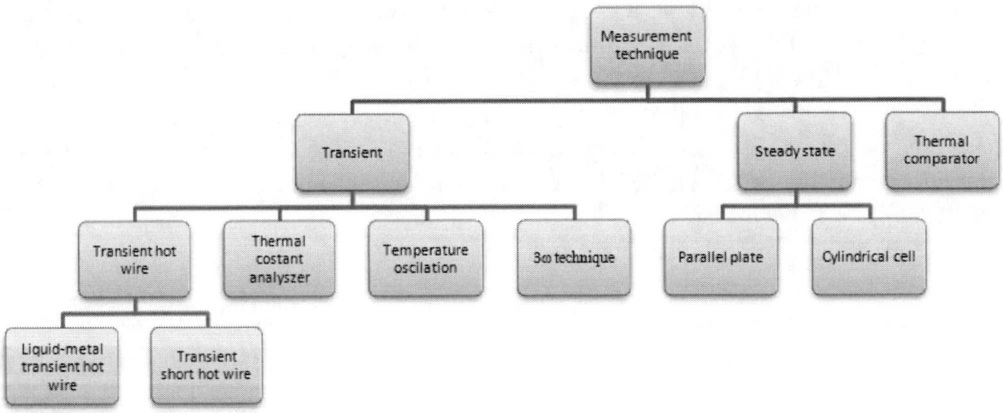


Figure 2. Various thermal conductivity measurement techniques for nanofluids [23].

Many researchers report different results on the enhancement levels; it ranges from less than 10% to more than 200%. Some of the published measurement data are presented in Table 2.

### *3.1.2. Viscosity*

Viscosity is one of the important thermophysical properties as it determines the pressure drop required and hence the pumping power. Several viscosity measurements are presented in Figure 3. Here several features are apparent; foremost among them is that the effective viscosity of the suspension increases considerably with particle volume fraction; albeit the percentage of increment is varied among different researchers. It is noted that the viscosity increase is higher than viscosity rise predicted by the Brinkman model; thus, it has raised serious concerns regarding reliability of using nanofluid heat transfer enhancement relative to the pumping power. Sasmito et al. [38] showed numerically that addition of small amount of nanoparticles up to 1% improves significantly the heat transfer performance in terms of the figure of merit, defined as heat transfer per unit pumping power; however, further addition tends to deteriorate heat transfer performance per unit pumping power.

**Table 2. Summary of some thermal conductivity measurement**

| Author | Nanoparticle | Concentration | Enhancement |
|---|---|---|---|
| Choi et al., 2003 [24] | SWCNTs | 9% | 250% |
| Llaguna et al., 2001 [25] | SWCNTs | 10% | 80% |
| Torii, 2009 [26] | Diamond, $Al_2O_3$, CuO and deionized water | 0.3 | 70% |
| Ali et al., 2010 [27] | Chromium in water, ethylene glycol (EG) and ethanol | 0.02% | 50% |
| Hojjat et al., 2011 [28] | C-$Al_2O_3$ in CMC solution | 1.5% | 4.21% |
| | $TiO_2$ in CMC solution | 1.5% | 7% |
| | CuO in CMC solution | 1.5% | 11.3% |
| Moosavi et al., 2010 [29] | ZnO in ethylence glycol and glycerol | 0.2% | 56% |
| Karthikeyan et al., 2008 [30] | CuO in distillate water + EG | 1% | 31.6% |
| Kwak et al., 2005 [31] | CuO in EG | <0.002% | 54% |
| Wang et al., 2009 [32] | $Al_2O_3$ in distillate water | 0.4% wt | 13% |
| | Cu in distillate water | 0.4% | 15% |
| Wei et al., 2009 [33] | $Cu_2O$ in distillate water | 0.01-0.05% | 24% |
| Lee et al., 2008 [34] | $Al_2O_3$ in distillate water | 0.01-0.3% | 0.05 – 2% |
| Murshed et al., 2008 [35] | $Al_2O_3$ in distillate water | 1% | 3-12% |
| | $Al_2O_3$ in EG | 0.5% | 2-10% |
| | $Al_2O_3$ in CTAB | 1% | 3-9% |
| Oh et al., 2008 [36] | $Al_2O_3$ in distillate water | 1-4% | 4.4-13.3% |
| | $Al_2O_3$ in EG | 1-4% | 1.9-9.7% |
| Kole et al., 2010 [37] | $Al_2O_3$ in car engine coolant | 3.5% | 10.41-11.25% |

## 3.2. Heat Transfer Performance

The heat transfer enhancement may be affected by several factors such as Brownian motion, molecular convection, dispersion, layering etc which will be explained in the next Section. Experiments show that thermal conductivity increase is not the only reason for heat transfer augmentation. Other factors may play role in heat transfer enhancement, e.g. particle interaction and fluctuations, especially at in high Peclet number may cause the change in flow structure. However, the proper and detailed mechanism of heat transfer enhancement has not been established to date. Increase in thermal conductivity might be offset by increase in viscosity and reduce the heat capacity which, in turn, reduce the figure of merit, albeit it may not be important in applications where the heat transfer performance alone is of paramount important.

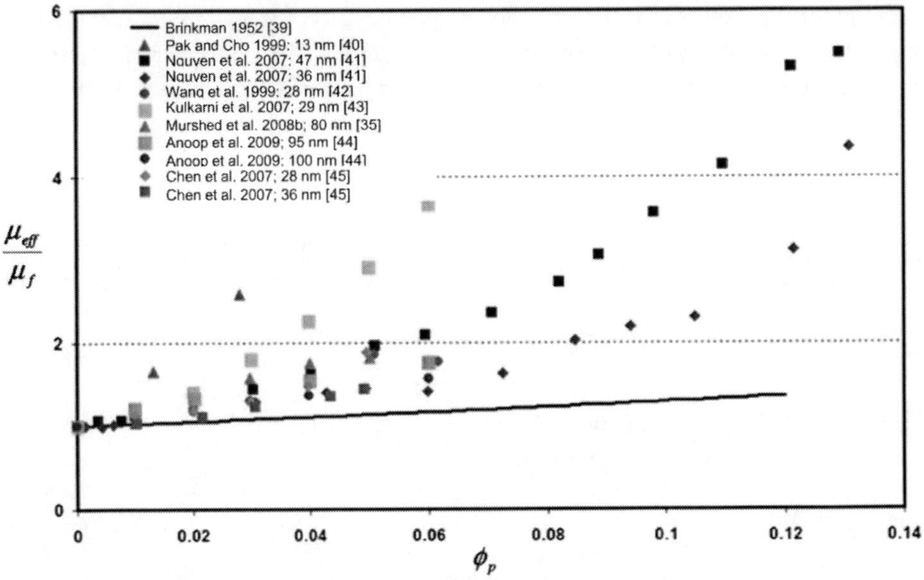


Figure 3. Viscosity data as function of volume fraction at ambient temperature [46].

As the heat transfer characteristics of nanofluids depend on several factors, e.g., properties of base liquid and nanoparticles, concentration, size, morphology, treatment etc., the general form of the Nusselt number correlation yields [47]:

$$Nu_{nf} = f\left(\mathrm{Re}, \mathrm{Pr}, \frac{k_p}{k_{bf}}, \frac{\left(\rho c_p\right)_p}{\left(\rho c_p\right)_{bf}}, \phi, \text{particle size and shape, flow structure}\right) \tag{1}$$

In addition, many researchers have proposed empirical correlations for convective heat transfer for nanofluids; these are summarized in Table 3.

## 4. Theoretical Aspects – Heat Transfer

### 4.1. Heat Transfer Enhancement – Mechanisms

Several mechanisms of heat transfer augmentation have been proposed by various researchers. This is well demonstrated by a recent analysis by Sergis and Hardalupas [18] who evaluated 131 published references based on the statistical analysis on mechanisms for heat transfer enhancement; they found that most of the researchers (33%) attribute the augmentation to Brownian motion, followed by interfacial layer theory (23%), shear thinning behaviour of flows (5%), aggregation and diffusion (2%), and other mechanisms. However, some of the proposed theories are physically inconsistent. Prasher et al. [58] showed an internally consistent physical theory to explain heat transfer augmentation of nanofluids: it employs very basic principles, and explains that most nanofluid behaviour is not strange or anomalous at all – although it may or may not be correct, it seems more plausible and physically consistent when compared with many other explanations and theories. Several other possible micro- or macro-scale explanations can still be investigated in the future.

**Table 3. Some empirical correlations for convective heat transfer of nanofluids [48]**

| Author | Correlation | Remarks |
|---|---|---|
| Pak and Cho, 1998 [40] | $Nu_{nf} = 0.021\,\mathrm{Re}_{nf}^{0.8}\,\mathrm{Pr}_{nf}^{0.5}$ | $Al_2O_3$-water, $TiO_2$-water; φ = 0.99-2.78%; turbulent, constant heat flux; Re = $10^4$-$10^5$. |
| Li and Xuan, 2002 [49] | $Nu_{nf} = 0.4328\left(1 + 11.258\phi^{0.754} Pe_p^{\ 0.218}\right)\mathrm{Re}_{nf}^{0.333}\,\mathrm{Pr}_{nf}^{0.4}$ | Cu-water; φ = 0.3-2%; laminar, constant heat flux; Re > 800. |
| Xuan and Li, 2003 [50] | $Nu_{nf} = 0.0059\left(1 + 7.6286\phi^{0.6886} Pe_p^{\ 0.001}\right)\mathrm{Re}_{nf}^{0.9238}\,\mathrm{Pr}_{nf}^{0.4}$ | Cu-water; φ = 0.3-2%; tubulent, constant heat flux; Re = (1-2.5) × $10^4$. |
| Yang et al., 2005 [51] | $Nu_{nf} = a\,\mathrm{Re}_{nf}^{b}\,\mathrm{Pr}_{nf}^{1/3}\left(D / L\right)^{1/3}\left(\mu_w / \mu_b\right)^{-0.14}$ | Graphite-oil; φ < 2.5%; laminar; Re = 5-110. |
| Maiga et al., 2005 [52] | $Nu_{nf} = 0.086\,\mathrm{Re}_{nf}^{0.55}\,\mathrm{Pr}_{nf}^{0.5}$ for constant heat flux<br>$Nu_{nf} = 0.28\,\mathrm{Re}_{nf}^{0.35}\,\mathrm{Pr}_{nf}^{0.36}$ for constant temperature | $Al_2O_3$-water, $Al_2O_3$-EG; φ < 10%; laminar; Re < 1000. |
| Maiga et al., 2006 [53] | $Nu_{nf} = 0.085\,\mathrm{Re}_{nf}^{0.71}\,\mathrm{Pr}_{nf}^{0.35}$ | $Al_2O_3$-water; φ < 10%; turbulent, constant heat flux; Re = (1-50) × $10^4$ |
| Buongiorno, 2006 [54] | $Nu_{nf} = \dfrac{\left(f / 8\right)\left(\mathrm{Re}_{nf} - 1000\right)\mathrm{Pr}_{nf}}{1 + \delta_v^{+}\left(f / 8\right)^{1/2}\left(\mathrm{Pr}_v^{2/3} - 1\right)}$ | Turbulent flow of nanofluids |
| Anoop et al., 2009 [55] | $Nu_{nf} = 4.36 + \left[a x_{+}^{-b}\left(1 + \phi^{c}\right)\exp^{-d x_{+}}\right]\left[1 + e\left(d_p / d_{ref}\right)^{-f}\right]$<br>$a = 6.219 \times 10^{-3},\ b = 1.1522,\ c = 0.1533,\ d = 2.5228$<br>$e = 0.57825,\ f = 0.2183,\ d_{ref} = 100nm,$<br>$x_{+} = x / \left(D\,\mathrm{Re}\,\mathrm{Pr}\right)$ | $Al_2O_3$-water; φ < 1.8%; laminar; Re = 500-2000. |
| Daungthongsuk and Wongwises, 2010 [56] | $Nu_{nf} = 0.074\,\mathrm{Re}_{nf}^{0.707}\,\mathrm{Pr}_{nf}^{0.385}\,\phi^{0.074}$ | $TiO_2$-water; φ =0.2-2%; turbulent; Re = 3000 – 18,000. |
| Vajjha et al., 2010 [57] | $Nu_{nf} = 0.065\left(\mathrm{Re}_{nf}^{0.65} - 60.22\right)\left(1 + 0.0169\phi^{0.15}\right)\mathrm{Pr}_{nf}^{0.542}$ | CuO, $SiO_2$, $Al_2O_3$; φ < 6% (CuO, $SiO_2$), φ < 10% ($Al_2O_3$); turbulent; Re = 3000 – 16,000. |

#### *4.1.1. Brownian Motion*

The random motion of nanoparticles transports energy directly from nanoparticles to and from the suspending fluid. The nanoparticles provide larger surface area for molecular collisions; the higher the momentum of nanoparticles due to higher mass concentrations, the higher the transfer of thermal energy at greater distances inside the base-fluid before they release it in a colder side. In addition, the micro-convection effect, which is due to fluid mixing around nanoparticles, play significant role to the heat transfer enhancement [59]. However, there are discrepancies and discordant observations in existing experiments and simulations, some support the role of Brownian motion (for example Xie et al. [60], Das et al. [61], Chopkar et al. [62], Li and Patterson [63], Shima et al. [64], Mintsa et al. [65]) while others provide evidence to the contrary (Keblinski et al. [66], Keblinski and Thomin [67] and Evans et al. [68]).

It is important to note that the motion of nanoparticles is influenced not only by the effect of Brownian motion, but also by other inter-particle forces such as van der Waals and electrostatic interactions between neighbouring particles, and by transport phenomena such as thermophoresis and frictional drag. Savithiri et al. [69] evaluated the significance of each inter-particle force through a scaling analysis; they found that the Brownian motion (one cannot properly speak of an inter-particle Brownian 'force') dominates thermal transport in smaller particles below 10 nm and at small (<1%) volume fractions; heat transport by friction, i.e. by drag between nanoparticles and the adjacent fluid is found to dominate in smaller particles below 10 nm and at high volume concentrations (>1%) The effect of thermopherosis and Magnus forces is found to increase with the particle size and concentration. As mentioned above, the inter-particle forces depend on the chemical and physical properties of suspension (base fluid and nanoparticle) as well as particle morphology, distribution, and treatment during preparation; therefore, it would be very challenging to accurately estimate the effect of Brownian motion alone on thermal conductivity [70]. In addition, most of colloidal particles (including nanofluids) are surrounded by electrical double layers (EDLs). This mechanism proposes an alteration of the strength of intermolecular interaction forces that affect the mean free path of the nanoparticles which, in turn, enhance the heat transfer [71].

#### *4.1.2. Interfacial Liquid Layering*

Yu et al. [72] showed that liquid molecules form layered structures around solid surfaces; their further observation, Yu and Choi [73], suggested that the nanolayers have a larger effective thermal conductivity than that of the base fluid, which might be responsible for thermal conductivity enhancement. Furthermore, molecular dynamics (MD) simulations revealed the important role of the liquid-solid interaction strength: a wetting liquid with strong interaction results in a strong liquid-layering over several atomic distance, thus, smaller interfacial thermal resistance; while a non-wetting liquid with weak interaction gives weak liquid layering and high thermal resistance [74]. Lee [75] proposed a postulate to calculate the thickness and thermal conductivity of the nanolayer by considering formation of an electric double layer around the nanoparticle. They showed via simulation that the thickness depends on several factors, e.g., dielectric constant, ionic strength, temperature, total charged surface density, ion density in the electrical double layer, pH value and thermal conductivity of nanoparticles and base fluid. Eapen et al. [76] showed, using MD simulation, that the liquid layering has significant role to the thermal conductivity enhancement; they noted that an increase of 55% on thermal conductivity was obtained at particle volume fraction 3.5%. In contrast, Xue et al. [77] showed using MD simulation that nanolayers have no effect on thermal transport in a simple monoatomic liquid. They also highlighted that the results might be different for water. Hence, more research is needed to understand the liquid layering effect for other more complicated base fluids with molecular chains.

#### *4.1.3. Particle Aggregation*

In nanofluid suspensions, nanoparticles may aggregate to form chain structures. The occurrence of this chain assembly is speculated to provide a faster path for heat diffusion in nanofluids. Philip et al. [78] conducted an experiment on $Fe_3O_4$ nanofluid by manipulating particle aggregation structures with variable strength of external magnetic fields; they found that stronger magnetic field yields longer particle chain structures in nanofluids which, in turn, results in large thermal conductivity enhancement (up to 300% at volume fraction 0.82%). This provides evidence that particle aggregation can play a significant role in thermal conductivity enhancement [79].

#### *4.1.4. Phonon Transfer*

One of the proposed mechanisms for heat transfer augmentation in nanofluid is attributed to specialised phonon and electron interaction and scattering at nanoscales / ballistic heat transport. Heat is considered to be transmitted by phonons, which are the quanta of energy in each mode of vibration. As the atoms in a solid are closely coupled together, an increase in temperature will be transmitted to the other parts, which may enhance the heat transfer in nanofluid [80].

#### *4.1.5. Thermophoretic Forces*

Thermophoretic forces on nanoparticles arise from the presence of temperature gradients in the fluid causing the concentration of the nanoparticles to change around heating and cooling sides relative to the mean value which induces particle migration and hydrodynamic diffusion. The consequence of this nanoparticle redistribution is the alteration of the heat transfer, which, in turn, enhances the thermal conductivity [81]. This can also be considered as part of force due to Brownian motion.

#### *4.1.6. Near Field Radiation*

At the nanoscales (near field radiation), there is an infrared radiation emission and absorption augmentation. This is postulated to enhance heat transfer between the heating surface, the nanoparticles, basefluid molecules and between the nanoparticles themselves [82].

#### *4.1.7. Non-Newtonian Rheology*

In nanofluid convective heat transfer, it is believed that the nanofluid exhibits non-Newtonian characteristic with shear thinning behaviour. As the shear rate of the nanofluids increases along the walls, the viscosity decreases at the solid boundaries of a flowing nanofluid; thus, the thermal boundary layer width is reduced which is further mirrored by increased heat transfer between the wall and the liquid, Ding et al. [83]. Hwang et al. [84] proposed that flattening of velocity profile due to viscosity can be one of the heat transfer enhancement mechanism. Viscosity change of nanofluids leads to a more uniform velocity profiles for convective heat transfer over pipes and ducts than the expected parabolic velocity profile. The increased near wall velocity is believed to provide an increased in the convective heat transfer. However, some researches also find that at low nanoparticle concentration (less than ~10%), the suspension are typically so dilute that non-Newtonian rheology of any kind is unlikely.

## 4.2. Modeling of Nanofluids Heat Transfer

### *4.2.1. Thermophysical Properties*

As stated previously, addition of small amount of nanoparticles to the base fluid can modify the thermophysical properties of the nano-suspension. Various theoretical models have been proposed to try to capture such modifications.

#### Thermal Conductivity

A wide range of theoretical studies have been reported in the literature in attempt to model thermal conductivity enhancement in nanofluids. Published results suggest that neither agreement nor unified explanation regarding the large discrepancy in the thermal conductivity model results – even for the same base fluid and nanoparticle size as well as concentration. Currently, there are no theoretical models that can accurately predict the thermal conductivity of nanofluids. In general, thermal conductivity models can be categorized into two groups: static and dynamic models. The former are based on the Maxwell and Hamilton–Crosser model, which presume immobile nanoparticles in the host fluid in which a conduction-based model predict the thermal transport properties; while the latter are based on the idea that nanoparticles possess arbitrary movement in the fluid (Brownian motion) or indirectly through micro liquid convection due to particle movement. Various thermal conductivity models are summarized in Table 4.

#### Viscosity

Similar to thermal conductivity, several effective viscosity models which account for nanoparticle concentration have been proposed by various researchers. An accurate viscosity model is essential to predict the effectiveness of nanofluid heat transfer enhancement with respect to the pressure drop required. However, to date, there is no single unified effective viscosity expression – even for same nanofluid type and concentration. Most of the models are based on the extended Einstein model, Brinkmann model or semi-empirical model from curve fitting of experimental measurement. Koo and Kleinstruer [97] proposed effective viscosity model by considering static and dynamic part of viscosity due to Brownian motion; however, they did not validate their model against experimental data. Several effective viscosity models are tabulated in Table 5.

#### Specific Heat Capacity

The specific heat capacity of nanofluid can be determined from weighted average of the base-fluid and the nanopaticles, given by

$$c_{p,eff} = \frac{\phi_p c_{p,p} \rho_p + (1 - \phi_p) c_{p,f} \rho_f}{\rho_{nf}} \tag{2}$$

However, some authors prefer to use simpler expression as

$$c_{p,eff} = \phi_p c_{p,p} + (1 - \phi_p) c_{p,f} \tag{3}$$

**Density**

The density of nanofluids is typically based on the physical principle of the mixture rule, defined as

$$\rho_{eff} = (1 - \phi_p) \rho_f + \phi_p \rho_p \tag{4}$$

### *4.2.2. Convective Heat Transfer*

**Homogeneous Model (Single Phase Model)**

This model treats the nanofluids suspension as a homogeneous fluid with effective properties. The choice of the most appropriate correlation for the effective properties of suspension is essential for accurate prediction. However, there is no such a universal correlation that is able to accurately predict the heat transfer performance at any given condition for any combination of independent variables, e.g. nanoparticle material, concentration, temperature, nanopaticles diameter, pH treatment, surfactant used etc; many different correlation also may lead to contradictory results. In addition, this model is not able to predict nanoparticles separation, sedimentation and agglomeration which may arise due to some forces/conditions, e.g., secondary flow in curved geometry, buoyancy forces etc.

The model solves for single conservation equations of mass, momentum and energy, given by

$$\nabla \cdot (\rho_{eff} \mathbf{u}) = 0 \tag{5}$$

$$\nabla \cdot (\rho_{eff} \mathbf{u} \otimes \mathbf{u}) = -\nabla p + \nabla \cdot \left[ \mu_{eff} \left( \nabla \mathbf{u} + (\nabla \mathbf{u})^{\mathbf{T}} \right) \right] \tag{6}$$

$$\nabla \cdot (\rho_{eff} c_{p,eff} \mathbf{u} T) = \nabla \cdot (k_{eff} \nabla T) \tag{7}$$

**Table 4. Summary of theoretical model for effective thermal conductivity of nanofluids**

| Models | Equation | Remarks |
|---|---|---|
| Maxwell, 1881 [1] | $\frac{k_{eff}}{k_f} = \frac{k_p + 2k_f + 2\phi_p(k_p - k_f)}{k_p + 2k_f - \phi_p(k_p - k_f)}$ | Based on the conduction solution theory through a stationary, random suspension of spheres. |
| Bruggeman, 1935 [85] | $\frac{k_{eff}}{k_f} = \frac{(3\phi_p - 1)k_p / k_f + (3(1-\phi_p) - 1) + \sqrt{\Delta}}{4}$<br>$\Delta = \left[(3\varphi_p - 1) - 1\right]k_p / k_f + \left\{3(1-\phi_p) - 1\right\}^2 + 8k_p / k_f$ | Based on the differential effective medium (DEM) theory at high spherical particle concentrations. |
| Hamiltion-Crosser, 1962 [86] | $\frac{k_{eff}}{k_f} = \frac{k_p + (n-1)k_f + (n-1)\phi_p(k_p - k_f)}{k_p + (n-1)k_f - \phi_p(k_p - k_f)}$ | Based on the effective thermal conductivity of a two-component mixture, $n$=3 for spheres; $n$=6 for cylinders |
| Jeffrey, 1973 [87] | $\frac{k_{eff}}{k_f} = 1 + 3\eta\phi_p + \phi_p^2\left(3\eta^2 + \frac{3\eta^2}{4} + \frac{9\eta^3}{16}\frac{\kappa+2}{2\kappa+3} + \ldots\right)$<br>$\kappa = \frac{k_p}{k_f}, \eta = \frac{\kappa - 1}{\kappa + 2}$ | Based on the conduction solution theory through a stationary random suspension of spheres. High order terms represent pair interactions of randomly distributed spherical particles. |
| Davis, 1986 [88] | $\frac{k_{eff}}{k_f} = 1 + \frac{3(\kappa - 1)}{(\kappa+2) - \phi_p(\kappa - 1)}\left[\phi_p + f(\kappa)\phi_p^2 + 0(\phi_p^3)\right]$ | Green theorem was applied to the space occupied by the matrix material. |
| Lu and Lin, 1996 [89] | $\frac{k_{eff}}{k_f} = 1 + a\phi_p + b\phi_p^2$ | The effective conductivity model of composites containing aligned spheroids of finite conductivity was modelled with pair interaction. |

| Models | Equation | Remarks |
|---|---|---|
| Yu and Choi, 2003 [73] | $k_{eff} = \frac{k_{pe} + 2k_f + 2\phi_p(k_{pe} - k_f)(1+\beta)^3}{k_{pe} + 2k_f - \phi_p(k_{pe} - k_f)(1+\beta)^3} k_f$ <br> $k_{pe} = \frac{2(1-\gamma) + (1+\beta)^3(1+2\gamma)\gamma}{-(1-\gamma) + (1+\beta)^3(1+2\gamma)} k_p$ <br> $\beta = t / r_p, \gamma = k_{layer} / k_p$ | Modified Maxwell equation for spherical nanoparticles; it includes effect of liquid nano-layering. |
| Yu and Choi, 2004 [90] | $k_{eff} = \left(1 + \frac{n f_e A}{1 - f_e A}\right) k_f,\ A = \frac{1}{3}\sum_{j=a,b,c} \frac{(k_{pj} - k_f)}{k_{pj} + (n-1)k_f}$ <br> $f_e = \frac{f\sqrt{(a^2+t)(b^2+t)(c^2+t)}}{abc}$ | Modified Hamiltion-Crosser model for non spherical nanoparticles; it considers liquid nano-layering effect |
| Xue 2003 [91] | $9\left(1 - \frac{\phi_p}{\lambda}\right)\frac{k_{eff} - k_f}{2k_{eff} + k_f} + \frac{\phi_p}{\lambda}\frac{k_{eff} - k_{c,x}}{k_{eff} + B_{2,x}(k_{c,x} - k_{eff})} + \ldots$ <br> $\ldots + \frac{\phi_p}{\lambda} 4 \frac{k_{eff} - k_{c,y}}{2k_{eff} + (1 - B_{2,x})(k_{c,y} - k_{eff})} = 0$ | Based on Maxwell model and the average polarization theory and on the assumption that there is an interfacial shell between the nanoparticles and the basefluid. |
| Xue and Xu, 2005 [92] | $0 = \left(1 - \frac{\phi_p}{\alpha}\right)\frac{k_{eff} - k_f}{2k_{eff} + k_f} + \ldots$ <br> $\ldots + \frac{\phi_p}{\alpha}\frac{(k_{eff} - k_{shell})(2k_{shell} - k_p) - \alpha(k_p - k_{shell})(2k_{shell} + k_{eff})}{(2k_{eff} + k_{shell})(2k_{shell} + k_p) + 2\alpha(k_p - k_{shell})(k_{shell} - k_{eff})}$ | A modified Bruggeman model including the effect of the interfacial shells for spherical particles. |

**Table 4. (Continued)**

| Models | Equation | Remarks |
| --- | --- | --- |
| Xie et al., 2005 [93] | $\frac{k_{eff}-k_f}{k_f}=3\Theta\phi_T+\frac{3\Theta^2\phi_T^2}{1-\Theta\phi_T}$ <br> $\phi_T=4/3\pi\left(r_p+t\right)^3 N_p=\phi_p\left(1+\beta\right)^3, \beta=t/r_p$ | Based on Fourier's law of heat conduction for low particle loading with nanolayer consideration |
| Wang et al., 2003 [94] | $\frac{k_{eff}}{k_f}=\frac{\left(1-\phi_p\right)+3\phi_p\int_0^{\infty}\frac{k_{cl}(r)n(r)}{k_{cl}(r)+2k_f}dr}{\left(1-\phi_p\right)+3\phi_p\int_0^{\infty}\frac{k_f(r)n(r)}{k_{cl}(r)+2k_f}dr}$ | A fractal model based on effective medium approximation and fractal theory. |
| Xuan et al., 2003 [95] | $\frac{k_{eff}}{k_f}=\frac{k_p+2k_f-2\phi_p(k_f-k_p)}{k_p+2k_f+\phi_p(k_f-k_p)}+\frac{\rho_p\phi_p c_p}{2k_f}\sqrt{\frac{k_B T}{3\pi r_c\mu}}$ | Based on Maxwell theory but it includes effect of the Brownian motion, particle size, concentration and temperature. |
| Jang and Choi 2004 [96] | $k_{eff}=k_f\left(1-\phi_p\right)+k_p\phi_p+3C\frac{d_f}{d_p}k_f \mathrm{Re}_{dp}^2 \mathrm{Pr}\,\phi_p$ | Consider four modes enhancement: collisions between fluid molecules, thermal diffusion of nanoparticles, collision between nanoparticles due to Brownian motion and thermal interaction of dynamic nanoparticles with basefluid molecules |
| Prasher et al., 2005 [59] | $k_{eff}=\left(1+A\,\mathrm{Re}^m\,\mathrm{Pr}^{0.333}\,\phi_p\right)\times\left[\frac{k_p+2k_f+2\phi_p\left(k_p-k_f\right)}{k_p+2k_f-\phi_p\left(k_p-k_f\right)}\right]k_f$ | Based on Maxwell model which account for convection caused by Brownian motion from nanoparticle movement. |

| Models | Equation | Remarks |
|---|---|---|
| Koo and Kleinstreuer, 2005 [97] | $k_{eff} = k_{static} + k_{Brownian}$<br>$k_{static} = \frac{k_p + 2k_f + 2\phi_p(k_p - k_f)}{k_p + 2k_f - \phi_p(k_p - k_f)} k_f$<br>$k_{Brownian} = 5 \times 10^4 \beta \phi_p \rho_p c_p \sqrt{\frac{k_B T}{\rho_p D}} f(T, \phi_p)$ | Based on static Maxwell theory and dynamic Brownian motion effect; considered surrounding liquid motion with randomly moving nanoparticles. |
| Chon et al., 2005 [98] | $k_{eff} = 1 + 64.7\phi_p^{0.74} \left(\frac{d_f}{d_p}\right)^{0.369} \left(\frac{k_p}{k_f}\right)^{0.747} \times \mathrm{Pr}^{0.9955} \mathrm{Re}^{1.2321}$<br>$\mathrm{Pr} = \frac{\mu_f}{\rho_f \alpha_f}$<br>$\mathrm{Re} = \frac{\rho_f V_{Br} d_p}{\mu_f} = \frac{\rho_f k_B T}{3\pi \mu_f^2 l_f}$ | Based on experimental curve fitting; Reynolds number is based on the Brownian motion velocity. |
| Li 2008 (KKL model) [99] | $k_{eff} = \frac{k_p + 2k_f + 2\phi_p(k_p - k_f)}{k_p + 2k_f - \phi_p(k_p - k_f)} k_f + \ldots$<br>$\ldots + 5 \times 10^4 \phi_p \rho_p c_p \sqrt{\frac{k_B T}{\rho_p d_p}} g(T, \phi_p, d_p)$<br>$g(T, \phi_p, d_p) = \left(a + b\ln(d_p) + c\ln(\phi_p) + d\ln(\phi_p)\ln(d_p) + e\ln(d_p)^2\right)\ln(T) + \left(g + h\ln(d_p) + i\ln(\phi_p) + j\ln(\phi_p)\ln(d_p) + k\ln(d_p)^2\right)$ | Revised Koo and Kleinstreue 2005 model and replace the function of β and $f(T,\varphi_p)$ with a new $g$ function which captures the influence of particle diameter, temperature and volume fraction. |
| Kihm et al., 2011 [100] | $\frac{k_{eff}}{k_f} = \left[1 + C \cdot \frac{\phi_p{}^a k_b \rho_p c_p T^{1.5}}{k_f h \mu^{0.5} d_p^{0.5}} \exp(-3.8T_b / T) \cdot \left(\frac{c_f}{c_p}\right)^b\right]$ | Account for Brownian motion with correction to the micro convection velocity, nanoparticle heat dissipation as well as coagulation effect |

| Author | Expression | Remarks |
| --- | --- | --- |
| Kulkarni et al., 2007 [43] | $\ln \mu_{eff} = A\left(\frac{1}{T}\right) - B$<br>$A = 20587\phi_p^2 + 15857\phi + 1078.3$<br>$B = -107.12\phi_p^2 + 53.54\phi + 2.8715$ | Curve fiting from experimental CuO-water; $5\% < \varphi_p < 15\%$; $d_p = 29$ nm; $278 < T < 323$; shear rate = 100 1/s. |
| Koo and Kleinstruer, 2005 [97] | $\mu_{eff} = \mu_{eff} + \mu_{Brownian}$<br>$\mu_{Brownian} = \frac{k_{Brownian}}{k_l} + \frac{\mu_l}{\mathrm{Pr}_l}$<br>$k_{Brownian} = 5 \times 10^4 \beta \rho_f \phi_p \sqrt{\frac{k_B T}{\rho_p d_p}} f(\phi_p, T)$ | For CuO-water suspension; $1\% < \varphi_p < 4\%$; $300 < T < 325$. |
| Namburu et al., 2009 [105] | $\log(\mu_{eff}) = Ae^{-BT}$<br>$A = -0.29956\phi_p^3 + 6.738\phi_p^2 - 55.444\phi + 236.11$<br>$B = \left(-6.4745\phi_p^3 + 140.03\phi_p^2 - 1478.5\phi + 20341\right) / 10^6$ | Curve fitting from experimental $Al_2O_3$-EG+water mixture; $1\% < \varphi_p < 10\%$; $d_p = 53$ nm; $278 < T < 323$. |

**Table 5. Several effective viscosity models of nanofluids**

| Author | Expression | Remarks |
|---|---|---|
| Einstein 1906 [101] | $\mu_{eff} = \left(1 + 2.5\phi_p\right)\mu_f$ | Based on the phenomenologival hydrodynamic equations for infinitely dilute suspension of spheres with no interaction between spheres itself. |
| Brinkman 1952 [39] | $\mu_{eff} = \dfrac{1}{\left(1 + 2.5\phi_p\right)^{2.5}}\mu_f = \left(1 + 2.5\varphi_p + 4.375\phi_p^2 + ...\right)\mu_f$ | Extended Einstein model by considering effect of the adition of one solute-molecule to an existing solution. |
| Batchelor 1977 [102] | $\mu_{eff} = \left(1 + \eta\phi_p + k_h\phi_p^2\right)\mu_f = \left(1 + 2.5\phi_p + 6.2\phi_p^2\right)\mu_f$ | Based on reciprocal theorm in Stokes flow problem to obtain an expression for the bulk stress due to the thermodynamic forces considering Brownian motion. |
| Lundgren 1972 [103] | $\mu_{eff} = \dfrac{1}{1 + 2.5\phi_p}\mu_f = \left(1 + 2.5\phi_p + 6.25\phi_p^2 + ...\right)\mu_f$ | Based on Taylor series expression in terms of particle volumetric concentration. |
| Graham 1981 [104] | $\mu_{eff} = \left(1 + 2.5\phi_p\right)\mu_f + \left[\dfrac{4.5}{\left(h / r_p\right)\left(2 + h / r_p\right)\left(1 + h / r_p\right)^2}\right]\mu_f$ | A cell theory was used to derive the dependence of the zero-shear-rate viscosity on volume concentration for a suspension of uniform, solid, neutrally buoyant spheres |
| Maiga et al., 2005 [53] | $\mu_{eff} = \left(1 + 7.3\phi_p + 123\phi_p^2\right)\mu_f$ for $Al_2O_3$-water<br>$\mu_{eff} = \left(1 - 0.19\phi_p + 306\phi_p^2\right)\mu_f$ for $Al_2O_3$-EG | Least square curve fitting from experimental data |
| Buongiorno 2006 [54] | $\mu_{eff} = \left(1 + 39.11\phi_p + 533.9\phi_p^2\right)\mu_f$ for $d_p = 13$ nm<br>$\mu_{eff} = \left(1 + 5.45\phi_p + 108.2\phi_p^2\right)\mu_f$ for $d_p = 27$ nm | Curve fitting from experimental data of nanofluids $Al_2O_3$-water |
| Nguyen et al., 2007 [41] | $\mu_{eff} = \mu_f \times 0.904e^{0.148\phi_p}$ for $d_p = 47$ nm<br>$\mu_{eff} = \left(1 + 0.025\phi_p + 0.015\phi_p^2\right)\mu_f$ for $d_p = 36$ nm | Curve fitting from experimental data of nanofluids $Al_2O_3$-water |

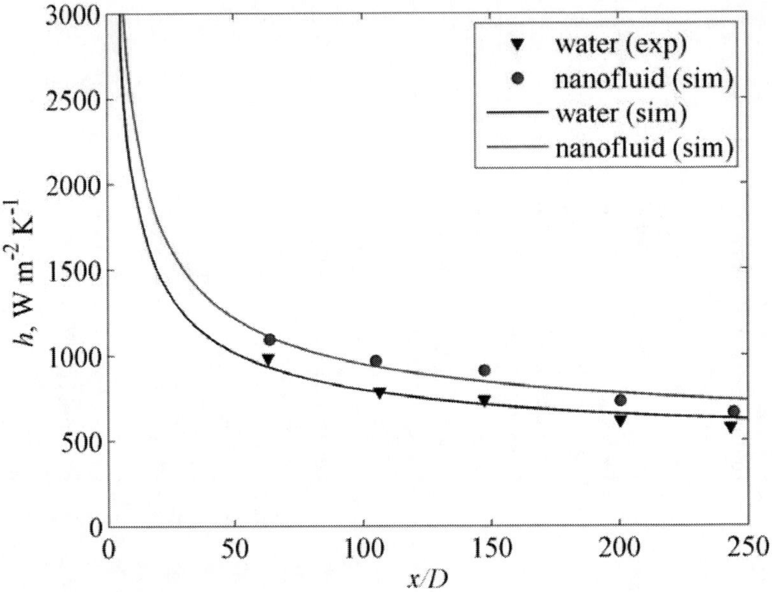


Figure 4. Comparison of experimental heat transfer coefficient by Anoop et al. [55] (experimental error of 4%) with numerical results from homogenous model by Sasmito et al. [38].

When developing mathematical models to predict nanofluid heat transfer, one needs to pay special attention to validation of the model due to the inherent complexity of coupled physical phenomena and interaction between base-fluid and nanoparticle. Sasmito et al. [38] validated their homogeneous model with experimental data by Anoop et al. [55]. The thermophysical properties used for the effective thermal conductivity are model based on Koo and Kleinstreuer [97], the effective viscosity model adopted from Namburu et al. [105]; whereas the effective specific heat capacity and density are based on Eqs. 2 and 4, respectively. The comparison between experimental heat transfer coefficient and numerical results are depicted in Figure 4; good agreement were obtained for local heat transfer coefficient for both water and nanofluids. Note that the choice of thermophysical model is essential in order to get good agreement.

**Two-Phase Model**

This approach is expected to capture phase interaction between the base liquid and solid nanoparticles, such as drag force, lift force, nanoparticles sedimentation, separation etc. There are a few studies that modelled nanofluids using two-phase approaches. Behzadmehr et al. [106], Mirmasoumi and Behzadmehr [107, 108], Akbarnia and Laur [109], Akbari et al. [110], Kalteh et al. [111], Kurowski et al. [112], Fard et al. [113] and Lotfi et al. [114] compared several two-phase flow model to predict heat transfer performance of nanofluids; they found that the two-phase model results are more precise than that of homogeneous one.

In general, there are three popular two-phase flow models reported in the literature, i.e., volume-of-fluid, mixture and Eulerian models.

**Volume of Fluid (VoF)**

The VoF model solves a single set of momentum equations for all phases and tracks the volume fraction all over the domain by solving additional scalar equations for the secondary phases. The total sum of the volume fractions for all the phases is equal to unity. In this method, all the physical properties are calculated by taking a weighted average of different phases based on their volume fraction. The single set of momentum equation is solved to get the velocity components, which are share by all phases which is identical to Eq. 6. Similarly, a shared energy equation is assumed to be in local thermal equilibrium and shared between all phases identical to Eq. 7. While the mass conservation equation is expressed as

$$\nabla \cdot \left(\phi_q \rho_q \mathbf{u}_q\right) = 0 \tag{8}$$

where $q$ refer to phases, $\sum_{q=1}^{n} \phi_q = 1$ and all properties are calculated by $N = \sum_{q=1}^{n} \phi_q N_q$ .

**Mixture Model**

The mixture model solves the continuity, momentum and energy equations for mixture as well as a volume fraction equation for the secondary phases. It then uses a correlation to calculate the relative velocity between the phases. The conservation equations for mass and momentum are

$$\nabla \cdot \left(\rho_m \mathbf{u}_m\right) = 0 \tag{9}$$

$$\begin{aligned}\nabla \cdot \left(\rho_m \mathbf{u}_m \otimes \mathbf{u}_m\right) = -\nabla p_m + \nabla \cdot \left[\mu_m \left(\nabla \mathbf{u}_m + \left(\nabla \mathbf{u}_m\right)^{\mathbf{T}}\right)\right] + \rho_m \mathbf{g} \\ + \nabla \cdot \left(\sum_{k=1}^{n} \varphi_k \rho_k \mathbf{u}_{dr,k} \otimes \mathbf{u}_{dr,k}\right)\end{aligned} \tag{10}$$

where the mixture velocity, density and viscosity are defined as

$$\mathbf{u}_m = \frac{\sum_{k=1}^{n} \phi_k \rho_k \mathbf{u}_k}{\rho_m} \tag{11}$$

$$\rho_m = \sum_{k=1}^{n} \varphi_k \rho_k \tag{12}$$

$$\mu_m = \sum_{k=1}^{n} \varphi_k \mu_k \tag{13}$$

The drift velocity of $k^{\text{th}}$ phase is

$$\mathbf{u}_{dr,k} = \mathbf{u}_k - \mathbf{u}_m \tag{14}$$

The conservation equation of energy is

$$\nabla \cdot \sum_{k=1}^{n} \left( \phi_k \rho_k c_{p,k} \mathbf{u}_k T \right) = \nabla \cdot \left( \left( \sum_{k=1}^{n} \phi_k k_k \right) \nabla T \right) \tag{15}$$

From the continuity equation for secondary phase *p*, the volume fraction for secondary phase *p* can be obtained using

$$\nabla \cdot \left( \phi_p \rho_p \mathbf{u}_m \right) = -\nabla \cdot \left( \phi_p \rho_p \mathbf{u}_{dr,p} \right) \tag{16}$$

The slip velocity between two phases is defined as the velocity of nanoparticles (*p*) relative to the velocity of the basefluid (*f*).

$$\mathbf{u}_{pf} = \mathbf{u}_p - \mathbf{u}_f \tag{17}$$

The drift velocity is related to the slip velocity as

$$\mathbf{u}_{dr,p} = \mathbf{u}_{pf} - \sum_{k=1}^{n} \frac{\phi_k \rho_k}{\rho_m} \mathbf{u}_{fk} \tag{18}$$

Assuming that the local equilibrium between phases is reached over short spatial length scale due to small size of nanoparticles, the slip velocity can be calculated from

$$u_{pf} = \frac{\rho_p d_p^2}{18 \mu_f f_{drag}} \frac{\left( \rho_p - \rho_m \right)}{\rho_p} \left( \mathbf{g} - \left( \mathbf{u}_m \cdot \nabla \right) \mathbf{u}_m \right) \tag{19}$$

where the drag force follows empirical relation developed by Schiller and Nauman [115]; however, this correlation was previously developed for much higher particle. Thus, more experimentation is required to determine drag related force in nanofluid.

$$f_{drag} = \begin{cases} 1 + 0.15\,\mathrm{Re}_p^{0.687} & \mathrm{Re}_p \le 1000 \\ 0.0183\,\mathrm{Re}_p & \mathrm{Re}_p > 1000 \end{cases} \tag{20}$$

**Eulerian Model**

In the Eulerian model, we solve two sets of conservation equations of mass, momentum and energy for basefluid and nanoparticles; the pressure is shared by all phases. The volume of each phase is calculated by integrating its volume fraction throughout the domain, whereas the summation of all the volume fractions is equal to unity. The conservation equation of mass is written as

$$\nabla \cdot \left(\phi_q \rho_q \mathbf{u}_q\right) = 0 \tag{21}$$

where $\mathbf{u}_q = \int_v \phi_q d\mathbf{u}$ and $\sum_{q=1}^{n} \phi_q = 1$ and $q$ indicates the phase. The conservation equation of momentum is

$$\nabla \cdot \left(\phi_q \rho_q \mathbf{u}_q \mathbf{u}_q\right) = -\phi_q \nabla p + \phi_q \nabla \cdot \left[\mu_q \left(\nabla \mathbf{u}_q + \left(\nabla \mathbf{u}_q\right)^{\mathbf{T}}\right)\right] + \phi_q \rho_q \mathbf{g} + \sum_{p=1}^{n} \mathbf{R}_{pq} + \mathbf{F}_{lift,q} \tag{22}$$

where $\sum_{p=1}^{n} \mathbf{R}_{pq}$ is the interaction forces between phases, defined as

$$\sum_{p=1}^{n} \mathbf{R}_{pq} = \sum_{p=1}^{n} S_{pq} \left(\mathbf{u}_p - \mathbf{u}_q\right) \tag{23}$$

$$S_{pq} = \frac{18 \mu_q \phi_p \phi_q \rho_q f}{\rho_p d_p^2} \tag{24}$$

$$f = \frac{C_D \operatorname{Re}}{24} \tag{25}$$

$$C_D = \begin{cases} \dfrac{24\left(1 + 0.15 \operatorname{Re}^{0.687}\right)}{\operatorname{Re}} & \operatorname{Re} \leq 1000 \\ 0.44 & \operatorname{Re} > 1000 \end{cases} \tag{26}$$

where Reynolds number is defined as

$$\operatorname{Re} = \frac{\rho_q \left|\mathbf{u}_p - \mathbf{u}_q\right| d_p}{\mu_q} \tag{27}$$

The lift force is calculated from

$$\mathbf{F}_{lift,q} = -0.5 \rho_p \phi_p \left(\mathbf{u}_p - \mathbf{u}_q\right) \times \left(\nabla \times \mathbf{u}_q\right) \tag{28}$$

The conservation of energy for Eulerian model is given as

$$\nabla \cdot \left(\phi_q \rho_q \mathbf{u}_q c_{p,q} T_q\right) = -\nabla \cdot \left(k_q \nabla T_q\right) - \boldsymbol{\tau}_q : \nabla \mathbf{u}_q + \sum_{p=1}^{n} \mathbf{Q}_{pq} \tag{29}$$

where $\mathbf{Q}_{pq} = h\left(\mathbf{u}_p - \mathbf{u}_q\right)$ is the heat exchange coefficient and $h = \left(6k_q\phi_q\phi_p\mathrm{Nu}_p\right) / \left(d_p^2\right)$. The $\mathrm{Nu}_p$ is calculated from Ranz and Marshal 1952 model

$$\mathrm{Nu}_p = 2 + 0.6\,\mathrm{Re}^{0.5}\,\mathrm{Pr}_q^{0.333} \tag{30}$$

and Prandtl number is defined as $\mathrm{Pr}_q = \left(c_{p,q}\mu_q\right) / k_q$.

Akbari et al. [110] compared local heat transfer for various nanofluids, as can be seen in Figure 5. It is noted that two-phase flow models result in better agreement with the experimental heat transfer coefficient by Wen and Ding [116]; while the single-phase model under-predicts the heat transfer rate. This deviation can be attributed to the selection of the effective thermophysical model. In this single-phase model, Akbari et al. [110] used thermal conductivity model proposed by Jang and Choi [96], whereas the effective viscosity is taken from Nguyen et al. [41]. Although the two-phase based models seem more advanced and more accurately predict the heat transfer performance compared to the homogeneous model, all the empirical relations for momentum and heat exchange between phases were originally derived for much larger particles (from micron to milimeter size solid particle); these interaction may or may not be correct for much smaller size (nano size) particles. Therefore more research is required to establish such correlation which is valid for nanofluids suspension.

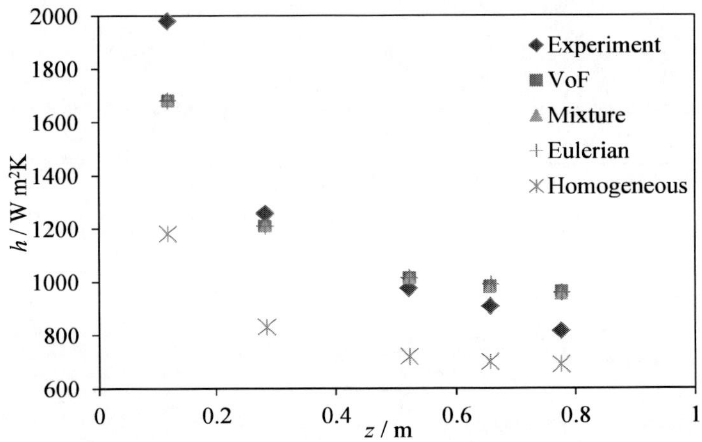


Figure 5. Comparison of various models for convective heat transfer coefficient with experimental data at Re 1050 and nanoparticle concentration 1% [110].

## Euler-Lagrange Model

Another approach to model convective heat transfer of nanofluids is the Eulerian-Lagrangian approach in which the basefluid is analysed in the Eulerian domain, whereas the nanoparticles are tracked in the Lagrangian one. Bianco et al. [117], He et al. [118], Savithiri et al. [69] attempted to model nanofluids convective heat transfer using Euler-Lagrange approach; their numerical results suggest that the Euler-Lagrange model is able to accurately predict heat transfer performance of nanofluids than that of homogeneous model.

Furthermore, the Euler-Lagrange approach is able to quantify the heat transfer enhancement due to Brownian motion, drag force, gravity force, thermophoretic force, Saffman's lift force, rotational force and Magnuss effect [69].

In this approach, the governing equations for conservation of mass, momentum and energy are identical to the homogeneous model (Eqs 5 -7) with additional term for momentum and heat exchange between basefluid and nanoparticles; while the nanoparticles are solved in the Lagrangian frame of reference, the equation of motion of a nanoparticles is given by:

$$m_p \frac{dv_p}{dt} = \sum \mathbf{F} \tag{31}$$

$$\sum \mathbf{F} = \mathbf{F}_{drag} + \mathbf{F}_{gravity} + \mathbf{F}_{Brownian} + \mathbf{F}_{thermophoresis} + \mathbf{F}_{Saffman} + \mathbf{F}_{rotational} + \mathbf{F}_{Magnus} \tag{32}$$

The drag force is the force generated in opposition to the direction of motion of a nanoparticle in a fluid. Drag force is proportional to the relative velocity between the basefluid and nanoparticle, expressed by

$$\mathbf{F}_{drag} = -\beta\left(\mathbf{u}_f - \mathbf{u}_p\right)m_p \tag{33}$$

where the interphase momentum exchange coefficient, $\beta$, is defined as

$$\beta = \frac{18\mu_{eff} C_D \mathrm{Re}_D}{\rho_p d_p^2 24} \tag{34}$$

The Reynolds number due to drag is defined as

$$\mathrm{Re}_D = \frac{\rho_p d_p \left|\mathbf{u}_f - \mathbf{u}_p\right|}{\mu_{eff}} \tag{35}$$

The drag coefficients are [69]

$$C_D = \begin{cases} \dfrac{24}{\mathrm{Re}_D}\left[1 + 0.15\,\mathrm{Re}_D^{0.687}\right] & \text{for spherical particle} \\ \dfrac{24}{\mathrm{Re}_D}\left[1 + b_1\,\mathrm{Re}_D^{b_2}\right] + \dfrac{b_3\,\mathrm{Re}_D}{b_4 + \mathrm{Re}_D} & \text{for non-spherical particle} \end{cases} \tag{36}$$

The gravitational force is proportional to the volume of the particle, and the relative density of nanoparticle and basefluid, expressed as

$$\mathbf{F}_{gravity} = -V_p\left(\rho_p - \rho_f\right)\mathbf{g} \tag{37}$$

The Brownian motion of nanoparticles within the suspension results from collisions between the nanoparticles and the molecules of the basefluid. It is a function of concentration gradient, surface area of the particle and the Brownian diffusion coefficient

$$\mathbf{F}_{Brownian} = \rho_p D_B \nabla\phi_p A_p \mathbf{u}_B \tag{38}$$

where the Brownian diffusion coefficient is defiend as [69]

$$D_B = \begin{cases} \dfrac{K_B T}{3\pi\mu_{eff} d_p} & \text{for spherical particle} \\ \dfrac{K_B T}{3\pi\mu_{eff} h}\left[\ln\left(\dfrac{h}{d_p}\right) - 0.662 + 0.917 - 0.05\left(\dfrac{h}{d_p}\right)^2\right] & \text{for non-spherical particle} \end{cases} \tag{39}$$

The Brownian velocity, $\mathbf{u}_B$, is a function of temperature and nanoparticle diameter

$$\mathbf{u}_B = \begin{cases} \dfrac{2K_B T}{\pi\mu_{eff} d_p^2} & \text{for spherical particle} \\ \dfrac{2K_B T}{\pi\mu_{eff} d_p^2 \left(A_{sp} / A\right)} & \text{for non-spherical particle} \end{cases} \tag{40}$$

Thermopherotic force defined as force that arises due to temperature gradient between basefluid and nanoparticles with the direction is opposite to the gradient, expressed as

$$\mathbf{F}_{thermophoresis} = \rho_p \mathbf{u}_T \phi_p A_p \mathbf{u}_T \tag{41}$$

where the thermophoretic velocity, $\mathbf{u}_T$, is given by [69]

$$\mathbf{u}_T = \frac{6\pi C_s \left(K + C_t \mathrm{Kn}\right)}{\left(1 + 3C_m \mathrm{Kn}\right)\left(1 + 2K + 2C_t \mathrm{Kn}\right)} \frac{\mu_{eff}}{\rho_{eff}} \frac{\nabla T}{T} \tag{42}$$

with $K=k_{eff}/k_p$ $C_s = 1.17$, $C_m = 1.14$ and $C_t = 2.18$. The Saffman's lift force due to free rotating particle moving in a shear flow is defined by

$$\mathbf{F}_{Saffman} = K_{Saffman} \mu_{eff} \left|\mathbf{u}_f - \mathbf{u}_p\right| r^2 \left(\frac{\left(8\mathbf{u}_p / D\right)}{\upsilon_{eff}}\right) \tag{43}$$

where $K_{Saffman} = 81.2$, $r$ is the particle radius, $\upsilon_{eff}$ is the kinematic viscosity and $D$ is the tube diameter. The particle rotational force due to motion around fixed axis is given by

$$\mathbf{F}_{rotational} = \frac{8\mu_{eff}\rho_p\mathbf{u}_p}{\rho_{eff}D}A_p \quad (44)$$

The Magnus force which arise under the effect of shear stress, a particle rotates about an axis perpendicular to the main flow direction. If a relative axial velocity exists between the particle and the fluid, a force perpendicular to the main flow direction will arise, expressed as:

$$\mathbf{F}_{Magnus} = m_p\left(\mathbf{u}_p - \mathbf{u}_{Magnus}\right) \quad (45)$$

The Magnus velocity is given by

$$\mathbf{u}_{Magnus} = 0.34\left(\frac{\left(\left(\mu_f + \mathbf{u}_p\right)/2\right)^{0.63} D}{\upsilon_{eff}}\right)\left(\frac{d_p}{D}\right)^{2.84}\left(\frac{z}{R}\right)\left(0.63 - \left(\frac{z}{R}\right)\right) \quad (46)$$

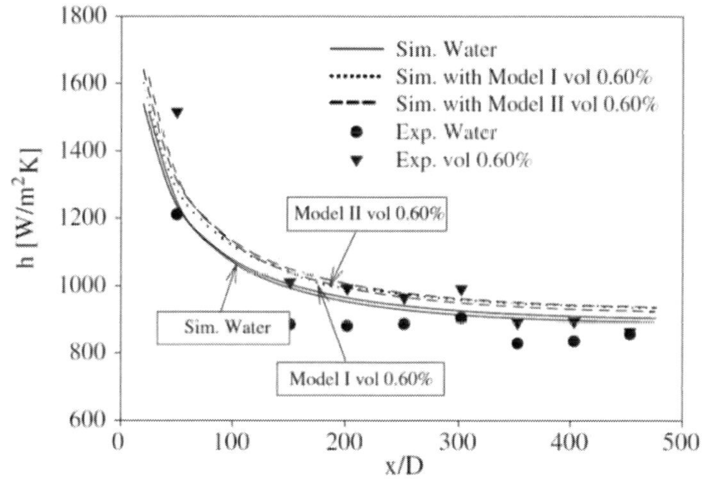


Figure 6. Comparison of heat transfer coefficient between experimental data, homogeneous model and Euler-Lagrange model for 0.6% nanofluid at q = 4000 W/m$^2$ at Re = 900. (Model 1: homogeneous model; Model II: Euler-Lagrange model).

He et al. [118] compared the local heat transfer coefficient between experimental data, homogeneous model and Euler-Lagrange model. They showed that the heat transfer coefficient for Euler-Lagrange model is somewhat higher than that of the homogenous model, as can be inferred from Figure 6. This is attributed the fact that the movement of nanoparticles and interactions between basefluid and nanoparticles are considered which enhances the heat transfer in some degree. They also concluded that the heat transfer coefficient is more affected by thermal conductivity than that of by the viscosity and Brownian force; the lift force and the thermophoretic force play small role on the heat transfer augmentation. In contrast, Savithiri et al. [69] showed by scaling analysis that the Brownian force dominates the mechanism at nanoparticle size bellow 10 nm and smaller volume fraction; whereas the drag force dominates in higher nanoparticle volume fraction. Hence, more work is required to clarify these phenomena in both experimental and theoretical aspect.

## CONCLUSION

This chapter presents an overview of the recent developments in nanofluids heat transfer from the preparation, characterization, theoretical aspects for thermophysical properties and convective heat transfer as well as stability of nanofluid from colloidal science point of view. Many interesting, complex and important phenomena have been reported to the literature. Researchers have given more attention to the thermal conductivity rather than heat transfer characteristics, suspension stability and cost. Although the use of nanofluids in a wide range of applications appears promising, the development of the field faces several challenges which need to be addressed in the near future:

- The lack of agreement between experimental results from different researchers for heat transfer enhancement leads to controversy on the reliability of the use of nanofluid for heat transfer enhancement. More precise and controlled experimentation is required to arrive at reliable conclusions.
- The choice of nanoparticle material is important. CNT and gold give rise to a higher heat transfer enhancement; however, the raw material is very expensive. This may not be suitable for low cost applications. Cheap, highly available and high heat transfer material can be a desired choice for future nanoparticles material; waste material from mining or minerals industry may have potential for nanofluids which needs to be explored further.
- Nanofluid preparation plays significant role on determining heat transfer characteristic and nanofluid stability. Cheap yet reliable and scalable preparation methods are required for successful real life applications to emerge.
- Several heat transfer enhancement mechanisms have been proposed by several researchers; however, no unified conclusion is possible. Further investigation is required in both experimental and theoretical work. The use of molecular dynamics simulation may help to understand the phenomena.
- Viscosity and thermal conductivity play important role in determining heat transfer enhancement. More precise observation is required to quantify heat transfer enhancement for both forced and natural convection with regard to the pressure drop.
- Two-phase flow modeling of nanofluids convective heat transfer requires special attention as most of the empirical correlation for drag force, slip force, lift force, Brownian force etc were derived for macro-size solid particle. It is therefore of interest to establish such correlation for nano-size particle suspension. Verification and validation of theoretical model with experimental counterpart is also required, not only for global heat transfer performance, but also for local physical phenomena.
- Currently most of research in nanofluids area is carried out by scientist. To ensure practical application, more engineering R&D is required.

## NOMENCLATURE

| | |
|---|---|
| $A$ | Surface area, $m^2$ |
| $c_p$ | Specific heat, J $kg^{-1}$ $K^{-1}$ |
| $D$ | Diameter / diffusion, m / $m^2s^{-1}$ |
| $d_p$ | Particle diameter, m |
| $\mathbf{F}$ | Forces, N |
| $h$ | Height, m |
| $k$ | Thermal conductivity, $Wm^{-1}K^{-1}$ |
| Kn | Knudsen number |
| $L$ | Length, m |
| $m$ | Mass flow rate, $kgs^{-1}$ |
| $p$ | Pressure, Pa |
| Pr | Prandtl |
| $r$ | Radius, m |
| Re | Reynolds |
| $T$ | Temperature, K |
| $\mathbf{u}$ | Velocity, $ms^{-1}$ |
| $V$ | Volume, $m^3$ |

### Greek

| | |
|---|---|
| $\beta$ | Interphase momentum exchange coefficient |
| $\mu$ | Viscosity, $kgm^{-1}s^{-1}$ |
| $\varphi_p$ | Nanoparticle volume fraction |
| $\rho$ | Density, $kgm^{-3}$ |

### Subscript

| | |
|---|---|
| eff | effective |
| f | basefluid |
| m | mixture |
| p | nanoparticle |
| ref | reference |

## REFERENCES


[1] Maxwell, J.C., 1881, *A Treatise on Electricity and Magnetism*, 2nd Ed; Clarendon Press: Oxford, UK,.

[2] Choi, S.U.S., 1995, "Enhancing Thermal Conductivity of Fluids with Nanoparticles," In *The Proceedings of the 1995 ASME International Mechanical Engineering Congress and Exposition*; San Francisco, USA, ASME; 1995, FED vol. 231/ MD vol. 66, pp 99-105.

[3] Wang, X.Q, and Mujumdar, A.S., 2008, "A Review on Nanofluids - Part I: Theoretical and Numerical Investigations," *Braz. J. Chem. Eng* 25, 613-630.

[4] Wang, X.Q, and Mujumdar, A.S., 2008, "A Review on Nanofluids - Part II: Experiments and Applications," *Braz. J. Chem. Eng,* 25, 631-648.

[5] Wang, X.Q, and Mujumdar, A.S., 2007, "Heat Transfer Characteristics of Nanofluids: a Review," *Int J Therm Sci,* 46, 1-19.

[6] Chandrasekar, M., and Suresh, S., 2009, "A Review on the Mechanisms of Heat Transport in Nanofluids," *Heat Transfer Eng.*, 30, 1136-1150.

[7] Daungthongsuk, W., and Wongwises, S., 2007, "A Critical Review of Convective Heat Transfer of Nanofluids," *Renew. Sustain. Energ. Rev.*, 11, 797-817.

[8] Trisaksri, V., and Wongwises, S., 2007, "Critical Review of Heat Transfer Characteristics of Nanofluids," *Renew. Sustain. Energ. Rev.*, 11, 512-523.

[9] Kakác, S., and Pramuanjaroenkij, A., 2009, "Review of Convective Heat Transfer Enhancement with Nanofluids," *Int. J. Heat Mass Transfer,* 52, 3187-3196.

[10] Wang, L.-Q., and Fan, J., 2010, 'Nanofluids Research: Key Issues," *Nanoscale Res. Lett.,* 5, 1241-1252.

[11] Godson, L., Raja, B., Lai, D. M., and Wongwises, S., 2010, "Enhancement of Heat Transfer Using Nanofluids—An Overview," *Renew. Sustain. Energ. Rev.*, 14, 629-641.

[12] Ghadimi, A., Saidur, R., Metselaar, H.S.C., 2011, "A Review of Nanofluid Stability Properties and Characterization in Stationary Conditions," *Int J Heat Mass Transfer,* 54, 4051-4068.

[13] Murshed, S.M.S., Leong, K.C., and Yang, C., 2008, "Thermophysical and Electrokinetic Properties of Nanofluids – A Critical Review," *Appl. Therm. Eng.*, 28, 2109-2125.

[14] Das, S.K., Choi, S.U.S., and Patel, H.E., 2006, "Heat Transfer in Nanofluids – A Review," *Heat Transfer Eng.*, 27, 3-19.

[15] Wen, D., Lin, G., Vafaei, S., and Zhang, K., 2009, "Review of Nanofluids for Heat Transfer Applications," *Particuology*, 7, 141-150.

[16] Li, Y., Zhou, J., Tung, S., Schneider, E., and Xi, S., 2009, "A Review on Development of Nanofluid Preparation and Characterization," *Powder Technol.*, 196, 89-101.

[17] Yu, W., France, D.M., Routbort, J.L., and Choi, S.U.S., 2008, "Review and Comparison of Nanofluid Thermal Conductivity and Heat Transfer Enhancements," *Heat Transfer Eng.,* 29, 432-460.

[18] Sergis, A., and Hardalupas, Y., 2011, "Anomalous Heat Transfer Modes of Nanofluids: A Review Based on Statistical Analysis," *Nanoscale Res. Lett.*, 6, 391.

[19] Jiang, L., Gao, L., and Sun, J., 2003, "Production of Aqueous Colloidal Dispersions of Carbon Nanotubes," *J Colloid Interface Sci.,* 1, 89-94.

[20] Assael, M.J. Metaxa, I.N., Arvanitidis, D.J, Christofilos, D., Lioutas, C., 2005, "Thermal Conductivity Enhancement in Aqueous Suspensions of Carbon Multi-Walled and Double-Walled Nanotubes in the Presence of Two Different Dispersants," *Int J Thermophys,* 26, 647-664.

[21] Xie, H., Lee, H., Youn, W., and Choi, M., 2003, "Nanofluids Containing Multi-Walled Carbon Nanotubes and Their Enhanced Thermal Conductivities," *J Appl Phys,* 94, 4967-4971.

[22] Hwang, Y., Lee, J.K., Lee, C.H., Jeong, Y.M., Cheong, S.I., Ahn, Y.C., and Kim, S.H., 2008, "Production and Dispersion Stability of Nanoparticles in Nanofluids," *Powder Technol.,* 186, 145-153.

[23] Paul, G., Chopkar, M., Manna, I., and Das, P.K., 2010, "Techniques for Measuring the Thermal Conductivity of Nanofluids: A Review," *Renew Sustain Energy Rev,* 14,1913-1924.

[24] Choi, E.S., Brooks, J.S., Eaton, D.S., Al-Haik, M.S., Hussaini, M.Y., Garmestani, H., Li, D., and Dahemen, K., 2003, "Enhancement of Thermal and Electrical Properties of Carbon Nanotube Polymer Composites by Magnetic Field Processing," *J Appl Phys,* 94, 6034-6039.

[25] Llaguno, M., Hone, J., Johnson, A., Fischer, J., 2001, "Thermal Conductivity of Single-Wall Carbon Nanotubes: Diameter and Annealing Dependence," In AIP conference proceeding, vol. 591, Woodbury, New York.

[26] Torii, S., 2010, "Turbulent Heat Transfer Behavior of Nanofluid in a Circular Tube Heated under Constant Heat Flux," *Advances in Mechanical Engineering,* 2010, 917612.

[27] Ali, F.M., Yunus, W.M.M.Y., Moksin, M.M., Aidin, Z., 2010, "Measurement Thermal Conductivity and Thermal Diffusivity of Chromium Nanofluids," *Australian J Basic Appl Sci,* 4, 3780-3793.

[28] Hojat, M., Etemad, S.Gh., Bagheri, R., Thibault, R., 2002, "Laminar Convective Heat Transfer of Non-Newtonian Nanofluids with Constant Wall Temperature," *Heat Mass Transfer,* 47, 203-209.

[29] Moosavi, M., Goharshadi, E.K., Youssefi, A., 2010, "Fabrication, Characterization, and Measurement of Some Physicochemical Properties of ZnO Nanofluids," *Int J Heat Fluid Flow,* 31, 599-605.

[30] Karthikeyan, N.R., Philip, J., Raj, B., 2008, "Effect of Clustering on the Thermal Conductivity of Nanofluids," *Mater Chem Phys,* 109, 50-55.

[31] Kwak, K., and Kim, C., 20056, "Viscosity and Thermal Conductivity of Copper Oxide Nanofluid Dispersed in Ethylene Glycol," *Korea Aust Rheol J,* 17, 35-40.

[32] Wang, X.J., and Li, X.F., 2009, "Influence of pH on Nanofluids' Viscosity and Thermal Conductivity," *Chin Phys Lett,* 26, 056601.

[33] Wei, X., Zhu, H., Kong, T., and Wang, L., 2009, "Synthesis and Thermal Conductivity of $Cu_2O$ Nanofluids," *Int J Heat Mass Transfer,* 52, 4371-4374.

[34] Lee, J.H., Hwang, K.S., Jang, S.P., Kim, J.H., Choi, S.U.S., and Choi, C.J., 2008, "Effective Viscosities and Thermal Conductivities of Aqueous Nanofluids Containing Low Volume Concentrations of $Al_2O_3$ Nanoparticles," *Int J Heat Mass Transfer,* 51, 2651-2656.

[35] Murshed, S.M.S., Leong, K.C., and Yang, C., 2008, "Investigations of Thermal Conductivity and Viscosity of Nanofluids," *Int J Therm Sci,* 47, 560-568.

[36] Oh, D.W., Jain, A., Eaton, J.K., Goodson, K.E., and Lee, J.S., 2008, "Thermal Conductivity Measurement and Sedimentation Detection of Aluminum Oxide Nanofluids by Using the 3@w Method," *Int J Heat Fluid Flow,* 29, 1456-1461.

[37] Kole, M., Dey, T.K., 2010, "Thermal Conductivity and Viscosity of $Al_2O_3$ Nanofluid Based on Car Engine Coolant," *J Phys D Appl Phys,* 43, 315501.

[38] Sasmito, A.P., Kurnia, J.C., and Mujumdar, A.S., 2011, "Numerical Evaluation of Laminar Heat Transfer Enhancement in Nanofluid Flow in Coiled Square Tubes," *Nanoscale Res Lett,* 6, 376.

[39] Brinkman, H.C., 1952, "The Viscosity of Concentrated Suspensions and Solutions," *J Chem Phys,* 20, 571.

[40] Pak, B.C., and Cho, Y.I., 1999, "Hydrodynamic and Heat Transfer Study of Dispersed Fluids with Submicron Metallic Oxide Particles," *Exp Heat Transfer,* 11, 151-170.

[41] Nguyen, C.T., Desgranges, F., Roy, G., Galanis, N., Mafe, T., Boucher, S., and Mintsa, H.A., 2007, "Temperature and Particle-size Dependent Viscosity Data for Water-Based Nanofluids - Hysteresis Phenomenon," *Int J Heat Fluid Flow,* 28, 1492-1506.

[42] Wang, X., Xu, X., and Choi, S.U.S., 1999, "Thermal Conductivity of Nanoparticle-Fluid Mixture," *J Thermophys Heat Transfer,* 13, 474-480.

[43] Kulkarni, D.P., Das, D.K., Patil, S.L., 2007, "Effect of Temperature on Rheological Properties of Copper Oxide Nanoparticles Dispersed in Propylene Glycol and Water Mixture," *J Nanosci Nanotechnol,* 7, 2318-2322.

[44] Anoop, K.B., Kabelac, S., Sundararajan, T., and Das, S.K., 2009, "Rheological and Flow Characteristics of Nanofluids: Influence of Electroviscous Effects and Particle Agglomeration," *J Appl Phys,* 106, 034909.

[45] Chen, H., Ding, Y., He, Y., and Tan, C.H., 2007, "Rheological Behaviour of Ethylene Glycol Based Titania Nanofluids," *Chem Phys Lett,* 444, 333-337.

[46] Khanafer, K., and Vafai, K., 2011, "A Critical Synthesis of Thermophysical Characteristics of Nanofluids," *Int J Heat Mass Transfer,* 54, 4410-4428.

[47] Wong, K.V., and Leon, D.O., 2010, "Heat Transfer Mechanisms and Clustering in Nanofluids," *Advances in Mechanical Engineering,* 2010, 519659.

[48] Sarkar, J., 2011, "A Critical Review on Convective Heat Transfer Correlations of Nanofluids," *Renew Sustain Energy Rev,* 15, 3271-3277.

[49] Li, Q., and Xuan, Y., 2002, "Convective Heat Transfer and Flow Characteristics of Cu-Water Nanofluid," *Sci China, Series E: Technol Sci,* 45, 408-416.

[50] Xuan, Y., and Li, Q., 2003, "Investigation on Convective Heat Transfer and Flow Features Of Nanofluids," *J Heat Transfer,* 125, 151-155.

[51] Yang, Y.Z., Zhang, G., Grulke, E.A., Anderson, W.B., and Wu, G., 2005, "Heat Transfer Properties of Nanoparticle-in-Fluid Dispersions (Nanofluids) in Laminar Flow," *Int J Heat Mass Transfer,* 48, 1107-1116.

[52] Maiga, S.E.B., Palm, S.J., Nguyen, C.T., Roy, G., and Galanis, N., 2005, "Heat Transfer Enhancement by Using Nanofluids in Forced Convection Flows," *Int J Heat Fluid Flow,* 26, 530-546.

[53] Maiga, S.E.B., Nguyen, C.T., Galanis, N., Roy, G., Mare, T., Coqueux, M., 2006, "Heat Transfer Enhancement in Turbulent Tube Flow using Al2O3 Nanoparticle Suspension," *Int J Numer Method Heat Fluid Flow,* 16, 275-292.

[54] Buongiorno, J., 2006, "Convective Transport in Nanofluids," *J Heat Transfer,* 128, 240-250.

[55] Anoop, K.B., Sundararajan, T., and Das, S.K., 2009, "Effect of Particle Size on the Convective Heat Transfer in Nanofluid in the Developing Region," *Int. J Heat Mass Transfer,* 52, 2189-2195.

[56] Daungthongsuk, W., and Wongwises, S., 2010, “An experimental Study on the Heat Transfer Performance and Pressure Drop of $TiO_2$-Water Nanofluids Flowing under a Turbulent Flow Regime,” *Int J Heat Mass Transfer,* 53, 334-344.
[57] Vajjha, R.S., Das, D.K., and Kulkarni, D.P., 2010, “Development of New Correlations for Convective Heat Transfer and Friction Factor in Turbulent Regime for Nanofluids,” *Int J Heat Mass Transfer, 53,* 4607-4618.
[58] Prasher, R., Phelan, P.E., Bhattacharya, P., 2006, “Effect of Aggregation Kinetics on the Thermal Conductivity of Nanoscale Colloidal Solutions (Nanofluid),” *Nano Lett,* 6, 1529-1534.
[59] Prasher, R., Bhattacharya, P., Phelan, P.E, 2005, Thermal Conductivity of Nanoscale Colloidal Solutions (Nanofluids), *Phys Rev Lett,* 94, 1-4.
[60] Xie, H., Wang, J., Xi, T., Liu, Y., and Ai, F., 2002, “Thermal Conductivity Enhancement of Suspensions Containing Nanosized Alumina Particles,” *J Appl Phys,* 91, 4568-4572.
[61] Das, S.K., Putra, N., Thiesen, P., and Roetzel, W., 2003, “Temperature Dependence of Thermal Conductivity Enhancement for Nanofluids,” *ASME J Heat Transfer,* 125, 567-574.
[62] Chopkar, M., Das, P.K., Manna, I., 2006, “Synthesis and Char- Acterization of Nanofluid for Advanced Heat Transfer Applications,” *Scr Matter,* 55, 549-552.
[63] Li, C.H., and Peterson, G.P., 2007, “Mixing Effect on the Enhancement of the Effective Thermal Conductivity of Nanoparticle Suspensions (nanofluids),” *Int J Heat Mass Transfer*, 50, 4668-4677.
[64] Shima, P.D., Philip, J., and Raj, B., 2009, “Role of Microconvection Induced By Brownian Motion of Nanoparticles in the Enhanced Thermal Conductivity of Stable Nanofluids,” *Appl Phys Lett,* 94, 223101.
[65] Mintsa, H.A., Roy, G., Nguyen, C.T., and Doucet, D., 2009, “New Temperature Dependent Thermal Conductivity Data for Water-Based Nanofluids,” *Int J Therm Sci,* 48, 363-371.
[66] Keblinski, P., Phillpot, S.R., Choi, S.U.S., and Eastman, J.A., 2002, “Mechanisms of Heat Flow in Suspensions of Nano-sized Particles (Nanofluids),” *Int J Heat Mass Transfer,* 45, 855-863.
[67] Keblinksi, P., and Thomin, J., 2006, “Hydrodynamic Field around a Brownian Particle,” *Phys Rev E,* 73, 010502.
[68] Evans, W., Fish, J., and Keblinski, P., 2006, Role of Brownian Motion Hydrodynamics on Nanofluid Thermal Conductivity,” *Appl Phys Lett.* 88, 093116.
[69] Savithiri, S., Pattamatta, A., and Das, S.K., 2011, “Scaling Analysis for the Investigation of Slip Mechanisms in Nanofluids,” Nanoscale *Res Lett,* 6, 471.
[70] Fan, J., and Wang, L., 2011, “Review of Heat Conduction in Nanofluids,” *ASME J Heat Transfer,* 133, 0408011.
[71] Jung, J Y.J., and Yoo, Y., 2009, “Thermal Conductivity Enhancement of Nanofluids in Conjunction with Electrical Double Layer (EDL),” *Int J Heat Mass Transfer,* 52, 525-528.
[72] Yu, C.J., Richter, A.G., Datta, A., Durbin, M.K., and Dutta, P., 1999, “Observation of Molecular Layering in Thin Liquid Films Using X-Ray Reflectivity,” *Phys Rev Lett,* 82, 2326-2329.

[73] Yu, W., and Choi, S.U.S., 2003, "The Role of Interfacial Layers in the Enhanced Thermal Conductivity of Nanofluids: A Renovated Maxwell Model," *J Nanopart Res,* 5, 167-171.

[74] Xue, L., Keblinski, P., Phillpot, S.R., Choi, S.U.S., and Eastman, J.A., 2003, "Two Regimes of Thermal Resistance at a Liquid-Solid Interface," *J Chem Phys,* 118, 337-339.

[75] Lee, D., 2007, "Thermophysical Properties of Interfacial Layer in Nanofluids," *Langmuir,* 23, 6011-6018.

[76] Eapen, J., Li, J., Yip, S., 2007, "Beyond the Maxwell Limit: Thermal Conduction in Nanofluids with Percolating Fluid Structures," *Phys Rev E,* 76, 062501.

[77] Xue, L., Keblinski, P., Phillpot, S.R., Choi, S.U.S., and Eastman, J.A., 2004, "Effect of Liquid Layering at the Liquid–Solid Interface on Thermal Transport," *Int J Heat Mass Transfer,* 47, 4277-4284.

[78] Philip, J., Shima, P.D., and Raj, B., 2008, "Evidence for Enhanced Thermal Conduction Through Percolating Structures in Nanofluids," *Nanotechnology,* 19, 305706.

[79] Liao, J., Zhang, Y., Yu, W., Xu, L., Ge, C., Liu, J., and Gu, N., 2003, "Linear Aggregation of Gold Nanoparticles in Ethanol," *Colloid Surf A Phsycochem Eng Aspects,* 223, 177-183.

[80] J. Avsec, The combined analysis of phonon and electron heat transfer mechanism on thermal conductivity for nanofluids, *Int J Heat Mass Transfer* 51 (2008) 4589-4598.

[81] Louge, M., and Chen, X., 2008, "Heat Transfer Enhancement in Suspensions of Agitated Solids. Part III: Thermophoretic Transport of Nanoparticles in the Diffusion Limit," *Int J Heat Mass Transfer,* 51, 5130-5143.

[82] Fu, C.J., and Zhang, Z.M., 2006, "Nanoscale Radiation Heat Transfer for Silicon at Different Doping Levels," *Int J Heat Mass Transfer,* 49, 1703-1718.

[83] Ding, Y., Alias, H., Wen, D., William, R.A., 2006, "Heat Transfer of Aqueous Suspensions of Carbon Nanotubes (CNT Nanofluids)," *Int J Heat Mass Transfer,* 49, 240-250.

[84] Hwang, K.S., Jang, S.P., and Choi, S.U.S., 2009, "Flow and Convective Heat Transfer Characteristics of Water-Based Al2O3 Nanofluids in Fully Developed Laminar Flow Regime," *Int J Heat Mass Transfer,* 52, 193-199.

[85] Bruggeman, D.A.G., 1935, "Berechnung Verschiedener Physikalischer Konstanten von Heterogenen Substanzen," *Ann Phys Leipzig,* 24, 636-679.

[86] Hamilton, R.L., 1962, "Thermal Conductivity of Heterogeneous Two- Component Systems," *Ind Eng Chem Fundam,* 1, 182-191.

[87] Jeffrey, D.J., 1973, "Conduction through a Random Suspension of Spheres," *Proc Roy Soc,* A355, 355-367.

[88] Davis, R.H., 1986, "The Effective Thermal Conductivity of a Composite Material with Spherical Inclusions," *Int J Thermophys,* 7, 609-620.

[89] Lu. S., and Lin, H., 1996, "Effective Conductivity of Composites Containing Aligned Spheroidal Inclusions of Finite Conductivity," *J Appl Phys,* 79, 6761-6769.

[90] Yu, W., and Choi, S.U.S., 2004, "The Role of Inter facial Layers in the Enhanced Thermal Conductivity of Nanofluids: A Renovated Hamilton-Crosser Model," *J Nanopart Res,* 6, 355-361.

[91] Xue, Q., 2003, "Model for Effective Thermal Conductivity of Nanofluids," *Phys Lett A,* 307, 313-317.

[92] Xue, Q., and Xu, W.M., "A Model of Thermal Conductivity of Nanofluids with Interfacial Shells," *Mater Chem Phys,* 90, 298-301.

[93] Xie, H., Fujii, M., and Zhang, X., 2005, "Effect of Interfacial Nanolayer on the Effective Thermal Conductivity of Nanoparticle-Fluid Mixture," *Int J Heat Mass Transfer,* 48, 2926-2932.

[94] Wang, B.X., Zhou, L.P., and Peng, X.F., 2003, "A Fractal Model for Predicting the Effective Thermal Conductivity of Liquid with Suspension of Nanoparticles," *Int J Heat Mass Transfer,* 46, 2665-2672.

[95] Xuan, Y. Li, Q., and Hu, W., 2003, "Aggregation Structure and Thermal Conductivity of Nanofluids," *AIChE J,* 49, 1038-1043.

[96] Jang, S.P., and Choi, S.U.S., 2004, "Role of. Brownian Motion in the Enhanced Thermal Conductivity of Nanofluids," *Appl Phys Lett,* 84, 4316-4318.

[97] Koo, J., and Kleinstreuer, C., 2005, "Laminar Nanofluid Flow in Microheat-Sinks," *Int J Heat Mass Transfer,* 48, 2652-2661.

[98] Chon, C.H., Kihm, K.D., Lee, S.P., and Choi, S.U.S., 2005, "Empirical Correlation Finding the Role of Temperature and Particle Size for Nanofluid (Al2O3) Thermal Conductivity Enhancement," *Appl Phys Lett,* 87, 153107.

[99] Li, J., 2008, Computational Analysis of Nanofluid Flow in Microchannels with Applications to Micro-Heat Sinks and bio-MEMS, *PhD Thesis NC State University,* Raleigh, NC, United States.

[100] Kihm, K.D., Chon, C.H., Lee, J.S., and Choi, S.U.S., 2011, "A New Heat Propagation Velocity Prevails over Brownian Particle Velocities in Determining the Thermal Conductivities of Nanofluids," *Nanoscale Res Lett.* 6, 361.

[101] Einstein, A., 1906, "Eine neue Bestimmung der Molekul dimension," *Ann Phys Leipzig,* 19, 289-306.

[102] Batchelor, G., 1977, "The Effect of Brownian Motion on the Bulk Stress in a Suspension of Spherical Particles," *J Fluid Mech,* 83, 97-117.

[103] Lundgren, T., 1972, "Slow Flow through Stationary Random Beds and Suspensions of Spheres," *J Fluid Mech,* 51, 273-286.

[104] Graham, A.L., 1981, "On the Viscosity of Suspensions of Solid Spheres," *Appl Sci Res,* 37, 275-286.

[105] Namburu, P.K., Das, D.K., Tanguturi, K.M., and Vajjha, R.S., 2009, "Numerical Study of Turbulent Flow and Heat Transfer Characteristics of Nanofluids Considering Variable Properties," *Int J Therm Sci,* 48, 290-302.

[106] Behzadmehr, A., Saffar-Avval, M., and Galanis, N., 2007, "Prediction of Turbulent Forced Convection of a Nanofluid in a Tube with Uniform Heat Flux Using a Two Phase Approach," *Int J Heat Fluid Flow,* 28, 211-219.

[107] Mirmasoumi, S., and Behzadmehr, A., 2008, "Numerical Study of Laminar Mixed Convection of a Nanofluid in a Horizontal Tube Using Two-phase Mixture Model," *Appl Therm Eng,* 28, 717-727.

[108] Mirmasoumi, S., and Behzadmehr, A., 2008, "Effect of Nanoparticles Mean Diameter on Mixed Convection Heat Transfer of A Nanofluid in a Horizontal Tube," *Int J Heat Fluid Flow,* 29, 557-566.

[109] Akbarnia, A., and Laur, R., 2009, "Investigating the Diameter of Solid Particles Effects on a Laminar Nanofluid Flow in a Curved Tube Using a Two Phase Approach," *Int J Heat Fluid Flow,* 30, 706-714.

[110] Akbari, M., Galanis, N., and Behzadmehr, A., "Comparative Analysis of Single and Two-Phase Models for CFD Studies of Nanofluid Heat Transfer," *Int J Therm Sci.* 50, 1343-1354.

[111] Kalteh, M., Abbasi, A., Saffar-Avval, M., and Harting, J., 2011, "Eulerian–eulerian Two-Phase Numerical Simulation of Nanofluid Laminar Forced Convection in a Microchannel," *Int J Heat Fluid Flow,* 32, 107-116.

[112] Kurowski, L., Chmiel-Kurowska, K., and Thullie, J., 2009, "Numerical Simulation of Heat Transfer in Nanofluids," *Computer Aided Chem Eng,* 26, 967-972.

[113] Fard, M.H., Esfahany, M.N., and Talaie, M.R., 2010, "Numerical Study of Convective Heat Transfer of Nanofluids in a Circular Tube Two-Phase Model versus Single-Phase Model," *Int Commun Heat Mass Transfer,* 37, 91-97.

[114] Lotfi, R., Saboohi, Y., and Rashidi, A.M., 2010, "Numerical Study of Forced Convective Heat Transfer of Nanofluids: Comparison of Different Approaches," *Int Commun Heat Mass Transfer,* 37, 74-78.

[115] Schiler, L., Naumann, A.Z., 1935, "Über die grundlegenden Berechungen bei der Schwerkraftaufbereitung," *Ver Deutsch Ing.*, 77, 318-320.

[116] Wen, D., and Ding, Y., 2004, "Experimental Investigation into Convective Heat Transfer of Nanofluids at the Entrance Region under Laminar Flow Conditions," *Int J Heat Mass Transfer,* 47, 5181-5188.

[117] Bianco, V., Chiacchio, F., Manca, O., and Nardini, S., 2009, "Numerical Investigation of Nanofluids Forced Convection in Circular Tubes," *Appl Therm Eng,* 29, 3632-3642.

[118] He, Y., Men, Y., Zhao, Y., Lu, H., and Ding, Y., 2009, "Numerical Investigation into the Convective Heat Transfer of TiO2 Nanofluids Flowing through a Straight Tube under the Laminar Flow Conditions," *Appl Therm Eng,* 29, 1965-1972.

In: Nanofluids
Editor: Yuwen Zhang
ISBN: 978-1-62618-165-6


*Chapter 3*

# FORMULATION AND THERMOPHYSICAL PROPERTIES OF NANOFLUIDS

***S. Suresh[1,*] and M. Chandrasekar[2]***

[1]Department of Mechanical Engineering, National Institute of Technology, Tiruchirappalli, India

[2]Department of Mechanical Engineering, University College of Engineering BIT Campus, Tiruchirappalli, India

## ABSTRACT

In this chapter, the formulation of nanofluids and experimental investigations on thermophysical properties together with the mechanisms proposed for the alteration in their values or characteristics due to the addition of nanoparticles are discussed. This chapter describes the two kinds of methods which have been frequently employed in producing nanofluids namely single-step method and the two-step method. Important thermophysical properties of nanofluids like thermal conductivity, viscosity, density, specific heat capacity and surface are also explained in this chapter.

## 1. INTRODUCTION

Generally, the thermal conductivity of solid particles is typically higher than that of liquids, seen from Table 1 [1-5]. Hence theoretically, all solid nanoparticles with high thermal conductivity can be used as additives of nanofluids.

[*] Email: ssuresh@nitt.edu.

**Table 1. Thermal conductivity of solids and liquids [1-5]**

| Material | | Thermal conductivity(W/mK) |
|---|---|---|
| Metallic solids | Cu | 401 |
| | Al | 237 |
| | Ag | 428 |
| | Au | 318 |
| | Fe | 83.5 |
| Nonmettalic solids | $Al_2O_3$ | 40 |
| | CuO | 76.5 |
| | Si | 148 |
| | SiC | 270 |
| | MWCNT | 3000 |
| | SWCNT | 6000 |
| Base fluids | H2O | 0.613 |
| | Ethylene glygol (EG) | 0.253 |
| | Engine oil (EO) | 0.145 |

The following nanoparticles that have been often used to prepare nanofluids

1) Metallic particles (Cu, Al, Fe, Au, and Ag)
2) Nonmetal particles ($Al_2O_3$, CuO, $Fe_3O_4$, $TiO_2$, and SiC)
3) Carbon nanotube and
4) Nanodroplet.

The base fluids commonly used are

1) Water
2) Engine oil
3) Acetone
4) Decene and
5) Ethylene glycol.

## 2. PREPARATION OF NANOFLUIDS

Preparation of nanofluids is the key step in the use of nanoparticles for heat transfer applications. Two kinds of methods have been employed in producing nanofluids. One is single-step method and the other is two-step method.

## 2.1. Single-Step Method

In a single-step method, synthesis and dispersion of nanoparticles into the fluid take place simultaneously. For example, Argonne national laboratory developed a one-step nanofluid production system in which nanoscale vapor from metallic source material can be directly dispersed into low vapor pressure fluids [1]. This novel one-step process was developed to overcome the van der Waals forces between the nanoparticles and produced stable suspensions of Cu nanoparticles without any dispersants. However, only low vapor pressure fluids are compatible with the process. This limits the application of this method. Lo et al. [2] prepared copper dioxide nanofluids by a single-step method called submerged arc nanoparticle synthesis system (SANSS). Another one-step physical process is wet grinding technology with bead mills [3]. One-step chemical methods for producing stable metallic nanofluids have been developed using a chemical reduction method [4, 5]. Also, Ag-water nanofluids were produced using one-step optical laser ablation in liquid [6]. In one-step method, the processes of drying, storage, transportation, and dispersion of nanoparticles are avoided. Hence the agglomeration of nanoparticles is minimized and the stability of fluids produced by this method is expected to be high.

## 2.2. Two-Step Method

Two-step method for preparing nanofluids involves dispersing of nanoparticles into base liquids. Nanoparticles, nanofibers or nanotubes used in this method are first produced as a dry powder by inert gas condensation, chemical vapor deposition, mechanical alloying or other suitable techniques, and the nano sized powder is then dispersed into the base fluid in a second processing step. This step by step method isolates the preparation of the nanofluids from the preparation of nanoparticles. As a result, agglomeration of nanoparticles may take place in both steps, especially in the process of drying, storage, and transportation of nanoparticles. The agglomeration will result in the settlement and clogging of heat transfer passages. Simple techniques such as ultrasonic agitation or the addition of surfactants to the fluids are often used to minimize particle aggregation and improve dispersion behavior. Since nanopowder synthesis techniques have already been scaled up to industrial production levels by several companies, there are potential economic advantages in using two-step synthesis methods that rely on the use of such powders. Most researchers purchase nanoparticles in powder form and mix them with the base fluid. However, these nanofluids are not stable, although stability can be enhanced with pH control or with surfactant addition. Some researchers purchase commercially available nanofluids. But these nanofluids contain impurities and nanoparticles whose size is different from vendor specifications. Although the two-step process works fairly well for oxide nanoparticles, it is not as effective for metallic nanoparticles [7]. The stabilization of the nanofluids prepared is an important problem that has to be addressed.

# 3. Thermophysical Properties of Nanofluids

## 3.1. Thermal Conductivity of Nanofluids

A number of techniques have been used for nanofluids. Figure 1 provides a summary of the available measurement techniques. Out of all the techniques, the transient hot-wire method has been used most extensively as shown in Figure 2 [8].

Anomalous enhancement observed in thermal conductivity of nanofluids has made nanofluid a prospective coolant for many applications and has been received considerable attention by the researchers. The effective thermal conductivities of $Al_2O_3$/water nanofluids with low volume concentrations from 0.01 to 0.3% were measured by Lee et al. [9]. The measured thermal conductivities of the dilute $Al_2O_3$/water nanofluids increase nearly linearly with volume concentration and a maximum enhancement of 1.44% at volume concentration of 0.3% was observed. These results show that the enhancement in the effective thermal conductivity of very dilute $Al_2O_3$/water nanofluids was not remarkable.

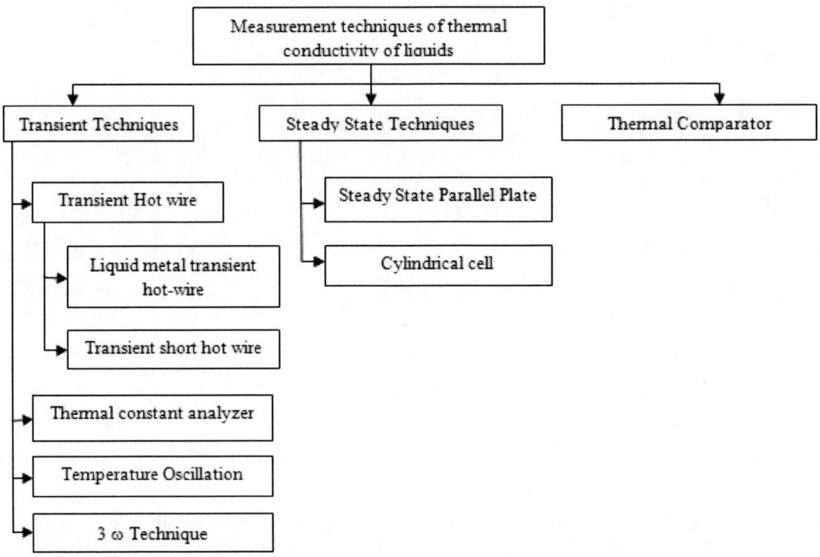


Figure 1. Different thermal conductivity measurement techniques for nanofluids [8].

Chandrasekar et al. [10] presented an experimental investigation of determination of effective thermal conductivity of $Al_2O_3$/water nanofluid. The thermal conductivity enhancements of current $Al_2O_3$/water nanofluids are 1.64%, 3.28%, 3.43%, 7.52% and 9.7%, correlated to volume concentrations of 0.33%, 0.75%, 1%, 2% and 3%. A linear relationship between the thermal conductivity enhancement and volume concentration was attributed to large regions of particle-free liquid with high thermal resistances created by highly agglomerated nanoparticles. They have also indicated that the relationship between the thermal conductivity enhancement and volume concentration is usually nonlinear for nanoparticles with a high aspect ratio (such as nanotube, nanorod, etc.) or nanoparticle alignment Paul et al. [11] synthesized Al-5wt% Zn nanoparticles by mechanical alloying and are dispersed to the tune of 0.01–0.10 % volume concentration in ethylene glycol (base fluid) following a careful mixing protocol. Thermal conductivity of the nanofluids and base fluid has been measured using the transient hot-wire method. It is observed that thermal conductivity of the nanofluids strongly depend on the concentration, particle size, fluid temperature and stability of dispersed nanoparticles in the base fluid. A maximum of 16% enhancement in thermal conductivity has been recorded at a volume concentration of 0.10%. It was pointed out that Al-5wt% Zn alloy is a new material that can be used as a dispersoid for nanofluid preparation. The level of enhancement recorded in this study at a much lower concentration surpasses earlier studies with comparable systems.

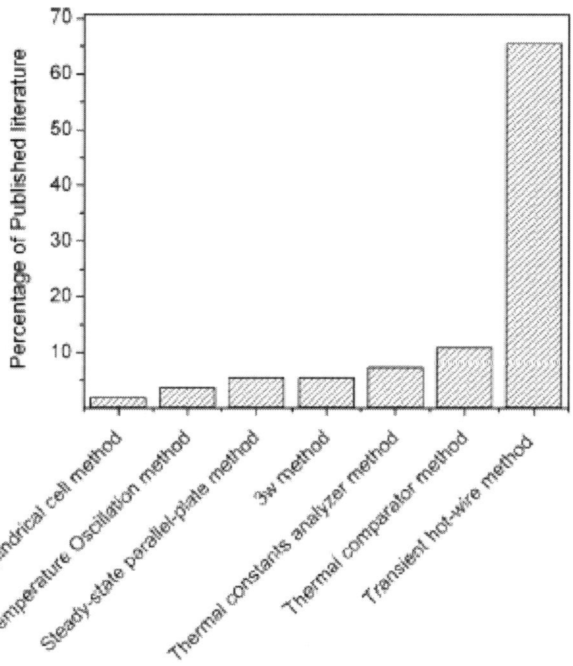


Figure 2. Comparison of the thermal conductivity measurement techniques for nanofluids [8].

Duangthongsuk and Wongwises [12] reported the thermal conductivity of $TiO_2$/water nanolfuids with volume concentration of 0.2-2 %. It was clearly shown that the use of nanoparticles dispersed in base liquid gave greater thermal conductivity than the base fluid, about 3-7 % for the volume concentration range between 0.2-2%.

Li et al. [13] experimentally investigated the effect of two types of magnetic nanoparticle on the thermal conductivity of magnetic fluids. They used $Fe_3O_4$/water and Fe/water magnetic fluid. The average particle diameter for $Fe_3O_4$ is 20 nm while that of Fe is 26 nm. Compared with pure water, the thermal conductivity of Fe/water magnetic fluid increased by about 14.9% at 5% volume concentration. They also illustrated that the thermal conductivity of Fe/water magnetic fluid increases with the strength of the external magnetic field and the thermal conductivity of water experienced almost no variation under the same external magnetic field. In addition they demonstrated that the application of the magnetic field enhances energy transport process in the magnetic fluid. For the Fe/water magnetic fluid of

51% volume concentration, its thermal conductivity was 0.667 W/m K under the field strength of 35 G while the thermal conductivity of the sample fluid was 0.833 W/m K under the field strength of 240 G.

Yu et al. [14] observed that the thermal conductivity of kerosene based $Fe_3O_4$ nanofluids increased almost linearly with volume concentration of the nanoparticles. When the volume concentration was 1%, the enhancement value was up to 34%, which was higher than that of $Fe_3O_4$ aqueous nanofluid. Kerosene based $Fe_3O_4$ nanofluids prepared did not showed the time-dependence of the thermal conductivity characteristic, which indicated a good suspension stability of the nanofluids prepared by them. In order to investigate the effect of temperature on the enhancement of the thermal conductivity, the thermal conductivities were measured in the temperature range of 10-60°C. The results indicated that the absolute thermal conductivities increased with the increasing temperature with the enhanced ratios being almost constant. Thus it was concluded that the thermal conductivities of the nanofluids track the thermal conductivities of the base liquid.

Thermal transport properties of ethylene glycol based nanofluids containing low volume concentration diamond nanoparticles (DNP/EG nanofluid) have been reported by Yu et al. [15]. The thermal conductivity enhancement ratio of DNP–EG nanofluid at a volume concentration of 1% was up to 17.23%. Moreover, it was found that the thermal conductivity enhancement decreased with elapsed time at pH= 7. While for the stable DNP/EG nanofluids at pH= 8.5, there was no obvious thermal conductivity decrease for 6 months. Hence they illustrated that the thermal conductivity enhancement was correlated with the pH values of the suspensions and the pH value has a direct effect on the stability of nanofluids. It was also mentioned that the thermal conductivity enhancement values DNP/EG nanofluids are much larger than those containing metallic oxide.

Thermal conductivity enhancements of nanodiamond particles suspended in pure deionized water with different volume concentrations in the range from 0.8% to 3% have been measured by Yeganeh et al. [16]. The highest observed enhancement in the thermal conductivity was 7.2% for a volume fraction of 3% at a temperature of 30°C. The thermal conductivity increased by about 9.8% when the temperature was raised to 50°C.

Jahanshahi et al. [17] measured the thermal conductivity of $SiO_2$/water nanofluid having volume concentrations ranging from 1 to 4% with a transient hot wire method. The thermal conductivity of nanofluid increased linearly with increasing particle volume concentration. The measured thermal conductivity of $SiO_2$ nanofluid was 3.23% and 23% at a volume fraction of 1 and 4%, respectively.

The degree of thermal conductivity enhancement of the nanofluid (with respect to the base fluid) as a function of volume concentration and size of gold nanoparticle has been determined using the transient hot-wire technique by Paul et al. [18]. They observed that the degree of enhancement increases with increase in concentration and decrease in size of nanoparticles. The maximum enhancement recorded was 48% at 0.00026% volume concentration and 21 nm average particle size.

Experimental investigations have been carried out by Vajjha and Das [19] for determining the thermal conductivity of three nanofluids containing $Al_2O_3$, CuO and ZnO nanoparticles dispersed in a base fluid of 60:40 (by mass) ethylene glycol and water mixture. Particle volumetric concentration tested was up to 10% and the temperature range of the experiments was from 25°C to 80°C. They observed an increase in the thermal conductivity of nanofluids compared to the base fluids with an increasing volume concentration of nanoparticles. The thermal conductivity also increased substantially with an increase in temperature. At room temperature, maximum thermal conductivity enhancement observed with nanofluids containing $Al_2O_3$, CuO and ZnO nanoparticles were 35%, 32% and 17% respectively.

Substantial thermal conductivity enhancements were seen for ethylene glycol (EG) based copper nanofluids by Yu et al. [20]. The enhancement ratios of the Cu/EG nanofluids with the volume concentration of 0.3% were 5%, 11% and 33% at 10 °C, 25 °C and 50 °C respectively. The corresponding enhancement ratios with the volume concentrations of 0.5% were 8%, 16% and 46% at 10 °C, 25 °C and 50 °C respectively. Since the thermal conductivities strongly depend on the temperature of fluid, and the enhancement ratios increased with the increasing temperatures. Hence they suggested Brownian motions of Cu nanoparticles would play the key role on determining thermal conductivity enhancement of nanofluids.

Habibzadeh et al. [21] studied the stability and thermal conductivity of nanofluids of $SnO_2$/water nanofluids at weight concentrations of 0.012, 0.018 and 0.024%. The thermal conductivities of nanofluids are enhanced with increase of the $SnO_2$ nanoparticles weight concentration. The maximum thermal conductivity enhancement of 7% is observed at the 0.024% weight concentration of the suspension. It was shown that the thermal conductivity enhancement depends upon the pH value, nanoparticles concentration and temperature. The results also show that there are more surface charges at a pH value of 8, at which the thermal conductivity is higher.

The thermal conductivity of ethylene glycol (EG) and propylene glycol (PG) based AlN nanofluids were measured by Yu et al. [22]. AlN nanoparticles, with an average diameter of about 50 nm were used. However, the measured average particle size in the formulated nanofluids is found to be much larger than that of the primary particles. The average particle sizes for AlN-EG and AlN-PG nanofluids were 165 and 169 nm respectively which revealed that the nanoparticles formed clusters. Substantial increases in thermal conductivity were noted for the nanofluids investigated, with thermal conductivity enhancement ratios of 38.71% and 40.2% for EG and PG based nanofluids respectively at 10% volume concentration of nanoparticles. The existence of approximately linear relationship between the thermal conductivity enhancement ratio and the volume fraction of AlN nanoparticles was reported. They also demonstrated that the increase in thermal conductivity enhancement ratios of AlN/EG and AlN/PG nanofluids was not appreciable in the temperature range of 10–60°C and they track the thermal conductivity trends of the base fluids.

Thermal conductivity of transformer oil based nanofluid containing $Al_2O_3$ powders (spherical and rod shaped) and AlN powders were measured by Choi et al. [23]. The thermal conductivity of suspended spherical-shape $Al_2O_3$ in transformer oil showed an enhancement of more than 20% at 4% volume concentration of nanoparticles. But the thermal conductivity enhancement is greatly higher for AlN-oil nanofluid compared with $Al_2O_3$ fluids. For AlN nanoparticles at a volume concentration of 0.5%, thermal conductivity is enhanced by 8%. It is concluded that the rod-shaped particle is favorable for the heat transfer compared with the spherical shaped particle.

Sharma et al. [24] experimentally investigated the enhancements of the thermal conductivity of silver/Ethylene glygol nanofluid as a function of the concentration of silver.

There was approximately 10% improvement in thermal conductivity of nanofluids when the nanofluid concentration is about 1000 ppm. On the other hand, when the concentration of silver nanoparticles is increased to 5000 ppm and 10,000 ppm, the thermal conductivity increased to 16% and 18% respectively. As the increase in thermal conductivity with increasing volume concentration of silver is not significant, it is concluded that the thermal conductivity with 10,000 ppm concentration did not show significant improvement as compared with 1000 ppm and 5000 ppm concentration. Hence they concluded that the high concentration of particles seems to hinder the enhancement effect due to the aggregation of silver particles. They also observed the enhancements in thermal conductivities of Ag/ethylene glygol nanofluids to decrease as time lapsed. After 30 days of preparation, the thermal conductivity of 1000 and 5000 ppm silver nanofluids decreased slightly from 10% and 16% to 9% and 14%, respectively. On the other hand, the thermal conductivity of 10,000 ppm nanofluid decreased from18% to 14% after 30 days. It is reported that the silver particles were aggregated in early stage of preparation (up to 15 days), which leads to the increase in the size of silver particles.

Thermal conductivity of Al–Cu/EG nanofluid was presented by Chopkar et al. [25]. The results indicated that the nanofluids containing a small volume fraction of nanoparticles have significantly higher thermal conductivity (up to 2 times) that of the base liquids without nanoparticles. This was attributed to the higher thermal conductivity and composition of the solid dispersoids. They observed a sharp increase between 0.75 and 1.5% volume concentration of particles. The increase was over 200% with merely 1.5% volume concentration of particles. Though the conductivity ratio increased with volume concentration, the stability of the nanofluid was adversely affected due to sedimentation and inhomogeneity beyond a volume concentration of 1.5 %.

Nanofluids containing $Al_2O_3$, ZnO, $TiO_2$ and MgO nanoparticles were prepared with a mixture of 55% distilled water and 45% ethylene glycol as base fluid and the thermal conductivity were measured by Xie et al. [26]. At a volume concentration of 1%, the increase in thermal conductivities of nanofluids containing $Al_2O_3$, ZnO, $TiO_2$, and MgO nanoparticles were 4%, 3.5%, 2% and 4.5% respectively. At the same volume fraction and flow condition, the enhancements in heat transfer coefficients of $Al_2O_3$ and ZnO nanofluids were 40% and 18% respectively while that of $TiO_2$ and MgO were 10% and 252% receptively. But from the well established heat transfer theories, for a fully developed inner-tube laminar flow with constant wall temperature, the enhancements in heat transfer coefficients would be the same to the thermal conductivity enhancement ratio. Hence, enhancements in heat transfer coefficients would not exceed 5%. As the experimental results showed intriguingly higher increase in heat transfer coefficients as several tens even hundreds percent, they concluded that the single phase correlation with nanofluid properties was not able to predict nanofluids heat transfer coefficients and increasing thermal conductivity was not the only mechanism responsible for the heat transfer enhancement and other factors such as dispersion, chaotic movement of nanoparticles, Brownian motion and particle migration might be the reasons for dramatic increase in heat transfer coefficients.

Chopkar et al. [27] synthesized $Al_2Cu$ and $Ag_2Al$ nanoparticles by mechanical alloying and then prepared nanofluids by dispersing about 0.2–1% volume concentration of these nanoparticles in water and ethylene glycol. They also characterized the size/microstructure of nanoparticles by X-ray diffraction and transmission electron microscopy and measured the thermal conductivity of nanofluids. The results indicated that the nanofluids records 50–150%

improvement in thermal conductivity. It was also noted that water based nanofluids recorded greater ability to transfer or conduct heat than that by ethylene glycol based nanofluids with comparable volume fraction of $Al_2Cu$ nanoparticles. However, ethylene glycol based nanofluid seems to possess greater stability of dispersion (resistance to sedimentation and clogging) than that of the water based one due to higher viscosity of the former than the latter. Both experimental results and analytical study indicate that the degree of enhancement strongly depends on identity/composition, size, volume fraction and shape (aspect ratio) of the dispersed nanoparticles.

Wei et al. [28] have recently developed a single-step chemical method known as chemical solution method (CSM) to vary and manipulate nanofluid microstructures effectively through adjusting synthesis parameters like the reactant concentration and pH value. They applied this method to synthesize two kinds of novel cuprous-oxide ($Cu_2O$) nanofluids namely (i) suspensions of spherical $Cu_2O$ nanoparticles in water and (ii) suspensions of octahedral $Cu_2O$ nanoparticle in water. For both types of nanofluids, the measured thermal conductivity showed a high nonlinearity to both the molar concentration and the temperature. The measured data revealed that the conductivity enhancement with spherical $Cu_2O$ nanoparticles is higher than that with octahedral $Cu_2O$ nanoparticles Variation in the $CuSO_4$ molar concentration has led to a change in nanofluid thermal conductivity through the change in nanofluid microstructure. An extraordinary conductivity enhancement of 24% was noticed for both types of nanofluids.

Again with CSM, Wei et al. [29] synthesized $CuS/Cu_2S$ nanofluids of either core-shell or hollow structure and presented experimental evidence that nanofluid thermal conductivity can be either increased or decreased by the presence of nanoparticles. This was attributed to a phenomenon predicted by the recent thermal-wave theory of nanofluids. However, the conductivity ratio was larger than one for most nanofluids with hollow structure. Depending on factors like material properties of nanoparticles and base fluids, nanoparticle's geometrical structure and their distribution in the base fluids, interfacial properties and dynamic processes on particle–fluid interfaces, the heat diffusion (at nano scale level) and thermal-wave (at macro scale level) may either enhance or counteract each other. Consequently, the heat conduction may be enhanced or weakened by the presence of nanoparticles.

Li et al. [30] focused on the effect of pH and sodium dodecyl benzene sulfonate (SDBS) surfactant on the thermal conductivity of $Cu/H_2O$ nanofluids. Their results showed that the thermal conductivity enhancements of $Cu/H_2O$ nanofluids are highly dependent on the weight fraction of nanoparticle, pH values and SDBS surfactant concentration of nano-suspensions. The particle size distributions of $Cu/H_2O$ nano-suspensions in the absence and presence of SDBS surfactant showed that there are obvious variations in the particle size characteristics between two samples. The average particle sizes obtained are 6770 nm and 207 nm in the absence and presence of SDBS surfactant respectively. Therefore, it was concluded that the stabilization of $Cu/H_2O$ suspension with SDBS surfactant is better. Therefore, they recommended the combined treatment with both the pH and chemical surfactant to improve the thermal conductivity for practical applications of nanofluid. For Cu nanoparticles, at a weight fraction of 0.1%, thermal conductivity was enhanced by up to 10.7% with an optimal pH value and SDBS concentration for the highest thermal conductivity.

Hwang et al. [31] conducted thermal conductivity measurements on various nanofluids containing nanoparticles such as multi-walled carbon nanotube (MWCNT), copper oxide and silicon dioxide dispersed in various base fluids such as water, ethylene glycol and mineral oil.

They observed that MWCNT nanofluid has the highest thermal conductivity enhancement among water based nanofluids whereas $SiO_2$ nanofluid has the lowest one. From this result, it was shown that the thermal conductivity enhancement of nanofluid depends on the suspended particles. Also, it was shown that MWCNT/oil nanofluid has higher thermal conductivity enhancement than MWCNT/water. Similarly it was shown that CuO/ethylene glycol nanofluid has higher thermal conductivity enhancement than CuO/water. These results imply that higher thermal conductivity enhancement can be obtained for base fluid of lower thermal conductivity. From the results, it was concluded that thermal conductivity enhancement of nanofluid depends on the thermal conductivity of base fluids and suspended particles.

Paul et al. [8] presented the results of thermal conductivity ratio for nanofluids containing $ZrO_2$ and $TiO_2$ nanoparticles dispersed in base fluid water and ethylene glycol. The results showed significant increase in the thermal conductivity ratio with increase in volume percent for both $ZrO_2$ and $TiO_2$ based nanofluids. The increase for ethylene glycol based nanofluids in both cases is more compared to water based nanofluids as the former is a more viscous fluid. It was also observed that the ethylene glycol based nanofluids are more stable than the water based nanofluids. The increase in thermal conductivity ratio is almost linear with increase in volume percent and reaches about 200% for ethylene glycol based $ZrO_2$ nanofluids.

Godsen et al. [32] presented an experimental investigation on the thermal conductivity and viscosity of silver-deionized water nanofluid. Nanofluid is studied for 0.3, 0.6, and 0.9% of volume concentrations and for temperatures between 50°C and 90°C. A minimum and maximum enhancement of 27% at 0.3 % volume concentration and 80% at 0.9 % volume concentration were observed at an average temperature of 70 °C. The effect of Brownian motion and thermophoresis on the thermo-physical properties was also discussed. They demonstrated that for every increase in concentration of 0.3%, the Brownian velocity increase by 40–60% and the thermopheretic velocity increase by 40–80%. Hence they concluded that out of these two mechanisms, thermophoretic velocity may play a significant role than the Brownian velocity in thermal conductivity enhancement of nanofluids.

Thermal conductivity of nanofluids containing carbon nanotube (CNT) has also been the focus of numerous investigators. Investigation on measurements confirmed conductivities of about 3000W/mK for multiwalled carbon nanotubes (MWCNTs) [33] and above 2000W/mK for single-walled carbon nanotubes (SWCNTs) [34]. Thermal conductivity of carbon nanotubes comprising magnetically sensitive metal oxides in heat transfer nanofluids were investigated by Hong et al. [35]. They observed that upon extended exposure to the magnetic field, the mixture slowly vibrated making the nanoparticles to straighten and align with respect to the magnetic field. The aligned nanoparticle chains appeared to be continuous and unbroken, forming a combination of aligned particles and clusters. Time dependent thermal conductivity experiments indicate that the alignment process dominates the thermal conductivity enhancement as opposed to micro convection. Thermal conductivity of stable silicone oil based nanofluids containing CNT has been reported by Chen and Xie [36]. Thermal conductivity results of the CNT nanofluids reveal that the collective effects involving straightness ratio, aspect ratio, and aggregation play a key role in the thermal conductivity of CNT nanofluids.

**Table 2. Summary of the experimental works on thermal conductivity of various nanofluids**

| Researcher | Nanofluid | Particle Size (nm) | Volume concentration (%) | Maximum enhancement (%) |
|---|---|---|---|---|
| Lee et al. [9] | $Al_2O_3$/water | 30±5 | 0.01 to 0.3 | 1.44 |
| Chandrasekar et al. [10] | $Al_2O_3$/water | 43 | 0.33- 3 | 9.7 |
| Paul et al. [11] | $Al_{95}Zn_{05}$/EG | 50-80 | 0.01–0.10 | 16 |
| Duangthongsuk and Wongwises [12] | $TiO_2$/water | 21 | 0.2-2 | 7 |
| Li et al. [13] | Fe/water | 26 | 1-5 | 14.9 [@]<br>25 [#] |
| Yu et al. [14] | $Fe_3O_4$ /Kerosene | 15 | 0.1-1 | 34 |
| Yu et al. [15] | Diamond/EG | 5-10 | 0.1-1 | 17.23 |
| Yeganeh et al. [16] | Diamond/Water | 10 | 0.8-3 | 7.2 at 30°C<br>9.8 at 50° C |
| Jahanshahi et al. [17] | $SiO_2$/water | 12 | 1-4 | 23 |
| Paul et al. [18] | Au/water | 21 | 0.6 $x10^{-4}$-$2.6x10^{-4}$ | 48 |
| Vajjha and Das [19] | $Al_2O_3$/ EG and water | 53 | 1-10 | 35 |
| | CuO / EG and water | 29 | 1-6 | 32 |
| | ZnO/ EG and water | 29 | 2&4 | 17 |
| Yu et al. [20] | Cu/EG | 5-10 | 0.3 & 0.5 | 46 at 50° C |
| Habibzadeh et al. [21] | $SnO_2$/water | 4.3-5.3 | 0.012-0.024 (wt %) | 7 |
| Yu et al.[22] | AlN/EG<br>AlN/PG | 50 | 1-10 | 38.71<br>40.2 |
| Choi et al. [23] | $Al_2O_3$/ Transformer oil | 13/2x100-1000 | 0.5-4 | 20 |
| | AlN/ Transformer oil | 50 | 0.5 | 8 |
| Sharma et al. [24] | Ag/EG | 100-500* | 1000-10,000 ppm | 18 |
| Chopkar et al. [25] | Al–Cu/EG | 20–40 | 0.2-2 | 120 |
| Xie et al. [26] | $Al_2O_3$ /DW-EG | 30±5 | 1 | 4 |
| | ZnO/DW-EG | | | 3.5 |
| | $TiO_2$ /DW-EG | | | 2 |
| | MgO/DW-EG | | | 4.5 |
| Chopkar et al. [27] | $Al_2Cu/H_2O$ | 20-80 | 2.5 | 150 |
| | $Al_2Cu$/EG | | | 120 |
| | $Ag_2Al$/ $H_2O$ | | | 175 |
| | $Ag_2Al$/ EG | | | 140 |
| Wei et al. [28] | $Cu_2O/H_2O$ (spherical) | 200.5 | 0.5 | 24 |
| | $Cu_2O/H_2O$(Octahedral) | -- | 0.02 (molar volume concentration) | |
| Wei et al. [29] | $CuS/Cu_2S$/Ammonium acetate | 100 | 0.03 (molar volume concentration | 20 |
| Li et al. [30] | $Cu/H_2O$ | 207 | 0.1 (weight fraction) | 10.7 |
| Hwang et al. [31] | $MWCNT/H_2O$ | (10-30 m) x (10-50)μm | 1 | 7 |
| | MWCNT/oil | | 0.5 | 8.5 |
| | $CuO/H_2O$ | 33 | 1 | 5 |
| | CuO/EG | 33 | 1 | 9 |
| | $SiO_2/H_2O$ | 12 | 1 | 3 |
| Godson et al. [32] | $Ag/H_2O$ | < 100 | 0.3, 0.6, 0.9 | 30 at 50° C |

[@] In the absence of magnetic field; [#] In the presence of magnetic field; Time dependent size.

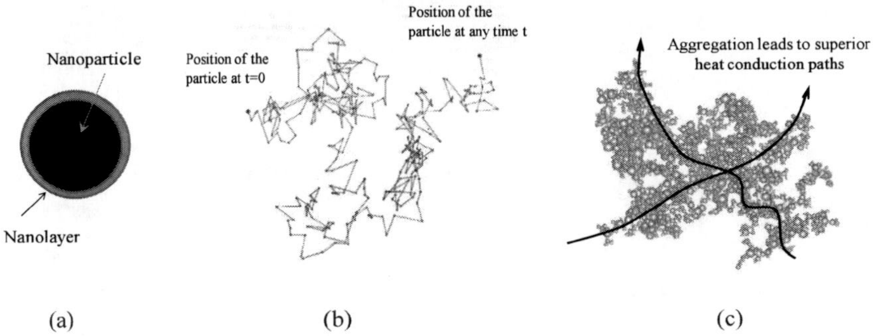


Figure 3. Conceptual schematic of (A) Nanolayer (B) Brownian motion (C) Clustering of nanoparticles.

Wen and Ding [37] investigated the effect of temperature on the thermal conductivity of MCNTs (20–60 nm in diameter and a few tens of micrometers in length)/water nanofluids. For temperatures lower than 30°C, an approximately linear dependence of thermal conductivity enhancement on temperature was obtained. Ding et al. [38] also showed that the effective thermal conductivity increases with increasing temperature in CNT/water suspensions.

Liu et al. [39] measured the thermal conductivities of nanofluids containing CNTs dispersed in ethylene glycol and synthetic engine oil. The increase of thermal conductivity is up to 12.4% for CNT/ethylene glycol suspensions at 1% volume concentration and 30% for CNT–synthetic engine oil suspensions at 2% volume concentration. It was concluded that the thermal conductivity is highly dependent on important factors such as the structure of the CNTs, clustering, temperature, etc.

Murshed et al. [40] measured the effective thermal conductivity of nanofluids containing titania nanorods of 10 nm diameter and 40 nm length and observed an enhancement of about 30 % in the effective thermal conductivity. Chen et al. [41] also carried out experiments to investigate the effective thermal conductivity of aqueous titanate nanofluids. Their results show a small thermal conductivity enhancement of about 3% at 25°C and about 5% at 40°C for 2.5% weight concentration of nanofluid. They demonstrated that the aspect ratio of the titanate nanotubes of approximately 10 is not sufficiently large to give a high thermal conductivity enhancement and the shape factor imposes more effect on the convective heat transfer.

The Summary of the experimental works on thermal conductivity of various nanofluids is given in Table 2.

The review on the thermal conductivity of nanofluids reveals that a long list of physical phenomenon has been proposed to explain the experimentally observed increase in thermal conductivity of nanofluids. The list includes volume concentration, size of the nanoparticles, higher specific surface area, effect of temperature that causes Brownian motion, magnetic field causing controlled nanoparticle particle alignment, formation of clusters or aggregates and the pH value favoring stability of the suspension. Earlier, Keblinski et al. [42] proposed four possible mechanisms for enhanced thermal conductivity of nanofluids namely (i) Brownian motion of the nanoparticles (ii) Nanolayering of the liquid at the liquid/particle interface (iii) the nature of heat transport in the nanoparticles, and (iv) Clustering of nanoparticles. As a result, several theoretical models have been proposed by taking into account the effect of Brownian motion [44-46], nanolayer [47-50] and clustering [51-53]. The conceptual schematic of Nanolayer, Brownian motion and clustering of nanoparticles is shown in Figure 3.

### 3.2. Viscosity of Nanofluids

While the thermal conductivity of nanofluids is important for heat transfer applications, viscosity is also important in designing nanofluids for flow and heat transfer applications because the pressure drop and the resulting pumping power depend on the viscosity. Compared to the works on thermal conductivity of nanofluids, only lesser investigations have been reported on the rheological behavior of nanofluids

The effective viscosities of the $Al_2O_3$/water nanofluids were measured using a viscometer of oscillation type by Lee et al. [9]. The effective viscosities of $Al_2O_3$/water nanofluids with low concentrations from 0.01 to 0.3% were measured in the temperature range from 21 to 39°C. They showed the effective viscosities of $Al_2O_3$/water nanofluids significantly decreases with increasing temperature and increase with increasing volume concentration. They also observed that the alumina nanofluids have a nonlinear relation between their viscosity and volume concentration. Nonlinear behavior was attributed to the particle–particle interactions and the increased surface of well dispersed nanoparticles in the nanofluids.

A maximum increase in viscosity of $Al_2O_3$/water nanofluids was 2.36 times that of water at 5% volume concentration as observed by Chandrasekar et al. [10]. The results show that the relative viscosity increase was almost linear up to 2% volume concentration. However, at volume concentrations more than 2%, the increase in relative viscosity shows a nonlinear relationship with volume concentration. This was attributed to the hydrodynamic interactions between particles which become important as the disturbance of the fluid around one particle interacts with that around other particles at higher volume concentrations.

Duangthongsuk and Wongwises [12] experimentally determined the viscosity of $TiO_2$-water nanofluids as a function of particle volume concentration and temperature. The results indicate that the viscosity of nanofluids significantly increases with decreasing nanofluid temperature and also increases with increasing particle volume concentration. It is obvious that the viscosity of nanofluids is higher than the base fluid by about 4–15%.

Li et al. [13] measured the viscosity of the magnetic fluid under the applied magnetic field. The measured results have indicated that the viscosity of the magnetic fluids increases with the concentrations of the suspended magnetic particles and the surfactant. The strength and orientation of the externally applied magnetic field play remarkable roles in affecting the viscosity of the magnetic fluid. This parameter increased with the strength of the external magnetic field. For the same magnetic fluid, its viscosity in the presence of a magnetic field perpendicular to the main flow is larger than that in the presence of a parallel field for the same field strength

Yu et al. [14] conducted rheological study on nanofluids containing AlN nanoparticles suspended in ethylene glycol and propylene glycol. They demonstrated that for AlN fraction below 5%, the two nanofluids behaved as Newtonian liquids, though beyond 5% volume concentration, the nanofluids demonstrated the shear-shinning behavior, which was intensified for lower temperatures and lower shear rates. The AlN volume fraction $\phi = 0.05$ is considered a critical concentration of rheological behavior. Differentiation between Newtonian or non-Newtonian behaviors of the nanofluids depends strongly on the kinds and the shapes of nanoparticles, their volume fractions and temperature.

Godsen et al. [32] presented an experimental investigation on viscosity of silver-deionized water nanofluid. Nanofluid consisted of silver nanoparticles of 0.3, 0.6, and 0.9% of volume concentrations and studied for temperatures between 50°C and 90°C. A minimum and maximum enhancement of 8% at 0.3% volume concentration and 28% at 0.9% volume concentration respectively were observed at 50°C. Viscosity are correlated as quadratic functions of volume concentration. The viscosity ratio for nanofluid with 0.3% volume concentration increased from 1.06 to 1.2, and for nanofluid with 0.9% volume concentration, the ratio increased from 1.23 to 1.43 for the temperature range between 50°C and 90°C

Kang et al. [54] measured the viscosities of UDD (ultra dispersed diamond)/ethylene glycol, silver/water, and silica/water nanofluids. They found that the viscosity increase was 50% for UDD/EG nanofluid, 30% increase for silver/water and 20% increase for silica/water nanofluids at volume concentrations of 1%, 2% and 3% respectively.

Prasher et al. [55] demonstrated that the viscosity of alumina/propylene glycol (PG) nanofluids was independent of shear rate, proving that the nanofluids are Newtonian in nature and increases as nanoparticle volume concentration increases. They found a 30% increase in viscosity at 3% volume concentration and attributed this increase to aggregation of the nanoparticles in the nanofluid with the size of the aggregates around three times the size of the individual nanoparticles.

Namburu et al. [56] presented an experimental investigation of rheological properties of copper oxide nanoparticles suspended in 60:40 (by weight) ethylene glycol and water mixture. Nanofluids of volume concentration ranging from 0% to 6.12% were tested. The experiments were also carried over temperatures ranging from -35°C to 50°C to demonstrate their applicability in cold regions. For the particle volume concentrations tested, nanofluids exhibited Newtonian behavior. The viscosity of nanofluids increased with increase in volume concentration of nanoparticles. They reported that the viscosity of 6.12% volume concentration nanofluid is about four times the value of the base fluid at -35°C. The viscosity of copper oxide nanofluids decreased exponentially with increase in the temperature.

Schmidt et al. [57] presented experimental shear and longitudinal viscosity data on two nanofluid systems namely $Al_2O_3$/decane and $Al_2O_3$/isoparaffinic polyalphaolefin (PAO). Their shear viscosity data exhibited an enhancement over the Einstein model and the longitudinal viscosity data indicate that the nanoparticles do not aggregate in these nanofluid systems. When the suspensions are extremely dilute, they show no signs of sedimentation, the results indicated that aggregation may not be responsible for the disagreement between effective medium models and the data in all nanofluid systems, and that the models based on Brownian dynamics or other nanoscale phenomena should be considered.

As organic fluids like EG are more favorable for the dispersion of nanosized particles compared to water, experimental work of Xie et al. [58] showed enhancement ratios of the viscosity of ethylene glycol (EG) based $Al_2O_3$ suspensions smaller than those of water based suspensions, indicating the significant influence of the base fluid on the viscosity of the nanofluids. They also studied the dependence of the viscosity on pH values. The isoelectric point is determined to be 9.2 for alumina nanoparticles. When pH is far from this, the nanoparticles are well dispersed because of the very large repulsive forces among the nanopaticles. The repulsion among nanoparticles decreases when pH value is close to isoelectric point. This causes coagulation or aggregation of nanoparticles and the viscosity increases for a nanofluid with pH value close to isoelectric point

The effect due to temperature and particle volume concentration on the dynamic viscosity for the $Al_2O_3$/water nanofluid has been experimentally investigated by Nguyen et al. [59]. They found that, in general, nanofluid dynamic viscosity increases considerably with particle volume concentration but decreases with a temperature increase. Their results have revealed the existence of a critical temperature beyond which the particle suspension properties seem to be drastically altered, which, in turn, has triggered a hysteresis phenomenon. The hysteresis phenomenon has raised serious doubts regarding the reliability of using nanofluids for heat transfer enhancement purposes.

Chen et al. [60] carried out experiments on the rheological behaviour of EG based titanate nanotubes (TNT) nanofluids containing weight concentrations of 0.5, 1, 2, 4 and 8 % TNT at 20–60°C. Their results showed a very strong shear thinning behaviour of the TNT nanofluids and big influences of particle concentration and temperature. The slopes of the shear viscosity–shear rate curves at the low shear rate region were estimated to be 0.61 and 0.86 for 4% and 8% weight concentrations respectively. It was also shown that the shear viscosity of nanofluids increases with increasing TNT concentration for a given shear rate. It is observed that the relative viscosity at low shear rate region increases sharply with increasing temperature, indicating stronger shear thinning behaviour of the EG/TNT nanofluids at higher temperatures. However, the relative viscosities at high shear rates converge to the same constant for all the temperatures investigated. These results suggest different mechanisms for the rheological behaviour at low and high shear regions. Hence they categorized the rheological behaviour of the EG/TNT nanofluids into four regimes like (i) dilute nanofluids (ii) semi-dilute nanofluids (iii) semi-concentrated nanofluids and (iv) concentrated nanofluids. The demarcation of these regimes depends on the effective particle concentration i.e., the aggregate size.

Anoop et al. [61] considered three types of nanofluids, viz., alumina/water, alumina/ethylene glycol and copper oxide/ethylene glycol to study their rheological properties. Both water-based as well as ethylene glycol-based nanofluids exhibited Newtonian behavior. The experiments revealed that with an increase in the particle concentration, the viscosity ratio increases and this increase was more predominant for water-based nanofluids than for ethylene glycol-based nanofluids. The increase in viscosity was attributed to the combined effects of electroviscous forces (due to the presence of an electrical double layer with electrostatic stabilization) and particle aggregation.

The dispersion and stability of nanofluids obtained by dispersing $Al_2O_3$ nanoparticles (obtained from different sources) in water were analyzed by Pastoriza-Gallego et al. [62]. They evaluated the differences arising from different dispersion techniques, the resulting particle size distribution and time stability among the different samples. It has been found that the influence of particle size in viscosity is very large and must be taken into account for any practical application. These viscosity differences were rationalized by considering a theory describing the aggregation state of the nanofluid. The effect of considering variable and constant aggregate size to predict nanofluid viscosity was also investigated by them. They showed that the assumption of constant aggregate size yielded poorer estimations at low concentration while the assumption of variable aggregate size would be desirable at higher concentrations.

Yu et al. [63] demonstrated that when volume concentration of ZnO/EG nanofluid is less than 2%, its viscosity is independent of shear rate from 20 to 100 $s^{-1}$ in the range of 20–60 °C. While for the ZnO/EG nanofluid with volume concentrations more than 3%, the shear-shinning behaviour was observed. These facts show that ZnO-EG nanofluids with volume concentrations less than 3% had non-Newtonian behaviors. It is concluded that the ZnO/EG nanofluids with the higher volume concentration exhibited different rheological behaviors from those with lower volume concentration.

Rheological behavior of agglomerated silver nanoparticles suspended in diethylene glycol over a wide range of volume concentrations was studied by Tamjid et al. [64]. The nanoparticle suspensions generally exhibited a yield pseudoplastic behavior. The strong dependency of relative viscosity on the volume concentration suggested that the particle interactions (or more specifically, the attractive interparticle potential) become more pronounced with increasing volume concentration. The critical solid loading was estimated to be about 11%.

A linear relation between shear stress and shear rate of Diamond/EG nanofluids was noticed by Yu et al. [15] and demonstrated the Newtonian behavior of such nanofluids. Their experiments revealed that with the increase of temperature, the viscosity of nanofluids decreases rapidly. The reason for decrease in viscosity with the increase in temperature was attributed to the weakening effect on the inner-particle/intermolecular forces.

Pastoriza-Gallego et al. [65] experimentally studied nanofluids composed by cupric oxide (CuO) nanoparticles dispersed in water, in a concentration range from 0.05% to 10% in weight fraction. Two different sets of samples were considered for viscosity measurements, one of them obtained from dispersion of commercial dry nanopowder, and the other from dispersion of synthesized (in situ) dry nanopowder.

Kole and Dey [66] presented the experimental results on the viscosity of nanofluid prepared by dispersing alumina in commercial car coolant. They showed that while the pure base fluid display Newtonian behavior over the measured temperature, it transforms to a non-Newtonian fluid with addition of a small amount of alumina nanoparticles and it behaves as a Bingham plastic with small yield stress. Their results also showed that viscosity of the nanofluid increases with increasing nanoparticle concentration and decreases with increase in temperature. They also demonstrated that the expression considering the influence of Brownian motion of nanoparticles in the base fluid could predict fairly the nanofluid viscosity. The same research group has also presented the experimental results on viscosity of CuO/gear oil nanofluid. The observations presented were same as that of their previous work expect that it was demonstrated that the expression considering aggregation could predict fairly the nanofluid viscosity.

Abareshi et al. [67] was the first to report the rheological properties of nanofluids of α-$Fe_2O_3$ nanoparticles and glycerol. The experimental results showed that the viscosity of α-$Fe_2O_3$/glycerol nanofluids increases with increasing the particle volume concentration, decreases with increasing temperature and the nanoflluid exihibited non-Newtonian behavior at lower temperatures. They also established that the nanofluid has some yield stress due to addition of nanoparticles to a base fluid.

Tsai et al. [68] investigated the effect of viscosity of the base fluid on the thermal conductivity of nanofluids in which $Fe_3O_4$ nanoparticles are suspended in the base fluid composed of diesel oil and polydimethylsiloxane. Viscosity of the base fluid is varied by changing the volumetric fractions between both fluids. The measured thermal conductivity of nanofluids gradually approaches the value predicted by the Maxwell equation by increasing the viscosity. It is demonstrated that the viscosity of nanofluids does affect the thermal conductivity of nanofluids, and the Brownian motion of suspended particles could be an important factor that enhances the thermal conductivity of nanofluids.

**Table 3. Summary of the experimental investigation on the viscosity of various nanofluids**

| Researcher | Nanofluid | Particle Size (nm) | Volume concentration (%) | Maximum Relative viscosity |
|---|---|---|---|---|
| Lee et al. [9] | $Al_2O_3$/ $H_2O$ | 30±5 | 0.01 to 0.3 | 1.029 |
| Chandrasekar et al. [10] | $Al_2O_3$/ $H_2O$ | 43 | 0- 5 | 2.36 |
| Duangthongsuk and Wongwises [12] | $TiO_2$/ $H_2O$ | 221 | 0.2-2 | 1.15 |
| Li et al. [13] | $Fe_3O_4$/$H_2O$ | 20 | 0-.2.83 | 2 |
| | Fe/$H_2O$ | 26 | 0-4 | 11.25* |
| Yu et al.[15] | AlN/EG | 50 | 1-10 | 38.71 |
| | AlN/PG | | | 40.2 |
| Godson et al. [32] | Ag/$H_2O$ | < 100 | 0.3, 0.6, 0.9 | 1.06-1.43 |
| Namburu et al. [56] | CuO/ EG and $H_2O$ | 29 | 1-6.12 | 4 |
| Schmidt et al. [57] | $Al_2O_3$/Decane | 40 | 0-1 | 1.85 |
| | $Al_2O_3$/PAO | | | 1.7 |
| Nguyen et al. [58] | $Al_2O_3$/ $H_2O$ | 36 | 0-13 | 4.5 |
| | | 47 | | 5.5 |
| Chen et al. [60] | TNT/EG | 10 X 100 | 0.1-1.8 | 1.7 |
| Anoop et al. [61] | $Al_2O_3$/ $H_2O$ | 95,100 | 0-6 | 1.6-1.8 |
| | $Al_2O_3$/EG | 100 | | 1.22 |
| | CuO/EG | 152 | | 1.24 |
| Pastoriza-Gallego et al. [62] | $Al_2O_3$/ $H_2O$ | 40-50 | 0.13 to 2.9 | - |
| | | <50 | | 1.3 |
| | | <20 | | 1.8 |
| Yu et al. [63] | ZnO/EG | 10-20 | 0.2-5 | 1.15* |
| Tamjid et al. [64] | Ag/DEG | 40 | 0.11–4.38 | 2.04* |
| Yu et al. [15] | Diamond/EG | 5-10 | 1 | 1.01 |
| Pastoriza-Gallego et al. [65] | CuO/ $H_2O$ | 23–37 | 0. 16 – 1.7 | 1.14* |
| | | 11 ± 3 | | 1.81 |
| Kole and Dey [66] | $Al_2O$ /car coolant | <50 | 0.1-1.4 | 2.4 |
| | CuO/gear oil | 40 | 0.5-2.5 | 3 |
| Abareshi et al. [67] | α-Fe2O3-glycerol | 5 | 0.1-0.75 | 1.37 |
| Timofeeva et al.[70] | SiC/ $H_2O$ | 16,29,66 &90 | 4 | 1.8+ |
| | SiC/ EG +$H_2O$ | | | 1.55+ |

* Calculated from the viscosity data; + for 6 nm.

Wang et al. [69] reported the effects of viscosity of base liquid on the thermal conductivity of nanofluids from the view of Brownian motion. They found that the thermal conductivity of nanofluids decreases with increasing viscosity and approaches a constant value. Based on the previously reported work [107], the authors [108] observed that the viscosity around 100 cP is a critical value of demarcation. Below 100 cP of viscosity, the Brownian motion is active, so the thermal conductivity of nanofluids increases. Over 100 cP of viscosity, the Brownian motion becomes inactive and thermal conductivity decreases.

Timofeeva et al. [70] presented the experimental data for the viscosity of nanofluids with SiC particles suspended in EG/$H_2O$ mixture with a 50/50 volume ratio. It was observed that at the same temperature, particle concentration and pH, the viscosity decreases with the increase in the average particle size. This lower viscosity is highly desirable for heat transfer applications to minimize the pumping power penalties. Comparing the viscosity increase in analogous water- and EG/$H_2O$-based suspensions reveals a lesser viscosity increase in the EG/$H_2O$ nanofluids. The difference in the viscosity increase is more pronounced at smaller particle sizes. Hence it is reported that the viscosity increase should be independent of the viscosity of the base fluid and only proportional to the particle volume concentration. The observed phenomena can be related to the difference in the structure and thickness of the diffuse fluid layers around the nanoparticles in various base fluids, which affects the effective volume concentration and ultimately the viscosity of the suspension. The summary of the experimental investigation on the viscosity of various nanofluids is given in Table 3.

## 3.3. Density of Nanofluid

Research work on the density, specific heat and surface tension of nanofluids is very limited compared to that on thermal conductivity and viscosity and hence only few of important works are presented in this article.

In the absence of experimental data, the density of the nanofluids has been reported to be consistent with the mixing theory [71] given by

$$\rho_{nf} = (1-\phi)\rho_{bf} + \phi\rho_s \tag{1}$$

where

$\rho_{nf}$ = density of nanofluid

$\rho_{bf}$ = density of base fluid

$\rho_s$ = density of solid particles

$\phi$ = volume concentration

Sommers and Yerkes [72] measured the density of the $Al_2O_3$/propanol nanofluid at room temperature using two methods and compared them. In the first method, a hydrometer was used to measure the specific gravity of a fluid sample. In the second method, a fluid sample of known volume was taken and then weighed on a high precision balance. Data collected using these two methods were then averaged and a nearly linear relationship between density and particle concentration was observed. Figure 4 shows the comparison of measured and calculated densities which indicate that the differences between them increase with increase in concentration of particles. However the difference is less than 5% when the weight concentration is upto 5%. In the absence of substantial works on density of nanofluids, it is obvious that further research works are clearly required to check the validity of equation based on mixing theory in predicting the density of nanofluids.

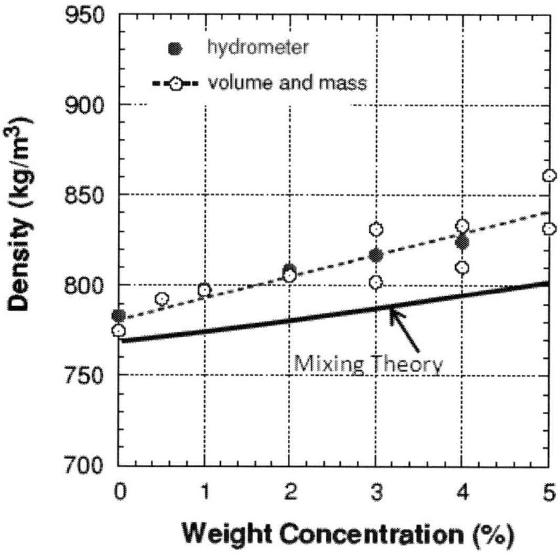


Figure 4. Comparison between experimental and theoretical values of density of nanofluid.

### 3.4. Specific Heat of Nanofluid

In the absence of available experimental data, the following two models have been extensively applied in the experimental and numerical nanofluid investigations to find the specific heat of nanofluid. The first model is one which is analogous to the mixing theory and the specific heat of a nanofluid is expressed as

$$c_{p,nf} = (1-\phi)c_{p,bf} + \phi c_{p,s} \tag{2}$$

where the subscripts nf, bf, and s refer to the nanofluid, base fluid and nanoparticle respectively.

Because of its simplicity, Eq. (2) has been used in the assessment of the heat transfer performance of nanofluids by researchers [73-75].

The second model is based on thermal equilibrium mechanism and the specific heat of a nanofluid is expressed as

$$\rho_{nf} c_{p,nf} = (1-\phi)\rho_{bf} c_{p,bf} + \phi \rho_s c_{p,s} \tag{3}$$

Eq. (3) has also been adopted in nanofluid investigations [76, 77]. In the case that no experimental data are available, it has been believed that both expressions can be considered equivalent and either one may be used to estimate nanofluid specific heat. However, the deviation between the two models is too large to the extent it cannot be ignored as shown in Figure 5.

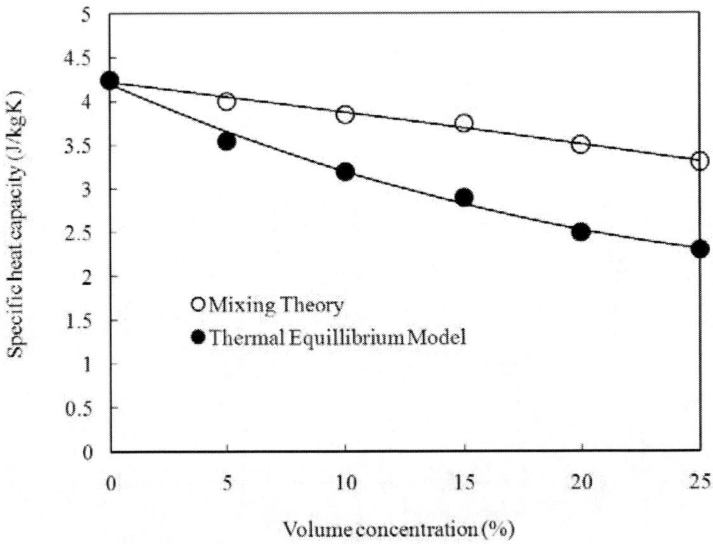


Figure 5. Comparison between values of specific heat capacity of nanofluid obtained by two different models.

Zhou and Ni [78] have presented an experimental investigation of the specific heat of water-based $Al_2O_3$ nanofluid with a differential scanning calorimeter. Their result indicates that the specific heat of nanofluid decreases gradually as the nanoparticle volume concentration increases. The relationship between them exhibits good agreement with the prediction from the thermal equilibrium model while the simple mixing model fails to predict the specific heat of nanofluid.

Shin and Baneerjee [79] observed the average enhancement on the specific heat capacity to be 14.5%. They pointed that the theoretical models based on thermal equilibrium given in Eq. 3 cannot explain the anomalous enhancement of the specific heat capacity at such low concentrations of the nanoparticles (0.6% by volume). This result implies that alternate transport mechanisms exist for nanofluids, thus distinguishing nanofluids from a simple mixture of two materials.

### 3.5. Surface Tension

As surface tension is known to be an important parameter in boiling, it was measured for the $Al_2O_3/H_2O$ fluid using the conventional ring method by Das et al. [80]. They observed that the variation of surface tension with particle volume concentration which is extremely nominal to have any surfactant effect on the boiling process.

Prakash Narayan et al. [81] have explained that the surface tension will definitely have a significant influence on boiling process due to its dependency of bubble departure and equilibrium conditions at the interfacial. At room temperature, surface tension does not show a significant change. Logically, the existence of active particles at the liquid-vapor interface is likely to have an influence on the value of the effective surface tension at near saturation temperatures when phase change is in inception. However, it is difficult to believe that this effect alone can bring about such a large amount of influence on the boiling process. Moreover, whatever is the effect of surface tension, it is not likely to be reversed for different cases of the same fluid and hence cannot explain the anomalous behavior of enhancement of boiling in one range and deterioration in another range. Hence it is mentioned that further comprehensive investigations are needed to understand the effect of surface tension on boiling heat transfer

Shi et al. [82] conducted experiments with iron (Fe) and alumina ($Al_2O_3$) nanoparticles boiled on a copper block. The authors concluded that Fe particles showed more enhancement than $Al_2O_3$ particles and that enhancement was mostly due to increases in thermal conductivity and lowered surface tension.

Kumar and Milanova [83] demonstrated that single-walled carbon nanotube (CNT) suspensions in a boiling environment can extend the saturated boiling regime and postpone catastrophic failure of the material if the surface tension of the nanofluid is carefully controlled. Hence they concluded that the heat transfer behavior is dictated not only by the deposition of nanotubes on the heated wire but also by the surface tension, which is a precursor to deposition and postponement of critical heat flux (CHF).

It is demonstrated by Murshed et al. [84] that the burnout heat flux of nanofluid strongly depends nonlinearly on its relaxation of surface tension with base fluid. Similar to the CHF, the burnout heat flux (BHF) of nanofluid also reaches its highest value at NaDBS and CNT concentration ratio of 1:5. The highest BHF value of nanofluid is about 265% enhancement over that for the base fluid at a relaxation surface tension of approximately 25 mN/m. The BHF of deionized water is found to be 1500 kW/m$^2$. The observed increment of relative BHF of nanofluid (i.e. $(BHF)_N/(BHF)_W$) was attributed to the Marangoni effect which results from the surface tension differences. Thus, the pool boiling heat transfer behavior of surfactant-added nanofluid is not only dictated by the deposition of nanotubes on heater surface, but also by the relaxation of surface tension of nanofluid, which is a precursor to the deposition and overall onsets of both the CHF and BHF.

Zhu et al. [85] measured the surface tension of $Al_2O_3/H_2O$ using the maximum bubble pressure method. With the maximum addition of 1 g/L $Al_2O_3$ nanoparticles, the surface tension of water increased to a maximum of about 5%. The surface tension of the nanofluids with 0.1 g/L is increased by about 3.4% compared to that of water. Considering the effect of the surface tension, they found that the CHF of nanofluids can be increased by about 1.4% compared to that of water. However, a maximum increase of 120% in CHF for nanofluids compared with that of water was observed in their study and hence they concluded that CHF cannot be related to surface tension of nanofluids,

Moosavi et al. [125] prepared ZnO nanofluids by dispersing ZnO nanoparticles in the EG and glycerol as the base fluids. The surface tension ratio of suspensions containing solid particles increased with increasing volume fraction of the solid particles. Since nanofluids are heterogeneous mixtures of nanoparticles dispersed in a base fluid, increasing in surface tension ratio in nanofluids was attributed to the presence of nanoparticles in the surface of base fluid.

## Conclusion

In this chapter the formulation of nanofluids and experimental investigations on thermophysical properties together with the mechanisms proposed for the alteration in their values or characteristics due to the addition of nanoparticles are discussed. The following section describes the summary of this chapter.

From the experimental works on thermal conductivity of nanofluid, it becomes apparent that smaller sized nanoparticles could bring Brownian motion at higher temperatures and the microstructure of the nanofluid provide details regarding the alignment of the nanoparticles or the cluster formation but there is no sufficient experimental technique to prove the presence of nanolayer and the value of its thickness. On the other hand, measurement of pH together with zeta potential will provide the information regarding stability of the prepared nanofluid. It is the fact that if the pH value is maintained away from the isoelectric point (pH value where the repulsive forces between the particles becomes zero), the particles will be well dispersed in the fluid avoiding sedimentation and improving the stability and thermal conductivity of nanofluid. It is also obvious that with the help of magnetic field it becomes easy to control the structure of nanoparticles suspended in the magnetic fluids. Thus, the future experimental works on the investigation of thermal conductivity of nanofluid must provide an experimental evidence to justify the reason attributed for enhancement in thermal conductivity.

It is learnt that the viscosity of the nanofluid may increase linearly or nonlinearly due to the addition of nanoparticles as the particle interactions become more prominent with increasing volume concentration. The viscosity of nanofluid decreases with the increase in temperature which was attributed to the weakening effect on the inner-particle/intermolecular forces. The fact, namely, the decrease in viscosity with increase in temperature will be favorable for its application in thermal management systems. However the phenomenon of hysteresis reported by Nguyen et al. [59] is required to be considered to ensure reliable and successful application of nanofluid in heat transfer systems. Furthermore, the review on the experimental nanofluid viscosity data indicates that nanofluid may exhibit Newtonian or non Newtonian behaviors which were coupled with the level of volume concentration. It is also revealed that nanofluid with smaller sized nanoparticles shows a significantly larger viscosity due to effects of the electric double layer repulsion. Recent findings of Wang et al. [69] about the decrease of thermal conductivity with increase in viscosity of nanofluid and the limiting viscosity value to decide whether the Brownian motion is active or inactive has created premises for future research works.

The literature survey also shows that the experimental works on density of the nanofluid is scant and it could be well predicted using the classical mixing theory. On the other hand, specific heat of nanofluid were calculated using mixing theory and thermal equilibrium model, and it was believed that either one of the theory can be used in predicting the specific heat of nanofluid. Later it was shown that the prediction from the thermal equilibrium model exhibits good agreement while the simple mixing model fails to predict the specific heat of nanofluid. Thus the thermal equilibrium model is used extensively by the researchers to calculate the specific heat of the nanofluid.

The property, namely, surface tension becomes pivotal in deciding the heat transfer performance of boiling processes as it offer a situation of variation in surface wettability. As the heat transfer coefficient in boiling heat transfer is inversely proportional to surface tension, a decrease in surface tension with nanofluid compared to base fluid is mandatory to improve boiling heat transfer. The literature search shows that the researchers have reported increase as well as decrease in surface tension of naofluid and also believe that the variation of surface tension is extremely nominal to bring any significant effect on the boiling process. Under these circumstances, it is learnt that the more experimental research works on surface tension of nanofluids are required to study the effect of variation of surface tension on boiling heat transfer and to create a reliable data base.

## REFERENCES


[1] Eastman, J.A., Choi, S.U.S., Li, S., Thompson, L.J., Lee, S., 1996, "Enhanced Thermal Conductivity through the Development of Nanofluids," Materials Research Society Symposium Proceedings, Boston.

[2] Lo, C.H., T.T. Tsung, L.C. Chen, C.H. Su and H.M. Lin, 2005, "Fabrication of Copper Oxide Nanofluid Using Submerged Arc Nanoparticle Synthesis System (SANSS)," *Journal of Nanoparticle Research*, 7, 313–320.

[3] Chopkar, M., Das, P.K., and Manna, I., 2006, "Synthesis and Characterization of Nanofluid for Advanced Heat Transfer Applications," *Scripta Materialia*, 55, 549–552.

[4] Liu, M.S., Lin, M.C.C., Tsai, C.Y., and Wang, C.C., 2006, "Enhancement of Thermal Conductivity with Cu for Nanofluids Using Chemical Reduction Method," *International Journal of Heat and Mass Transfer*, 49, 3028–3033.

[5] Zhu, H.T., Zhang, C.Y., Liu, S.Q., Tang, Y.M., and Yin, Y.S., 2006, "Effects of Nanoparticle Clustering and Alignment on Thermal Conductivities of $Fe_3O_4$ Aqueous Nanofluids," *Applied Physics Letters*, 89, 023123-1-3.

[6] Phuoc, T.X., Soong, Y., and Chyu, M.K., 2007, "Synthesis of Ag-Deionized Water Nanofluids using multi-beam laser ablation in liquids," *Optics and Lasers in Engineering*, 45, 1099-1106.

[7] Choi, S.U.S., 2009, "Nanofluids: From Vision to Reality through Research," *Journal of Heat Transfer,* 131, 033106-1-9.

[8] Paul, G., Chopkar. M., Manna, I., and Das, P.K., 2010, "Techniques for Measuring the Thermal Conductivity of Nanofluids: A Review," *Renewable and Sustainable Energy Reviews*, 14, 1913–1924.

[9] Lee, J.H., Hwang, K.S., Jang, S.P., Lee, B.H., Kim, J.H., Choi, S.U.S., and Choi, C.J., 2008, "Effective Viscosities And Thermal Conductivities Of Aqueous Nanofluids Containing Low Volume Concentrations of $Al_2O_3$ Nanoparticles," *International Journal of Heat and Mass Transfer*, 51, 2651–2656.

[10] Chandrasekar, M., Suresh, S., Chandra, B.A., 2010, "Experimental Investigations and Theoretical Determination of Thermal Conductivity and Viscosity of $Al_2O_3$/Water Nanofluid," *Experimental Thermal and Fluid Science,* 34, 210–216.

[11] Paul, G., Philip, J., Raj, B., Das, P.K., and Manna, I., 2011, "Synthesis, Characterization, and Thermal Property Measurement of Nano-$Al_{95}Zn_{05}$ Dispersed Nanofluid Prepared by a Two-Step Process," *International Journal of Heat and Mass Transfer*, 54, 3783-3788.
[12] Duangthongsuk, W., and Wongwises, S., 2009, "Measurement of temperature-dependent thermal conductivity and viscosity of $TiO_2$-water nanofluids," *Experimental Thermal and Fluid Science,* 33, 706–714.
[13] Li, Q., Xuan, Y., and Wang, J., 2006, "Experimental Investigations on Transport Properties of Magnetic Fluids," *Experimental Thermal and Fluid Science*, 30, 109–116.
[14] Yu, W., Xie, H., Chen, L., and Li, Y., 2010, "Enhancement of Thermal Conductivity of Kerosene-based Fe3O4 Nanofluids Prepared via Phase-transfer Method," *Colloids and Surfaces A: Physicochemical and Engineering Aspects*, 355, 109–113.
[15] Yu, W., Xie, H., Li, Y., Chen, L., and Wang, Q., 2001, "Experimental Investigation on The Thermal Transport Properties of Ethylene Glycol Based Nanofluids Containing Low Volume Concentration Diamond Nanoparticles," *Colloids and Surfaces A: Physicochemical and Engineering Aspects*, 380, 1–5.
[16] Yeganeh, M., Shahtahmasebi, N., Kompany, A., Goharshadi, E.K., Youssefi, A., and Siller, L., 2010, "Volume fraction and temperature variations of the effective thermal conductivity of nanodiamond fluids in deionized water," *International Journal of Heat and Mass Transfer*, 53, 3186–3192.
[17] Jahanshahi, M., Hosseinizadeh, S.F., Alipanah, M., Dehghani, A., and Vakilinejad, G.R., 2010, "Numerical Simulation of Free Convection Based on Experimental Measured Conductivity in a Square Cavity Using Water/$SiO_2$ Nanofluid," *International Communications in Heat and Mass Transfer*, 37, 687–694.
[18] Paul, G., Pal, T., and Manna, I., 2010, "Thermo-physical property measurement of nano-gold dispersed water based nanofluids prepared by chemical precipitation technique," *Journal of Colloid and Interface Science,* 349, 434–437.
[19] Vajjha, R.S., and Das, D.K., 2009, "Experimental determination of thermal conductivity of three nanofluids and development of new correlations," *International Journal of Heat and Mass Transfer*, 52, 4675–4682.
[20] Yu, W., Xie, H., Chen, L., Li, Y., 2010, "Investigation on the Thermal Transport Properties of Ethylene Glycol-based Nanofluids Containing Copper Nanoparticles," *Powder Technology*, 197, 218–221.
[21] Habibzadeh, S., Beydokhti, A.K., Khodadadi, A.A., Mortazavi, Y., Omanovic, S., Niassar, M.S., 2010, "Stability and Thermal Conductivity of Nanofluids of Tin Dioxide Synthesized via Microwave-Induced Combustion Route," *Chemical Engineering Journal*, 156, 471–478.
[22] Yu, W., Xie, H., Li, Y., and Chen, L., 2001, "Experimental Investigation on Thermal Conductivity and Viscosity of Aluminum Nitride Nanofluid," *Particuology,* 9, 187–191.
[23] Choi, C., Yoo, H.S., Oh, J.M., 2008, "Preparation and Heat Transfer Properties of Nanoparticle-in-transformer Oil Dispersions as Advanced Energy-Efficient Coolants Investigation on Thermal Conductivity and Viscosity of Aluminum Nitride Nanofluid," *Current Applied Physics*, 8, 710–712.

[24] Sharma, P., Baek, I.H., Cho, T., Park, S., and Lee, K.B., 2011, "Enhancement of Thermal Conductivity of Ethylene Glycol Based Silver Nanofluids," *Powder Technology*, 208, 7–19.

[25] Chopkar, M., Das, P.K., Manna, I., 2006, "Synthesis and Characterization of Nanofluid for Advanced Heat Transfer Applications," *Scripta Materialia*, 55, 549–552.

[26] Xie, H., Li, Y., and Yu, W., 2010, "Intriguingly High Convective Heat Transfer Enhancement of Nanofluid Coolants in Laminar Flows," *Physics Letters A*, 374, 2566–2568.

[27] Chopkar, M., Kumar, S., Bhandari, D.R., Das, P.K., and Manna, I., 2007, "Development and Characterization of $Al_2Cu$ and $Ag_2Al$ Nanoparticle Dispersed Water and Ethylene Glycol Based Nanofluids," *Materials Science and Engineering B*, 139, 141–148.

[28] Wei, X., Zhu, H., Kong, T., and Wang, L., 2009, "Synthesis and Thermal Conductivity of $Cu_2O$ Nanofluids," *International Journal of Heat and Mass Transfer*, 52, 4371–4374.

[29] Wei, X., Kong, T., Zhu, H., and Wang, L., 2010, "CuS/Cu2S Nanofluids: Synthesis and Thermal Conductivity," *International Journal of Heat and Mass Transfer*, 53, 1841–1843.

[30] Li, X.F., Zhu, D.S., Wang, X.J., Wang, N., Gao, J.W., and Li, H., 2008, "Thermal Conductivity Enhancement Dependent pH and Chemical Surfactant for Cu-$H_2O$ Nanofluids," *Thermochimica Acta*, 469, 98–103.

[31] Hwang, Y., Park, H.S., Lee, J.K., and Jung, W.H., 2006, "Thermal Conductivity and Lubrication Characteristics of Nanofluids," *Current Applied Physics*, 6S1, 67–71.

[32] Godson, L., Raja, B., Mohan, Lal, D., and Wongwises, S., 2010, "Experimental Investigation on the Thermal Conductivity and Viscosity of Silver-Deionized Water Nanofluid," *Experimental Heat Transfer*, 23, 317–332.

[33] Kim, P., Shi, L., Majumdar, A., and McEuen, P.L., 2001, "Thermal Transport Measurements of Individual Multiwalled Nanotubes," *Physical Review Letters*, 87, 215502/1–4.

[34] Yu, C., Shi, L., Yao, Z., Li, D., and Majumdar, A., 2006, "Thermal Conductance and Thermopower of an Individual Single-wall Carbon Nanotube," *Nano Letters*, 5, 1842–6.

[35] Hong, H., Luan, X., Horton, M., Li, C., and Peterson, G.P., 2011, "Alignment of Carbon Nanotubes Comprising Magnetically Sensitive Metal Oxides in Heat Transfer Nanofluids," *Thermochimica Acta*, doi:10.1016/j.tca.2011,07.025.

[36] Chen, L., and Xie, H., 2009, "Silicon Oil Based Multiwalled Carbon Nanotubes Nanofluid with Optimized Thermal Conductivity Enhancement," *Colloids and Surfaces A: Physicochemical and Engineering Aspects*, 352, 136–140.

[37] Wen, D., and Ding, Y., 2004. "Effective Thermal Conductivity of Aqueous Suspensions of Carbon Nanotubes (Carbon Nanotube Nanofluids)," *Journal of Thermophysics and Heat Transfer,* 18(4), 481–485.

[38] Ding, Y., Alias, H., Wen, D.S., Williams, R.A., 2006, "Heat Transfer of Aqueous Suspensions of Carbon Nanotubes (CNT Nanofluids)," *International Journal of Heat and Mass Transfer*; 49, 240–250.

[39] Liu, M.S., Ching-Cheng Lin, M., Huang, I.T., and Wang, C.C., 2006, "Enhancement of Thermal Conductivity with Carbon Nanotube for Nanofluids," *International Communications in Heat and Mass Transfer*, 32 (9), 1202–1210.

[40] Murshed, S.M.S., Leong, K.C., and Yang, C., 2006, "Enhanced Thermal Conductivity of $TiO_2$-Water Based Nanofluids," *International Journal of Thermal Sciences* 44, 367–373.

[41] Chen, H., Yang, W., He, Y., Ding, Y., Zhang, L., Tan, C., Lapkin, A.A., and Bavykin, D.V., 2008, "Heat Transfer and Flow Behaviour of Aqueous Suspensions of Titanate Nanotubes (Nanofluids)," *Powder Technology*, 183, 63–72.

[42] Keblinski, P., Phillpot, S.R., Choi, S.U.S., and Eastman, J.A., 2002, "Mechanism of Heat Flow in Suspension of Nano-Sized Particles (Nanofluids)," *International Journal of Heat and Mass Transfer* 45, 855–863.

[43] Jang, S.P., and Choi, S.U.S., 2007, "Effects of Various Parameters on Nanofluid Thermal Conductivity," *Journal of Heat Transfer*, 129, 618–623.

[44] Prasher, R., Bhattacharya, P., and Phelan, P.E., 2006, "Thermal Conductivity of Nanoscale Colloidal Solutions (Nanofluids)," *Physical Review Letters*, 94, 025901.

[45] Koo, J., and Kleinstreuer, C., 2004, "A New Thermal Conductivity Model for Nanofluids," *Journal of Nanoparticle Research,* 6, 577–588.

[46] Kumar, D.H., Patel, H.E., Kumar, V.R.R., Sundararajan, T., Pradeep, T, and Das, S.K., 2004. "Model for Heat Conduction in Nanofluids," *Physical Review Letters* 93, 144301.

[47] Xuan, Y., Li, Q., and Hu, W., 2003, "Aggregation Structure and Thermal Conductivity of Nanofluids," *AIChE Journal,* 49, pp.1038–1043.

[48] Xie, H., Fujii, M., and Zhang, X., 2006, "Effect of Interfacial Nanolayer on the Effective Thermal Conductivity of Nanoparticle–Fluid Mixture," *International Journal of Heat and Mass Transfer*, 48, 2926–2932.

[49] Xue, Q., and Xu, W.M., 2006, "A Model of Thermal Conductivity of Nanofluids with Interfacial Shells," *Materials Chemistry and Physics*, 90, 298–301.

[50] Yu, W., and Choi, S.U.S., 2003. "The role of interfacial layers in the enhanced thermal conductivity of nanofluids: A renovated Maxwell model," *Journal of Nanoparticle Research,* 5, 167–171.

[51] Wang, B.X., Zhou, L.P., and Peng, X.F., 2003. "A Fractal Model for Predicting the Effective Thermal Conductivity of Liquid with Suspension of Nanoparticles," *International Journal of Heat and Mass Transfer* 46, 2665–2672.

[52] Prasher, R., Phelan, P.E., and Bhattacharya, P., 2006, "Effect of Aggregation Kinetics on the Thermal Conductivity of Nanoscale Colloidal Solutions (Nanofluid)," *Nano Letters*, 6, 1529–1534.

[53] Feng, Y., Yu, B., Xu, P., and Zou, M., 2007, "The Effective Thermal Conductivity of Nanofluids Based on the Nanolayer and the Aggregation of Nanoparticles," *Journal of Physics D: Applied Physics*, 40, 3164–3171.

[54] Kang, H.U., Kim, S.H., and Oh, J.M., 2006, "Estimation of Thermal Conductivity of Nanofluid Using Experimental Effective Particle Volume," *Experimental Heat Transfer*, 19, 181–191.

[55] Prasher, R., Song, D., Wang, J., and Phelan, P.E., 2006, "Measurements of Nanofluid Viscosity and Its Implications for Thermal Applications," *Applied Physics Letters*, 89, 133108-1-3.

[56] Namburu, P.K., Kulkarni, D.P., Misra, D., and Das, D.K., 2007, "Viscosity of Copper Oxide Nanoparticles Dispersed in Ethylene Glycol and Water Mixture," *Experimental Thermal and Fluid Science*, 32, 397–402.

[57] Schmidt, J., Chiesa, M., Torchinsky, D.H., Johnson, J.A., Boustani, A., McKinley, G.H., Nelson, K.A., and Chen, G., 2008, "Experimental Investigation of Nanofluid Shear and Longitudinal Viscosities," *Applied Physics Letters*, 92, 244107.

[58] Xie, H., Chen, L., and Wu, Q., 2008, "Measurements of the Viscosity of Suspensions (Nanofluids) Containing Nanosized $Al_2O_3$ Particles," *High Temperatures-High Pressures*, 37, 127–135.

[59] Nguyen, C.T., Desgranges, F., Galanis, N., Roy, G., Mare, T., Boucher, S., and Mintsa, H.A., 2008, "Viscosity Data for $Al_2O_3$/Water Nanofluid-Hysteresis: Is Heat Transfer Enhancement Using Nanofluids Reliable?" *International Journal of Thermal Sciences* 47, 103-111

[60] Chen, H., Ding, Y., and Lapkin, A., 2009, "Rheological Behaviour of Nanofluids Containing Tube / Rod-Like Nanoparticles," *Powder Technology*, 194, 132–141.

[61] Anoop, K.B., Kabelac, S., Sundararajan, T., and Das, S.K., 2009, "Rheological and Flow Characteristics of Nanofluids: Influence of Electroviscous Effects and Particle Agglomeration," *Journal of Applied Physics*, 106, 034909.

[62] Pastoriza-Gallego, M.J., Casanova, C., Páramo, R., Barbés, B., Legido, J.L., and Piñeiro, M.M., 2009, "A Study on Stability and Thermophysical Properties (Density and Viscosity) of $Al_2O_3$ in Water Nanofluid," *Journal of Applied Physics*, 106, 064301.

[63] Yu, W., Xie, H., Chen, L., and Li, Y., 2009, "Investigation of Thermal Conductivity and Viscosity of Ethylene Glycol Based ZnO Nanofluid," *Thermochimica Acta*, 491, 92–96.

[64] Tamjid, E., and Guenther, B.H., 2010, "Rheology and Colloidal Structure of Silver Nanoparticles Dispersed in Diethylene Glycol," *Powder Technology*, 197, 49–53.

[65] Pastoriza-Gallego, M.J., Casanova, C., Legido, J.L., and Pineiroa, M.M., 2011, "CuO in Water Nanofluid: Influence of Particle size and Polydispersity on Volumetric Behaviour and Viscosity" *Fluid Phase Equilibria*, 300, 188–196.

[66] Kole M, Dey TK, 2011, "Effect of aggregation on the viscosity of copper oxide-gear oil nanofluids," *International Journal of Thermal Sciences*, 50, 1741-1747.

[67] Abareshi, M., Sajjadi, S.H., Zebarjad, S.M., and Goharshadi, E.K., 2011, "Fabrication, Characterization, and Measurement of Viscosity of $\alpha$-$Fe_2O_3$-Glycerol Nanofluids," *Journal of Molecular Liquids*, 163, 27–32.

[68] Tsai, T.H., Kuo, L.S., Chen, P.H., and Yang, C.T., 2008, "Effect of Viscosity of Base Fluid on Thermal Conductivity of Nanofluids," *Applied Physics Letters*, 93(23), 233121.

[69] Wang B, Wang X, Lou W, Hao J, 2011, "Ionic liquid-based stable nanofluids containing gold nanoparticles," *Journal of Colloid and Interface Science*, 362, 5–14.

[70] Timofeeva, E.V., Yu, W., France, D.M., Singh, D., and Routbort, J.L., 2011, "Base Fluid and Temperature Effects on the Heat Transfer Characteristics of SiC in Ethylene Glycol/$H_2O$ and $H_2O$ Nanofluids," *Journal of Applied Physics*, 109, 014914.

[71] Pak, B.C., and Cho, Y., 1998, "Hydrodynamic and Heat Transfer Study of Dispersed Fluids with Submicron Metallic Oxide Particle," *Experimental Heat Transfer*, 11, 151–170.

[72] Sommers, A.D., and Yerkes, K.L., 2010, "Experimental Investigation into the Convective Heat Transfer and System-Level Effects of $Al_2O_3$-Propanol Nanofluids," *Journal of Nanoparticle Research*, 12, 1003–1014.

[73] Xuan, Y., and Roetzel, W., 2006, "Conceptions for Heat transfer Correlation of Nanofluids," *International Journal of Heat and Mass Transfer*, 43, 3701-3707.

[74] Gosselin, L, and De Silva, A.K., 2004. "Combined Heat Transfer and Power Dissipation Opimization of Nanofluid Flows," *Applied Physics* Letters, 85, 4160–4162.

[75] Lee, J., and Mudawar, I., 2007, "Assessment of the Effectiveness of Nanofluids for Single Phase and Two-Phase Heat Transfer in Micro-channels," *International Journal of Heat and Mass Transfer*, 50, 452-463.

[76] Avsec, J., Oblak, M., 2007, "The Calculation of Thermal Conductivity, Viscosity and Thermodynamic Properties for Nanofluids on the Basis of Statistical Nanomechanics," *International Journal of Heat and Mass Transfer*, 50(19), 4331-4341.

[77] *Buongiorno, J., 2006,* "Convective Transport in Nanofluids," *Journal of Heat Transfer, 128, 240-*250.

[78] Zhou, S.Q., and Ni, R., 2008, "Measurement of the Specific Heat Capacity of Water-Based $Al_2O_3$ Nanofluid," *Applied Physics Letters*, 92, 093123.

[79] Shin, D., and Banerjee, D., 2011, "Enhancement of Specific Heat Capacity of High-temperature Silica-nanofluids Synthesized in Alkali Chloride Salt Eutectics for Solar Thermal-energy Storage Applications," *International Journal of Heat and Mass Transfer*, 54, 1064–1070.

[80] Das, S.K., Putra, N., and Roetzel, W., 2003, "Pool Boiling Characteristics of Nanofluids," *International Journal of Heat and Mass Transfer*, 46, 851–862.

[81] Prakash, N.G., Anoop, K.B., and Das, S.K., 2007, "Mechanism of Enhancement/Deterioration of Boiling Heat Transfer Using Stable Nanoparticle Suspensions over Vertical Tubes," *Journal of Applied Physics*, 102, 074317.

[82] Shi, M.H., Shuai, M.Q., Chen, Z.Q., Li, Q., and Xuan, Y.M., 2007, "Study on Pool Boiling Heat Transfer of Nano-Particle Suspensions on Plate Surface," *Journal of Enhanced Heat Transfer*, 14 (3), 223–231.

[83] Kumar, R., and Milanova, D., 2009, "Effect of Surface Tension on Nanotube Nanofluids," *Applied Physics Letters*, 94, 0731079.

[84] Murshed, S.M.S., Milanova, D., and Kumar, R., 2009, "An Experimental Study of Surface Tension Dependent Pool Boiling Characteristics of Carbon Nanotubes-Nanofluids," Proceedings of 7th international ASME conference on nanochannels, microchannels and minichannels. 2009,

[85] Zhu, D., Wu, S., Wang, N., 2010, "Thermal Physics and Critical Heat Flux Characteristics of $Al_2O_3$– $H_2O$ Nanofluids," *Heat Transfer Engineering*, 31(14), 1213–1219.

[86] Moosavi, M., Goharshadi, E.K., and Youssefi, A., 2010, "Fabrication, Characterization, and Measurement of Some Physicochemical Properties of ZnO Nanofluids," *International Journal of Heat and Fluid Flow*, 31, 599–605.

In: Nanofluids
Editor: Yuwen Zhang
ISBN: 978-1-62618-165-6


*Chapter 4*

# THERMAL CONDUCTIVITY OF NANOFLUIDS AND HEAT PROPAGATION VELOCITIES

***Kenneth D. Kihm*** * ***and Chan Hee Chon*** †
Department of Mechanical, Aerospace, and Biomedical Engineering
The University of Tennessee, Knoxville, TN, US

## ABSTRACT

The empirical correlation shows that the base fluid temperature plays an important role for effective thermal conductivity enhancement of nanofluids with decreasing nanoparticle sizes. Furthermore, it shows that the Reynolds number, function of temperature, has a more dominant role than other factors such as base fluid thermal conductivity, particle size, and diffusivity. Despite that the Brownian velocity of nanoparticles has been traditionally used for the Reynolds number in modeling, it is too slow to properly describe the effective thermal conductivity enhancement of nanofluids for many cases. The more comprehensive model based on a faster heat transfer velocity, in the same order as the speed of sound, agrees better with wide range of experimental data.

**Keywords**: Nanofluid, effective thermal conductivity, heat propagation velocities

## 1. INTRODUCTION

Nanofluid as a mixture of nanoscale particles suspended in liquid [1] is perceived as an extended scope of earlier efforts to study multiphase systems containing microscale particles embedded in solid materials [2-4] or dispersed in liquids [5].

---

* Corresponding author, E-mail: kkihm@utk.edu, jointly appointed as World Class University (WCU) Professor of Multiscale Mechanical Design Division, Seoul National University, Seoul 151-742, Korea.
† Present affiliation: Director for R&D Center of Infopia Co. Ltd, 891 Hogye-dong, Dongan-gu, Anyang, Kyunggi 431-080, Korea.

Nanoparticles have many attractive characteristics to lend to the idea of a nanofluid. First, nanoparticles are nearly free from the sedimentation because of the minimal gravitational pull over their surface drag. The particle sedimentation depends on particle size, base fluid viscosity, and density difference between particle and base fluid:

$$U_s = \frac{2}{9}\frac{d_p^2}{\mu_{BF}}(\rho_p - \rho_{BF})g \tag{1}$$

where $U_s$ is the particle sedimentation velocity, $d_p$ is the particle diameter, $\mu_{BF}$ is the dynamic viscosity of the base fluid, $g$ is the gravitational acceleration, and $\rho_p$ and $\rho_{BF}$ are densities of particle and base fluid, respectively. The particle diameter $d_p$ dominates in Eq. (1) and the sedimentation speed $U_s$ approaches to zero for nanometer-sized particles.

In addition, the surface area of a nanometer-sized particle is $10^6$ times larger than that of a micron-sized particle. This feature greatly enhances the heat conduction, which is an interfacial phenomenon between the particles and the surrounding fluid. The smaller the particle, the greater the capacity for enhancing the effective heat transfer of nanofluids.

Since the first publication on measurements of the thermal conductivity of water suspensions of $Al_2O_3$ nanoparticles using the transient hot-wire technique in 1999 [6], a number of successive studies have extended the original findings by examining the parametric variations affecting the conductivity enhancement [7-23]. These experimental studies revealed the parametric importance of thermal conductivity enhancement, including the volume concentration of nanoparticles and their sizes, clustering or aggregation effect, pH effect, surfactant effect, and the base fluid temperature. More recently, Chon et al. [24] have constructed an experimentally extrapolated equation that predicts the nanofluid conductivity in terms of these related parameters.

In parallel to these advances in experimental investigations, theoretical predictions of the effective thermal conductivity enhancement of nanofluids have been continually published [25-30]. Table 1 shows the chronological presentation of published theories predicting conductivities, either for particle-embedded solid materials or for nanofluids. The first attempt at mathematical modeling dates back to 1873 by Maxwell [2], who presented effective thermal conductivity for a heterogeneous solid material, consisting of spherical solid particles of thermal conductivity, $k_p$ embedded in a continuous solid phase with thermal conductivity $k_{BF}$. The volume concentration $f$ of the embedded spheres is taken to be sufficiently small, such that the spheres do not interact thermally and the effect of the particle size is assumed negligible. In 1962, Hamilton and Crosser [3] extended Maxwell's model and incorporated a modification for non-spherical particles by the empirical shape factor $n$ ($n = 3$ for sphere).

A number of alternative models have been proposed with the use of the Brownian motion-induced micro-convection in a nanofluid. By adding the second term to the Maxwell model, Xuan et al. proposed a model incorporating the Brownian motion of nanoparticles in 2003 [25]. A year later, Jang and Choi [26] introduced the Brownian-motion-driven convection model and attempted to describe the temperature-dependency of nanofluid thermal conductivity. They assumed the Nusselt number (Nu) to be the product of Reynolds number (Re) and Prandtl number (Pr), i.e., $Nu = Re^2Pr^2$, based on the postulation of Reynolds number of the order of unity. However, this assumption is invalid because it is incorrect to neglect the first two terms, i.e., lower degree terms of RePr, in the expression for the Nusselt number that

Acrivos and Taylor [31] have derived for heat transfer from a spherical particle at low values of the Reynolds number.

Furthermore, in 2005, Prasher et al. [28] developed a model combining the Maxwell-Garnett model [33] incorporating both the Kapitza resistance effect of particles with the surrounding medium and the effect of the Brownian motion-induced convection. Later, they expanded their theoretical prediction for nanofluid thermal conductivity by adding aggregation conductivity contributions for the convection enhancements [29]. However, they assumed a less justifiable Brownian velocity of nanoparticles as $V_{Brownian} = \sqrt{18k_bT/\pi\rho_p d_p^3}$ based on the kinetic theory of gas, which is limited for fine particles suspended in a dilute gas (Boltzmann constant $k_b = 1.3807 \times 10^{-23} J/K$, the base fluid temperature $T$, the nanoparticle densify $\rho_p$, and its diameter $d_p$). Quite possibly because of this conflict, their model fits only to a subset of their experimental data, e.g., agrees fairly well with $Al_2O_3$ nanofluid data, but fails to fit to CuO nanofluid data.

**Table 1. Historical development of nanofluidic thermal conductivity models**

| Author | Thermal conductivity enhancement, $k_{eff}/k_{BF}$ |
|---|---|
| Maxwell [2] (1873) | $\frac{k_{eff}}{k_{BF}} = \left[\frac{k_p + 2k_{BF} - 2f(k_{BF} - k_p)}{k_p + 2k_{BF} + f(k_{BF} - k_p)}\right]$ |
| Hamilton & Crosser [3] (1962) | $\frac{k_{eff}}{k_{BF}} = \left[\frac{k_p + (n-1)k_{BF} - (n-1)f(k_{BF} - k_p)}{k_p + (n-1)k_{BF} + f(k_{BF} - k_p)}\right]$ |
| Xuan et al. [25] (2003) | $\frac{k_{eff}}{k_{BF}} = \left[\frac{k_p + 2k_{BF} - 2f(k_{BF} - k_p)}{k_p + 2k_{BF} + f(k_{BF} - k_p)}\right] + f\frac{\rho_p C_p}{2k_{BF}}\sqrt{\frac{2k_bT}{3\pi d_p \mu}}$ |
| Jang & Choi [26] (2004) | $\frac{k_{eff}}{k_{BF}} = \left[(1-f) + \frac{\beta k_p}{k_{BF}} f + 3C_\alpha \frac{d_{BF}}{d_p} Re_{d_p}^2 \mathrm{Pr} f\right]$, $\beta$=0.01 |
| Kumar et al. [27] (2004) | $\frac{k_{eff}}{k_{BF}} = \left[1 + C_\beta\left(\frac{2k_bT}{\pi\mu d_p^2}\right)\frac{f d_{BF}}{k_{BF}(1-f)d_p}\right]$ |
| Prasher et al. [28,29] (2005,2006) | $\frac{k_{eff}}{k_{BF}} = (1 + C_\gamma \mathrm{Re}^m \mathrm{Pr}^{0.333} f) + \frac{[k_a + 2k_{BF}] + 2f_a(k_a - k_{BF})}{[k_a + 2k_{BF}] - f(k_a - k_{BF})}$ |
| Patel et al. [30] (2008) | $\frac{k_{eff}}{k_{BF}} = 1 + \frac{k_p \pi}{k_{BF}}\left(6 + \frac{\alpha_{BF} \cdot \pi\mu d_p}{C_\delta \cdot 2k_bT}\right)^{-1}\left(\frac{6f}{\pi}\right)^{\frac{1}{3}}$ |

[$\alpha_{BF}$ is the diffusivity of base fluid ($m^2/s$), $C_p$ is the specific heat of particle (J/kg·K), and $C_\alpha$, $C_\beta$, $C_\gamma$, $C_\delta$ and $m$ are empirical constants. Suggested constants [26,27,29] are $18\times10^6$ for $C_\alpha$, 2.9 to 3.0 for $C_\beta$, 40,000 for $C_\gamma$, and 2.4 to 2.75 for $m$.]

The effect of the Brownian motion-induced microconvection remains controversial among different research groups. Eapen et al. [34] strongly argued that microconvection around randomly moving nanoparticles does not influence the thermal conductivity of the nanofluid. In 2007, Das group proposed a nanofluid thermal conductivity model based on a cell model [30]. Their cell model tried to explain the nonlinear dependence of thermal conductivity of nanofluids on particle volume fraction. However, their empirical constants were defined only to fit to their experimental data. In fact, their model constants did not show consistency for an identical $Al_2O_3$ nanofluid.

The kinetic principle well describes the thermal conductivity of gas, as the gas molecules are assumed to be freely moving due to their relatively lean distributions [35]. For liquids, however, their stronger intermolecular forces make it necessary to modify the kinetic theory. In addition, the molecular collision velocities of gases are too low to explain liquid thermal conductivities that are at least one order of magnitude higher than the gas conductivities. Hence, the thermal conductivities of denser liquids are conjectured to be more properly expressed by the faster sound propagation in the case of liquids, and by the phonon velocity in the case of solids.

This chapter presents an overview of the effective thermal conductivities of nanofluids and a new concept of heat propagation velocity to more appropriately elaborate the nanofluid thermal conductivities than the Brownian motion of nanoparticles alone. In Section 2, the transient hot-wire technique for thermal conductivity measurement is described for its working principle as well as the system construction. Section 3 presents the comprehensive empirical correlation of the effective thermal conductivity with varying temperature and particle sizes of different nanofluids. In Section 4, new heat propagation velocity of the same order as the sound propagation speed is proposed and extensively examined in determining the effective thermal conductivity of nanofluids, in contrast to the Brownian velocity that has been traditionally believed to play the role.

## 2. Transient Hot-Wire Technique

The thermal conductivity $k$ is equivalent to heat flux (W/m$^2$) per unit temperature gradient $\nabla T$, i.e., $k = \frac{q}{A\nabla T}$. The working principle of the transient hot wire technique was introduced by Schieirmacher [36] and then tested by Van der Held and Van Drunen [37]. Very fine platinum wire is centered in the vertical test cell filled with a test fluid. The wire is used as a temperature sensor as well as a heat source [38]. Because nanofluids are electrically conductive, the use of a bare platinum wire is not adequate but, an electrically insulated platinum wire works properly [39].

Considering a long and thin wire of radius $r$ under the heating power $Q$, the transient surface temperature $\theta$ at any given time $t$ is given by [40]:

$$\theta = \frac{Q}{4\pi k} \cdot \ln\left(\frac{4\alpha_t t}{r^2 C_\gamma}\right) \quad (2)$$

where $C_\gamma = \exp \gamma$, with Euler constant $\gamma = 0.577216$, $\alpha_t$ is the thermal diffusivity expressed as $\alpha_t = k/\rho C_p$, $\rho$ is the density of the surrounding fluid, and $C_p$ is the specific heat. Therefore, the temperature rise $\theta_2 - \theta_1$ measured over a time interval $t_2 - t_1$ is given by:

$$\theta_2 - \theta_1 = \frac{Q}{4\pi k} \cdot \ln\left(\frac{t_2}{t_1}\right) \quad (3)$$

And thus, the thermal conductivity is given by:

$$k = \frac{Q}{4\pi(\theta_2-\theta_1)} \cdot \ln\left(\frac{t_2}{t_1}\right) = \frac{Q}{4\pi} \cdot \frac{1}{S} \quad (4)$$

which $S$ is the slope of the linear portion of the $\theta$ - $ln\, t$ curve, 1.e.,

$$S = \frac{\theta_2-\theta_1}{\ln(t_2)-\ln(t_1)}. \quad (5)$$

The transient hot-wire thermal conductivity measurement system consists of three parts [41]: the voltage divider circuitry, the data acquisition, and the miniature test chamber of approximately 10 cc in volume (Figure 1). The voltage divider is a series of two resistors with a known resistance, $R_s$, and an unknown platinum wire (electrically insulated) resistance, $R_w$, which varies with temperature providing the heating power $Q = i^2R_w$. A 50-μm diameter platinum wire works as a heating wire as well as a thermometer.

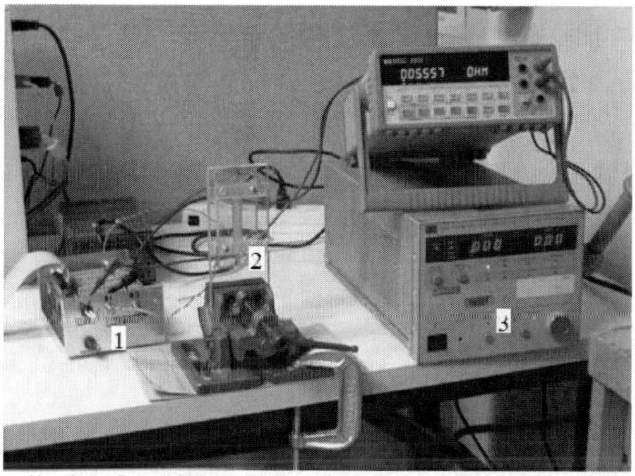


**(1: signal amplifier, 2: test chamber, 3: DC power supply)**

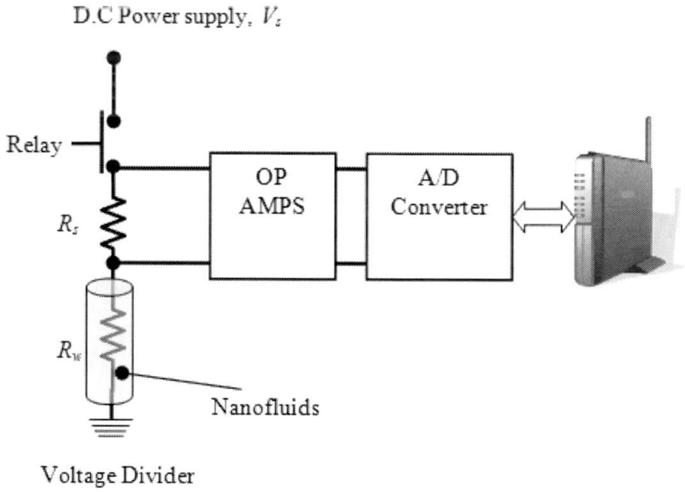


Figure 1. Experimental setup for nanofluid thermal conductivity measurement: (a) the overall experimental setup, and (b) electrical circuit layout with the voltage divider.

Switching the power supply to the voltage divider initiates the voltage change in the hot wire, and the time varying voltage is recorded by NI PCI-6033 analog-digital (A/D) converter with a resolution of 1.22mV at a sampling rate of 50Hz for 10 seconds. The measured voltage change is converted to a resistance change and the heating current through the wire and the temperature variation of the wire can be calculated by the temperature-resistance relationship of the platinum wire.

## 3. EFFECTIVE THERMAL CONDUCTIVITY OF NANOFLUIDS

This section presents a systematic experimental result to delineate the effect of temperature and particle size on the effective thermal conductivity of nanofluids. Three nanofluid samples consist of three differently sized $Al_2O_3$ nanoparticles: (1) 11-nm nominal diameter, (2) 47-nm nominal diameter, and (3) 150-nm nominal diameter. For enhanced dispersion of nanoparticles in the base fluid, ultrasonic vibration is used. The nanofluid temperature ranges (21°C - 71°C) is controlled by placing the test chamber inside a circulating thermal bath with ± 0.01°C accuracy.

Based on Buckingham-Pi theorem with a linear regression for the experimental data, an empirical correlation for the nanofluid effective thermal conductivity ($k_{eff}$), normalized by the base fluid thermal conductivity ($k_{BF}$), is given with 95% confidence level by [24]:

$$\frac{k_{eff}}{k_{BF}} = 1 + 64.7 f^{0.746} \left(\frac{d_{BF}}{d_p}\right)^{0.369} \left(\frac{k_p}{k_{BF}}\right)^{0.7476} \mathrm{Pr}^{0.9955} \mathrm{Re}^{1.2321} \tag{6}$$

where $f$ denotes the volume concentration of nanoparticles of diameter $d_p$ with thermal conductivity of $k_p$, suspended in the base fluid of molecular diameter of $d_{BF}$ with thermal conductivity of $k_{BF}$. The Prandtle number is defined as $\mathrm{Pr} = \mu/\rho_{BF}\alpha$ and the Reynolds number is defined as $\mathrm{Re} = \frac{\rho_{BF} V_{Br} d_p}{\mu} = (\rho_{BF} k_b T)/(3\pi\mu^2 l_{BF})$ with $l_{BF}$ denoting the mean free path of the base fluid. The mean free path of a molecule, that is, the average distance it can travel between collisions is derived as $l_{BF} = 1/n\sigma$ where $n$ represents the molecular number density, which can be regarded as negligibly weak function of temperature for aquatic water for the tested range, and $\sigma$ denotes the collision cross section that is equivalent to four times of the geometrical cross section of a molecule. For improved accuracy, the derivation should be modified to account for the motion of the target molecules as, $l_{BF} = 1/\sqrt{2}n \cdot 4\pi d_{BF}^2$ [42].

The temperature dependence of the dynamic viscosity of the base fluid (water) is given by [43]:

$$\mu = 2.414 \times 10^{-5} \cdot 10^{\frac{247.8}{T-140}} \tag{7}$$

The Brownian velocity of nanoparticles based on the Einstein diffusion theory is given by [44]:

$$V_{Br} \equiv \frac{k_b T}{3\pi\mu d_p l_{BF}} = \frac{k_b}{3\pi d_p l_{BF}} \cdot \frac{T}{2.414 \times 10^{-5} \cdot 10^{\frac{247.8}{T-140}}} \tag{8}$$

The correlation Eq. (6) can now be rewritten exclusively in terms of nanoparticle diameter and suspension temperature as:

$$\frac{k_{eff}}{k_{BF}} = 1 + C_1 \cdot \left( \frac{\Pr(T)^{0.9955} T^{1.2321}}{d_p^{0.369} k_{BF}(T)^{0.7476} 10^{\frac{247.8}{T-140}}} \right) \quad (9)$$

where $C_1$ represents all specified or invariant experimental parameters.

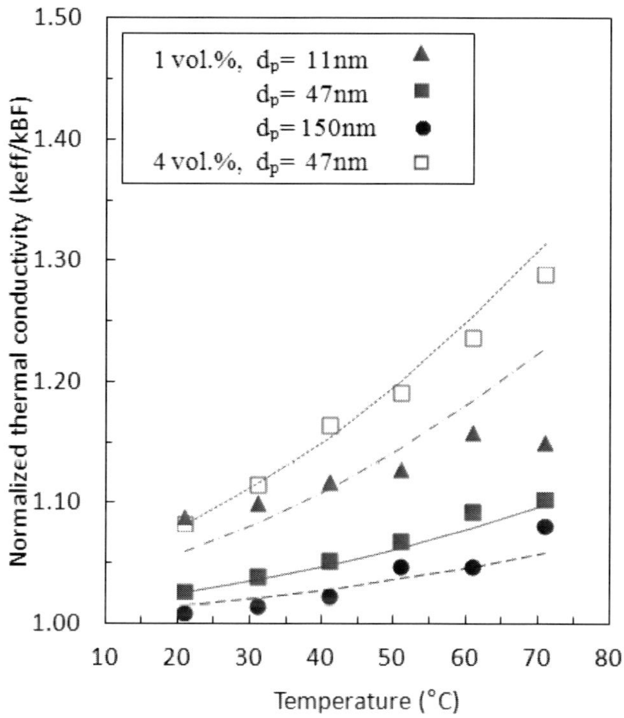


Figure 2. Temperature dependence of the thermal conductivity enhancement of three different $Al_2O_3$ nanofluids with 11-nm, 47-nm, and 150-nm sized nanoparticles at 1 and 4 vol.% concentration, normalized by the thermal conductivity of the base fluid at the specific temperature [24].

Figure 2 shows the measured $Al_2O_3$ nanofluid thermal conductivity normalized by the base fluid thermal conductivity at each specified temperature (symbols) and the experimental correlation of Eq (9) (curves) for different nanoparticle sizes and volume concentration. Nanofluid thermal conductivity increases with increasing nanofluid temperature and with decreasing nanoparticle sizes.

Figure 3 shows the temperature dependence of the three primary parameters in Eq. (6), $k_{BF}$, Pr, and Re with the subscript o being at the reference temperature of 21°C, for the case of 47-nm nanoparticles at 1 vol.%. Reynolds number that represents the mobility of nanoparticles shows dominating temperature dependence whereas Pr shows slightly decreasing dependency and $k_{BF}$ shows practically no dependence on temperature.

It is shown that the Reynolds number is considered as a dominant mechanism for the temperature dependence of thermal conductivity of nanofluid. The Brownian velocity depends upon the base fluid viscosity, which in turn depends on the base fluid temperature

(Eq. (8)). By approximating the weak temperature dependence of $k_{BF}$ and Pr as invariant in Eq. (9), the empirical correlation can explicitly show the effects of nanoparticle size and nanofluid temperature as:

$$\frac{k_{eff}}{k_{BF}} = 1 + C_2 \cdot \left(\frac{1}{d_p}\right)^{0.369} \left(\frac{T^{1.2321}}{10^{\frac{247.8}{T-140}}}\right) \tag{10}$$

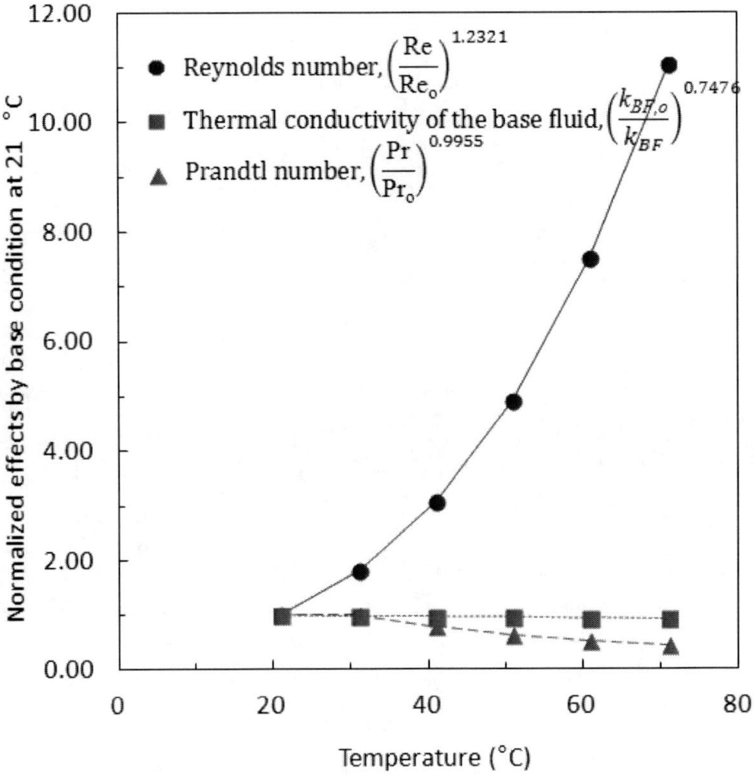


Figure 3. Comparison of temperature dependencies of the three primary parameters ($k_{BF}$, Pr, and Re) of the empirical correlation (Eq. (6)) for the case of the $Al_2O_3$ nanofluid sample with 47-nm nanoparticles at 1 vol.% concentration [24].

In conclusion, the systematic experiments for thermal conductivity measurements using three different batches of fairly monodipersed $Al_2O_3$ samples, together with the analytical approaches using Buckingham-Pi theorem, provides an empirical correlation equation, Eqs. (9) and (10), to comprehensively describe the effect of individual physical parameters on nanofluid thermal conductivity. According to the empirical correlation, the base fluid temperature is most dominant factor for a given particle size.

## 4. Prevailing Heat Propagation Velocity in Nanofluids

To predict the effective thermal conductivity of nanofluids, diverse theoretical approaches were conducted. However, the widely applied prediction such Brownian particle velocities are often found not enough to describe the substantial enhancement of the effective thermal conductivities of nanofluids. Therefore, an alternative insight is presented concerning more appropriate heat propagation velocity scales in predicting the effective thermal conductivities of nanofluids [45].

The enhanced thermal conductivity of nanofluids containing highly conductive metal or oxide metal particles, such as $Al_2O_3$ or CuO, is believed to be attributed to the interaction of nanoparticles with the base fluid molecules. Hence, the thermal conductivity enhancement, $\Delta k_{enh}$, by the molecular interaction with nanoparticles is given by [35]:

$$\Delta k_{enh} = \left(k_{eff} - k_{BF}\right) \sim f \cdot \rho_p \cdot C_p \cdot V_{ht} \cdot l_{ht} \tag{11}$$

where the heat propagation velocity $V_{ht}$ represents the heat propagation rate by the vibration of base fluid molecules. The heat travel distance $l_{ht}$ in the above is defined as the freely traveled distance of heat energy during the interaction of base fluid molecules and nanoparticles as [25]:

$$l_{ht} = C_3 \cdot \sqrt{\frac{T}{\mu d_p}} \tag{12}$$

where $\mu$ is the dynamic viscosity of the base fluid and $C_3$ is a proportional constant. Note that this estimation is consistent with the conjectures of the characterisitc heat propagation velocity being on the scale of the sound propagation velocity in a colloidal medium [46, 47].

In a stationary liquid, individual molecules are constantly moving, and their motions are largely confined within a "cage" formed by the closely-packed neighboring molecules [48]. This virtual cage is conceived by the energy barrier of height $\Delta\tilde{G}_0^+/\tilde{N}$ where $\Delta\tilde{G}_0^+$ represents the molar free energy of activation for escaping the cage and $\tilde{N}$ denotes the molar Avogadro number. The molecular vibrational frequency $v$ is given by:

$$v = \frac{k_b T}{h} exp\left(-\Delta\tilde{G}_o^+/RT\right) \tag{13}$$

where $h$ and $R$ are the Planck constant and the specific gas constant, respectively, and $T$ is the fluid temperature. The free energy of activation, $\Delta\tilde{G}_0^+$, is assumed to be constant for a specified fluid and also assumed to be directly related to the internal energy of vaporization at the normal boiling point [49]. The internal energy is given from Trouton's rule [50] as $\Delta\tilde{U}_{vap} \approx \Delta\tilde{H}_{vap} - RT_b \cong 9.4RT_b$:

$$\Delta\tilde{G}_o^+ \approx 0.408\Delta\tilde{U}_{vap} \cong 3.8RT_b \tag{14}$$

where $\Delta\tilde{H}_{vap}$ is the enthalpy of vaporization at the normal boiling point $T_b$. Combining equations (12) and (13) gives an expression for the heat propagation velocity as:

$$V_{ht} = \lambda_{ht} \cdot \frac{k_b T}{h} exp(-3.8T_b/T) \quad (15)$$

where $\lambda_{ht}$ represents the heat propagation length scale. The heat propagation velocity can be estimated by examining the order-of-magnitudes of the involved parameters.

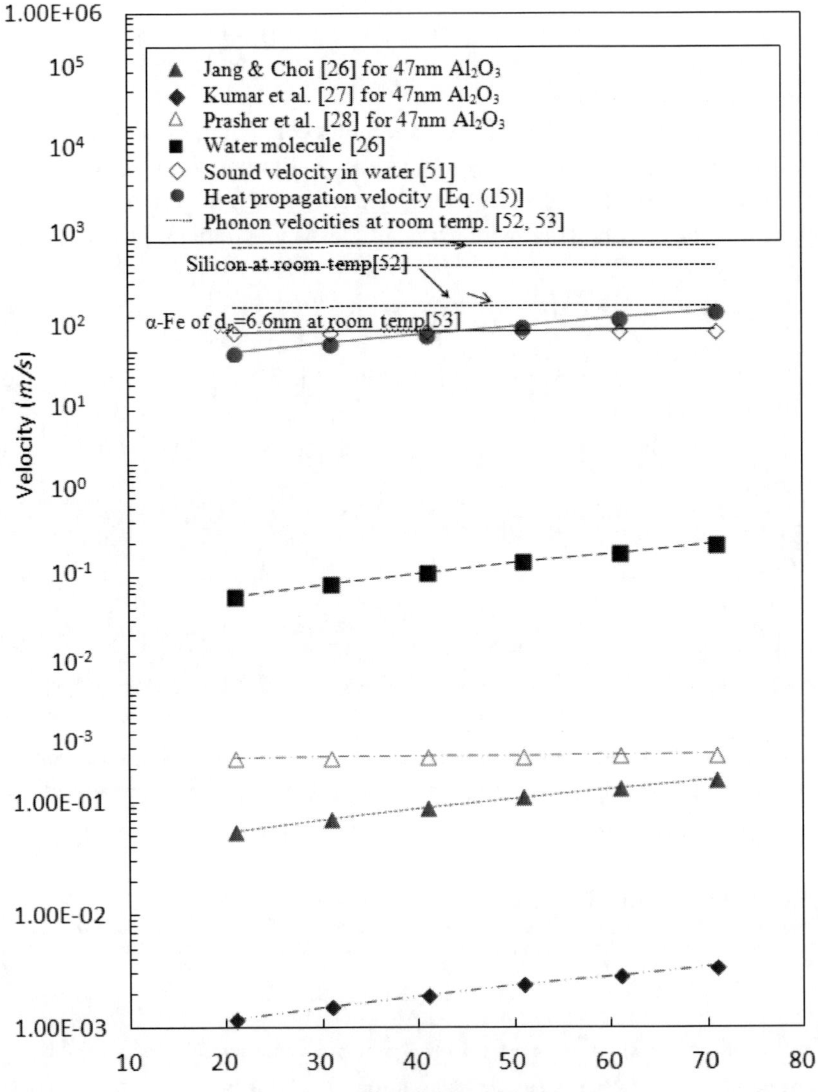


Figure 4. Temperature dependence of Brownian velocities of nanoparticles and water molecules [26, 27, 28], speed of sound in water [51], phonon velocities in selected solid mediums [52, 53], and the heat propagation velocity of Eq. (15).

Figure 4 compares widely different types of velocity scales that are considered relevant in describing effective thermal conductivities of nanofluids: three differently defined Brownian velocities for 47-nm $Al_2O_3$ nanoparticles [26-28], the Brownian velocity of the base fluid (water) molecules [26], the heat propagation velocity based on the currently proposed model

(Eq. (14)), the sound velocity in water [51], and phonon velocities for selected molecules and nanoparticles [52, 53]. Note that they are spanned as widely as $10^6$ in their magnitudes for the considered temperature range from 21 to 71°C.

Table 2 shows the functional expressions of these velocities and their calculated magnitudes. Despite that almost all reported studies used the Brownian velocities to manage correlating with the heat propagation scale in nanofluids, the heat propagation velocity of Eq. (15) is comparable to the sonic velocity in the base fluid that is several orders larger than the Brownian velocities. The Brownian velocities based on the nanoparticles are too slow to be compatible with the relatively faster heat conduction phenomena in liquids. Furthermore, the Brownian velocities of the base fluid of water molecules are also considered too slow to properly model the nanofluidic conductivities.

**Table 2. Differently defined Brownian velocities and heat propagation velocities [45]**

| Author | Velocity model | Calculated velocity (m/s) |
|---|---|---|
| Brownian velocity of nanoparticles [26] | $V_{Br} = \frac{k_b T}{3\pi\mu d_p l_{BF}}$ | 0.055 ~ 0.160 |
| Brownian velocity of nanoparticles [27] | $V_{Br} = \frac{2k_b T}{\pi\mu d_p^2}$ | 0.0012 ~ 0.0035 |
| Brownian velocity of nanoparticles [28] | $V_{Br} = \sqrt{\frac{18k_b T}{\pi\rho d_p^3}}$ | 0.249 ~ 0.270 |
| Brownian velocity of water molecules [26] | $V_{Br} = \frac{k_b T}{3\pi\mu d_{BF} l_{BF}}$ | 6.710 ~ 19.534 |
| Sound propagation velocity in water [53] | $V_s = \sqrt{\left(\frac{\partial p}{\partial \rho}\right)_{\text{isentropoic}}}$ | 1480 ~ 1555 |
| Heat propagation velocity [Eq. (15)] | $V_{ht} = \lambda_{ht}\frac{k_b T}{h} exp(-3.8T_b/T)$ | 950 ~ 2250 |

The heat propagation velocity, Eq. (15), shares the same order of magnitude with the speed of sound, and this is consistent with the conjecture of the sound velocity being used to describe thermal conductivities in liquid medium [22]. The phonon velocities shown for selected solid mediums of α–Fe and silicons in Figure 4 are expected to be faster than the heat propagation velocity in liquid because of the relatively higher heat conductivities in solid mediums.

Substituting Eqs. (12) and (15) into Eq. (11) gives:

$$\frac{k_{eff}}{k_{BF}} = 1 + C_4 \cdot \frac{f\rho_p C_p}{k_{BF}} \cdot \frac{\lambda_{ht} k_b T}{h} exp(-3.8T_b/T) \cdot \sqrt{\frac{T}{\mu d_p}} \tag{16}$$

where $C_4$ is a proportional constant. Further modifications of Eq. (16) can be implemented to account for both the nanoparticle heat dissipation to the surrounding licquid and the effect of nanoparticle coagulation. The finite heat dissipation time from nanoparticles into the base fluid medium is known to impede the thermal conductivity [54, 55], and the heat dissipation time increases with increasing nanoparticle heat capacity and decreases with increasing heat capacity of the surrounding fluid. In other words, nanoparticles with a higher heat capacity require longer heat dissipation time to the base fluid, and this results in slower thermal diffusion and lower effective thermal conductivity.

In addition, nanoparticle coagulation is inevitable to some extent, and the coagulation generally becomes more severe with increasing particle concentrations. According to Keblinski et al. [47], the surface to surface distance of nanoparticles is twice the particle size at 1 vol. %; however, it can decrease to half the particle size at 5 vol. %. The coagulation of nanoparticles can effectively decrease the volume concentration to $f^a$ with a < 1. Therefore, the effective thermal conductivity of nanofluids of Eq. (16) is modified by accommodating the reduced volume concentration as follows:

$$\frac{k_{eff}}{k_{BF}} = 1 + C' \cdot \frac{f^a \rho_p C_p k_b T^{1.5}}{k_{BF} \cdot h \cdot \mu^{0.5} d_p^{0.5}} \cdot exp(-3.8T_b/T) \cdot \left(\frac{C_{BF}}{C_p}\right)^b \quad (17)$$

where $C'$ is a modified constant and $C_{BF}$ is the base fluid specific heat. As $\lambda_{ht}$, a length scale to introduce the velocity from the frequency as shown in Eq. (15), cannot be calculated directly, it was determined by order analysis and merged into the constant $C'$. The exponents $a$ and $b$ are empirical constants that represent the effect of nanoparticle coagulation and of nanoparticle heat dissipation, respectively. A regression analysis of published experimental data by the authors [24] provides $a = 0.70$, $b = 1.5$, and $C' = 1.33\text{x}10^{-25}$ for the case of $Al_2O_3$ nanoparticles of three different sizes of 11-nm, 47-nm, 150-nm diameters, respectively suspended in water under various experimental conditions of a volume concentration range of 1 to 4 vol. % and a tested temperature range of 21 to 71°C.

Figures 5-a to c show the effective thermal conductivity based on the heat propagation velocity model for nanofluids, Eq. (17), in comparison with five previously published models [25-30], for the three different nanofluids. The symbols represent the corresponding experimental data for $Al_2O_3$ [24] and CuO [45]. For all three nanofluids with 47-nm $Al_2O_3$ at 1% and 4%, and 30-nm CuO at 1%, Xuan et al. [25] overestimated the Maxwell's model [2] for nanofluids. Jang and Choi's model [26] shows proximity with experimental data for up to about 50°C for 1 vol.% $Al_2O_3$ and 40 °C for 4 vol.% $Al_2O_3$, but substantially deviates thereafter. This deviation beyond a certain temperature is believed to be attributed to their incorrect postulation implied in determining the Nusselt number, as previously noted. For the CuO nanofluid, their model shows large discrepancies throughout the tested temperature range. Additionally, the model by Kumar et al. [27] wrongly postulates the mean free path of the base fluid and completely fails to predict nanofuidic thermal conductivities for the tested conditions.

Prasher et al. [28, 29] show fairly good agreement with the experiments for the Al2O3 nanofluid, as shown in Figures 5-a and 5-c. However, for the CuO nanofluid in Figure 5-b, their model underestimates the corresponding experimental data [24]. When completely different model parameters were imposed for CuO from that of Al2O3, the model agrees well with the data; however, the model then lacks comprehensiveness because different model parameters need to be determined for different types of nanofluids. Finally, Patel et al. [30]

agrees fairly well with the experimental data at higher concentrations (Figure 5-c) but overestimate the conductivities for low volume concentrations (Figures 5-a and b).

In contrast, the heat propagation model of Eq. (17) shows consistent agreement with the experimental data not only for both nanofluids but for all the tested conditions of temperatures and volume concentrations. Furthermore, Figure 6 demonstrates the comprehensiveness of this new model for effective thermal conductivities of nanofluids in comparison with published experimental data for both $Al_2O_3$ and CuO nanofluids from different leading groups.

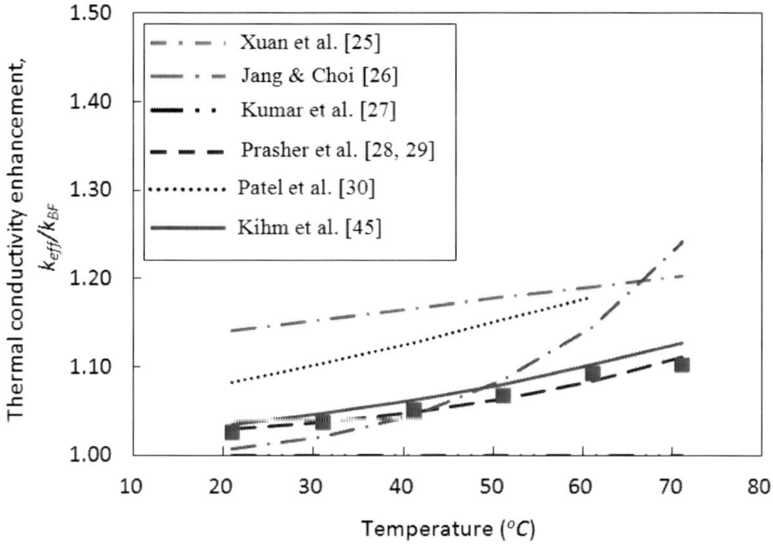


(A) 1 vol. % 47-nm sized $Al_2O_3$

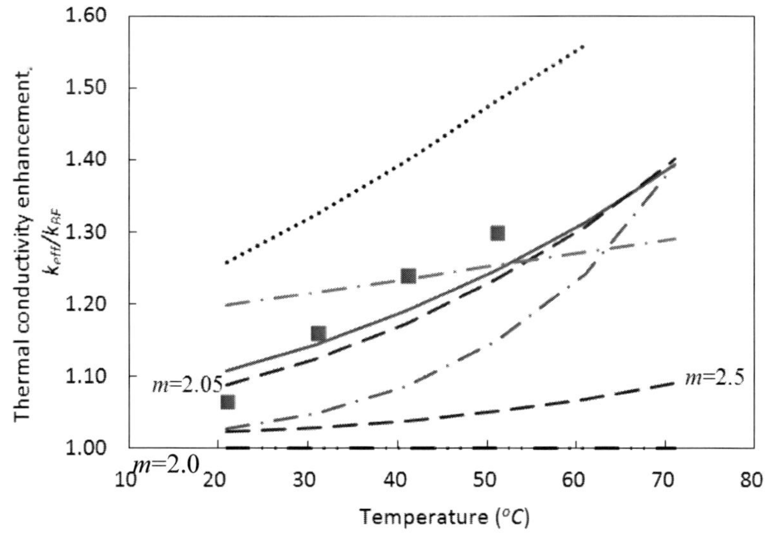


(B) 1 vol. % 30-nm sized CuO

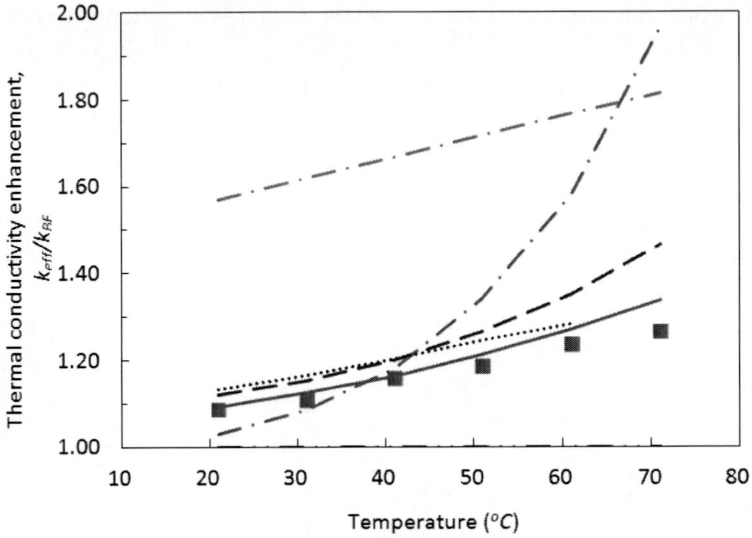


**(C) 4 vol. % 47-nm sized $Al_2O_3$**

Figure 5. Comparison of the published models [25-30, 45] for the thermal conductivities of nanofluids. The symbols represent the measured conductivities: (a) 1 vol. % $Al_2O_3$ nanofluid, (b) 1 vol. % CuO nanofluid, and (3) 4 vol. % $Al_2O_3$ nanofluid [24, 45].

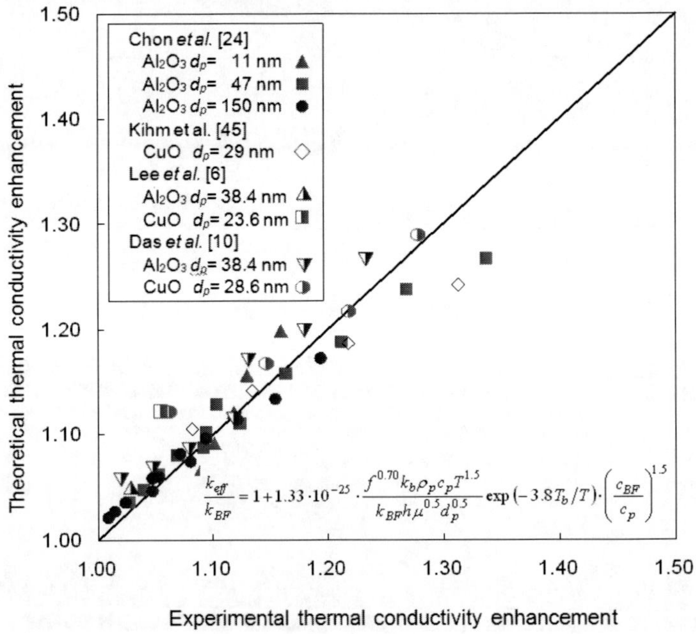


Figure 6. The extrapolated correlation of Eq. (17) is compared with corresponding experimental data for various $Al_2O_3$ and CuO nanofluid samples [6, 10, 24].

## CONCLUSION

Nanoparticles suspended in the nanofluid play important roles for the thermal conductivity enhancement such as high surface-to-volume ratio, thermally-driven particle motion in liquid as well as settlement-free condition because of the extremely small sizes. The effect of the base fluid temperature is considered as an important role for high thermal conductivity of the nanofluid. Experimental studies show that the Reynolds number, a function of temperature, has a more dominant role than other factors such as base fluid thermal conductivity, particle size, and diffusivity. The Brownian motion of nanoparticles in nanofluid is reported as a possible key mechanism for nanofluid thermal conductivity enhancement by many researchers in the past. However, the thermal conductivity enhancement by the Brownian particle motion is not sufficient to explain the comparably high thermal conductivity of the nanofluid. The new model based on the faster heat propagation velocity more appropriately describes the physics of the effective thermal conductivities and also shows substantially better agreement with wide range of experimental data for the effective thermal conductivities of various nanofluids.

## REFERENCES

[1] Choi, S.U.S., 1995, "Developments and Applications of Non-Newtonian Flows," ASME, New York, FED-Vol231, pp. 99-105.

[2] Maxwell, J.C., 1954, *A Treatise on Electricity and Magnetism*, 3rd ed., Dover, New York.

[3] Hamilton, R.L. Crosser, O.K., 1962, "Thermal Conductivity of Heterogeneous Two-Component Systems," *Industrial and Engineering Chemistry Fundamentals*, 1, 182-191.

[4] Bionnecaze, R.R., Brady, J.F., 1991, "The Effective Conductivity of Random Suspensions of Spherical Particles," Proceedings of the Royal Society London Series A (Mathematical and Physical Science) 432, 445-465.

[5] Adler, P.M., 1992, Porous Media: Geometry and Transports, Butterworth-Heinemann, Boston.

[6] Lee, S., Choi, S.U.S., Li, S., Eastman, J.A., 1999, "Measuring Thermal Conductivity of Fluids Containing Oxide Nano-Particles," *Journal of Heat Transfer*, 121(2), 280-289.

[7] Eastman, J. A., Choi, S.U.S., Li, S., Yu, W., Thompson, L.J., 2001, "Anomalously Increased Effective Thermal Conductivities of Ethylene Glycol-Based Nano-Fluids Containing Copper Nano-Particles," *Applied Physics Letters,* 78(6), 718-720.

[8] Choi, S. U. S., Zhang, Z.G., Yu, W., Lockwood, F.E., Grulke, E.A., 2001, "Anomalously Conductivity Enhancement in Nano-Tube Suspensions," *Applied Physics Letters*, 79(14), 2252-2254.

[9] Xie, H. Lee, H., Youn, W., Choi, M., 2002, "Thermal Conductivity Enhancement Of Suspensions Containing Nanosized Alumina Particles," *Journal of Applied Physics*, 91(7), 4568-4572.

[10] Das, S. K., Putra, N., Thiesen, P., Roetzel, W., 2003, "Temperature Dependence of Thermal Conductivity Enhancement for Nanofluids," *Journal of Heat Transfer* 125(4), 567-574.

[11] Patel, H.E., Das, S.K., Sundararajan, T., Nair, A.S., George, B., and Pradeep, T., 2003, "Thermal Conductivities of Naked and Monolayer Protected Metal Nanoparticle Based Nanofluids: Manifestation of Anomalous Enhancement and Chemical Effects," *Applied Physics Letters*, 83(14), 2931-2933.

[12] Wang, X., Xu, X. and Choi, S.U.S., 1999, "Thermal Conductivity of Nanoparticle-Fluid Mixture," *Journal of Thermophysics and Heat Transfer*, 13(4), 474-480.

[13] Kim, S.H., Choi, S.R., Kim, D., 2007, "Thermal Conductivity of Metal-oxide Nanofluids: Particle Size Dependence and Effect of Laser Irritation," *Journal of Heat Transfer*, 129(3), 298-307.

[14] Jang, S.P., and Choi, S.U.S., 2007, "Effects of Various Parameters on Nanofluid Thermal Conductivity," *Journal of Heat Transfer*, 129(5), 617-623.

[15] Mintsa, H.A., Roy, G., Nguyen, C.T., and Doucet, D., 2009, "New Temperature Dependent Thermal Conductivity Data for Water-Based Nanofluidics," *International Journal of Thermal Sciences*, 48(2), 363-371.

[16] Hwang, Y., Lee, J.K., Lee, C.H., Jung, Y.M., Cheong, S.I., Lee, C.G., Ku, B.C., and Jang, S.P., 2007, "Stability and Thermal Conductivity Characteristics of Nanofluids," *Thermochima Acta*, 455(1-2), 70-74.

[17] Hong, K.S., Hong, T.-K., Yang, H.-S., 2006, "Thermal Conductivity of Fe Nanofluids Depending on the Cluster Size of Nanoparticles," *Applied Physics Letters*, 88(3), 031901.

[18] Prasher, R. Evans, W., Meakin, P., Fish, J., Phelan, P., and Keblinski, P., 2006, "Effect of Aggregation on Thermal Conduction in Colloidal Nanofluids," Applied Physics Letters, 89(14), 143119.

[19] Karthikeyan, N.R., Philip, J., and Ray, B., 2008, "Effect of Clustering on The Thermal Conductivity of Nanofluids," *Materials Chemistry and Physics*, 109(1), 50-55

[20] Li, X.F., Zhu, D.S., Wang, X.J., Wang, N., Gao, J.W., and Li, H., 2008, "Thermal Conductivity Enhancement Dependent pH and Chemical Surfactant for Cu-$H_2O$ Nanofluids," Thermochimica Acta, 469(1-2), 98-103.

[21] Evans, W., Prasher, R. Fish, J., Meakin, P., Phelan, P., and Keblinski, P., 2008, "Effect of Aggregation and Interfacial Thermal Resistance on Thermal Conductivity of Nanocomposites and Colloidal Nanofluids," *International Journal of Heat and Mass Transfer*, 51(5-6), 1431-1438.

[22] Li, C.H., and Peterson, G.P., 2007, "The Effect of Particle Size on the Effective Thermal Conductivity of $Al_2O_3$-Water Nanofluids," Journal of Applied Physics 101(4), 044312.

[23] Timofeeva, E.V., Gavrilov, A.N., McCloseky, J.M., and Tolmachev, Y.V., 2007, "Thermal Conductivity and Particle Agglomeration in Alumina Nanofluid: Experiment and Theory," *Physical Review E,* 76(6), 061203.

[24] Chon, C.H., and Kihm, K.D., Lee, S.P., and Choi, S.U.S., 2005, "Empirical Correlation Finding the Role of Temperature and Particle Size for Nanofluid ($Al_2O_3$) Thermal Conductivity Enhancement," *Applied Physics Letters*, 87(15), 153107.

[25] Xuan, Y., Li, Q., and Hu, W., 2003, "Aggregation Structure and Thermal Conductivity of Nanofluids," *AIChE Journal*, 49(4), 1038-1043.

[26] Jang, S. P., and Choi, S. U. S., 2004, "Role of Brownian Motion in the Enhanced Thermal Conductivity of Nanofluids," *Applied Physics Letters*, 84(21), 4316-4318.
[27] Kumar, D. H., Patel, H.E., Kumar, V.R.R., Sundararajan, T., and Das, S.K., 2004, "Model for Heat Conduction in Nanofluids," *Physical Review Letters*, 93(14), 144301.
[28] Prasher, R., Bhattacharya, P., Phelan, P.E., 2005, "Thermal Conductivity of Nanoscale Colloidal Solution (Nanofluids)," *Physical Review Letters*, 94(2), 025901.
[29] Prasher, R., Phelan, P.E., and Bhattacharya, P., 2006, "Effect of Aggregation Kinetics on the Thermal Conductivity of Nanoscale Colloidal Solutions (Nanofluids)," *Nano Letters*, 6(7), 1529-1534.
[30] Patel, H.E., Sundararajan, T., and Das, S.K., 2008, "A Cell Model Approach for Thermal Conductivity of Nanofluids," *Journal of Nanoparticle Research*, 10(1), 87-97.
[31] Acrivos, A., and Taylor, T.D., 1962, "Heat and Mass Transfer from Single Spheres in Stokes Flows," *Physics of Fluids*, 5(4), 387-394.
[32] Keblinski, P., Eastman, J.A., Cahill, D.G., 2005, "Nanofluids for Thermal Transport," *Materialstoday*, 8, 36-44.
[33] Nan, C.W., Birringer, R., Clarke, D.R., and Gleiter, H., 1997, "Effective Thermal Conductivity of Particulate Composites with Interfacial Thermal Resistance," *Journal of Applied Physics*, 81(10), 6692-6699.
[34] Eapen, J., Williams, W.C., Buongiorno, J., Hu, L., Yip, S., 2007, "Mean-Field versus Microconvection Effects in Nanofluid Thermal Conduction," *Physical Review Letters*, 99(9), 095901.
[35] Carey, V.P., 1999, *Statistical Thermodynamics and Microscale Thermophysics*, Cambridge University Press, New York.
[36] Schleiermacher, A., 1888, "Über die wärmeleitung der gase," *Annalen der Physik* 270(8A), 623-646 (in German).
[37] Van der Held, E.F.M., and Van Drunen, F.G., 1949, "A Method of Measuring the Thermal Conductivity of Liquids," *Physica A*, 15(1), 865-881.
[38] Roder, H.M., 1981, "A Transient Hot-Wire Thermal Conductivity Apparatus for Fluids," *Journal of Research of the National Bureau of Standards*, 86(5), 457-483.
[39] Nagasaka, Y., and Nagashima, A., 1981, "Absolute Measurement of The Thermal Conductivity of Electrically Conducting Liquids by The Transient Hot-Wire Method," *Journal of Physics E: Scientific Instruments*, 14(12), 1435-1440.
[40] Carslaw, H.S., and Jaeger, J.C., 1986, *Conduction of Heat in Solids*, 2nd ed., Oxford University Press, Oxford.
[41] Chon, C.H., 2006, The Thermophysical Role of Nanoparticles in Nanofluidic Heat and Mass Transport, Ph.D. dissertation, University of Tennessee, Knoxville, TN.
[42] Vincenti, W. G., Kruger, C. H., 1965, *Introduction to Physical Gas Dynamics*, Krieger, New York.
[43] Fox, R.W., McDonald, A.T., Pritchard, P.J., 2004, *Introduction to Fluid Mechanics*, 6th ed., Wiley & Sons, New York.
[44] Einstein, A., 1956, *Investigation on the Theory of Brownian Movement*, Dover, New York.
[45] Kihm, K. D., Chon, C. H., Lee, J. S., Choi, S. U. S., 2011, "A New Heat Propagation Velocity Prevails over Brownian Particle Velocities in Determining the Thermal Conductivities of Nanofluids," *Nanoscale Research Letters*, 6(1), 361.

[46] Bird, R.B., Stewart, W.E., and Lightfoot, E.N., 2002, *Transport Phenomena*, 2nd ed., Wiley & Sons, New York.
[47] Keblinski, P., Phillpot, S.R., Choi, S.U.S., and Eastman, J.A., 2002, "Mechanisms of Heat Flow in Suspensions of Nano-sized Particles (Nanofluids)," *International Journal of Heat and Mass Transfer*, 45(4), 855-863.
[48] Glasstone, S., Laidler, K.J., Eyrinig, H., 1941, *Theory of Rate Processes*, McGraw-Hill, New York.
[49] Kincaid, J.F., Eyring, H., Stearn, A.E., 1941, "The Theory of Absolute Reaction Rates and Its Application to Viscosity and Diffusion in the Liquid State," *Chemical Reviews*, 28, 301-365.
[50] Digilov, R.M., and Reiner, M., 204, "Trouton's Rule for the Law of Corresponding States," *European Journal of Physics*, 25(1), 15-22.
[51] Young, H.D., and Freedman, R.A., 1996, *University Physics*, 9th ed., Addison-Wesley, Massachusetts.
[52] Balandin, A.A., 2002, "Nanoscale Thermal Management," *IEEE Potentials*, 21(1), 11-15.
[53] Pasquini, L., Barla, A., Chumakov, A.I., Leupold, O., Rüffer, R., Deriu, A., and Bonetti, E., 2002, "Size and oxidation effects on the vibrational properties of nanocrystalline α-Fe," *Physical Review B*, 66(7), 073410.
[54] Wilson, O.M., Hu, X., Cahill, D.G., Braun, P.V., 2002, "Cooloidal Metal Particles as Probes of Nanoscale Thermal Transport In Fluids," *Physical Review B*, 66(22), 224301.
[55] Ge, Z., Cahill, D.G., Braun, P.V., 2004, "AuPd Metal Nanoparticles as Probes of Nanoscale Thermal Transport in Aqueous Solution," *Journal of Physical Chemistry B*, 108(49), 18870-18875.

In: Nanofluids
Editor: Yuwen Zhang
ISBN: 978-1-62618-165-6


*Chapter 5*

# MAGNETIC NANOFLUIDS: SYNTHESIS, PROPERTIES AND APPLICATIONS

***G. P. Singh*[1] *and S. Ram*[2,*]**

[1]Centre for *Nanotechnology,* Central University of Jharkhand, Ranchi, Jharkhand, India
[2]Materials Science Centre, Indian Institute of Technology, Kharagpur, India

## ABSTRACT

Innovative designs of different kinds of stable ferrofluids (magnetic or ferroelectric nanoparticles (NPs) dispersed in a suitable carrier liquid) are described in this chapter with well-known chemical synthesis routes, and basic physics underlying behind tailoring functional properties for medicals, biosensors and other applications. Common nanostructured flocculates described in this chapter include magnetic oxides such as $\gamma$-$Fe_2O_3$, $Fe_3O_4$ and spinel ferrites $AB_2O_4$ or perovskites $ABO_3$, A = a divalent cation and B = a tri or tetravalent cation in general. Surface modified NPs dispersed and immobilized in different polymer and ionic fluids give a road map of understanding the formation and properties of different ferrofluids with stable properties at elevated temperatures for applications. The most fascinating properties of a practically useful ferrofluid are its stability and rheology with self-adjustable magneto-viscosity and magnetic/ferroelectric features upon perturbations at moderate magnetic or electric fields. Applications of ferrofluids primarily depend on the basic properties such as ability to manipulate the parent features subject to an applied perturbation field (such as magnetic, electric, optical, etc.), rheology, absorption capacity of electromagnetic energy at selective frequencies and heating-up, thermal conductivity, magnetocalory, biological activities, ferroelectrics, light-energy transfer, magnetic switching, and other perturbation sensitive features. Those cover diversified fields such as chemical and biochemical separations, magnetic target drug delivery, magnetic resonance imaging contrast agents, hyperthermia therapy, disease diagnosis, chromatography, electrochemical biosensors, optical detection and luminescence detection, heat sinker and dampers, multicolored pigments, magnetic switches and valves, phosphors, and many more in the time to come in the near future.

---

[*] Corresponding author: sram@matsc.iitkgp.ernet.in.

# 1. INTRODUCTION

## 1.1. Historical Background of Nanofluids

In the early part of the $20^{th}$ century, the efforts of the scientists were put into dealing with diverse phenomena of the matter. A conceptual paradigm of describing the concept of an ordered solid matter through a dense periodic lattice of atoms has been spectacularly successful in recitation pertinent properties and continued to underlie much of the ongoing research on developing novel materials for possible applications [1-6]. For instance, in the 1960s, a particular class of technologically useful materials appeared, named "soft materials" which is also known as a "*magnetic fluid*" or a "*ferrofluid*" in general, whose properties and behaviors are different to those of the conventional solids in many respects. This class of the materials led to a tremendous revolution in evolution of the condensed matter science from the solid state science. Usually, the soft materials are characterized by the certain facts like effortlessly mechanical deformation in the presence of an applied external parameter such as stress, an electromagnetic irradiation, or a gravitational field, etc. that may include fluids, liquid crystals, polymers, emulsions and colloids [5, 6]. With the high sensitivity and actuation, the applications of the soft materials are growing rapidly with versatile value added industrial interests. They are considered to be "smart materials" of using as soft building blocks in devising structural tools and devices for specific applications.

In the meantime, another pioneering progress achieved during this cutting age of the science and technology was the sprouting of the nanosciecne and technology which deal with a more precise control over the chemical and structural perfections in a material having self-confined dimension, usually at a scale of 1-100 nm depending upon its shape and surface topology [7-13]. Further, dispersion of such small particles, called nanoparticles (NPs), into an ionic or a polymer liquid leads to bridge an interface that tailors new properties in comparison to those available in the two structural parts. Such materials, usually called "nanofluids", have value added commercial impacts in cosmetics, camouflage, medicine, biosensors, heat-transfer, automotives, inkjet printing, or color dyes. Probably, it is a right time to take a glimpse at this new class of hybrid materials. In a case, such materials possesses magnetic and/or ferroelectric properties devise a smart magnetic fluid, called a "ferrofluid" in general, and that has outsmarted many other materials and heralded in an entirely new rode map in fabrication of many more emerging technologies, especially in the area of biomedicines, magnetic mirrors, spin-valves and different kinds of sensors.

The nature is full of natural nanofluids, like blood, a complex biological nanofluid where different NPs (at a molecular level) accomplish different functions in a leaving creature. Many natural processes in biosphere and atmosphere cover a wide spectrum of NPs with different kinds of the fluids. Many mining and manufacturing processes leave waste products which consist of NPs embedded selective fluids. A nomenclature "nanofluid" is originated from the concept of a colloid but in a real meaning it is marked different from conventional nanocolloids. As a separate discipline of the synthetic chemistry or materials science, the colloidal science was introduced as early as 1861 by the Scottish scientist Thomas Graham [14]. A colloid is a type of a chemical mixture that appears in a solution but it is actually a chemical mixture of a dispersed phase in a dispersion medium of a continuous phase. People extended this concept to nanocolloids in case the dispersed phase limits its size in a nanometer regime. On the other hand, a "nanofluid" is an engineered colloidal suspension of NPs in a base fluid, viz., an aqueous or nonaqueous solution, or an organic liquid.

As demonstrated in Figure 1(a), a nanofluid is a smart chemical dispersion of NPs at a much finer scale in an effectively dilute and rheological fluid with a stable chemical bonding of the NPs with the surrounding. Individual NPs represent single magnetic domains which are embedded and grafted in a molecular bed (a surfactant or an immobilizer) so that they behave to be locally immobilized immersing in the base liquid of a carrier. In Figure 1(a), the yellow balls represent NPs of a given material, the light blue background shows the base fluid, and the layered circles around the yellow balls represent the surface layers at the solid-liquid interfaces. In general, average size in the basic structure in a ferrofluid varies at a scale of a few nanometers (one nanometer is billionth of a meter, or one millionth the size of a pinhead, or thousandth the length of a typical bacterium, or the length of a few tens of atoms). Most of technological ferrofluids often comprise 5% magnetic NPs and 10% surfactant with the rest 85% in volume-wise of the carriers [15-17]. The common surfactants used in preparing commercial ferrofluids are oelic acid, tetramethyl ammonium hydroxide, citric acid, or soy lecithin. In a nanofluid, the size and morphology of NPs as well as the nature of the surface-interfaces in the individual particles devise unique characteristics of greatly enhanced Gibbs free-energy (ΔG), momentum and mass transfer, reduced tendency for sedimentation and erosion of the NPs surfaces, as well as largely improved surface enhanced properties. Till a few years back, the term "nanofluid" was being used basically for traditional heat-transfer fluids in which NPs of selective materials were dispersed in specific kinds of liquids in order to increase average values of the thermal conductivity and the heat-transfer rate.

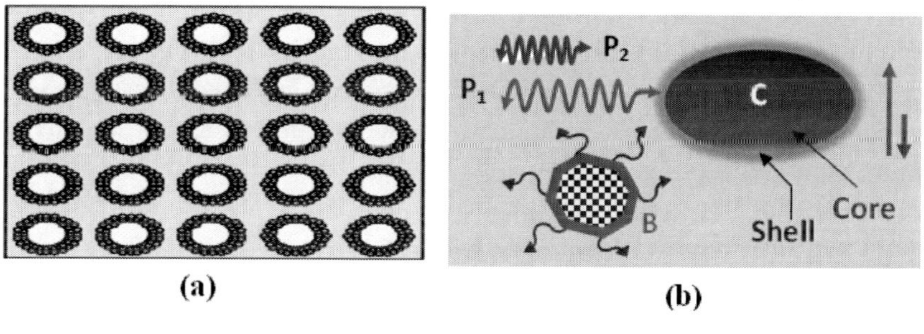


Figure 1. (a) A model of magnetic nanofluid consisting of NPs encapsulated in a surface immobilizer in a core-shell structure and (b) a core C with a bonded surface layer (shell) in a specific structure B.

In a model structure in Figure 1(b), the NPs in a nanofluid embedded in a bonded surface layer which bound them rigidly in a way that they stay separated apart in a distinct core-shell structure. A core C bonds a shell B in a specific structure through the surface atoms in the two gradients with a stable interface. The surface interface plays a crucial role in localizing the charge carriers, spins, phonons (also photons if available in the effect) in it in a correlated spectrum of the different phenomena in a kind of a hybrid composite system. In a simple case, the dynamics of the core and shell can be described in terms of two different mechanical and/or electromagnetic waves $P_1$ and $P_2$ that determine how they exchange couple each other. The two waves mix-up and ultimately respond collectively to the effects of an applied parameter in an application. In minimizing the total Gibbs free-energy, the spins in the shell order opposite to those in the core (at magnetic field H = 0). As a result, each entity can be treated as a tiny magnet, or a capacitor, depending on its chemistry, local structure, and effective dimension. Obviously, a nanofluid is not limited to devise merely a nanofluid device but it rather opens a new era to synthesize and fabricate novel ferroelectric and optical nanofluids and derivatives (including films and surface coatings) for versatile applications.

In a rheological nanofluid, an advantage with a stable surfactant is that it restricts the NPs (flocculates) not to agglomerate even when a strong magnetic field gradient is applied either during measuring the properties or applications. The selection of a surfactant capable to form a stable nanofluid can be made on the basis of its structure and properties as follows.

1) It should bond and/or adhere to the carrier type rigidly,
2) It shod be able to disperse in a base liquid of the carrier,
3) It should be able to overcome the attractive van der Waals and magnetic forces between NPs as the flocculates,
4) It should not be volatile, and
5) It should be biocompatible.

Other than these five general points, the thermal stability and thermal conductivity of a ferrofluid are highly important parameters for the choice of the surfactants on the case to case basis as per the requirements in specific devices [17-19]. Thermodynamically, as a solute behavior in a dilute medium, the NPs diffuse and interact with solvent molecules or additives if any used to modulate rheology and other functional properties in a product, a diagnostic tool, or a device. Further, in the presence of magnetic gradient fields, such nanofluids get viscosity enhanced on an apparently increased density. This property is used in damper applications [17, 18]. In fact, a ferrofluid hardly displays a magnetic hysteresis loop with significant coercivity ($H_c$) and remnant magnetization ($M_r/M_s$, with $M_s$ the saturation magnetization) even if the ferromagnetic NPs are dispersed in a small dosage below a critical value in an adequate liquid. Traditionally, such magnetic NPs are described as "superparamagnetic" entities due to pretty high low-field magnetic susceptibility [15-19] by a few orders of magnitude over the virgin paramagnetic state. Effective size, shape and surface topology in a kind of NPs are known to tailor the surface magnetism, hysteresis loop, and other physical and chemical behaviors in an assembly of randomly oriented magnetic dipoles before a magnetic field is applied. As a result, using the microstructure dependent features of the NPs united on a proper surface coating (of a rigidly bonded surface layer with a surfactant) in order to fulfill a colloidal thermal stability of a ferrofluid, we can tailor selective properties in an assembly of such magnetic NPs dispersed in a base liquid of nonpolar and polar carriers. Long-term thermal stability of concentrated ferrofluids in strong magnetic fields, e.g., in rotating seals or bearings imposes severe necessities on dispersion/stabilization of magnetic NPs in different organic carriers [18, 19]. Water-based ferrofluids are a special category of polar magnetic fluids with discerning features of particle interactions and assembling processes.

Ferrofluids consisting of several kinds of magnetic NPs such as $Fe_3O_4$, $\gamma$-$Fe_2O_3$, FePt [20-26], FePd [27], CoPt [28, 29], CoO [30], and $CoFe_2O_4$ [31], MnPt [32], Gd-based complexes [33-37], $CrO_2$ [38], and their inorganic-organic hybrid nanocomposites [39-45] have been developed for different applications. Amongst them, $Fe_3O_4$ magnetite and $\gamma$-$Fe_2O_3$ maghemite have great potential for various biomedicals, rotating seals, magnetic mirrors, and different kinds of sensor applications. Control of the surface chemistry of the magnetic NPs is a big challenge to devise nontoxic ferrofluids for bio-related applications. For examples, pristine shaped NPs tend to coalesce into effectively too large assemblies, because of their large contact surface area and strong dipole-dipole magnetic interactions, and that ultimately stifles required superparamagnetic properties. A coalescence of the NPs can be minimized by a surface modification of them by using a surface coating of an adequate material (which may be a metallic, a nonmetallic, or a polymeric ligand). A bonded surface layer if grafted on such NPs not only inhibits their coalescence in a stable nanofluid, but also improves their solubility in the liquid phase, biocompatibility, nonspecific adsorption to cells, and bioconjugation. Several approaches to surface modifying NPs by using small molecules including biomolecules have been developed in devising water-soluble NPs. Two well-known approaches in use for surface modifying NPs in a nanofluid are as follows.

***Post Reaction with a Soluble Ligand***

In this approach, separately prepared NPs by a suitable method are chemical treated with a soluble ligand in a base liquid, which can be an ionic or a polymer solution. As described in detail elsewhere, it involves direct adsorption [46-50], addition of a second layer [51-53], a ligand exchange [35, 54-57], a functional ceramic coating [58-62], or an ionic inclusion [63].

***In Situ Reaction with a Stabilizer***

In this process, formation of the NPs and a surface passivation reaction of the later are carried out in-situ in a dilute solution at an elevated temperature. Reactive NPs with nascent surfaces readily react with a stabilizer from the surrounding so that a stable surface layer is obtained easily on the growing NPs. Most common stabilizing ligands used for this purpose include d-mannose [64], 2-pyrrolidone [65], and poly(ethylene oxide) diacids (HOOC-PEG-COOH, with PEG = polyethylene glycol) [66], and dextran (Dex) [67], etc. The author's group [68-74] has explored poly(vinyl alcohol) (PVA), poly(vinyl pyrrolidone) (PVP), poly(vinylidene fluoride) ($PVF_2$) in devising series of optical nanofluids with noble metal NPs in aqueous as well as nonaqueous media. In these examples, the polymer molecules PVA, PVP, as well as $PVF_2$ serve both a reducing agent (to deduce the metal of NPs from a metal salt) as well as a surface stabilizer in a situ-reaction in a hot condition with ultrasonic irradiation. A combination of ultrasonic irradiation and microwave heating is highly capable in preparing stable nanofluids with surface stabilized metal and metal oxide NPs.

## 1.2. Core-Shell Structure

A modification of surfaces in a nanostructure includes surface engineering morphology and surface structure in the basic geometrical entities. As an illustration, Figure 2(a) displays how a bonded interface-surface layer on model NPs mediates to bond them in small groups through macroscopic interaction between the NPs in a rheological liquid in the form of a nanofluid. The NPs encaged singles in a polymer in (a) a core-shell structure can assemble in small assemblies by chemical bonding through the polymer molecules in the forms of (b, c) a single or (d) a multicore. A polymer layer which is not grown so thick intimately to the core in a model assembly (b) is grown distinctly in a prolonged reaction in a model assembly (c).

A single shell displayed on the NPs in the assembly (b) is grown into a multishell structure in a due reaction with polymer molecules in the model assembly (c). As also proposed by Horak et al. for encapsulated magnetic NPs in several polymer molecules for a biological separation [75], NPs with a bonded polymer surface layer in a multishell can grow slowly in a number of morphologies. Known morphologies from the microscopic images include a "raspberry" morphology (c, d) and a brush or a hair like complex morphology (e) in strong polymer chains attach to a magnetic core strongly and ultimately diverge in bridging to flocculates present around within a range of the interactions.

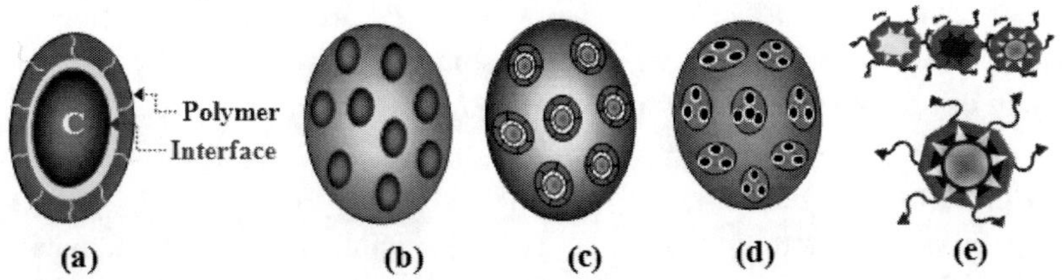


Figure 2. Morphologies in surface modified NPs with a polymer; (a) a core-shell, (b) a multicore-shell assembly, (c, d) a hetero-coagulated raspberry with single NPs (c) and assemblies (d), and (e) a brush-like hybrid structure.

A polymer molecule in bonds to magnetic NPs with more than one functional moiety in a multidentate ligand not only gives a colloidal stability to flocculates of them but rather lead to tailor magnetic, optical, electronic, and biological properties for wide spread applications. In particular, biological applications include magnetic resonance imaging (MRI), targeted drug delivery, biological separation, hyperthermia, protein immobilization, and biosensors [17-19]. Most of the synthetic and bio-based polymer molecules being highly transparent in the visible range of the electromagnetic spectrum hardly hamper the biological processes. As described schematically in the models in Fig 2(d, e), a polymer medium when cross-links NPs dispersed in a network like structure can breed the mechanical toughness of the resultant structure. Further, polymer inspired soft assemblies and cross-linked gels in pretty small effective sizes, often called microgels or nanogels, help to encapsulate and immobilize therapeutics and imaging agents in for planed targeted delivery at a controlled rate. Biocompatible materials of the interest include NPs of nontoxic metals (e.g., titanium, copper, silver, gold, etc.) and simple or mixed transition metal oxides (including the cermets).

Low mechanical stability due to a poor miscibility (or chemical bonding) of the NPs (flocculates) in a base liquid of a solvent is an open technical problem in devising an engineering nanofluid. When the NPs are being dispersed in a dispersion medium (a polar solvent or an electrolyte), the nascent NPs surfaces in general charge through one or more of the following processes and/or mechanisms:

1) Preferential adsorption of ions from the medium,
2) Dissociation of the surface charged species,
3) Isomorphic substitution of the ions,
4) Accumulation or depletion of electrons at the nascent surfaces, and
5) Physical adsorption of charged species onto the surfaces.

The van der Waals force acting between the flocculates dispersed in a dilute medium of the interest in a ferrofluid is a weak characteristic force and it becomes significant only at a very short separation between the flocculates. The Brownian motion ensures the NPs colliding with one another all the time. Thus, a combination of the van der Waals attraction force and the Brownian motion would render either a desired dispersion of the NPs or adversely formation of an assembly from such NPs at the outset of the first kinds of the predominant interactions. Because of these reasons, even isolated metal or oxide NPs tend to be fairly unstable in a liquid medium. Hence, a special surface engineering has to be taken to inhibit unwanted coalescence and precipitation of such NPs. As briefed above, a common strategy is of using a suitable stabilizer during preparing a ferrofluid following in situ method. It suppresses due particle coalescence and promotes the functionalized properties. In general, the stabilization of NPs of a material can be done in two ways of electrostatic and steric stabilizations.

The electrostatic stabilization, a kinetic stabilization method, is described in lyophobic colloids by the DLVO theory, named after the work of Derjaguin, Landau, Verwey, and Overbeek [76, 77]. Though this theory has been proposed initially for infinitely flat surfaces with no variation in the surface charge density it applies well for nonplanar surfaces. According to it, a colloid contains an electrical double layer on the NPs surfaces. In two NPs stay separated apart by a distance between the surfaces larger than the combined thickness of two electric double layers in the NPs, there is no particle-particle interaction in lack of overlap of the diffusion double layers. This is demonstrated with model diagrams in a book written by Derjaguin et al., and Verwey et al. [76, 77]. In a case, the two electric double layers overlap on the NPs move closer, a repulsive force is developed. With the gap reduces, the repulsion increases to a maxima at the distance between the NPs surfaces equals to the distance between the repulsive barrier and the surface. A major limitation with an electrostatic stabilization is that it is difficult to apply to a multiple phase system. Since different solids develop different surface charges and electric potentials it is not easy to disperse them evenly in a nanofluid.

Such problem does not arise in a steric stabilization, a thermodynamic method, in which NPs are easily redispersible in a nanofluid. In general, polymers are used as a particle stabilizer in synthesizing nanofluids following a soft chemical route. Compared to the electrostatic stabilization, a polymer stabilization used in this method offers an advantage of the synthesis of NPs, particularly when a fairly narrow particle size distribution is required for applications. A polymer layer which bonds on the reactive nascent surfaces in growing NPs serves as a diffusion barrier layer to the growing structure, resulting in a diffusion-limited growth process in the subsequent growth step of the sample. Interactions between the polymer and the solid surface in a growing structure are limited to the adsorption process of the reactive polymer molecules onto the reactive surfaces in a reaction medium. As observed in selective systems [68-74, 78], the adsorption can be either by forming chemical bonds between the surface entities on a growing structure and the polymer molecules, or just by a physical adsorption on the surfaces. There is no restriction whether one or multiple bonds form between the two kinds of the entities of the NPs and the polymer molecules. Depending on the interactions in the polymer and the nascent NPs in a nanofluid, a polymer chosen in such reactions can be grouped into (i) a anchored polymer (which irreversibly bonds to the solid surface in a growing particle by one end only, and typically are diblock polymers) and

(ii) an adsorbing polymer (which adsorbs weakly at the random points along the polymer backbone).

### 1.3. Distribution of Up-to-Date Information on Ferrofluids

In this investigation, our main focus is to review recent advances in design of different kinds of the ferrofluids with established synthesis routes, and basic physics underlying behind tailored properties for different applications, especially pertaining to biomedical diagnosis, treatments, and medical devices. Common examples of flocculates are limited to magnetic NPs such as $\gamma$-$Fe_2O_3$, $Fe_3O_4$ and derivatives of spinel ferrites and orthoferrites $ABO_3$, which in general formulate a combination of a divalent A and a tetravalent cation B. Surface modified NPs dispersed in different polymer and ionic fluids give a road map of understanding the chemistry and physics of formation and properties of typical ferrofluids.

## 2. Types of Ferrofluids and Assemblies

Magnetic and rheological behaviors of a ferrofluid are determined from how it responses in an external magnetic or electromagnetic field. A flocculate being used in effectively small dosages devise functional properties on controlled interactions through the receptive medium and quite differently those from usual magnetorheological fluids. Subject to a magnetic field, a conventional magnetorheological fluid rises-up in viscosity enormously so as it freezes like a solid, while a rather dilute ferrofluid containing lesser flocculates (NPs) can retain its fluidity even in sufficiently large magnetic fields ($H \geq 10$ kOe) enough to reach the magnetic saturation. Ferrofluids are usually optically isotropic in nature but upon applying a critical magnetic field can exhibit an induced birefringence [79]. Similar birefringences also arise when a ferrofluid wets a substrate in thin ferrofluid layers [80]. A surface layer used to coat magnetic NPs affect the birefringences. Depending on ionic features of the surface coating, the ferrofluids can be classified into two major groups (i) surfacted ferrofluid and (ii) ionic ferrofluid.

### 2.1. Surfacted Ferrofluid

In this kind of surfacted ferrofluids, the material which is used in coating over the NPs behaves as a stable surfactant like amphiphilic molecules, oleic acid, aerosol sodium di-2-ethylhexyl-sulfosuccinate, etc. As a covalent molecular species, the surfactant bonds to the NPs so as they stay separated apart in a reasonably dispersed structure in a susceptible base liquid used. They exert a steric repulsion between the core NPs and acts as a physical barrier to keep a bonded particle distance apart from neighboring NPs in a stable sample. For instance, Figure (3) shows how a model magnetic particle bonds a surfactant on it in (a) a single layer and (b) a double layer, feasible in a polymer based and a water based ferrofluid respectively. In a nonpolar medium of an oily liquid, often a single layer of a surfactant breeds an external hydrophobic layer so that a polar head of the surfactant species is bonded to the particle surface and the carbonic chain stands in an intimate contact to the fluid carriers. In a realistic example of a nanofluid, such surface protected NPs can assemble by bonding further through bonded surface species on the NPs in different shapes and sizes. Four simple structures are proposed in Figure 3(c to f). The process follows the basic size and shape in the NPs (as the building blocks) and the way the host shares and propagates chemical bonding to cross-link the later via it in an extended co-branched structure. Ultimately, a cross-linked equilibrium structure attains and determines the rheology, mechanical stiffness, magnetic properties, optical properties, and heat-transfer; all these depend on a correlated structure that exchanges energy so as it propagates through the sample. Phenomenologically, the optimal features can arise only on the optimal structures (can be different for the different properties) between the NPs, assemblies and average host grafted through the surfactant species.

Figure 3. Grafting surfactant species on a nanoparticle in (a) a single and (b) a double bonded overlayer in a dilute ferrofluid, with possible assemblies in figures (c to f). ● Head group in a surface layer.

A technical advantage with a polar liquid (or solution) used as a base medium in a nanofluid is that it builds-up a hydrophilic layer around the NPs with a bonded surface layer so that the polar heads of a surfactant species can be cationic, anionic, or a nonionic in general. These geometrical entities carry over localized electrostatic charges in the forms of the basic building blocks of an assembly (flocculate) in a nanofluid. The sample thus owns a significant in-built average dielectric field that affects the magnetic and other functional properties out of a magnetodielectric nanofluid. As described at several places in the literatures [81-83], such ferrofluids are commercially available for use in different forms in different kinds of devices. Further, the magnetodielectric features can be retained on casting the sample in the forms of films or other engineering shapes for other applications.

## 2.2. Ionic Ferrofluids

In an ionic ferrofluid, the shell in the individual NPs contains ionic species in the forms of self-localized charges, magnetic spins, photons and phonons in general of a multifunctional material. Concurrence of these four kinds of the properties can make the sample a highly sensitive probe for pertinent applications. For illustration, Figure 4(a, b) represents two typical examples with an acid and an alkaline nature of the surface species in the core-shell NPs that share the charge distribution over the NPs surfaces embedded within the carriers. As a result, an induced ionic cloud builds-up on the NPs surfaces so as it maintains the ionic and mechanical equilibria of a stable sample. This is a necessary condition of bridging a colloidal system stable so as it holds the NPs suspended within the carriers. Magnetic NPs mostly $\gamma$-

$Fe_2O_3$ and $AB_2O_4$ ferrites ($A$ = Mn, Co, Zn, Cu, Ni, and B = Fe) are used in preparing stable ferrofluids. An acid-alkaline reaction with the NPs, prepared via a chemical precipitation method, gives a sample of electrically charged surfaces of the NPs with a bonded surface layer. Ferrofluids with $Fe_3O_4$ NPs present pH dependent stability in alkaline and acidic hosts [84]. Low polarizing cations in alkaline host (pH > 9), or low polarizing anions in an acidic host (pH < 5), immobilize the flocculates. Flocculates coalesce on charges neutralize at intermediate pH. As a result, an aqueous nanofluid hardly stands stable in absence of a surfactant. Depeyrot et al. [85] studied magnetic and magneto-optical properties in Cu and Zn ferrite nanofluids, synthesized via an electric double layered magnetic fluid (EDL-MF) using a static magneto optical birefringence. Another example, $NiFe_2O_4$ NPs, co-precipitated from a precursor solution using low-polarizable counterions, were dispersed in a polar host like water so as they adsorb amphoteric hydroxyl groups and set-up electrostatic particle-particle repulsion. The repulsion protects them from coalescence under magnetic and van der Waals interactions [86, 87].

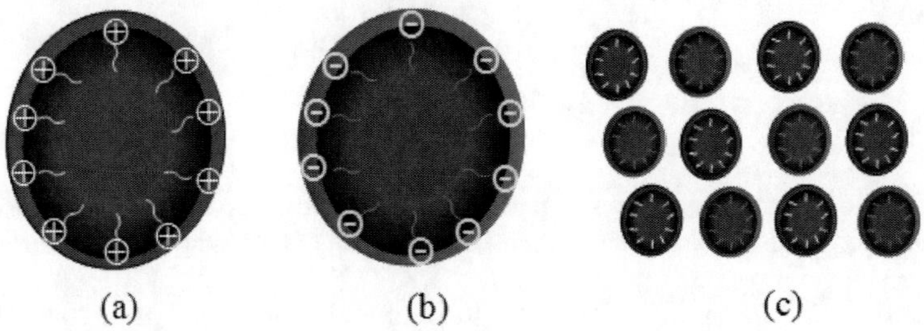

Figure 4. (a) Positively and (b) negatively charged surface distributions over magnetic NPs with an acidic and an alkaline nature in a ferrofluid, (c) a model distribution of two kinds of the entities in a mixed sample.

The two kinds of the common ferrofluids described in Figure 4(a, b) often use an aqueous complex with either a very small pH of ~2, or an effectively very large value ~12. The pH value controls the average sign of the surface charges on the NPs. At small pH of the solution under 7 (acidic), the ferrofluid has positively charged NPs, while at larger pH above 7 (basic), the ferrofluid has negatively charged NPs. The pH governs surface charge density ($\sigma_{sc}$) of NPs in a ferrofluid. A $\sigma_{sc}$-value of the order of $20\mu C/cm^2$ is found in a ferrofluid consisting of $\gamma$-$Fe_2O_3$ NPs with hydroxolligands in water [87], i.e., tailored roughly by an order of magnitude from a value $2\mu C/cm^2$ shown in a $NiFe_2O_4$ ferrofluid [88]. The $\sigma_{sc}$-value is measured at the point of a zero charge (defined as the value of the pH at which the NPs are not electrically charged, viz., the ferrofluid precipitates). Further, it is possible to devise a mixed solution of two kinds of the nanofluids as demonstrated in Figure 4(c). In this case, the electrostatic interactions between the oppositely charged entities can tune a finer dispersion of them via the carriers and in turn closer exchange interactions and multifunctional properties. Magnetic and dielectric entities when combine in a single system in this way give rise to a class of "magnetoelectric nanofluids". In principle, ferromagnetic and ferroelectric features favor each other.

## 3. Stability and Flocculate Dimension in Ferrofluid

The stability of a ferrofluid with and without applying a field H determines its functions in a particular application. Besides the sample magnetization, it depends on kinetic energy in the NPs and on the balance between attractive (van der Waals and dipole-dipole) and repulsive (steric and electrostatic) interactions between the NPs through a dispersive rheological medium used to uphold the NPs of a flocculate and to render their mobility in a nanofluid. A typical nanofluid comprises a numerical density of NPs of ~ $10^{23}$ $m^{-3}$. Dispersed NPs in an ideal ferrofluid have an effective hydrocyanic diameter ($D_h$), which relates to the thermal energy and dipole-dipole pair energy in a simple empirical relation [89],

$$D_h \leq \left( \frac{72 k_B T}{\pi \mu_o \mu^2} \right)^{1/3}, \tag{1}$$

where $k_B$, $T$, $\mu_o$ and $\mu$ are the Boltzmann's constant, the absolute temperature, the permeability of the free space and the magnetic moment respectively. For example, using the experimental values $\mu$ = 400 *kA/m*, known from a ferrofluid of $\gamma$-$Fe_2O_3$ NPs in the presence of hydroxoligands in water [89], in this Eq. 1 gives a value $D_h$ = 10 *nm*. The knowledge of the effective $D_h$ value is important in understanding magnetism and magnetization behavior of a ferrofluid and in turn its ability as a magnetic probe for various biological and other application.

### 3.1. Attractive Interactions in Flocculates

In a magnetic nanofluid, two major types of attractive interactions incur in between the magnetic NPs in part of the flocculates. They are normally known as the van der Waals-London and the magnetic dipole-dipole interactions. As described by Scholten in a reference book [90], the van der Waals-London interaction, $U_{Aw}$, between two particles of a model spherical shape of diameter $D$, separated by a distance $r$, is expressed by a model relation,

$$U_{AW} = -\frac{A}{6}\left[ \frac{2}{\alpha^2 - 4} + \frac{2}{\alpha^2} + \ln\left( \frac{\alpha^2 - 4}{\alpha^2} \right) \right], \tag{2}$$

where $\alpha = 2r/D$ and $A$ is the Hamaker constant (A ~ $10^{-19} J$ is estimated for spinel ferrite particles [89]). As it is well described in the literature [87-90], this is a short-range interaction and its value increases with increasing the effective particles size. The interaction energy between two magnetic dipoles, $\mu_1$ and $\mu_2$, separated by a distance $r_1$ and $r_2$, is given by:

$$U_{Ad} = \frac{\mu_o}{4\pi r^2}\left[ \mu_1 \cdot \mu_2 - 3\left( \mu_1 \cdot \frac{r_1}{r_2} \right)\left( \mu_2 \cdot \frac{r_1}{r_2} \right) \right], \tag{3}$$

where $r_1$ and $r_2$ are relative positions of the particles. In an ionic aqueous ferrofluid containing $MnFe_2O_4$ NPs ($D_h$ = 10.7 *nm)*, average magnetic moment of particles is about $10^{-19}Am^2$ (or $10^{-16}$ emu) per particle. A close matching with the experiment values confer validity of these mathematical relations in formulating the magnetic interactions between the flocculates in a magnetic fluid. The effect of a bonded surface layer grafted on the NPs in terms of its volume fraction and magnetic/dielectric features is yet to develop in a more realistic correlation in the microstructure, microscopic interactions and related properties.

## 3.2. Repulsive Interactions in Flocculates

As can be seen from a model in Figure 4(a, b), an ionic ferrofluid, which comprises the same kinds of the electronic charges on the NPs, experiences long-range electrostatic interactions between the NPs that give to repulsive interactions. Such interactions keep a colloidal stability of the sample very rigidly in a balance with the up-thrust that acts on the flocculates (NPs) in a liquid and the conceive forces between the entities. On the other hand, in the surfacted ferrofluids, there are steric repulsion forces of a short range nature. Here, let us limit our discussion on repulsive interactions in the case of an ideal ionic ferrofluid. As the bulk of this material is electronically neutral, there are counter-ions in the bulk of the suspension in order to compensate the net surface charged entities. Such counter ions are driven to the surfaces on the individual NPs and ultimately share the electrostatic repulsion. An estimation of the electrostatic repulsion between the charged NPs thus has to take into the account this complex distribution of the counter ions. In a double layer model [91], the first layer comprises the counter ions (called the Stern layer), of a thickness of a subnanometer, that involves the particle and the second diffuse layer lying over it. The two layers are separated by a Helmholtz plane. The diffuse layer is characterized by a Debye length, $l_D \equiv k^{-1}$, which is defined conventionally as the distance from the Helmholtz plane at which the electrostatic potential is 37% smaller than the value at the plane. As a result, following the pioneering work of Israelachvili on a ferrite based ferrofluid [91], the interaction between two electronically charged particles (spherical) of diameter *D*, separated by a distance *r*, is described by a simple relation which is expression as,

$$U_R = \frac{D\pi\sigma^2}{\varepsilon_o\varepsilon_r k^2}\exp\left[-k(r-D)\right], \tag{4}$$

where $\sigma_{sc} = \varepsilon_o\varepsilon_r k\psi_o$ is the surface charge density, $\varepsilon = \varepsilon_o\varepsilon_r$ is the dielectric permittivity of the flocculate carrier, and $\psi_0$ is the surface potential of the charged particle at the Helmholtz plane, with a typical value $\psi_0 \leq 50\ mV$. Hasmonay [88] has described steric repulsions in model examples in his Ph. D. Thesis. It turns out that this repulsion energy is linearly dependent on absolute temperature T. For a sample consisting of spherical NPs with a bonded surface layer of thickness $\delta$ and molecular density $\zeta$ *(*molecules per $nm^2$), one gets

$$\frac{U_{st}}{k_B T} = \frac{\pi D^2\zeta}{2}\left[2 - \frac{l+2}{t}\ln\left(\frac{1+t}{1+l/2}\right) - \frac{l}{t}\right], \tag{5}$$

where $l = 2s = D$ and $t = 2\delta/D$ with $s = r - D$ as the separation between the flocculates.

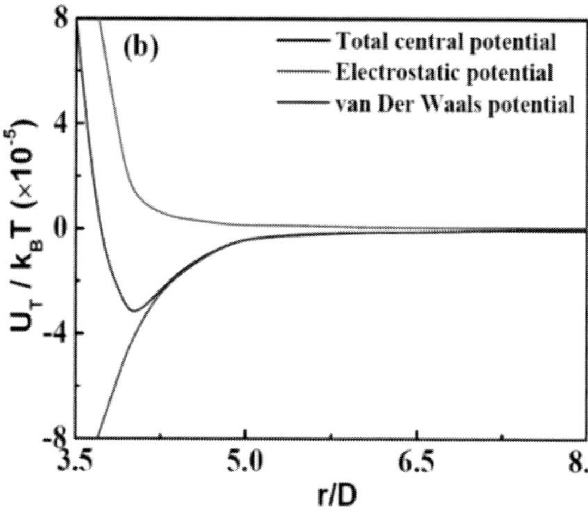


Figure 5. Interparticle central potential plotted against r/D value in the vicinity of (a) the cutoff potential and (b) the secondary minima in a ferrofluid [91].

### 3.3. Total Interaction Potential

A combination of the above three eqs. 2, 3 and 4 describes average behavior of the total interaction potential between flocculates in a ferrofluid $U_T$ as a function of average interparticle distance $r$. This is portrayed in Figure 5(a) for a typical ferrofluid. At short interparticle distances, $r/D = 1$, there is a cutoff net potential due to the hard-sphere interactions (not shown in the figure). Evidently, in increasing $r$, but still in the vicinity of $r/D \approx 1$, the force is dominated by the van der Waals attractions. Upon increasing the $r$-value further above $r/D \approx 1$, the shielded Coulomb repulsion dominates over the resultant value. This is the case in an ionic ferrofluid, or in an example of a prominent steric repulsion in a surfacted ferrofluid.

As plotted in Figure 5(b), still larger $r$-values trace out one secondary minimum at $r/D \approx 4$. The most stable ferrofluids are designed in a way that the interparticle average distance to the nearest neighbors is approximately equal to that corresponding to the secondary minima. Larger values of $\psi_0$ (possible at lower pH) lead to reasonably higher potential barriers and smaller values of the depth of the secondary minima. The data in Figures 5 were calculated with an arbitrary value $\psi_0 = 60\ mV$ (at pH~ 4.5).

## 4. Synthesis and Fabrication of Ferrofluids

In a ferrofluid, a career liquid bond to the flocculate NPs so that they behave to be surface stabilized but can flow along with it as a mobile fluid. Within the last one decade, special effort has been devoted to synthesis of magnetic NPs stabilized in various shapes in the forms of colloids in way that they devise new hybrid properties. Such magnetic NPs retain high specific surface area and high effective magnetization as required for applications of ion-exchange separation [92-98], drug delivery [99-103], MRI [104-109], targeted gene therapy, or biosensors [110-114]. A ferrofluid fabrication involves two main steps (a) synthesis of the NPs and (b) functionalizing the NPs in a nonpolar or a polar liquid. Well-established methods in use of synthesis of requisite magnetic NPs [19, 115-118] are as follows:

1) Co-precipitation
2) Reverse micelle
3) Seed-mediated growth
4) Solvothermal decomposition
5) Sol-gel chemistry
6) Ultrasound irradiation
7) Hydrothermal process
8) Microwave plasma synthesis
9) Laser induced pyrolysis, and
10) Controlled combustion of inorganic-organic complexes.

Here, we restrict our discussion on the first three methods which promptly control shape and size of magnetic NPs useful for ferrofluids as follows.

### 4.1. Synthesis of Shape Controlled Magnetic NPs

As a versatile method, the co-precipitation process is applied to produce magnetic oxide NPs of varied sizes, usually 3 to 20 nm, at moderate temperature from a solution precursor. For example, magnetite $Fe_3O_4$ can be produced in large scale by this process. A minor variation of the initial $Fe^{3+}$:$Fe^{2+}$ molar ratio from 2:1 to 1.5:1 in the presence of KOH, NaOH, or $NH_4OH$ tunes fine precipitates form a precursor solution. Part of the excess $Fe^{2+}$ converts to $Fe^{3+}$ during the processing so as it balances the stoichiometry. A hydroxide co-precipitates in an average $Fe_3O_4$ composition by a base solution [119]. An oxidation $Fe^{2+} \rightarrow Fe^{3+}$ causes merely a small decrease in magnetization (to utmost 10%) and it hardly affects colloidal stability of the magnetic NPs in a base carrier liquid in a ferrofluid. A technical difficulty is that this method is not able to control a homogeneous microstructure of the final sample.

Executing a better control over size/shape in magnetic NPs following a liquid precursor, a newer process known as "Reverse micelle method" is introduced. Numerous reports are available on preparing oxides as well as metals/alloys with practically homogeneous NPs with this method [117, 120-125]. In general, reverse micelles form of microemulsions with a dilute solution precursor wherein two immiscible liquids coexist with a stable microscopic interface. In principle, it is a three-component system comprising of a solvent, a metal salt, and a surfactant. A prerequisite metal salt or a complex is taken in a solution of an adequate concentration in a pertinent solvent followed by a surfactant is admixed with mechanical stirring (or ultrasonication) at an elevated temperature. For example, $Fe_3O_4$ NPs of ~5 nm or still smaller sizes are prepared as follows. In a small batch in a laboratory, a rheological precursor containing a salt solution (reddish color) of 2 mmol $Fe(acac)_3$, 10 mmol 1,2-hexadecanediol, 2 mmol of oleic acid, and 2 mmol oleylamine with 20 ml of phenyl is heated at 200°C for 30 min in a specific reactor in argon before raising the temperature gradually to

265°C and reflux further over another 30 min, forming fine slurries which are dispersed in the mother solution of a blackish-brown apparent color. After the reaction, the sample is cooled down to room temperature and ethanol is added in order to precipitate out $Fe_3O_4$ NPs, a characteristic blackish color, which could be separated by centrifuging.

A recovered sample in this way was dispersed in hexane in the presence of oleic acid and oleylamine, and centrifuged at a speed of 4500 rpm for 15 min to separate away any undispersed species. Finally, washing the later in ethanol gives $Fe_3O_4$ NPs (~4 nm size) of a powder free from byproduct impurities. A quite possible is that a surfactant used in this method is not compatible to carriers when forming a ferrofluid. In an easy way to circumvent this microscopic problem, the surfactant is either etched out from the NPs, and a compatible surfactant is deployed, or a new surfactant is coated over the parent surfactant in a kind of a two-surfactant molecular layers.

Surface stabilized magnetic NPs can be grown in tunable sizes/shapes by a seed mediated chemical growth in a liquid host under specific conditions. Being widely applied to grow nanostructural metals, oxides, or composites [120-122], this method appears a quite efficient in making size selective monodispersed NPs along with LaMer's method through fast supersaturated-burst nucleation [123] and Finke's method via slow, continuous nucleation and fast, autocatalytic surface growth [124]. In this process, pretty small $Fe_3O_4$ NPs prepared from reverse micelles, as described above, are used as a seed to grow-up to stipulated values with functional properties for pertinent applications. Besides the initial seed size/shape, rheology of the host with the reaction species present and dispersion of the seeds deliver a topotactic growth depending on the experiment conditions. As example, $Fe_3O_4$ NPs used as seeds could be grown easily in the 10-20 nm range depending upon the starting value of a few nanometers [125]. In a typical batch, when a 62 mg mass of 8 nm $Fe_3O_4$ NPs with 2 mmol of $Fe(acac)_3$, 2 mmol of oleic acid, and 2 mmol oleylamine, 10 mmol stearyl alcohol heated at 300 °C for 30 min in the procedure described above for 4 nm sample had grown to an average 12 nm $Fe_3O_4$ NPs. In another experiment, a 15 mg mass of 12 nm $Fe_3O_4$ seeds had grown to an average 16 nm upon heating in the same conditions. Figure 6(a-d) compares TEM images from $Fe_3O_4$ NPs prepared in different sizes from the seeds of different sizes. In all the four micrographs, the NPs are in round shaped discs of sharp sizes monodispersed in a core-shell structure.

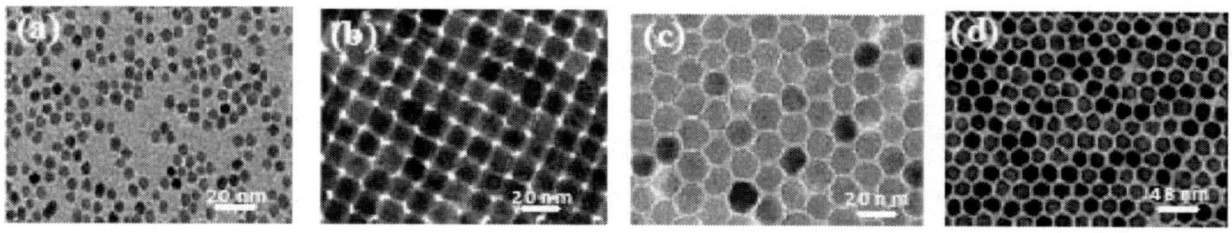


Figure 6. $Fe_3O_4$ NPs (discs) prepared in different sizes (a) 6 nm, (b) 10 nm, (c) 12 nm, and (d) 16 nm of average diameters [125].

## 4.2. Surface Functionalization of Magnetic NPs With Polymers

Surface functionalization of magnetic NPs is a process to enhance performance in known applications and to explore newer applications in different disciplines including the treatment of contaminated water, bioseparation, therapeutic index, controlled release and drug targeting due to increase specificity, evade the immune system, enhance their safety, selectivity and targetability, and improve in vivo efficacy and limit cytotoxicity [126-134]. Known surface modification processes for magnetic NPs include

1) Radical polymerization
2) Layer-by-layer surface polymerization
3) Self-assembly and self-association, and
4) Heterogeneous polymerization.

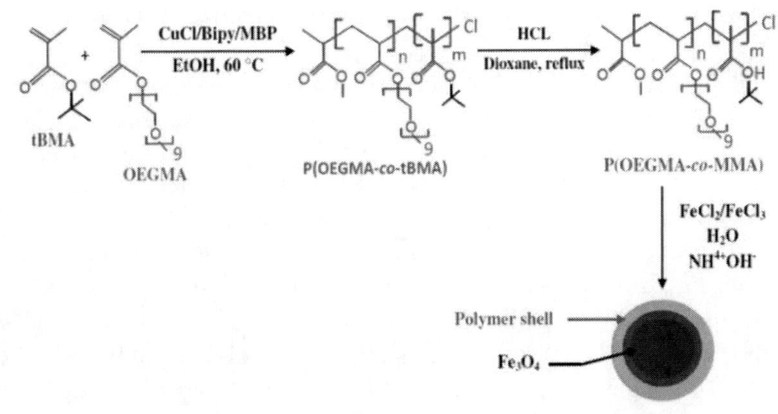


Figure 7. Schematic diagram of synthesis steps in functionalized $Fe_3O_4$ with a polymer POEOMA-b-PMAA layer [138].

They involve physical adsorption and/or addition of a second layer, leading to induce different morphologies of magnetic NPs encapsulated or bonded with a polymer as a surfactant, as demonstrated in Figure (2). The physical adsorption approach is achieved by adding a stabilizer (a copolymer), either during or after preparation of the NPs. A copolymer typically consists of two different blocks, viz., an anchoring block, which enables the adhesion to the surfaces of the NPs and the other a water-soluble block, which renders NPs dispersible in a carrier liquid and biocompatible direct modification with the polymer used. Now, let us describe the above four methods in detail with selective examples.

#### *4.2.1. Radical Polymerization Process*

This polymerization process is used in functionalizing magnetic NPs in a colloidal solution and in a core-shell structure by anchoring block co-polymers onto the NPs surfaces. It unites two processes of a controlled radical polymerization (CRP) and a reversible addition-fragmentation chain transfer (RAFT) polymerization. In the CRP method, a controlled block co-polymer is prepared of an average molecular weight ($M_w$) with a narrow distribution (<0.3 deviation) via atom radical transfer process [135-137]. For example, a POEOMA-b-poly(methacrylic acid) (POEOMA-b-PMAA) block co-polymer is produced by hydrolyzing poly(oligo(ethylene oxide) monomethyl ether methacrylate)-b-poly(t-butyl methacrylate) (POEOMA-b-PtBMA) in an acidic condition. This block co-polymer can be anchored on magnetic NPs through in-situ or ex-situ process. The PMAA block is anchored onto the NPs surfaces so as POEOMA block renders assembly in the surface coated NPs, a water dispersible and biocompatible material. Size of such core-shell NPs can be tuned in a 10-25 nm range by varying the copolymer concentration. Polymer surface stabilized magnetic NPs in this way have long-term colloidal stability useful for MRI contrast agents [138]. Figure 7 demonstrates formation of a model surface functionalized magnetic particle $Fe_3O_4$ with a block copolymer of a compatible surface layer in a typical core-shell structure.

The RAFT provides synthesis of a tri-block co-polymer, e.g., one consisting of poly(acrylic acid) (PAA), poly(N-isopropylacrylamide) (PNIPAM), and POEOMA blocks [139,140]. A block copolymer PAA-b-PNIPAM-b-POEOMA so obtained is post treated to the surface modified magnetic NPs such as $Fe_3O_4$ described above. The later perceives a voluminous change in response to the temperature and pH. For instance, $Fe_3O_4$ NPs treated with PAA41-b-PNIPAM150-b-POEOMA90 tri-block co-polymer decrease in size from 70 to 45 nm on cooling from 39 to 25°C. This quality is highly useful in hyperthermia [141]. Well-controlled poly(2-acetoacetoxyethyl methacrylate) (PAAEM)-b-POEOMA is also prepared by this process. Co-precipitating $Fe^{3+}$ and $Fe^{2+}$ in the presence of a PAAEM-b-POEOMA block copolymer gives surface stabilized $Fe_3O_4$ NPs with di-block copolymer, in which pendent acetoxy groups anchor to the NPs surfaces [142].

Other polymerization methods explored for preparing well controlled block polymers and functionalized magnetic NPs consist of pendent COOH or epoxy groups (anchoring block) that encapsulate the NPs. Examples include anionic polymerization and sequential photo-cross linking in polyisoprene-b-poly(2-cinnamoylethyl methacrylate)-b-PAA [143,144], or polycondensation in PEO-b-poly(COOH-containing urethane)-b-PEO [145]. Other examples are ring-opening polymerization and subsequent hydrolysis for PEO-b-poly(aspartic acid) [146] and ring-opening metathesis polymerization (ROMP) for di-block copolymer of bicycle {2,2,1} hept-5-ene-2-carboxylic acid xiranylmethyl ester [147]. Biopolymers are modified with functional anchoring groups to NPs for MR cancer imaging. They include dopamine-conjugated hyaluronic acid (HA) [148], polypeptide with a sequence of GGGGYSAYPDSVPMMSK (a targeting ligand for ovarian cancer cells) virus [149], dextran (Dex) [150], Dex derivatives modified with TAT-derived peptide [151], dopaminc-plus-human serum albumin [152], and dendrimers [153].

#### *4.2.2. Layer-by-Layer Surface Polymerization Process*

In a simple case, this process grafts a single anchoring group by a single layer polymerization from ATRP (atomic transfer radical polymerization) initiated species (halides or amphiphilic block copolymers by a second layer polymerization) on magnetic NPs, as demonstrated in Figure 2(e) with a model brush-like surface features. A single layer polymerization, which simply grafts an adhesive surface layer in a "grafting from" method, utilizes often a mono-dentate polymer with a single anchoring group at the end of polymer chain so as it bonds intimately on the magnetic NPs. It more strongly adheres to the NPs than a multi-dentate polymer with pendent anchoring groups in the CRP method. The process can be executed in two ways. The one approach deals with a physical adsorption of acid functionalized halides on the magnetic NPs. Functionalized halides include 3-chloropropionic acid [154], 2-bromo-2-methyl propionic acid [155], or 10-carboxydecanyl-2-bromo-2-methyl-thiopropanoate [156], which initiates the ATRP of styrene and OEOMA in bulk or in an organic solvent, yielding coated magnetic NPs with a hydrophobic styrene or a water-soluble OEOMA. In the other approach, a covalent network grafts on NPs as an immobilizer via silanization, e.g., 2-(4-chlorosulfonylphenyl) ethyltrichlorosilane for PMMA [157], [11-(2-bromo-2-methyl)- propionyloxy]undecyltrichlorosilane for polystyrene [158], or [4-(chloromethyl)phenyl]trichlorosilane for polyOEOMA [159]. Also a "grafting onto" method yields similar coated NPs by a single polymer layer. The design of polymers requires mono-dentate that has a terminal anchoring group at the end of polymer, or made of hyper-branched architectures with functionalities introduced in the focal points. Thiol-terminated polystyrene [160] and phosphonate-functionalized PEO are typical examples [161].

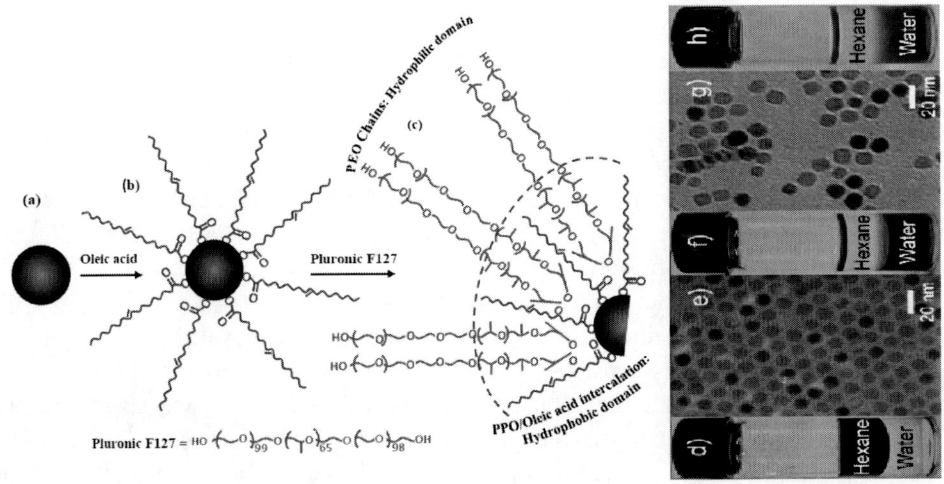


Figure 8. Schematic illustration for preparing water soluble stabilized (a) $Fe_3O_4$ NPs with (b) oleic acid and (c) an amphiphilic PEO-b-PPO-b-PEO F127 block copolymer, digital pictures (d, f, and h) show the colloidal stability of NPs, and TEM images in (e) hexane and (g) water [162].

The second layer polymerization process grafts an amphiphilic block copolymer onto the magnetic NPs immobilized with a parent hydrophobic ligand such as oleic acid as described in section 4.1. A block co-polymer, e.g., pluronic F127, a PEO-b-PPO-b-PEO [162-164], consists of a hydrophobic portion that intercalates the hydrophobic stabilizing ligands on the NPs and a hydrophilic portion that promotes water-solubility. As illustrated in Figure 8 (a-h), oleic acid-stabilized $Fe_3O_4$ NPs are prepared by thermal decomposition of iron oleate ($Fe(acac)_3$) in dioctyl ether and then an aqueous pluronic F127 solution is added. Upon drying the oil in water micro-emulsion yields F127-stabilized $Fe_3O_4$ NPs. Round shaped discs of NPs have sharp distribution with an average 4-12 nm diameter and that stand highly stable in a ferrofluid in hexane as well as in water for several months. This illustrates essence of grafting a second polymer layer on magnetic NPs in immobilizing them in a stable nanofluid. An amphiphilic graft copolymer is prepared by a reaction of terminal amine-functionalized PNIPAM (PNIPAM-NH2) with poly(maleic anhydride-altoctadecene). Magnetic NPs stabilized with resulting copolymer exhibit a dramatic volume change in response to the temperature [165], and it can be utilized for biological sensors, hyperthermia and other applications.

#### *4.2.3. Self-Assembly and Self-Association Process*

Self-assembly process let to design amphiphilic block co-polymers that self-assemble in aqueous medium, forming core/shell micellar NPs, wherein the hydrophobic core serves as a carrier for the magnetic NPs and anticancer drugs and the hydrophilic shell assists particle immobility. Ionic interaction between NPs and anionic- or cationic polymers cause self-association. The surfaces of hydrophobic magnetic NPs and anticancer drugs embedded in micellar particles after immobilizing oleic acid through self-assemblies are attached with targeting ligands for multifunctional nanomedicines. For example, polyester-based amphiphilic block copolymers are generally prepared by ring opening polymerization (ROP) of d,l-lactide (LA) and ε-caprolactone (CL). Here, a well-defined COOH-terminated PEO-b-PLA amphiphilic block copolymer self-assembles in the presence of hydrophobic magnetic NPs and doxorubicin (Dox) of a core/shell structure, and then functionalizes with therapeutic antibodies (targeting species to tumor) for an ultrasensitive MRI probe. A well-controlled PLA-b-POEOMA amphiphilic block copolymer is prepared by a combination of ROP and ATRP from 2-hydroxyethyl-2'-methyl-2'-bromopropionate, a double headed initiator (Fig 3b). A copolymer which self-assembles in the presence of hydrophobic $Fe_3O_4$ NPs is functionalized with folates for cancer cell targeting [166]. Maleimide-terminated PEO-b-PLA and methoxy terminated PEO-b-PLA self-assemble in the presence of DOX and NPs through a mixed micellization. Resulting maleimide functionalized NPs react with RGD tripeptide that can target integrin $\alpha_v\beta_3$ on tumor endothelial cells [167]. Other such copolymers of the interest are folateen coded PEO-b-PCL [168], PEO-modified polystyrene [169, 170], and hydrophobically modified poly-α-β-(N-2-dimethylaminoethyl L-aspartamide) [171].

Examples of self-association from ionic interaction between magnetic NPs and anionic- or cationic polymers include ionic pairs between (i) positively charged magnetic NPs and negatively charged head-groups of polymers such as poly(amidoamine) (PAMAM) dendrimers [172], (ii) negatively charged magnetic NPs and posisively charged head groups of polymers such as poly(trimethylammonium ethyl acrylate methylsulfate)-b-PAAm [173], or (iii) vesicle aggregates cross-linked by positively charged magnetic NPs [174]. A positively charged poly(l-lysine) (PLL) mixed with negatively charged magnetic NPs modified with citrate ions results in self-complex coacervates, and then post cross-linking with glutaraldehyde (GA) cross-linkers yields PLL-magnetic NPs hybrid micro-gels [175, 176]. Further, self-complexation through hydrophobic interactions is also explored, e.g., inclusion interactions of β-cyclodextrins with oleic acid-immobilized magnetic NPs [177]. An emulsification/solvent evaporation technique is used in preparing polymeric NPs based on PLA homo-polymers embedded with hydrophobic magnetic NPs. For example, a hydrophobic PLA dissolved in a volatile organic solvent is mixed with a surfactant in aqueous solution, forming oil in water emulsion. Evaporating solvent yields stable PLA-based NPs [178-181].

#### *4.2.4. Heterogeneous Polymerization*

A heterogeneous polymerization reaction of hydrophilic or water-soluble monomers is being used to prepare magnetic NPs embedded in a polymer bed in a nearly spherical shape as well as cross-linked microgels, nanogels, or hydrogels for medicines and biomedical applications. Spherical NPs are especially important in tuning magnetic coercivity on controlled shape anisotropy in the ideal single magnetic domains. Reactions give inverse (i) mini-emulsion polymerization, (ii) dispersion polymerization, and (iii) precipitation polymerization. Polymerizing hydrophobic monomers such as styrene and MMA in the presence of hydrophobic magnetic NPs in a conventional emulsion [182], a mini-emulsion [183-188], or a suspension [189] is found to produce promptly magnetic hydrophobic NPs.

Inverse mini-emulsion polymerization is a water in-oil polymerization process in which aqueous droplets of water-soluble monomers are dispersed with aid of an oil-soluble surfactant in an organic medium. Mechanical stirring promotes dispersion in inverse emulsion process, while sonification helps inverse mini-emulsion polymerization. Upon adding a radical initiator initiates polymerization within the aqueous droplets and form colloidal NPs. Incorporation of a multifunctional cross-linker leads to cross-link a microgel or a nanogel (a submicron-sized microgel). Several reports demonstrate the use of inverse mini-emulsion polymerization in preparing hydrophilic water soluble polymeric particles [190-194] and cross-linked microgels/nanogels [195-200]. Recently, a unique method utilizing controlled ATRP in inverse mini-emulsion is introduced in preparing functionalized biodegradable nanogels for targeted drug delivery, color pigments, and other applications [201-210].

The PAAm-based microgels, 60-160 nm diameter, are a good example of a magnetogel as deduced by this method. An aqueous solution in AAm, N,N'-methylenebisacrylamide (MBAm), and PMAA-supported ferrofluid was admixed to a solution of Span 80 (sorbitol monooleate) in cyclohexane. An inverse mini-emulsion of PAAm-microgels (100 nm diameter) polymerized when adding a free radical initiator retains a magnetic part as much as 13 wt% consistent to the initial dosage [211]. An in situ formation of magnetic NPs in the presence of poly(AAm-co-MAA) gives similar microgels. It involves reducing magnetic species such as Fe(II) ions, for example, with hydrazine, and the COOH groups (from polymer) anchore Fe atoms during the reaction. Swollen magnetic poly(AAm-co-MAA) microgel measures 9-30 μm diameters in the dynamic light scattering (DLS) [212]. Inverse emulsion/micro-emulsion polymerization also give well-defined polymer embedded magnetic NPs. For instance, aqueous droplets of MAA, 2-hydroxyethyl methacrylate (HEMA), and $Fe_3O_4$ were dispersed in sodium dioctylsulfosuccinate in toluene. Copolymerizing monomers yields hydrophilic polymer beads embedding $Fe_3O_4$. However, poor dispersibility gives a maximum 3.3 wt% $Fe_3O_4$ loading [213]. More $Fe_3O_4$ could be loaded if functionalized with a double-hydrophilic block copolymer PEO-b-PMAA. In synthesis, a water-soluble monomer mixture and $Fe_3O_4$ functionalized with PEO-b-PMAA, HEMA, and MAA were immersed in poly(ethene-co-butene)-b-PEO in decane. When adding an organic initiator magnetic microspheres (50-250 nm average diameter) form with a loading of SIONPs up to 18 wt% [214].

A dispersion polymerization yields micrometer-size polymer microspheres with a narrow size distribution in a single step. Selection of reaction medium is important in that although the monomers are soluble, but the polymer which results on the reaction becomes insoluble along with the magnetic NPs. While heating the polymerization mixture, initiator decomposes to form oligomer radicals. Oligomeric chain no longer stays dissolved in the medium when grows to a critical length. They associate growing NPs, which aggregate and at the same time adsorb the stabilizer forming primary mature particles with a magnetic core. As described elsewhere [215-217], under the conditions when no new NPs nuclei turn-up, the primary NPs grow and reach a uniform size. For example, HEMA was polymerized in the presence of iron oxide needles or cubes (100–500 nm size) in a mixture of toluene/2-methylpropan-1-ol, with magnetic microspheres (1-2 μm diameter) stabilized with acetate butyrate cellulose. EGDMA is added to promote cross-linked microspheres [218]. A similar process is applied for preparing magnetic microsphere of poly(glydicyl methacrylate) (PGMA) [219-221] and PAAm [222] stabilized with a linear polymer such as poly(vinyl pyrrolidone) (PVP) molecules. Magnetic microspheres of poly(styrene-co-GMA) with as much as 72 wt%

magnetic NPs are available [223]. Thermoresponsive polymeric microgels loaded with SIONPs and anticancer drugs are of interest for multifunctional cancer therapies because such NPs can be used for magnetic drug targeting followed by simultaneous hyperthermia and drug release. Following the work of Purushotham et al. [224], a schematic overview of the dox-composite NPs synthesis procedure is given in Figure (9), wherein the composite is synthesized first by polymerizing NIPAM monomers in the presence of $\gamma$-$Fe_2O_3$ NPs of 43 nm average diameter (Figure 9a).

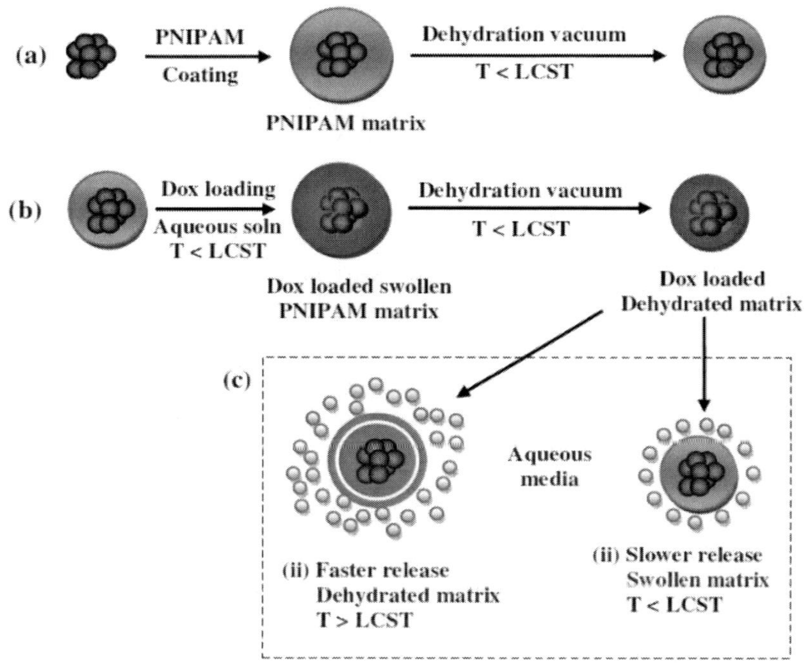


Figure 9. Schematic overview of surface modified (a) composite magnetic NPs, (b) drug loading, and (c) drug release processes [224].

For storage, the polymer composite $\gamma$-$Fe_2O_3$ NPs are dehydrated in vacuum below the LCST (lower critical solution temperature above which they suffer from a reversible volume shrinkage). The dox-composite forms upon dehydrated composite NPs are swollen below the LCST in an aqueous solution with doxorubicin IONPs with 14, 19, and 43 nm diameters were synthesized by high temperature decomposition (Figure 9b). The drug loaded NPs are again dehydrated for storage and other applications. Thermo-responsiveness of the polymer-$\gamma$-$Fe_2O_3$ composite NPs exhibits facile loading and release of drugs at temperatures below and above the LCST around 34°C. The sample shows Fickian diffusion release kinetics; the maximum dox release at 42°C after 101 h was 41%. In vitro simultaneous hyperthermia and drug release of therapeutically relevant quantities of dox were achieved. For example, 14.7% of loaded dox was released in 47 min at hyperthermia temperatures [224].

### 4.3. Functionalizing Magnetic NPS with Ceramic-Polymer Hybrids

A ceramic-polymer hybrid can graft a reasonably more sustainable bed required to embed and coat magnetic NPs with functionalized properties. Well-known biocompatible ceramics like $SiO_2$, $ZrO_2$, $Al_2O_3$, and derivatives are recommended to explore bonding over selective magnetic NPs (metals, alloys, or oxides) in a stable core-shell structure that broaden applications as hybrd composite NPs in biosensors, florescent probes, and diagnostic medical devices. A sol-gel process is an efficient chemical method to prepare such ceramic NPs (beds) and to incorporate magnetic NPs therein through a proper precursor such as tetraethyl orthosilicate (TEOS) in a case of the $SiO_2$ based polymer composites. For illustration, a sample of silica encapsulated magnetic NPs is further treated with a suitable polymer, forming a multifunctional inorganic-organic hybrid. Preparation of a hairy hybrid composite consisting of a magnetic core, silica shell, and functional polymer brushes is illustrated in Figure (10). The synthetic process starts with a sol–gel reaction in the presence of $Fe_3O_4$ NPs, for example, and fluorescein isothiocyanate (FITC), a fluorescent dye, producing $Fe_3O_4$ NPs encapsulated with silica shell, in which FITC is covalently incorporated. The sample so obtained is functionalized with methacrylate by reacting with 3-methacrylo-xypropyltrimethosysilane (MPS) and then with chlorine via a radical cross linking polymerization of ethylene glycol, dimethacrylate (EGDMA) and vinyl benzyl chloride. The chlorine groups serve as an initiator for ATRP of OEOMA, forming a hybrid composite with biocompatible POEOMA in the form of nanobrushes, as described by Li et al. [225].

Silica encapsulated $Fe_3O_4$ superparamagnetic NPs, prepared by the sol-gel reaction of TEOS, when encapsulated in polyphosphazene yield a pomegranate-like core/shell nanostructure [226]. As shown from (a) AFM and (b, c) TEM images in Figure (10), after the sol-gel reaction, average size of the $Fe_3O_4$-$DySiO_2$ core-shell NPs has increased considerably up to 73 nm (~8 nm the initial value). Interesting to note is that the parent disc shape does not change much upon immobilizing in a ceramic-polymer bed and upon the growth. A distinct shell, ~32 nm thickness, is well resolved in a $Fe_3O_4$-$DySiO_2$-PMAA composite particle (Fig. 10c). The dysprosium adds fluorescent properties, with a strong light-emission at 518 nm wavelength, in a multifunctionalized magnetic-optical hybrid nanocomposite particle [227].

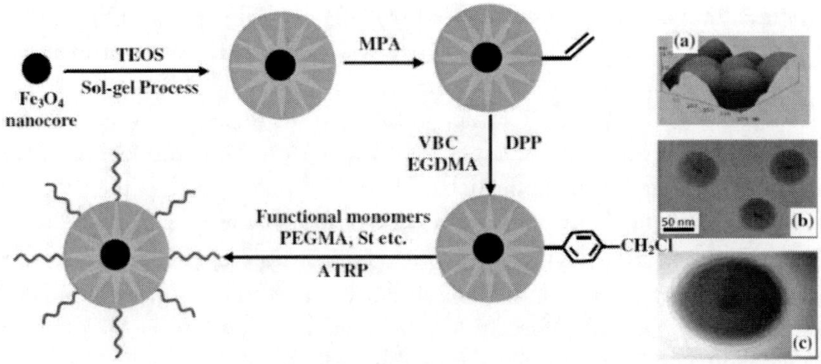


Figure 10. Synthesis of surface modified magnetic NP with silica and polymer shell, (a) an AFM and (b, c) TEM images of $Fe_3O_4$-$DySiO_2$-PMAA [225].

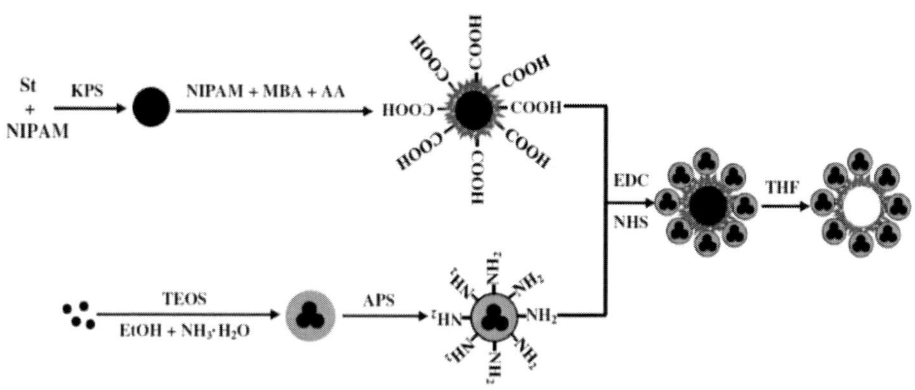


Figure 11. Schematic of preparation procedure of cross-linked PNIPAM capsules surface-anchored with silica coated magnetic $Fe_3O_4$ NPs [230].

A second synthetic approach involves prepare copolymers bearing trimethoxysilyl groups capable to cross-link on the magnetic NPs in a hybrid core-shell structure. Typical copolymers consisting of poly(3-trimethoxysilyl)propyl methacrylate (PEPMA) and poly(N-acryloxysuccinimide) (PNAS) [228] are often used in devising such composite magnetic NPs with functional properties for applications. A Cy5.5 fluorescent dye is further employed to functionalize the PNAS block for early in vivo tumor detection by dual features of magnetic resonance and fluorescence imaging [229]. In a third approach to derive hybrid composite magnetic NPs, silica coated $Fe_3O_4$ of isolated NPs when reacted with 3-aminopropyltriethoxysilane (APS) create amino-functionalized silica coated $Fe_3O_4$ NPs with a double core-shell structure, viz., a large effective surface area in single magnetic domains. The later is mixed and reacted with an acid-functionalized core/shell microgel so as it results in a poly(styrene-co-NIPAM) core with a cross-linked polymer PNIPAM shell. As demonstrated in Figure (11), a removal of the core poly(styrene-co-NIPAM) by dissolving it in an organic solvent such as tetrahydrofuran yields cross-linked PNIPAM of nanocapsules surface-anchored with silica-coated magnetic $Fe_3O_4$ [230], useful for drug delivery, light imaging, and other multifunctional medicinal applications.

High quality of silica based polymer composite microgels is illustrated with typical AFM images in Figure 12(a-d) from pioneering work by Zhang and Wang on preparing silica based magnetic composite NPs with selective compositions and functionalized properties [230]. For illustration, (a) the height image and (b) phase image of poly(styrene) seeds, for example, indicate that the seed microspheres are narrowly dispersed in size/shape and exhibit a raspberry–like structure, demonstrating involvement of a two-step-polymerization process which occurs and results in the formation of a rich polymer PNIPAM shell on the magnetic NPs of iron oxides. The subsequent AFM images (c, d) in this figure present the short-growth composite NPs and the corresponding phase image. Compared to the parent poly(styrene) seeds, the short-growth NPs are reasonably larger. The phase image confers the core-shell structure is substantially stable on dry samples of a hybrid ceramic-polymer composite. A reasonably hard core stands out from a softer poly(NIPAM-co-AA) shell collapsed on the mica substrate for every microsphere. Other details of synthesis, properties and applications on the poly(NIPAM-co-AA) composites with poly(styrene) seeds in the form of stable core-shell microspheres are described in the original article of Zhang and Wang [230] and in a review article on iron-oxide based superparamgentic polymer nanomaterials compiled recently by Oh and Park [231].

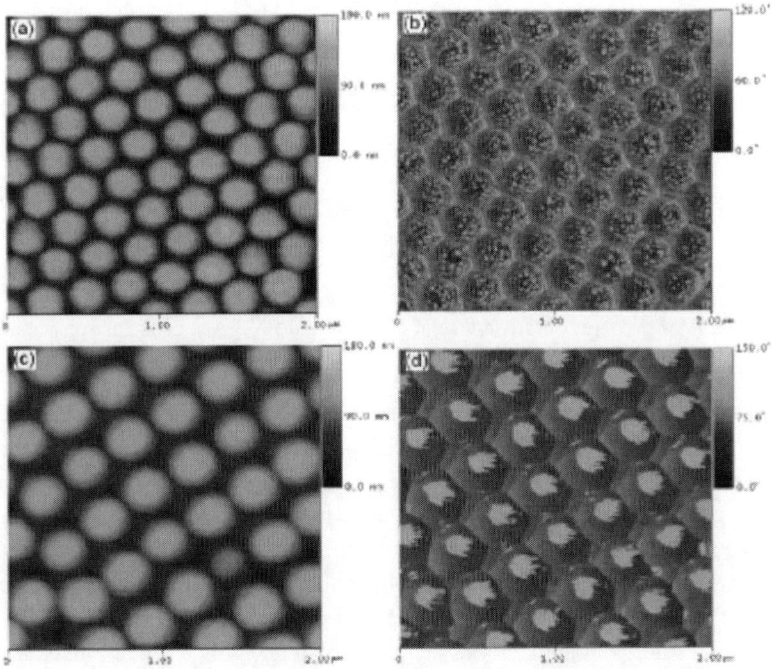


Figure 12. AFM images of poly(styrene) seeds and composites with poly(NIPAM-co-AA) of core-shell microspheres; (a) height and (b) phase images of the poly(styrene) seeds with (c) height and (d) corresponding phase images of the core-shell microspheres [230].

## 5. FUNCTIONALIZED PROPERTIES OF FERROFLUIDS

One of the most fascinating physical properties of a practically useful ferrofluid for pertinent applications is its rheology with unusually large distortion or manipulation of its basic structural units of flocculates when exposed to a magnetic and/or electric field. For instance, when a ferrofluid is brought into contact with a magnetic field, the magnetic moments of the particles reorient along the field lines readily that depends on its magnetic viscosity. The magnetization of the ferrofluid responds immediately to the changes in the applied magnetic field, and when the applied field is removed, the moments tend to randomize quickly to the initial state so that the net magnetization vanishes [17-19]. In a gradient field the whole fluid responds as a homogeneous single phase magnetic liquid which moves to the region of highest magnetic flux density. This means that a ferrofluid can be precisely positioned and controlled by an external magnetic field. The forces upholding the magnetic fluid in place are proportional to the gradient of the external field and the magnetization value of the fluid. This means that the retention force of a ferrofluid can be adjusted by changing either the magnetization of the fluid or the magnetic field in the region.

The performance and the operating life of various ferrofluids in general depends on the following primary properties

1) Stability of ferrofluids in magnetic/electric fields
2) Structural properties
3) Magnetic properties
4) Magneto-viscosity properties.

## 5.1. Stability of Ferrofluids

As described above in section 1.0, a ferrofluid consists of dispersion and suspension of pretty small magnetic or ferroelectric particles, usually 5-10 nm effective size, in the form of a stable colloidal suspension in a base liquid carrier (water, oil or diester) in such a way that the particles remain dispersed indefinitely despite of rather long range magnetic/electric forces and short range van der Waals forces of attraction [17, 19]. Due to the short range forces, the dispersed particles intend to coalesce in small assemblies that are very detrimental for the performance of the fluids for commercial applications. The first criterion for a feasible colloid is thus of course the stability against to sedimentation under gravity and in strong magnetic fields. In constructing a workable ferrofluid, the energy scale for gravitational and magnetic interactions must be much smaller than the kinetic energy $k_B$T in the individual particles or flocculates. In a simplified case, this condition can be written as $k_B T > \mu H$ [17]. Reconstructive cohesive and viscoelastic features as per an applied shear force or a magnetic/electric field modulate the fluid rheology with a Newtonian or non-Newtonian behavior. In order to prevent unwanted flocculate aggregation, the basic particles are covered with a surfactant, which bonds over the NPs a stable protective surface layer, usually a long chain polar polymer molecule. The surface layer bonds and adheres to the surface of the particles so that they do not recombine and grow bigger and also do not coalesce in hard aggregates by the entropic repulsion. In the case of the long range magnetic/electric forces, the magnetic/electric energy is a minimum for two particles in contact as described above with eq. 1. Hence, the surfactant and the liquid carrier properties must be concurrent to overcome the attractive van der Waals and magnetic forces between the NPs dispersed (and often cross-linked) through the carrier liquid of bonded molecules.

Nevertheless, it is quite likely that a ferrofluid becomes more stable in the NPs bond in small groups in the form of hard assemblies or clusters of large mass and in turn large effective magnetic moment of the specimen on average. It depends on the volume fraction of the magnetic NPs which varies usually between 2 and 10% by volume. However, the maximum NPs uploading is limited because the NPs begin to separate from a cross-linked structure with the carrier liquid and as a result the viscosity ($\eta$) does not stand so stable especially in an applied magnetic field. Eventually, the ferrofluid loses its viscoelastic features. An empirical relationship in the viscosity and NPs dosage can be expressed as [232]

$$\eta / \eta_o = 1/\left(1 - 2.5\phi + b\phi^2\right) \quad (6)$$

with

$$b = (2.5\phi_c - 1)/\phi_c^2, \quad (7)$$

where $\phi_c$ is the critical volume fraction of NPs above which the "fluid" suffers from a phase separation, and $\eta_o$ is the carrier viscosity. $\phi_c = 0.74$ for hexagonal close packed spheres. The concentration gradient $\nabla n$ for NPs in a gravitational field is written as [232]

$$\nabla n = \pi D^3 (\rho_s - \rho_o) gn / 6k_B T, \quad (8)$$

where $\rho_o$ is the density of the carrier liquid, $\rho_s$ is the density of the NPs, and $n$ is the number of NPs per unit volume at temperature T. A similar concentration change arises in a magnetic field gradient $\nabla H$, and in this case an equivalent expression to eq. (8) can be written as

$$\nabla n = nD^3 \pi \mu_o M_s \nabla H / 6k_B T, \quad (9)$$

The factor $\nabla n/n$ is important in determining stability of the sample. If $\nabla n$ is appreciable, the viscosity of the ferrofluid increases in the field gradient leading to bulk when $\phi \rightarrow \phi_c$. A criterion for the fluid stability can be derived in terms of an $\nabla n$ value. This is considered as a limiting value for a particular ferrofluid application and it depends on what change in the flocculate NPs density is acceptable. Thus $D$, $M_s$, and $\nabla H$ are the vital variants which may be adjusted to tailor the critical stability. A ferrofluid thus may be regarded to be practically stable in conditions of small $\nabla H$, while unstable on a larger $\nabla H$. So, a ferrofluid for a particular application must be chosen on the basis of the field gradients within the device. Phenomenological, the same may not be suitable for others where the field gradients are effectively larger. Required features can be readjusted on the case to case basis by using smaller NPs with suitable polymer additives.

## 5.2. Structural Properties

The structural properties of a nanofluid cover its basic microstructure, the local structures of the flocculates and the carrier liquid phase, how the basic components mix-up (or interact) each other in a coupled system, or how their average microstructure readjust-itself to cope with an external perturbation such as a magnetic or electric field, shear force, or even a radiation. For example, a structural pattern of a magnetic fluid is usually studied by applying an external magnetic field either parallel or perpendicular to a sample sealed in a container (often a glass cell with a thickness of tens of micrometer). Such a thin sample of a nanofluid is considered as a magnetic fluid film for a practical purpose. Under parallel magnetic field, the inhomogeneous magnetic fluid exhibits a randomly oriented needle like chain structure (Figure 13a), while a homogeneous fluid forms a one dimensional (1D) periodic long chain structure (Figure 13b) [233, 234]. The chain length is growing as the field strength increases in a result of the NPs grow in small assemblies [235]. A random needle chain like structure in an inhomogeneous magnetic fluid occurs due to a phase separation dependent growth of the assemblies in a result of thermodynamic instability of the NPs dispersion [17, 236]. Furthermore, the 1D chain structure varies as per the magnetic field {Figure 13(b)}. A field dependent measured spacing between two neighboring chains and the averaged chain width is shown in Figure (14) and both decrease as the field increases.

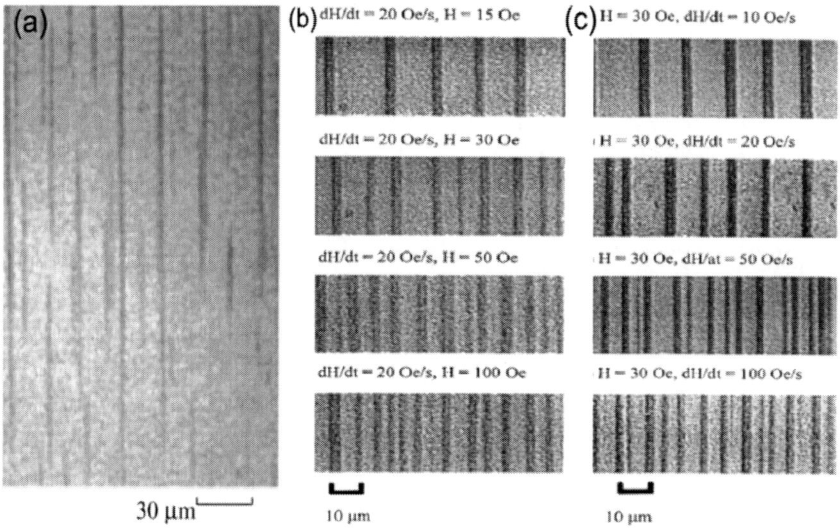


Figure 13. (a) Needle and (b) periodic like long chain structural patterns formed in a magnetic fluid film (5.0 μm thickness and 22.4% volume fraction of $Fe_3O_4$ NPs in kerosene) at constant sweep rate, and (c) at different sweep rates under a parallel field H [234, 235].

In place of needle like chains in the above example, a magnetic fluid rearrange in the forms of a cylindrical magnetic column structure in a case the field is passing through the film perpendicular to it. As the H increases, the columns increase in numbers with a broaden distribution of the size in the fluid film [237]. At this stage, the columns are located randomly within a film. Each column acts in an isolated manner, and the interactive force between a column and its surrounding columns cannot be balanced. As shown in Figure 15(a), such structural pattern is called a disordered column pattern [234]. If H is increased further, the NPs combine and form new columns and ultimately construct in a granular hexagonal ordered structure as can be seen from a typical five- and six-fold coordinated columns along the grain boundaries pattern shown in Figure 15(b) [238] after a pioneering work of Hong et al. on magnetic $Fe_3O_4$ fluid films. Figure 15(c) plots average distance between two neighboring fluid columns against H-value in a $Fe_3O_4$ fluid film from the work of Hong et al. [238]. The solid and dashed lines are the theoretical curves from the column-dipole behavior with interaction between the first nearest neighboring columns and up to the second neighboring columns, respectively (6.0 μm film thickness and 19.4 vol% NPs). An ordered hexagonal structure of magnetic columns in the ferrofluids results in novel optical properties of a new class of the multifunctional materials called "magnetochromics" [239]. A novel implication of this result is that such a hexagonal structure of magnetic columns can act as a 2D optical grating, opening new optical phenomena with field dependence optical properties for diverse applications.

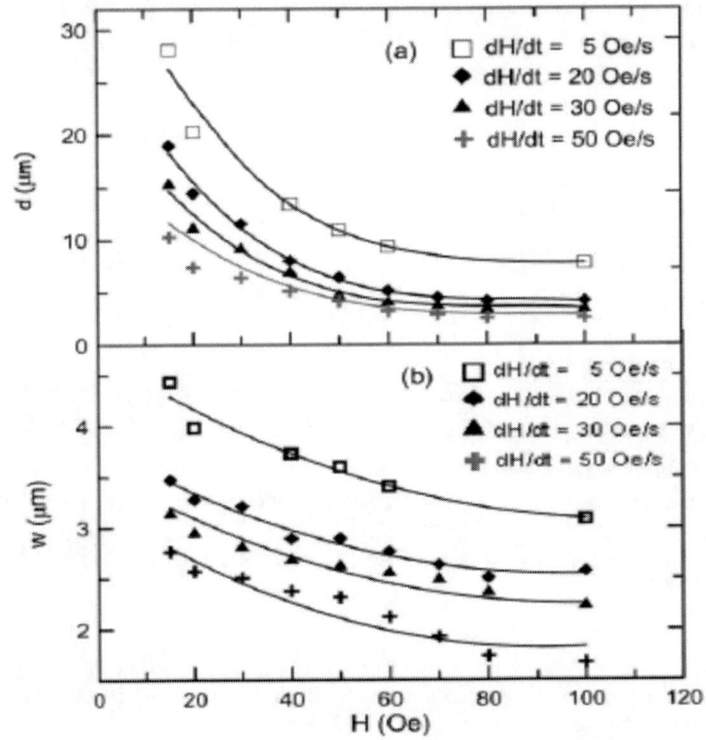


Figure 14. (a) The spacing *(d)* between long chains and (b) the chain width (w) as a function of field H at various sweep rates in a magnetic fluid film as in Figure (13) [235].

## 5.3. Magnetic Properties

A fundamental property of the magnetic fluids in the presence of a non-homogeneous magnetic flux (B) is that the flocculates converge in the region where the field intensity H is a maximum. It occurs due to the magnetic moments ($\mu$) in the NPs, or rigid assemblies, rotate along to the applied field where the minimum energy ($U = - \mu \cdot B$) is conserved. The rotation of the $\mu$ within a ferrofluid depends on two distinct mechanisms, namely, the *Brownian rotation and* the *Neél rotation.* In case of a *Brownian rotation*, the rotation of the magnetic NPs inside the liquid carrier occurs even in absence of B due to molecular collisions, with relaxation time ($\tau$) of $\tau_B = 3V\eta/k_BT$ [89,241], where $V$ is the particle's volume and $\eta$ the liquid's viscosity, while the rotation of the $\mu$ with respect to the particle is known as the "*Neél rotation*". The $\tau$-value for this kind of a local rotation is expressed as $\tau_N = f^{-1}\exp(K_aV/k_BT)$, and it is strongly dependent on the NPs volume and the sample temperature, where $f$ is the Larmor frequency and $K_a$ is the anisotropy constant of the particle. In the case of $Fe_3O_4$ NPs, $K_a = 1.1\times10^4 Jm^{-3}$ at room temperature and the $\tau_N$-value increases from $4\times10^{-9}$ *s* to $7\times10^{-5}$ *s* upon increasing the particle's diameter from 10 *nm* to 20 *nm*. When the relaxation is dominated by the Brownian rotation, the *μ-value* can be considered as blocked within a particle. Reorientation of the magnetic dipoles in an applied H then occurs by rotation of the entire particle. At temperatures far below the *freezing temperature* ($T_F$) or *blocking temperature* ($T_B$) of the carrier fluid, the NPs are presumably immobilized and, consequently, relaxation of the $\mu$ occurs by rotation of the dipoles within the particles (the Neél relaxation) and this $\tau_N$ stands larger than the typical observation time. Below $T_B$ temperature, a particle is no longer superparamagnetic, but the average magnetic fluid still behaves to be a superparamagnetic because the particle, and so as the $\mu$, continues to be quasi-free to undergo a local rotation. An equation of dynamic motion for $\mu$ is too general to be applicable for the cases of superparamagnetic and non-superparamagnetic NPs, as well as a mixed phase of the two kinds of the macroscopic phases dispersed each other in a hybrid composite structure. Strong microscopic interaction between them through the medium modifies the parent characters.

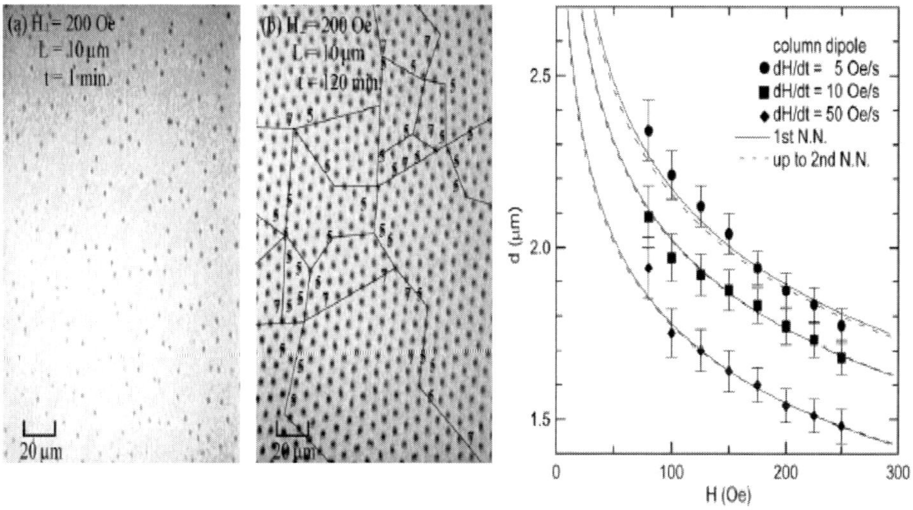


Figure 15. (a) Disordered and (b) ordered hexagonal column patterns of a magnetic fluid film (10 μm thickness) after subjecting into a perpendicular 200 Oe field [234] and (c) the column distance against H at different sweep rates [238].

Figure (16) compares the zero-field cooled (ZFC) and field cooled (FC) dc magnetization curves measured on $Fe_3O_4$ of 14 nm particles at H = 100 Oe (a) before and after dispersed in (b) hexane and (c) dodecane in the form of a magnetic nanofluid. The ZFC curve in Figure 16(a) exhibits a broad maximum at $T_B \sim$ 200 K in a polydispersed nature of the $Fe_3O_4$ NPs together with a broad distribution in size and anisotropy axes in the individual NPs [241, 242]. The flocculates dispersed in a carrier liquid show a distinct cups at ~186 K in hexane (Figure 16b), while at ~270 K in dodecane (Figure 16c). The cusp at 270 K in the dodecane ferrofluid primarily attributes to the freezing of the carrier liquid (dodecane, $T_F$ ~264 K) while that of 186 K in the hexane ferrofluid is a combined effect of both the carrier liquid freezing (hexane, $T_F$ ~178 K) and the field dependent particle blocking. Moreover, an inflection shown in dodecane ferrofluid at ~181 K refers to a magnetic anomaly in the blocking effect of the magnetic NPs. Since when the carrier liquid freezes stops abruptly the local physical motion of the NPs and alignment of the magnetic moments to the field direction, the cups at 186 and 270 K are sharper compared with that of 200 K, intrinsic of the blocking process of the spins within the NPs, that is due to a competition between thermal energy ($k_B$T) and anisotropy energy ($K_aV$). The cusp at 186 K is sharper in hexane because the freezing and blocking concur in a narrow window.

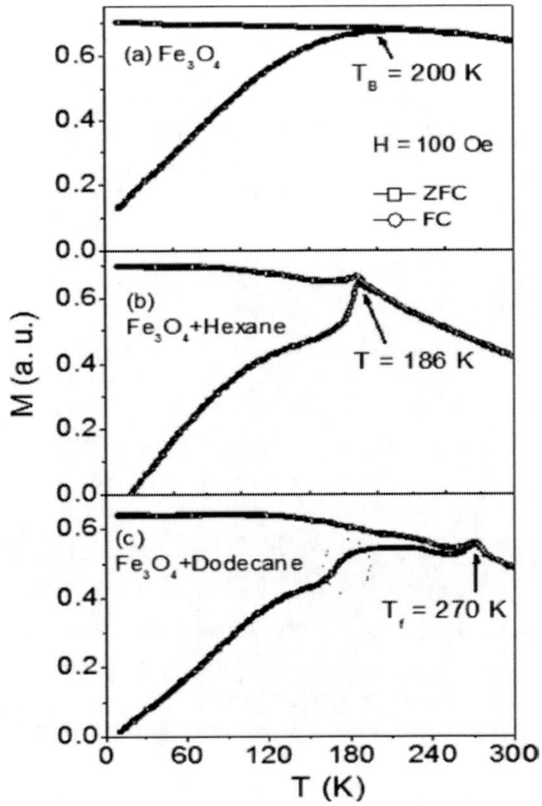


Figure 16. Temperature dependent ZFC-FC curves of $Fe_3O_4$ NPs (a) before and after dispersed in (b) hexane and (c) dodecane [244].

The ac susceptibility thermogram provides another probe of characterizing the freezing and blocking processes in a magnetic fluid. For illustration, Figure (17) compares temperature dependence ZFC dc magnetization and ac complex susceptibility measured from a $Fe_3O_4$ ferrofluid in dodecane over 0 - 300 K. A spin-glass-transition like peak (270 K) and a $T_b$ point (181 K) shown in the ZFC magnetization more clearly appear in two distinct $\chi''(T)$ relaxation peaks very similar to the features reported in a sample $Fe_3O_4$ NPs by Zhang et al. [243]. In general, three characteristic states, viz., a liquid phase, a mixed (m) phase, and a frozen solid {or solid state (s)} can concur in the flocculates or solvent in a ferrofluid. For temperatures above the pour point ($T_{pour}$), a liquid phase exists as the system flows like a fluid, whereas a solid phase turns-up at low temperatures below $T_{s\text{-}m}$, and a mixed phase exists in the temperature range between $T_{s\text{-}m}$ and $T_{pour}$, where $T_{s\text{-}m}$ is the transition temperature from the s-m phase. Logically, the first $\chi''$ peak (at high temperature) lies in the $T_{s\text{-}m} < T < T_{pour}$ range (in the mixed state) while the second $\chi''(T)$ peak (at lower temperature) to the frozen state [243]. As a result, the second $\chi''(T)$ peak (Figure 17) ascribes to the $T_b$ point, while the first one ascribe to the host freezing. Nevertheless, the first peak reflects the magnetic behavior in the mixed state with the second one representing the magnetic behavior in the frozen state in the ZFC curve. The blocking in the frozen state significantly affects the interparticle dipole-dipole interaction, causing characteristic spin-glass-like dynamics.

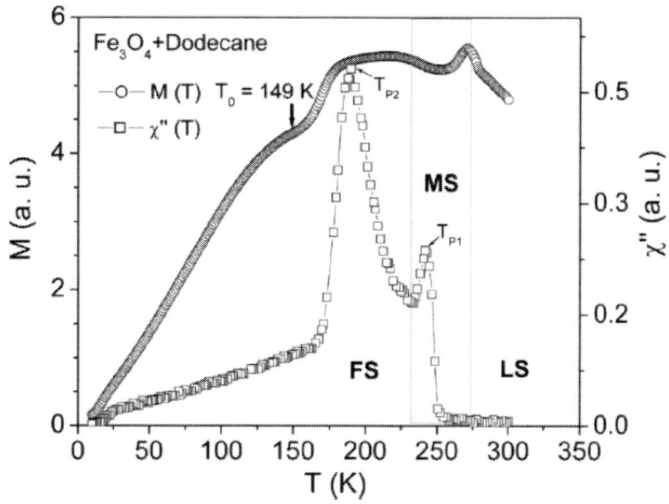


Figure 17. Temperature dependence ZFC magnetization and imaginary part of magnetic susceptibility $\chi''$ in a $Fe_3O_4$ ferrofluid in dodecane. LS: Liquid state, MS: mixed state; FS: frozen state, and $T_0$: the temperature at which the system enters a glassy state. $T_{p1}$ and $T_{p2}$ represent the $\chi''$ peaks in mixed and frozen states, respectively [244].

## 5.4. Magneto-Viscous Properties

A long ago in 1969, Rosensweig observed, for the first time, a marked change in viscosity in a concentrated $Fe_3O_4$ ferrofluid in water in the presence of a magnetic field [245]. Only a few months later McTaguem [246] carried out similar studies in an independent work on a highly dilute cobalt ferrofluid in a similar aqueous host. Both the groups observed an increase of viscosity with increasing H. Moreover, McTaguem's experiments also showed a dependence of the effect on the relative angle between the magnetic field direction and the vorticity of the flocculates. These phenomena can qualitatively be described by a model as follows. Assuming that the magnetic particles dispersed through the host species are not interacting one another much and that the magnetic moment is fixed within the particles, a shear flow would generate a local rotation of the particles in the sample and in-turn also of the magnetic moments ($\mu_i$) in the individual particles (marked as $i = 1 \rightarrow N$, with N the total number of the particles in the sample). The average axis of rotation of the particles is aligned with the vorticity of the flow as shown schematically in Figure (18). As soon as a magnetic field is applied to the system, the magnetic moments in the particles tend to align along the magnetic field. If the moments are fixed inside the particles, as assumed above, a rotation caused by the viscous friction would lead to a misalignment of average $\langle\mu_i\rangle$ from the H direction. Such misalignment gives rise to a magnetic torque which counteracts a mechanic torque due to a shear flow of the flocculates. Thus, a free rotation of the particles is hindered and, macroscopically, the viscosity rises-up. If the fluid's vorticity is aligned with the field, the rotation of the particles cannot reinforce a <$\mu_i$> misalignment from the H direction, and thus no magnetic torque would counteract the free rotation of the magnetic particles. As a result, no increase in the viscosity should take place in this relative configuration of field and vorticity.

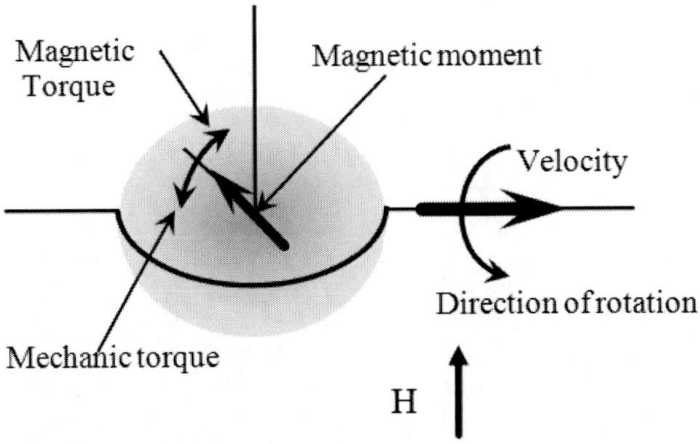


Figure 18. A model illustration of the mechanism describing an increase in the viscosity in a ferrofluid with an applied magnetic field [247].

Considering the above model assumption, the rotational ferrofluid's viscosity $\eta_r$ in a ferrofluid can be written in a simple form [17]

$$\eta_r = \frac{3}{2}\eta_o V \frac{\alpha - \tanh\alpha}{\alpha + \tanh\alpha}\left\langle \sin^2 \beta \right\rangle, \tag{10}$$

where $\eta_o$ denotes the viscosity of the fluid for vanishing magnetic fields, $V$ the volume concentration of the magnetic NPs including their surfactant, $\beta$ the angle between vorticity and magnetic field $H$, and $\alpha = \mu_o mH/k_BT$ the ratio of magnetic and thermal energy of the NPs with magnetic moment $m$ ($\langle....\rangle$ devotes the spatial average). In weak magnetic fields, $\eta_r$ increases proportional to $H^2$, while in large magnetic fields, the $\eta_r$ -value saturates as

$$\eta_r^{\max} = \frac{3}{2}\eta_o V . \tag{11}$$

Thus, a commercial standard ferrofluid with 7 vol% of magnetic NPs, a mean particle diameter of about 10 nm and a surfactant layer thickness of 2 nm results in a value $V$ = 19.2 vol% and it implies a maximum relative change of viscosity $\eta_r/\eta_o$ ~ 30%. This estimate is based on the assumption that all magnetic NPs contribute to the average viscosity. The magnetic relaxation behavior of the NPs sets-up this boundary condition. In this case, the magnetic NPs can either align to the field by a rotation of the whole particle, a process called the "Brownian relaxation", or by a change of the direction of the magnetic moments inside the particles, so-called the "*Neél* relaxation process". In a typical example, Figure (19) displays size distribution (Figure 19a) of $Fe_3O_4$ NPs and its effect on the field dependence viscosity (Figure 19b) in a derived magnetic nanofluid. The magnetic NPs with a diameter above a critical value of nearly 13 nm, where the relaxation behaviour of the NPs becomes Brownian, are present in a small volume concentration of less than 1 vol%. As a result, a maximum change in the shear viscosity calculated following a model relation (11) should be pretty small only, i.e., ~ 2%.

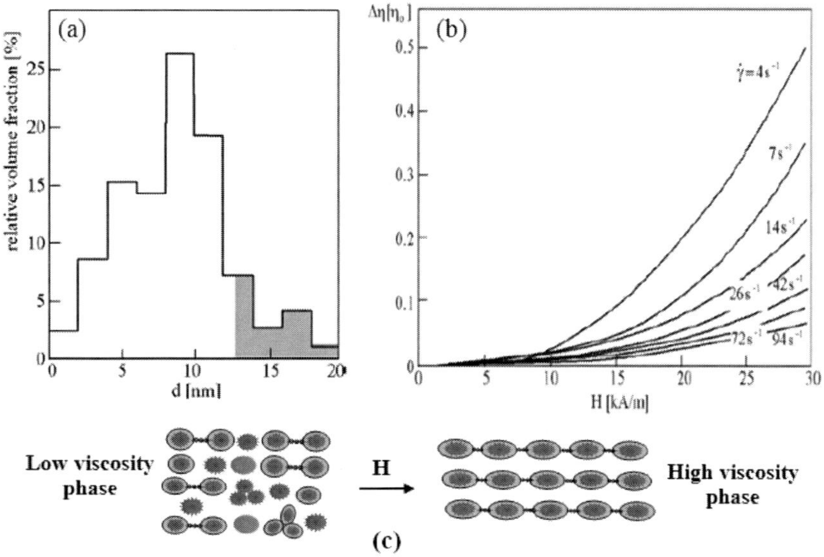


Figure 19. (a) Particle size distribution of $Fe_3O_4$ and (b) change of viscosity of $Fe_3O_4$ dispersed in an oil carrier of a ferrofluid as a function of H at different shear rates [247], with (c) a model describing how the magnetic NPs (blue-green stars) bond anistropic long chains through the host molecules (yellowish-brown) in a viscous ferrofluid.

In a poor agreement, the experimental observation portrayed in Figure 19(b) on a ferrofluid of 7 vol% magnetic $Fe_3O_4$ NPs (size distribution in Figure 19b) in an oil carrier liquid illustrates a much larger viscosity change of more than 100% with a moderate H = 30 kA $m^{-1}$ (or 30 kOe). Even a correction of V of that on magnetization the magnetic NPs expand in volume significantly cannot lead to raise $\eta_r/\eta_o$ so heavily. Assuming a revised V = 24 % (arbitrary) on the NPs expand by ~ 25% thus hardly gives ~ 5% increase in the initial $\eta_r/\eta_o$ value. Thus, the difference between the two values indicates that the field induces a structural variant that promotes the viscosity so largely. Quite possibly, the NPs cross-link through the host molecules in an anisotropic shape as small assemblies that breed high viscosity in a sufficiently large magnetic field. Here, total volume of the assemblies should be taken as the effective flocculate volume in relation (11). Model shapes proposed in Figure 19(c) brief how such assemblies grow with the magnetic NPs bond one another through the host molecules. The magnetic NPs which do not constitute a cross-linked structure are called as the "dead species" that do not impart a high viscosity state. Logically, the shear rate suppresses the final $\eta_r/\eta_o$ value because it simply compresses the flocculates bonded via the host molecules in model anisotropic shapes.

# 6. Application of Ferrofluids

A distinguishing feature of variety of ferrofluids made available today offers ample applicability in versatile areas of science, medicines, camouflage, and small technological devices. Most applications of the magnetic (or also ferroelectric) fluids are based primarily on the following major properties such as ability to manipulate with an applied perturbation field (such as magnetic, electric, optical, etc.), rheology, absorption capacity of electromagnetic energy at selective frequencies and heating-up, thermal conductivity, magnetocalory, biological activities, ferroelectrics, light-energy transfer, magnetic switching, and other perturbation sensitive features. Well-known applications of the ferrofluids are diversified in various fields such as chemical and biochemical separations, magnetic target drug delivery, magnetic resonance imaging contrast agents, hyperthermia therapy, disease diagnosis, chromatography, electrochemical biosensors, optical detection and luminescence detection, heat sinker and dampers, multicolored pigments, magnetic switches and valves, phosphors, and several others [110, 230, 248-255].

## *6.1. Biotechnological Applications*

Small magnetic NPs dispersed in a biocompatible liquid carrier tailor several attractive features useful for biomedicines with not only several auspicious characteristics but desirably controlled responses. A basic advantage is that the size of the magnetic NPs in a fluid comparable to the biological entities such as a cell (10-100μm), a virus (20-450 nm), a protein (5-50 nm) or a gene (2 nm wide and 10-100 nm long) are facile to coat over biological molecules in order to make them interact with or bind to a biological entity for controllable 'tagging' or "addressing". Further, the magnetic nature makes it feasible for the manipulation by an external magnetic field gradient. This 'action at a distance', combined with the intrinsic penetrability of a controlled magnetic dosage into human tissue, opens-up many applications involving the transport and/or immobilization of the magnetic NPs, or of magnetically tagged biological entities in a fluid and derivatives. In this way, they can be made to deliver a package such as an anticancer drug, or a cohort of radionuclide species, to a targeted region of a living body, such as a tumor. A third point, the resonating response character in the presence of time-varying magnetic field also opens a path to the transfer of energy from the applied perturbation field to the NPs tagged with the biological entities, as described in details elsewhere [256-259]. For example, the magnetic $Fe_3O_4$ NPs can be made to heat-up locally that leads to their use as hyperthermia agents, delivering toxic amounts of thermal energy to targeted bodies such as tumors; or as chemotherapy and radiotherapy enhancement agents, where a moderate degree of tissue warming results in more effective malignant cell destruction. These, and many other potential applications, are made available in biomedicine as a result of the unique physical properties on different kinds of magnetic NPs which are not toxic themselves in nature. Here, we brief selective applications of such pertinent ferrofluids for bio-isolation, MR imaging (or dual imaging with optical imaging based on fluorescence), targeted drug delivery, and hyperthermia for example.

### *6.1.1. Bio-isolation*

A magnetic responsive nature of a ferrofluid that adsorbs functionalized magnetic NPs as an absorbent in combination with a magnetic separation technique offers an expedient bio-isolation process because of selective manipulation and separation properties in the presence of other suspended species in a rheological fluid. In this process, functionalized adsorbents bind selective target species directly out of a crude biological process without a perturbation field, and then in an applied magnetic field they intend to assemble in specific configurations under the influence of the inter-particle macroscopic attractive interactions [248-259]. When an applied field is removed, the absence of any magnetic memory allows soft NPs agglomerates to redisperse with very fast adsorption and desorption processes so as entrained contaminants relieve over the open sample surfaces. This technique allows to bypass several stages of sample pretreatment process especially centrifugation, filtration and membrane separation before its application on a sample of packed bed chromatography columns. As described in the original research articles [98, 260-262], it is pretty fast, gentle, and easy to scale-up process that demonstrates its credibility in a wide range of disciplines which include

1) Purification and isolation of protein and enzymes,
2) Cell isolation,
3) Nucleic acid isolation.

#### 6.1.1.1. Purification and Isolation of Proteins and Enzymes

In biosciences and biotechnology, the isolation, separation and purification of proteins are generally done by the chromatography, electrophoresis, and ultrafiltration and precipitation techniques [263-268]. The magnetic NPs embedded in a pertinent fluid (ferrofluid) functionalized with suitable target protein mixed with the complex analytes at adequately low concentration is a simple and inexpensive procedure for effective detection and accurate quantification of a specific protein from a given sample to ascertain the DNA sequence and analyze disease diagnosis, biological threat defense, and drug development. In this process, a ferrofluid used possesses either an immobilized affinity or hydrophobic ligand, or ion exchange groups, which were used to mix with the target protein. For example, the magnetic NPs functionalized with a nucleic acid, chemi-luminescence enzyme, carboxyl groups, and epichlorohydrin coated with starch are used for separating recognition duplexes upon amplification, free thyroxine in human serum, DNA from chicken erythrocytes, amylases from soybean proteins, respectively [264, 265]. It is also widely used as supports to separate the target enzymes from the reaction medium and easily be removed from the systems using an external magnetic field. Further, the magnetic NPs are also considered as efficient carriers, binders and anchors to obtain His-tagged protein directly from total cell lysate. Hence, biofriendly ferrofluids are promising a novel purification system for 6xHis-tagged recombinant proteins and enzymes because of its high efficiency, easiness of operation, cost-effectiveness, biocompatibility, and versatility.

#### 6.1.1.2. Nucleic Acid Isolation

Nucleic acid isolation based on magnetic separation using a suitable ferrofluid has played an important role in a molecular biology for clinical diagnosis in comparison with traditional methods. It is a quick and efficient method and widely applied to the extraction and detection of nucleic acids from crude cells [262-265]. A ferrofluid to be used in these applications must have a good compatibility and sufficient magnetization at moderate applied magnetic field of few kOe at ambient temperature. Various types of magnetic nanocomposites such as magnetic hydrophilic polymer NPs, magnetic colloids (latexes) and surface modified magnetic NPs are used to devise a pertinent ferrofluid for the separation of the target nucleic acid either by capturing on the solid flocculates or by retaining those in the ferrofluid base liquid [264, 266-268]. On other hand, a magnetic separation module combined with a chemagic viral DNA and RNA kit using special magnetic beads could also be used to extract selective viral nucleic acids. Consequently, a safety of blood products can be maintained. As described in some research articles [269-271], nucleic acids can also be selectively separated from complex mixtures (human genomic DNA) using streptavidin-coated magnetic beads. In fact, magnetic beads have great potential for the separation of nucleic acids and as a platform for optimized DNA hybridization.

The ferrofluids consisting of anionic magnetic NPs like $Fe_3O_4$ and other ferrites with hydrophilic cationic poly[(*N*-isopropylacrylamide)-*co*-styrene] in a core–shell structure have great future for promptly extracting cells, proteins, DNA, RNA, bacteria and viruses in a very short time-period of less than 5 min because of adequately large specific surface area, good magnetic properties, high colloidal stability and chemical stability in the flocculates. As a matter of fact, using a suitable ferrofluid can reduce significantly the identification time of diagnostically important genes for a target organism with high specificity and sensitivity.

#### 6.1.1.3. Cell Isolation

Cell isolation is often advantageous to separate out specific biological entities from their native environments for the production of concentrated samples of these entities or to free an infected sample there from. It is a two-step process. In the first step, a magnetic particle is attached to a desired biological entity, and then in the second step, the magnetic NPs are pulled together with their "prey" out of the native environment by the action of a magnetic field gradient. Magnetically separable cells can be classified into two major categories of magnetic and non-magnetic cells. Magnetic cells can be magnetically separated via a direct isolation of target cells from a crude material such as blood, food, water, etc. [257-259, 271]. In the other case, the non-magnetic cells should firstly be coated with a magnetic layer using affinity ligands and then can be separated by using an external magnetic field. The surface of the magnetic particle is usually functionalized by an additional coating such as dextran, polyvinyl alcohol and many others [257-259]. Usages of cell isolation include clean-up bone marrow from cancer infected area, capturing dead and live bacteria such as *aeromonas, salmonella, pseudomonas, enterococcus, bacillus, staphylococcus* and *E. coli* from matrix, purification of the targeted cells, and removing of molecules and living cells flowing biological fluids without involving any washing steps [251, 272]. In this process, using surface modified $Fe_3O_4$ NPs, an average 94-98% of purified target cells could be recovered from co-culture systems of rat hepatocytes, human keratinocyte HaCaT cells and mouse embryonic stem cells. Additionally, the separation efficiency of approximately 95% is obtained in isolation of human hematopoietic progenitor cells from umbilical cord blood and peripheral blood based on magnetic nano-bacterial spheres.

### *6.1.2. Targeted Drug Delivery*

A ferrofluid drug delivery technique is considered as a very efficient tool of delivering a drug to localized disease sites in a body. It is based on the magnetic drug transport technique in which the drug can be either encapsulated in small packets in a ferrofluid or conjugated on the surface of the flocculates in a ferrofluid. When the ferrofluid carrier is intravenously administered, the accumulation takes place within selected regions to which the magnetic field is applied. The magnetic field helps to extravasation the accumulated ferrofluid carrier into the targeted regimes so that very high concentration of chemotherapeutic agents can be converged near the target site without any toxic effect [273, 274]. Accumulation of ferrofluid carrier depends on physiological parameters, e.g., particle size, surface characteristics, field strength, and blood flow rate, etc.

According to the mechanism of control of drug release, drug delivery systems can be divided into diffusion-controlled systems, chemically controlled systems, solvent-activated systems and modulated-release systems. Most drug delivery systems act by a combination of different mechanisms depending upon the ferrofluid used. Controlled release of therapeutic agents with carriers of magnetic NPs dispersed in a ferrofluid has received growing interest owing to its numerous advantages such as high delivery efficiency, site-specific therapy, and magnetic resonance [275-280]. For example, surface modified magnetic $Fe_3O_4$ NPs have been considered as a most suitable material for drug delivery due to low cytotoxicity, high surface area, tunable size, biocompatibility, and biodegradable characteristics. $Fe_3O_4$ NPs coated with rhodamine B through grafting gum Arabic show efficient delivery of dexamethasone drug by a photo-triggered-on-demand drug response in the near-infrared region [281]. An enhanced chemotherapy effect of doxorubicin is also observed in poly(vinyl pyrolidone) or poly(oxyethylene 25-propylene glycol stearate) coated with doxorubicin by combining with drug-conjugated magnetic $Fe_3O_4$ NPs [282]. Moreover, the $Fe_3O_4$ NPs could efficiently enhance the relevant drug infusion into cancer cells via internalization/endocytosis processes when the drugs are delivered in terms of reducing their clearance from joints because of surface energized species in small sizes [282, 283].

### *6.1.3. Magnetic Resonance Imaging*

Magnetic resonance imaging (MRI) is one of the most powerful non-invasive imaging methods being widely utilized in clinical medicines to distinguish the pathologic tissues from the normal tissues through the contrast imaging. It results in different signal intensities produced by each tissue in response to a particular sequence of applied radiofrequency (RF) pulses. This response is dependent on the proton density and magnetic relaxation times [284, 285]. In general, the tissue or organ-specific contrast agents consist of two components: a magnetic label capable of altering the signal intensity on the MR images and a target-group molecule having a characteristic affinity to a specific type of cell or receptor. Recently, developments of molecular and cellular imaging have focused on visualizing disease-specific biomarkers at the molecular and cellular levels, and the migration of cells labeled with magnetic resonance contrast agents and tissues [286, 287]. These systems in general include biocompatible magnetic NPs and are able to degrade into eliminable fragments after a drug release. A basic objective of such materials is to promote controlled and local drug delivery through the incorporation of magnetic NPs conjugated with target-specific biomolecules with coordinate bonding, which serves as a contrast agent for MRI by increasing the relaxation rates of water protons in tissues the agent accumulates [287]. Those have been used as MRI contrast agents due to their very large net positive magnetic susceptibility in the presence of an external magnetic field.

A clinical application of ferrofluids includes contrast-enhanced imaging of tumor, liver, spleen, and lymph nodes. The flocculates also serve as a perfusion agent. The use of magnetic NPs as imaging of the reticulo-endothelial system (RES) and perfusion agents has several advantages. Some of the advantages are: site-specific in vivo mapping, cellular tagging and in vivo transport, in vivo mapping of blood-brain barrier disruption and development of combined “drug systems” [288-292]. Site-specific in vivo mapping implies the attachment of site specific molecules, such as sugars and proteins, to the magnetic NPs. The fictionalization of surface of NPs for an MRI contrast agent depends on the type of the diseases. Thermally cross-linked $Fe_3O_4$ NPs and poly(ethylene-glycol) (PEG) (HOOC-PEG-COOH)-coated $Fe_3O_4$ nanocrystals conjugated with a cancer-targeting antibody anti-carcinoembryonic antigen (CEA) monoclonal antibody rch 24 (rch 24 mAb) is used for a tumor detection. $Fe_3O_4$, $\gamma$-$Fe_2O_3$ and similar oxide magnetic NPs surface modified with dextran or citrate are used for diagnostic of diverse range of diseases including cardiovascular and neurological disorders and cancer. Metallic magnetic NPs and metal complexes of manganese, iron and gadolinium are used as MRI contrast agents [293-295]. Gadolinium is bit toxic and ferric ions are not compatible to some regimes of human body such as gastrointestinal tract, which produces gastric irritation and slow clearance of the contrast agent.

The factors that influence MRI signal include particle size (also shape), surface coating material and its type, surface topology of the flocculates, average surface of the fluid, dopants if any used, different phases present, etc [249, 289, 296]. The signal sensitivity of MRI largely depends on the specificity and selectivity of the ligands used to conjugate onto the magnetic NPs surfaces to target the NPs to specific tissues. In this regard, many ligands have been investigated, such as antibody-antigen, herceptin, folic acid, nucleic acid, dextran, starch, dopamine or catechaldehyde and genes bearing functional groups such as amino and formyl groups [289, 291, 297-300]. As a result, MRI is currently the most accurate, noninvasive technique for detecting and treating human diseases such as ovarian, cervical and breast carcinoma, via dipole and exchange interactions. Multifunctional oxide magnetic NPs like $Fe_3O_4$ represent a key milestone in magnetic imaging due to their versatility in encapsulation by both hydrophilic or hydrophobic active moieties and conjugation of various targeting and bioresponsive moieties. They are highly safe to the human body with good selectivity and targetability [301-303].

### *6.1.4. Hyperthermia*

Hyperthermia (heat) treatment is the oldest technique for curing malignant tumors. Use of magnetic induction hyperthermia provides a better therapeutic modality for treating malignant tumors, leading to kill cancer not with drugs, but with targeted NPs heaters in the presence of an alternating magnetic field [302, 303]. Magnetic hyperthermia is an identified technology for treating tumors based on injecting a ferrofluid in the diseased localized regime of a living body. The ferrofluid heats-up the selected regime by absorbing electromagnetic energy at a particular frequency without heating the surrounding much at the same time. A basic principle involved is that the magnetic NPs dispersed in the fluid carrier can generate heat by a hysteresis loss, Neel relaxation loss, and Brownian loss upon applying an alternating magnetic field. This technique is applied for detecting and curing various types of tumors [302].

In particular, intracellular hyperthermia methods have been recommended with magnetic NPs, whereby the particles are concentrated at the tumor site and they are remotely heated using an applied magnetic field in order to achieve a requisite hyperthermic temperature (42-45°C). The normal cells survive safely at these temperatures. The cancer cells burn out at higher temperatures. [303-305]. Magnetic NPs with the Curie temperature in the range 43-46°C, low coercive field $H_c$ and high $M_s$ are suitable for hyperthermia therapy [250, 304]. The

$H_c$ and $M_s$ largely vary with size and shape of the NPs. Moreover, the NPs used should possess non-toxicity, biocompatibility, injectability, and high-level accumulation in the target tumor.

Recently, Johannsen et al. [305] and Wust et al. [306] found in the clinical trials that 10-30 nm $Fe_3O_4$ NPs have potential on mediated thermotherapy on prostate carcinoma, breast cancer and on the tolerance of intestinal tumors. Furthermore, $Fe_2O_3$ and $As_2O_3/Fe_2O_3$ NPs complexes used in hyperthermia therapy indicated that the heating characteristics of the ferrofluid used depend on different particle concentrations, and the complexes exerted radiofrequency-induced hyperthermia and drug toxicity to tumors without any liver or kidney damage [307, 308]. Various kinds of magnetic NPs are being used for hyperthermia that include dextran, silica, liposomes, cobalt ferrite, iron, magnetite, and hybrid Au/$Fe_3O_4$-loaded thermo-responsive micelles of amphiphilic block copolymer like poly[(*N*-iso-propylacrylamide)-*co*-acrylamide]-*block*-poly(ε-caprolactone) [308-311]. It is proved that the efficiency of heat dissipation from superparamagnetic NPs is enhanced further when the particles size distribution is narrow. Maximum heat dissipation is observed for a suspension with an average 14 nm particle diameter. In this condition, the heating efficiency of the dispersed magnetic suspension is improved markedly [309-311].

## 6.2. Deformable Mirror Technology

In adaptive optics, the conventional deformable mirror devised by using solid thin plates or membranes is immensely used for optical applications [312, 313]. They are very expensive and are typically limited to diameters of a few cm and strokes of a few microns, a major limitation for some applications like vision science. A liquid mirror offers an interesting alternative that does not suffer from these restrictions. For example, they can have smooth deviations from flatness measuring several centimeters. Supremacy of liquid optics arises from the fact that the surface in a liquid adapts an equipotential surface to a very high precision. Optical shop tests have shown that rotating mercury mirrors have excellent optical qualities [314, 315]. Liquid mirror telescopes (LMTs) are built and are in use to collect astronomical, space science and atmospheric sciences data [315, 316]. These liquid mirrors bear low cost, about two orders of less value than the conventional parabolic glass mirrors. Ferrofluids have low reflectivity and for many applications need to be coated with a reflective layer. This can be done with nano-engineered reflective liquids based on interfacial films of silver particles known as metal liquid-like films (MELLFs). The MELLFs have combined properties of metals and liquids which can be deformed easily, and are therefore widely adapted to applications in the field of liquid optics [89, 317, 318].

More recently, highly reflective Ag-NPs surface films have been spread on water, ethylene glycol, and several other hydrophilic liquids containing a ferrofluid in order to follow the surface contours as they are dynamically modified with an array of magnetic actuators in the base of the mirror [319, 320]. The use of ethylene glycol as a polar carrier liquid for ferrofluid has been proposed. Particle stabilization in polar carrier liquids, such as water or ethylene glycol, has been achieved by introducing surface charges. It is well known that the particle aggregation is prevented by electrostatic stabilization employing the surface adsorption of citrate or hydroxide ions to produce negatively charged NPs [321, 322]. Bilayers of ionic surfactants have also been reported to provide electrostatic stabilization through an outer layer of charged head groups surrounding the NPs [323]. Moreover, a difficulty is that the magnetic NPs stabilized with citrate (ferrofluid) are not compatible with the reflective surface films of Ag-NPs.

A ferrofluid composed of positively charged $Fe_3O_4$ NPs coated with 2-[2-(2-methoxyethoxy)ethoxy]acetic acid (MOEEAA) dispersed in ethylene glycol promises a highly stable magnetic response in a magnetic field. MOEEAA exhibits a strong affinity for the carrier liquid (ethylene glycol) due to an ethoxy group (-O-$CH_2$-$CH_2$-) within the chain. Furthermore, the carboxylate group ensures stable grafting onto the $Fe_3O_4$ NPs, increases stability, and prevents unwanted NPs agglomeration through creation of repulsive barrier in ethylene glycol liquid carrier. Moreover, the positive charge in MOEEAA coated $Fe_3O_4$ NPs is compatible with the Ag- NPs net positive charge. Hence, the MOEEAA based system with the deposition of surface films of Ag-NPs allows preparing a magnetically deformable liquid surface. Such mirrors exhibit optical quality surfaces and magnetic performance and that remain stable more than three months. The identification of a ferrofluid compatible with the Ag-NPs constitutes a key step in developing commercial mirrors based on these technologies [319-323].

## 6.3. Dynamic Sealing

The idea of controlling the leakage of a machine using a ferrofluid with a magnetic field has led to many innovative applications. Localizing a ferrofluid tailored by the field leads to develop excellent sealing materials. The basic seal components consist of a ferrofluid, a permanent magnet, two pole pieces and a magnetically permeable shaft. The magnetic structure, completed by the stationary pole pieces and the rotating shaft, concentrates magnetic flux in the radial gap under each pole. When a ferrofluid is applied to the radial gap, it assumes the shape of a "liquid O-ring" and produces a hermetic seal. This is achieved by a hole inside a magnet and a shaft made of a soft magnet. A groove in the shaft is filled-up with the ferrofluid, which is kept in place by the magnetic field, obstructing the passage of any impurity, but leaving the axle free to rotate, because the obstructing material used is a mobile liquid [89, 240].

Such magnetic sealants have several advantages over conventional mechanical sealants, (for example, low viscous drag, 100% torque transmission, high speed capability, non-contaminating, long and reliable life, very good leak tightness, wide temperature range. Such sealants are being used in a variety of machines including 100% leak-free, no-wear vacuum rotary seal for use in the manufacture of semiconductor wafers, vacuum processing applications, centrifuges, or computer hard disk drives that prevent contamination, increased memory capacity and improved processing throughput. Ferrofluids based hermetic seal has been developed for smooth operation of airborne targeting cameras, which are used even in harsh environments of military aircrafts. This property is exploited for separating mixtures of industrial scrap metals such as titanium, aluminum and zinc for sorting diamonds. Another promising application is to use magnetic fluid inks for high speed, cost effective and noise-free printers [324, 325].

### 6.4. Heat dissipation and Damping

A magnetisation in the range 4-40 $kAm^{-1}$ and a viscosity 1 mPa to 1 Pa make a ferrofluid practically useful [240, 324, 325]. High viscosity gives excellent damping features of a fluid which can be held in position with a magnetic field. Loudspeakers widely use variety of ferrofluids with the voice coil oscillates between the pole pieces of a circular magnet. The power handling capacity and the sound quality are major limitations. The voice coil can damage at high powers through overheating while the reverberations deteriorate the sound quality. One way of extracting the heat away from a device heats-up in the functioning, so that it turns-up not so hot, is by using a good heat conductor interfaced to the system with a heat sink of an open surface to dissipate the heat relieves. In some cases, the heat conductor may not be necessarily a solid otherwise it can block the device operating. This problem can be resolved if using a ferrofluid as a heat-exchanger. A ferrofluid kept in place with a magnet, which is fixed on the loudspeaker's horn, transfers heat from the coil heats-up during the functioning. Nowadays most of high power loudspeakers are equipped with some ferrofluids. A fluid present around the coil damps resonances produce unpleasant noise. The loudspeakers use a ferrofluids damper [81, 240].

In an inertia damper of a ferrofluid, the settling time of a stepper motor can be reduced by connecting the damper to the shaft of the stepper motor. The seismic mass which consists of a series of permanent ring magnets levitates in the ferrofluid, often hosted in an aluminium container. No support or bearings are required. As the shaft rotates and is bought to the rest, a frictional force the ferrofluid exerts on the seismic mass produces a torque which opposes the motion of the stepper motor assembly bringing it rapidly to rest without 'ringing'.

### 6.5. Magnetic Inks

Practically, a ferrofluid behaves as an ink whose magnetic response enables to use it to produce fluid jets that can be deflected by magnetic fields to form characters. More familiar high speed printing uses electrostatic fields for a similar purpose. Both the processes produce high quality prints at high speeds but the magnetic system reserves its own advantage that it requires pretty low voltage. Newer printing techniques are emerging of magnetic ink allows attract magnetized characters formed on a magnetic tape prior to printing. As described at several places [326, 327], the ink is then transferred onto paper by a direct contact and the characters erase at the end of the printing process. Another advantage with such ink is the magnetic properties retain even on the carrier evaporates out. Here, the characters can be reproduced with a magnetic reader on the remnant is significantly large. Use of magnetic ink in magnetic reading has open opportunity in the field of security printing, reading bank notes and bounce cheques.

## 7. Future Prospective of Ferrofluids

The science and technology of ferrofluids and derivatives have created great excitement and expectations within the last few years. Although a lot of work has been done on synthesis, assembly and stabilization of ferrofluids in different rheological media it is not yet so easy to compromise high quality with long lasting reproducible features for wide variety of technologies. The next decade is likely to witness major strides in the preparation, characterization and exploitation of various shapes of magnetic NPs and other nanounits, and their assemblies. There is plenty of scope in fabrication, implementation and commercialization of small items and devices that the scientists/engineers can nature from natural and synthetic nanofluids. Such technological developments are bound to deliver a tremendous impact on biotechnology, mechanical, chemical, energy, electronics and space industries. Medicines, health care, genomics, drug and gene delivery, biosensors, etc. are other important disciplines of utilizing such materials.

The biomedical applications, which until now have been based almost exclusively on experiments with small animals, are turning rapidly to uses on the mankind. Treating a cancer fast, for example, needs develop alternatives of traditional medicines with use of smart nanofluids for drug targeting, hyperthermia, or combining this technique with traditional radiotherapy, chemotherapy, or even surgery, is the subject of immense research in the future. In this view, a ferrofluid will be utilized for biocompatible coatings that give affinities to different kinds of the living cells. Yet limiting to preliminary research, the computer technology that produces very fast, powerful and not expensive computers, as well as very convenient softwares can be used in conjunction with realistic models for numerical simulation of predicting properties and qualities of ferrofluids to be tested in the laboratory. Stochastic molecular dynamics are in use for this purpose. Associated with the hydrodynamic forces, this approach can lead to very realistic predictions of ferrofluids with functional properties and applications.

## REFERENCES

[1] Friedel, D., 1922, "The Mesomorphic States of Matter," *Ann. Physique*, 18, 273-474.

[2] Critchley, J. P., 1970, "A Review of the Poly(azoles)," *Progr. Polymer Sci.*, 2, 47-49.

[3] De Gennes, P. G., 1979, *Scaling Concepts in Polymer Physics*, Cornell University Press, Ithaca, NY, Vol. 1, 13-17.

[4] Gray, G. W., 1962, *Molecular Structure and the Properties of Liquid Crystals*, Academic Press, NY.

[5] Lehmann, O., 1889, "On Flowing Crystals," *Z. Phys. Chem.*, 4, 462-472.

[6] Jannik, G., and Des, C., 1990, "Polymers in Solution," *J. Phys.: Condens. Matter.*, 2, 1-24.

[7] Chidsey, E. E. D., and Murry, R. W., 1986, "Electroactive Polymers and Macromolecular Electronics," *Science*, 231, 25-31.

[8] Krutschmer, W., Lamb, L. D., Fostiropoulos, K., and Huffman, D. R., 1990, "Solid $C_{60}$: A New Form of Carbon," *Nature*, 347, 354-358.

[9] Siegel, R. W., 1991, "Cluster-Assembled Nanophase Materials," *Annu. Rev. Mater. Sci.*, 21, 559-578.

[10] Arriagada, F. J., and Osse-asave, K. J., 1995, "Synthesis of Nanosize Silica in Aerosol OT Reverse Microemulsions," *Coll. Inter. Sci.*, 170, 8-17.

[11] Ram, S., 2001, "Self-confined Dimension of Thermodynamic Stability in Co-nanoparticles in *fcc* and *bcc* Allotropes with a Thin Amorphous $Al_2O_3$ Surface Layer," *Acta Mater.*, 49, 2297-2307.

[12] Mohanty, P., and Ram S., 2002, "Confined Growth of $Eu_2O_3$ Nanocrystals in a New Polymorph in Amorphous Mesoporous $Al_2O_3$," *Philos. Magn. B*, 82, 1129-1144.

[13] Gleiter, H., 2003, "Discrete Dislocation Plasticity Modeling of Short Cracks in Single Crystals," *Acta Mater.*, 51, 1-29.

[14] Graham, T., 1861, "Liquid Diffusion Applied to Analysis," *Philos. Trans. R. Soc.*, 151, 183-224.

[15] Blums, E., 1995, "New Applications of Heat and Mass Transfer Processes in Temperature-Sensitive Magnetic Fluids," *Brazilian J. Phys.*, 25, 112-117.

[16] Albrecht, T., Bührer, C., Fähnle, M., Maier, K., Platzek, D., and Reske, J., 1997, "First Observation of Ferromagnetism and Ferromagnetic Domains in a Liquid Metal," *Appl. Phys. A: Mater. Sci. & Process.,* 65, 215-220.

[17] Odenbach, S., 2003, "Ferrofluids: Magnetically Controllable Fluids and Their Applications," Editor: Odenbach, S., *Applied Rheology*, Springer, NY, Vol. 4, 03-18.

[18] Berkovsky, B. M., Medvedev, V. F., Krakov, and M. S., 1989, *Magnetic Fluids*, Moscow, Khimia, Russian, 240.

[19] Raj, K., 1996, "Magnetic Fluids and Devices: A Commercial Survey," Editors: Berkovski, B., and Bashtovoi, V., *Magnetic Fluids and Applications Handbook*, Begell House, NY, 657-751.

[20] Black, C. T., Murray, C. B., Sandstrom, R. L., and Sun, S., 2000, "Spin Dependent Tunneling in Self-Assembled Cobalt-Nanocrystal Superlattices," *Science*, 290, 1131-1134.

[21] Gu, H., Ho, P. L., Tsang, K. W. T., Wang, L., and Xu, B., 2003, "Using Biofunctional Magnetic Nanoparticles to Capture Vancomycin-Resistant Enterococci and Other Gram-Positive Bacteria at Ultralow Concentration," *J. Am. Chem. Soc.*, 125, 15702-15703.

[22] Chen, M., Liu, J. P., and Sun, S., 2004, "One-Step Synthesis of FePt Nanoparticles with Tunable Size," *J. Am. Chem. Soc.*, 126, 8394-8395.

[23] Yan, Q., Purkayastha, A., Kim, T., Kroger, R., Bose, A., and Ramanath, G., 2006, "Synthesis and Assembly of Monodisperse High-Coercivity Silica-Capped FePt Nanomagnets of Tunable Size, Composition, and Thermal Stability from Microemulsions," *Adv. Mater.*, 18, 2569-2573.

[24] Gao, J., Li, L., Ho, P. -L., Mak, G. C., Gu, H., and Xu, B., 2006, "Combining Fluorescent Probes and Biofunctional Magnetic Nanoparticles for Rapid Detection of Bacteria in Human Blood," *Adv. Mater.*, 18, 3145-3148.

[25] Srivastava, S., Samanta, B., Jordan, B. J., Hong, R., Xiao, Q., Tuominen, M. T., and Rotello, V. M., 2007, "Integrated Magnetic Bionanocomposites through Nanoparticle-Mediated Assembly of Ferritin," *J. Am. Chem. Soc.*, 129, 11776-11780.

[26] Kim, J., Rong, C., Liu, J. P., and Sun, S., 2009, "Dispersible Ferromagnetic FePt Nanoparticles," *Adv. Mater.*, 21, 906-909.

[27] Hou, Y., Kondoh, H., Kogure, T., and Ohta, T., 2004, "Preparation and Characterization of Monodisperse FePd Nanoparticles," *Chem. Mater.*, 16, 5149-5152.

[28] Tzitzios, V., Niarchos, D., Margariti, G., Fidler, J., and Petridis, D., 2005, "Synthesis of CoPt Nanoparticles by a Modified Polyol Method: Characterization and Magnetic Properties," *Nanotechnology,* 16, 287-291.

[29] Wang, Y., and Yang, H., 2005, "Synthesis of CoPt Nanorods in Ionic Liquids," *J. Am. Chem. Soc.*, 127, 5316-5317.

[30] Ghosh, M., Sampathkumaran, E. V., and Rao, C. N. R., 2005, "Synthesis and Magnetic Properties of CoO Nanoparticles," *Chem. Mater.*, 17, 2348-2352.

[31] Scarberry, K. E., Dickerson, E. B., McDonald, J. F., and Zhang, Z. J., 2008, "Magnetic Nanoparticle-Peptide Conjugates for In vitro and In vivo Targeting and Extraction of Cancer Cells," *J. Am. Chem. Soc.*, 130, 10258-10262.

[32] Lee, D. C., Ghezelbash, A., Stowell. C. A., and Korgel, B. A., 2006, "Synthesis and Magnetic Properties of Colloidal $MnPt_3$ Nanocrystals," *J. Phys. Chem. B*, 110, 20906-20911.

[33] Kim, J. S., Rieter, W. J., Taylor, K. M. L., An, H., Lin, W., and Lin, W., 2007, "Self-Assembled Hybrid Nanoparticles for Cancer-Specific Multimodal Imaging," *J. Am. Chem. Soc.*, 129, 8962-8963.

[34] Taylor, K. M. L., Kim, J. S., Rieter, W. J., An, H., Lin, W., and Lin, W., 2008, "Mesoporous Silica Nanospheres as Highly Efficient MRI Contrast Agents," *J. Am. Chem. Soc.*, 130, 2154-2155.

[35] Rowe, M. D., Chang, C. -C., Thamm, D. H., Kraft, S. L., Harmon Jr, J. F., Vogt, A. P., Sumerlin, B. S., and Boyes, S. G., 2009, "Tuning the Magnetic Resonance Imaging Properties of Positive Contrast Agent Nanoparticles by Surface Modification with RAFT Polymers," *Langmuir,* 25, 9487-9499.

[36] Cheng, Z., Thorek, D. L. J., Tsourkas, A., 2010, "Gadolinium-Conjugated Dendrimer Nanoclusters as a Tumor-Targeted T1 Magnetic Resonance Imaging Contrast Agent," *Angew. Chem. Int. Ed.*, 49, 346-350.

[37] Crossley, E. L., Aitken, J. B., Vogt, S., Harris, H. H., and Rendina, L. M., 2010, "Selective Aggregation of a Platinum-Gadolinium Complex within a Tumor-Cell Nucleus," *Angew. Chem. Int. Ed.*, 49, 1231-1233.

[38] Singh, G. P., and Ram, S., 2010, "Optical and Electron Paramagnetic Resonance Properties of Native $Cr_2O_3$ Surface over $CrO_2$," *J. Magn. & Magn. Mater.*, 322, 1484-1487.

[39] Kim, H., Achermann, M., Balet, P., Hollingsworth, J. A., and Klimov, V. I., 2005, "Synthesis and Characterization of Co/CdSe Core/Shell Nanocomposites: Bi-functional Magnetic-Optical Nanocrystals," *J. Am. Chem. Soc.*, 127, 544-546.

[40] Choi, J. -S., Jun, Y. -W., Yeon, S. -I., Kim, H. C., Shin, J. -S., and Cheon, J., 2006, "Biocompatible Heterostructured Nanoparticles for Multimodal Biological Detection," *J. Am. Chem. Soc.*, 128, 15982-15983.

[41] Gao, J., Zhang, B., Gao, Y., Pan, Y., Zhang, X., and Xu, B., 2007, "Fluorescent Magnetic Nanocrystals by Sequential Addition of Reagents in a One-Pot Reaction: A Simple Preparation for Multifunctional Nanostructures," *J. Am. Chem. Soc.*, 129, 11928-11935.

[42] Yang, Y., Chen, O., Angerhofer, A., Cao, Y. C., 2008, "On Doping CdS/ZnS Core/Shell Nanocrystals with Mn," *J. Am. Chem. Soc.*, 130, 15649-15661.

[43] Sun, X., Tabakman, S. M., Seo, W.-S., Zhang, L., Zhang, G., Sherlock, S., Bai, L., and Dai, H., 2009, "Separation of Nanoparticles in a Density Gradient: FeCo@C and Gold Nanocrystals," *Angew. Chem. Int. Ed.*, 48, 939-942.

[44] Zhou, T., Lu, M., Zhang, Z., Gong, H., Chin, W. S., and Liu, B., 2010, "Synthesis and Characterization of Multifunctional FePt/ZnO Core/Shell Nanoparticles," *Adv. Mater.*, 22, 403-406.

[45] Lee, Y., Garcia, M. A., Frey Huls, N. A., and Sun, S., 2010, "Synthetic Tuning of the Catalytic Properties of Au-$Fe_3O_4$ Nanoparticles," *Angew. Chem. Int. Ed.*, 49, 1271-1274.

[46] Xu, C., Xu, K., Gu, H., Zheng, R., Liu, H., Zhang, X., Guo, Z., and Xu, B., 2004, "Dopamine as a Robust Anchor to Immobilize Functional Molecules on the Iron Oxide Shell of Magnetic Nanoparticles," *J. Am. Chem. Soc.*, 126, 9938-9939.

[47] Kim, T., Reis, L., Rajan, K., and Shima, M., 2005, "Magnetic Behavior of Iron Oxide Nanoparticle-Biomolecule Assembly," *J. Magn. & Magn. Mater.*, 295, 132-138.

[48] Bertorelle, F., Wilhelm, C., Roger, J., Gazeau, F., Menager, C., and Cabuil, V., 2006, "Fluorescence-Modified Superparamagnetic Nanoparticles: Intracellular Uptake and Use in Cellular Imaging," *Langmuir*, 22, 5385-5391.

[49] Thomas, L. A., Dekker, L., Kallumadil, M., Southern, P., Wilson, M., Nair, S. P., Pankhurst, Q. A., and Parkin, I. P., 2009, "Carboxylic acid-stabilised Iron Oxide Nanoparticles for Use in Magnetic Hyperthermia," *J. Mater. Chem.*, 19, 6529-6535.

[50] Basly, B., Felder-Flesch, D., Perriat, P., Billotey, C., Taleb, J., Pourroy, G., and Begin-Colin, S., 2010, "Dendronized Iron Oxide Nanoparticles as Contrast agents for MRI," *Chem. Commun.*, 46, 985-987.

[51] Shen, L., Laibinis, P. E., and Hatton, T. A., 1999, "Bilayer Surfactant Stabilized Magnetic Fluids: Synthesis and Interactions at Interfaces," *Langmuir,* 15, 447-453.

[52] Shen, L., Stachowiak, A., Hatton, T. A., and Laibinis, P. E., 2000, "Polymerization of Olefin-Terminated Surfactant Bilayers on Magnetic Fluid Nanoparticles," *Langmuir,* 16, 9907-9911.

[53] Bucak, S., Jones, D. A., Laibinis, P. E., and Hatton, T. A., 2003, "Protein Separations Using Colloidal Magnetic Nanoparticles," *Biotechnol. Prog.*, 19, 477-484.

[54] Fried, T., Shemer, G., and Markovich, G., 2001, "Ordered Two-Dimensional Arrays of Ferrite Nanoparticles," *Adv. Mater.*, 13, 1158-61.

[55] Boal, A. K, Das, K., Gray, M., and Rotello, V. M., 2002, "Monolayer Exchange Chemistry of $\gamma$-$Fe_2O_3$ Nanoparticles," *Chem. Mater.*, 14, 2628-2636.

[56] Song, H. -T., Choi, J. -S., Huh, Y. -M., Kim, S., Jun, Y. -W., Suh, J. -S., and Cheon, J., 2005, "Surface Modulation of Magnetic Nanocrystals in the Development of Highly Efficient Magnetic Resonance Probes for Intracellular Labeling," *J. Am. Chem. Soc.*, 127, 9992-9993.

[57] Huh, Y. -M., Jun, Y. -W., Song, H. -T., Kim, S., Choi, J. -S., Lee, J. -H., Yoon, S., Kim, K. S., Shin, J. -S., Suh, J. -S., and Cheon, J., 2005, "In Vivo Magnetic Resonance Detection of Cancer by Using Multifunctional Magnetic Nanocrystals," *J. Am. Chem. Soc.*, 127, 12387-12391.

[58] Kohler, N., Fryxell, G. E., and Zhang, M., 2004, "A Bifunctional Poly(ethylene glycol) Silane Immobilized on Metallic Oxide-Based Nanoparticles for Conjugation with Cell Targeting Agents," *J. Am. Chem. Soc.*, 126, 7206-7211.

[59] Kohler, N., Sun, C., Wang, J., and Zhang, M., 2005, "Methotrexate-Modified Superparamagnetic Nanoparticles and Their Intracellular Uptake into Human Cancer Cells," *Langmuir,* 21, 8858-8864.

[60] Kohler, N., Sun, C., Fichtenholtz, A., Gunn, J., Fang, C., and Zhang, M., 2006, "Methotrexate-Immobilized Poly(ethylene glycol) Magnetic Nanoparticles for MR Imaging and Drug Delivery," *Small*, 2, 785-792.

[61] El-Boubbou, K., Gruden, C., and Huang, X., 2007, "Magnetic Glyco-nanoparticles: A Unique Tool for Rapid Pathogen Detection, Decontamination, and Strain Differentiation," *J. Am. Chem. Soc.*, 129, 13392-13393.

[62] El-Boubbou, K., Zhu, D. C., Vasileiou, C., Borhan, B., Prosperi, D., Li, W., and Huang, X., 2010, "Magnetic Glyco-Nanoparticles: A Tool to Detect, Differentiate, and Unlock the Glyco-Codes of Cancer via Magnetic Resonance Imaging," *J. Am. Chem. Soc.,* 132, 4490-4499.

[63] Bergemann, C., Muller-Schulte, D., Oster, J., Brassard, L., Lubbe, A. S., 1999, "Magnetic Ion-Exchange Nano- and Microparticles for Medical, Biochemical and Molecular Biological Applications," *J. Magn. & Magn. Mater.*, 194, 45-52.

[64] Horak, D., Babic, M., Jendelova, P., Herynek, V., Trchova, M., Pientka, Z., Pollert, E., Hajek, M., and Sykova, E., 2007, "D-Mannose-Modified Iron Oxide Nanoparticles for Stem Cell Labeling," *Bioconjugate. Chem.*, 18, 635-644.

[65] Li, Z., Chen, H., Bao, H., and Gao, M., 2004, "One-Pot Reaction to Synthesize Water-Soluble Magnetite Nanocrystals," *Chem. Mater.*, 16, 1391-1393.

[66] Hu, F., Wei, L., Zhou, Z., Ran, Y., Li, Z., and Gao, M., 2006, "Preparation of Biocompatible Magnetite Nanocrystals for In Vivo Magnetic Resonance Detection of Cancer," *Adv. Mater.*, 18, 2553-2556.

[67] Park, J.-H., Von Maltzahn, G., Zhang, L., Schwartz, M. P., Ruoslahti, E., Bhatia, S. N., and Sailor, M. J., 2008, "Magnetic Iron Oxide Nanoworms for Tumor Targeting and Imaging," *Adv. Mater.*, 20, 1630-1635.

[68] Biswas, S., Srivastava, V. K., Ram, S., and Fecht, H. -J., 2007, "Nanorods of Silver-Coated Magnetic $CrO_2$ Particles from a Polymer Template in Hot Water," *J. Phys. Chem. C*, 111, 7593-7598.

[69] Mishra, A., Tripathy, P., Ram, S., and Fecht H. -J., 2009, "Optical Properties in Nanofluids of Gold Nanoparticles in Poly(vinylpyrrolidone)," *J. Nanosci. & Nanotech.*, 9, 4342-4347.

[70] Mishra, A., Ram, S., and Ghosh, G., 2009, "D-Mannose-Modified Iron Oxide Nanoparticles for Stem Cell Labeling," *J. Phys. Chem. C*, 113, 6976-6982.

[71] Phule, A. D., Ram, S., and Tyagi, A. K., 2010, "Tailoring Absorption and Emission of Light in Dielectric Polymer Nanofluids of Reinforced Poly(vinylidene fluoride) with Silver Nanoparticles," *J. Appl. Phys.*, 107, 034316-4.

[72] Ram, S, Phule, A. D., and Tyagi, A. K., 2010, "Highly Fluorescent Silver Nanofluids with Ferroelectric Molecules," *Philos. Magn. Lett.*, 90, 781-791.

[73] Ram, S, and Fecht, H. J., 2011, "Modulating Up-Energy Transfer and Violet-Blue Light Emission in Gold Nanoparticles with Surface Adsorption of Poly(vinyl pyrrolidone) Molecules," *J. Phys. Chem. C,* 115, 7817-7828.

[74] Ram, S, Agrawal, L., Mishra, A., and Roy S.K., 2011, "Synthesis and Optical Properties of Surface Stabilized Gold Nanoparticles with Poly(Vinyl Pyrrolidone) Polymer Molecules of a Nanofluid," *Adv. Sci. Lett.*, 4, 3431-3438.

[75] Horak, D., Babic, M., Mackova, H., and Benes, M. J., 2007, "Preparation and Properties of Magnetic Nano- and Microsized Particles for Biological and Environmental Separations," *J. Sep. Sci.*, 30, 1751-1772.

[76] Derjaguin, B.V. and Landau, L. D., 1941, "Theory of the Stability of Strongly Charged Lyophobic Sols and of the Adhesion of Strongly Charged Particles in Solutions of Electrolytes," *Acta Phys. Chim. URSS*, 14, 633-662.

[77] Verwey, E. J. W., and Overbeek, J. T. G., 1948, *Theory of the Stability of Lyophobic Colloids*, Elsevier, Amsterdam, New York.

[78] Napper, D. H., 1983, *Polymeric Stabilization of Colloidal Dispersions*, Academic press, London.

[79] Cabuil,V., Bacri, J. C., Perzynsky, R., and Raikher, Yu., 1996, "Colloidal Stability of Magnetic Fluids," Editors: Berkovski, B., and Bashtovoi V., *Magnetic Fluids and Applications Handbook*, Begell House, NY, US, pp.33-56.

[80] Magalhães, M., Figueiredo Neto, A. M., Bee, A., and Bourdon, A., 2000, "Wetting of Glass Surfaces by Ionic Magnetic Fluids: Effect of the Concentration and the pH of The Solution and of the Size of the Magnetic Grains on a Surface Stabilized Birefringent Layer," *J. Chem. Phys.*, 113, 10246-10251.

[81] Wohfarth, E. P., 1980, *Ferromagnetic Materials*, North-Holland Publishing Company, Amsterdam, Vol. 2, 509-559.

[82] Croucher, M. D., and Winnik, M. A., 1990, *In an Introduction to Polymer Colloids*, Editors: Candau, F., Ottewill, R. H., Kluwer Academic, Amsterdam, 35-72.

[83] Ugelstad, J., Ellingsen, T., Berge, A., and Helgee, B., 1983, "Magnetic Polymer Particles and Process for the Preparation Thereof," Int. Pat. WO 83/03920.

[84] Massart, R., 1981, "Preparation of Aqueous Magnetic Liquids in Alkaline and Acidic Media," *IEEE Trans. Magn.*, 17, 1247-1248.

[85] Depeyrot, J., Da Silva, G. J., Alves, C. R., Sousa, E. C., Magalhães, M., Figueiredo Neto, A. M., Sousa, M. H., and Tourinho, F. A., 2001, "Static Magneto Optical Bifringence of New Electric Double Layered Magnetic Fluids," *Braz. J. Phys.*, 31, 390-397.

[86] Massart, R., Dubois, E., Cabuil, V., and Hasmonay, E., 1995, "Preparation and Properties of Monodisperse Magnetic Fluids," *J. Magn. & Magn. Mater.*, 149, 1-5.

[87] Hasmonay E., Bee, A., Bacri, J. -C., and Perzynski, R., 1999, "pH Effect on an Ionic Ferrofluid: Evidence of a Thixotropic Magnetic Phase," *J. Phys. Chem. B*, 103, 6421-6428.

[88] Dubois, E., Cabuil, V., Boué, F., and Perzynski, R., 1999, "Structural Analogy between Aqueous and Oily Magnetic Fluids." *J. Chem. Phys.*, 111, 7147-7160.

[89] Rosensweig, R. E., 1997, *Ferrohydrodynamics*, Dover.Publ.Inc., Mineola, NY, 1-31.

[90] Scholten, P. C., 1978, "Colloid Chemistry of Magnetic Fluids," Editor: Berkovski, B., *Thermomechanics of the Magnetic Fluids*, Hemisphere Publ. Corp., Bristol, PA, pp.1-26.

[91] Israelachvili, J. N., 1991, *Intermolecular and Surface Forces*, Academic, NY, USA, 291.

[92] Afkhami, A., and Norooz-Asl, R., 2009, "Removal, Preconcentration and Determination of Mo(VI) from Water and Wastewater Samples Using Maghemite Nanoparticles," *Colloids Surf. A*, 346, 52-57.

[93] Baldrian, P., Merhautov, V., Gabriel, J., Nerud, F., Stopka, P., and Hrub, M., 2006, "Decolorization of Synthetic Dyes by Hydrogen Peroxide with Heterogeneous Catalysis by Mixed Iron Oxides," *Appl. Catal. B*, 66, 258-264.

[94] Boyer, T., and Handsinger, P. C., 2005, "Bench-Scale Testing of a Magnetic Ion Exchange Resin for Removal of Disinfection by-Product Precursors," *Water Res.*, 39, 1265-1276.

[95] Stormer, M., Kleesiek, K., and Dreier, J., 2007, "High-Volume Extraction of Nucleic Acids by Magnetic Bead Technology for Ultrasensitive Detection of Bacteria in Blood Components," *Clin. Chem.*, 53, 104-110.

[96] Willner, I., Cheglakov, Z., Weizmann, Y., and Sharon, E., 2008, "Analysis of DNA and Single-Base Mutations Using Magnetic Particles for Purification, Amplification and Dnazyme Detection," *Analyst*, 133, 923-927.

[97] Mohapatra, S., Pal, D., Ghosh, S. K., and Pramanik, P., 2007, "Design of Superparamagnetic Iron Oxide Nanoparticle for Purification of Recombinant Proteins," *J. Nanosci. Nanotechnol.*, 7, 3193-3199.

[98] Ito, A., Jitsunobu, H., Kawabe, Y., Ijima, H., and Kamihira, M., 2009, "Magnetic Separation of Cells in Coculture Systems Using Magnetite Cationic Liposomes," *Tissue Eng. C*, 15, 413-423.

[99] Cormode, D. P., Skajaa, T., Fayad, Z. A., and Mulder, W. J. M., 2009, "Nanotechnology in Medical Imaging: Probe Design and Applications," *Arterioscler Thromb. Vasc. Biol.*, 29, 992-1000.

[100] Kaminski, K., Zazakowny, K., Szczubialka, K., and Nowakowska, M., 2008, "pH-Sensitive Genipin-Cross-Linked Chitosan Microspheres for Heparin Removal," *BioMacromolecules*, 9, 3127-3132.

[101] McGill, S. L., Cuylear, C. L., Adolphi, N. L., Osinski, M., and Smyth, H. D. C., 2009, "Magnetically Responsive Nanoparticles for Drug Delivery Applications Using Low Magnetic Field Strengths," *IEEE Trans Nanobiosci.*, 8, 33-42.

[102] Luo, Y. L., Fan, L. H., Gao, G. L., Chen, Y. S., and Shao, X. H., 2009, "Fe3O4/PANI/P(MAA-co-NVP) Multilayer Composite Microspheres with Electric and Magnetic Features: Assembly and Characterization," *J. Nanosci. & Nanotechnol.*, 9, 6439-6452.

[103] Mahmoudi, M., Simchi, A., Imani, M., Milani, A., and Stroeve, P., 2008, "Optimal Design and Characterization of Superparamagnetic Iron Oxide Nanoparticles Coated with Polyvinyl Alcohol for Targeted Delivery and Imaging," *J. Phys. Chem. B*, 112, 14470-14481.

[104] Taboada, E., Solanas, R., Rodriguez, E., Weissleder, R., and Roig, A., 2009, "Supercritical-Fluid-Assisted One-Pot Synthesis of Biocompatible Core(γ-Fe2O3)/Shell(SiO2) Nanoparticles as High Relaxivity $T_2$-Contrast Agents for Magnetic Resonance Imaging," *Adv. Funct. Mater.*, 19, 2319-2324.

[105] Doiron, A. L., Homan, K. A., Emelianov, S., and Brannon-Peppas, L., 2009, "Poly(Lactic-co-Glycolic) Acid as a Carrier for Imaging Contrast Agents," *Pharm. Res.*, 26, 674-682.

[106] Tang, T. Y., Muller, K. H., Graves, M. J., Li, Z. Y., Walsh, S. R., and Young, V., 2009, "Iron Oxide Particles for Atheroma Imaging Arterioscler," *Thromb. Vasc. Biol.*, 29, 1001-1008.

[107] Duan, L., Jia, S., Wang, Y., Chen, J., and Zhao, L., 2009, "Synthesis of $Fe_3O_4$ Polyhedra by Hydrothermal Method: Using l-Arginine as Precipitator," *J. Mater. Sci.*, 44, 4407-4412.

[108] Teodor, E., Litescu, S., Lazar, V., and Somoghi, R., 2009, "Hydrogel-magnetic Nanoparticles with Immobilized l-Asparaginase for Biomedical Applications," *J. Mater. Sci. Mater. Med.*, 20, 1307-1314.

[109] Wang, D., Duan, X., Zhang, J., Yao, A., Zhou, L., and Huang, W., 2009, "Fabrication of Superparamagnetic Hydroxyapatite with Highly Ordered Three-dimensional Pores," *J. Mater. Sci.*, 44, 4020-4025.

[110] Huska, D., Hubalek, J., Adam, V., Vajtr, D., Horna, A., and Trnkova, L., 2009, "Automated Nucleic Acids Isolation using Paramagnetic Microparticles Coupled with Electrochemical Detection," *Talanta*, 79, 402-411.

[111] Li, Y. G., Gao, H. S., Li. W. L., Xing, J. M., and Liu, H. Z., 2009, "In situ Magnetic Separation and Immobilization of Dibenzothiophene-desulfurizing Bacteria," *Bioresour. Technol.*, 100, 5092-5096.

[112] Veyret, R., Elaissari, A., Marianneau, P., Sall, A. A., and Delair, T., 2005, "Magnetic Colloids for the Generic Capture of Viruses," *Anal. Biochem.*, 346, 59-68.

[113] Jin, H., Lin, J.-M., Wang, X., Xin, T., Liang, S., and Li, Z., 2009, "Magnetic Particle-based Chemiluminescence Enzyme Immunoassay for Free Thyroxine in Human Serum," *J. Pharm. Biomed. Anal.*, 50, 891-896.

[114] Olsvik, O., Popovic, T., Skjerve, E., Cudjoe, K. S., Hornes, E., and Ugelstad, J., 1994, "Magnetic Separation Techniques in Diagnostic Microbiology," *Clin. Microbiol. Rev.*, 7, 43-54.

[115] Kang, Y. S., Risbud, S., Rabolt, J. F., and Stroeve, P., 1996, "Synthesis and Characterization of Nanometer-Size $Fe_3O_4$ and $\gamma$-$Fe_2O_3$ Particles," *Chem. Mater.*, 8, 2209-2211.

[116] Fried, T., Shemer, G., and Markovich, G., 2001, "Ordered Two-Dimensional Arrays of Ferrite Nanoparticles," *Adv. Mater.*, 13, 1158-1161.

[117] Moumen, N., and Pileni, M. P., 1996, "Control of the Size of Cobalt Ferrite Magnetic Fluid," *J. Phys. Chem.*, 100, 1867-1873.

[118] Liu, C., Zou, B., Rondinone, A. J., and Zhang, Z. J., 2000, "Reverse Micelle Synthesis and Characterization of Superparamagnetic $MnFe_2O_4$ Spinel Ferrite Nanocrystallites," *J. Phys. Chem. B*, 104, 1141-1145.

[119] Khalafalla, S. E., and Reimers, G. W., 1973, "Magneto-fluids and their Manufacture-Forming Non-magnetic Iron Suboxide Colloidal Dispersion," U. S. Patent-3 764 540.

[120] Brown, K. R., and Natan, M. J., 1998, "Hydroxylamine Seeding of Colloidal Au Nanoparticles in Solution and on Surfaces," *Langmuir,* 14, 726-728.

[121] Jana, N. R., Gearheart, L., and Murphy, C. J., 2001, "Evidence for Seed-Mediated Nucleation in the Chemical Reduction of Gold Salts to Gold Nanoparticles," *Chem. Mater.*, 13, 2313.

[122] Yu, H., Gibbons, P. C., Kelton, K. F., and Buhro, W. E., 2001, "Heterogeneous Seeded Growth: A Potentially General Synthesis of Monodisperse Metallic Nanoparticles," *J. Am. Chem. Soc.*, 123, 9198-9199.

[123] LaMer, V. K., and Dinegar, R. H., 1950, "Theory, Production and Mechanism of Formation of Monodispersed Hydrosols," *J. Am. Chem. Soc.*, 72, 4847-4854.

[124] Watzky, M. A., and Finke, R. G., 1997, "Transition Metal Nanocluster Formation Kinetic and Mechanistic Studies. A New Mechanism When Hydrogen Is the Reductant: Slow, Continuous Nucleation and Fast Autocatalytic Surface Growth," *J. Am. Chem. Soc.*, 119, 10382-10400.

[125] Sun, S., Zeng, H., Robinson, D. B., Raoux, S., Rice, P. M., Wang, S. X., and Li, G., 2004, "Monodisperse $MFe_2O_4$ (M = Fe, Co, Mn) Nanoparticles," *J. Am. Chem. Soc.*, 126, 273-279.

[126] Mody, V. V., Nounou, M. I., and Bikram, M., 2009, "Novel Nanomedicine-based MRI Contrast Agents for Gynecological Malignancies," Adv. Drug Deliv. Rev., 61, 795-807.

[127] Hu, F., Neoh, K. G., Cen, L., and Kang, E. T., 2006, "Cellular Response to Magnetic Nanoparticles "PEGylated" via Surface-Initiated Atom Transfer Radical Polymerization," *BioMacromolecules*, 7, 809-816.

[128] Nicolas, J., Ruzette, A. V., Farcet, C., Gerard, P., Magnet, S., and Charleux, B., 2007, "Nanostructured Latex Particles Synthesized by Nitroxide-mediated Controlled/Living Free-radical Polymerization in Emulsion," *Polymer*, 48, 7029-7040.

[129] Jiang, C., and Tsukruk, V. V., 2006, "Freestanding Nanostructures via Layer-by-Layer Assembly," *Adv. Mater.*, 18, 829-840.

[130] Guo, J., Yang, W., Wang, C., He, J., and Chen, J., 2006, "Poly(N-isopropylacrylamide)-Coated Luminescent/Magnetic Silica Microspheres: Preparation, Characterization, and Biomedical Applications," *Chem. Mater.*, 18, 5554-5562.

[131] Wang, X., Ding, X., Zheng, Z., Hu, X., Cheng, X., and Peng, Y., 2006, "Magnetic Molecularly Imprinted Polymer Particles Synthesized by Suspension Polymerization in Silicone Oil," *Macromol. Rapid. Commun.*, 27, 1180-1187.

[132] Pollert, E., Knizek, K., Marysko, M., Zaveta, K., Lancok, A., and Bohacek, J., 2006, "Magnetic Poly(glycidyl methacrylate) Microspheres Containing Maghemite Prepared by Emulsion Polymerization," *J. Magn. & Magn. Mater.*, 306, 241-247.

[133] Lu, S., and Forcada, J., 2006, "Preparation and Characterization of Magnetic Polymeric Composite Particles by Miniemulsion Polymerization," *J. Polym. Sci. A: Polym. Chem.*, 44, 4187-4203.

[134] Zeng, H., Lai, Q., Liu, X., Wen, D., and Ji, X., 2007, "Factors Influencing Magnetic Polymer Microspheres Prepared by Dispersion Polymerization," *J. Appl. Polym. Sci.*, 106, 3474-3480.

[135] Matyjaszewski, K., and Davis, T. P., 2002, *Handbook of Radical Polymerization*. John Wiley & Sons Inc., USA.

[136] Wang, J. S., and Matyjaszewski, K., 1995, "Controlled/"living" Radical Polymerization. Atom Transfer Radical Polymerization in the Presence of Transition-Metal Complexes," *J. Am. Chem. Soc.*, 117, 5614-5615.

[137] Matyjaszewski, K., and Xia, J., 2001, "Atom Transfer Radical Polymerization," *Chem. Rev.*, 101, 2921-2990.

[138] Lutz, J.-F., Stiller, S., Hoth, A., Kaufner, L., Pison, U., and Cartier, R., 2006, "One-Pot Synthesis of PEGylated Ultrasmall Iron-Oxide Nanoparticles and Their in Vivo Evaluation as Magnetic Resonance Imaging Contrast Agents," *BioMacromolecules*, 7, 3132-3138.

[139] Chiefari, J., Chong, Y. K., Ercole, F., Krstina, J., Jeffery, J., Le, T. P. T., Mayadunne, R. T. A., Meijs, G. F., Moad, C. L., Moad, G., Rizzardo, E., and Thang, S. H., 1998,

"Living Free-Radical Polymerization by Reversible Addition–Fragmentation Chain Transfer: The RAFT Process," *Macromolecules*, 31, 5559-5562.

[140] Perrier, S., and Takolpuckdee, P., 2005, "Macromolecular Design via Reversible Addition-Fragmentation Chain Transfer (RAFT)/xanthates (MADIX) Polymerization," *J. Polym. Sci. Part A: Polym. Chem.*, 43, 5347-93.

[141] Aqil, A., Vasseur S., Duguet, E., Passirani, C., Benoit, J. P., Jerome, R., and Jerome, C., 2008, "Magnetic Nanoparticles Coated by Temperature Responsive Copolymers for Hyperthermia," *J. Mater. Chem.*, 18, 3352-3360.

[142] Papaphilippou, P., Loizou, L., Popa, N. C., Han, A., Vekas, L., Odysseos, A., and Krasia-Christoforou, T., 2009, "Superparamagnetic Hybrid Micelles, Based on Iron Oxide Nanoparticles and Well-Defined Diblock Copolymers Possessing β-Ketoester Functionalities," *BioMacromolecules*, 10, 2662-2671.

[143] Underhill, R. S., and Liu, G., 2000, "Triblock Nanospheres and Their Use as Templates for Inorganic Nanoparticle Preparation," *Chem. Mater.*, 12, 2082-2091.

[144] Harris, L. A., Goff, J. D., Carmichael, A. Y., Riffle, J. S., Harburn, J. J., St Pierre, T. G., and Saunders, M. 2003, "Magnetite Nanoparticle Dispersions Stabilized with Triblock Copolymers," *Chem. Mater.*, 15, 1367-1377.

[145] Kumagai, M., Kano, M. R., Morishita, Y., Ota, M., Imai, Y., Nishiyama, N., Sekino, M., Ueno, S., Miyazono, K., and Kataoka, K., 2009, "Enhanced Magnetic Resonance Imaging of Experimental Pancreatic Tumor in vivo by Block Copolymer-coated Magnetite Nanoparticles with TGF-Beta Inhibitor," *J. Control Release*, 140, 306-311.

[146] Biswas, S., Belfield, K. D., Das, R. K., Ghosh, S., and Hebard, A. F., 2009, "Block Copolymer-Mediated Formation of Superparamagnetic Nanocomposites," *Chem. Mater.*, 21, 5644-5653.

[147] Lee, Y., Lee, H., Kim, Y. B., Kim, J., Hyeon, T., Park, H., Messersmith, P. B., and Park, T. G., 2008, "Bioinspired Surface Immobilization of Hyaluronic Acid on Monodisperse Magnetite Nanocrystals for Targeted Cancer Imaging," *Adv. Mater.*, 20, 4154-4157.

[148] Huh, Y. -M., Lee, E. -S., Lee, J.,-H., Jun, Y. -W., Kim, P. -H., Yun, C. -O., Kim, J. H., Suh, J.,-S., and Cheon, J., 2007, "Hybrid Nanoparticles for Magnetic Resonance Imaging of Target-Specific Viral Gene Delivery," *Adv. Mater.*, 19, 3109-3112.

[149] Nath, S., Kaittanis, C., Ramachandran, V., Dalal, N. S., and Perez, J. M., 2009, "Synthesis, Magnetic Characterization, and Sensing Applications of Novel Dextran-Coated Iron Oxide Nanorods," *Chem. Mater.*, 21, 1761-1767.

[150] Lewin, M., Carlesso, N., Tung, C. -H., Tang, X. -W., Cory, D., Scadden, D. T., and Weissleder, R., 2000, "Tat: Peptide-derivatized Magnetic Nanoparticles Allow in vivo Tracking and Recovery of Progenitor Cells," *Nat. Biotechnol.*, 18, 410-414.

[151] Xie, J., Wang, J., Niu, G., Huang, J., Chen, K., Li, X., and Chen, X., 2010, "Human Serum Albumin Coated Iron Oxide Nanoparticles for Efficient Cell Labeling," *Chem. Commun.*, 46, 433-435.

[152] Bulte, J. W. M., Douglas, T., Witwer, B., Zhang, S. -C., Strable, E., Lewis, B. K., Zywicke, H., Miller, B., van Gelderen, P., Moskowitz, B. M., Duncan, L. D., and Frank, J. A., 2001, "Magnetodendrimers Allow Endosomal Magnetic Labeling and in vivo Tracking of Stem Cells," *Nat. Biotechnol.*, 19, 1141-1147.

[153] Vestal, C. R., and Zhang, Z. J., 2002, "Atom Transfer Radical Polymerization Synthesis and Magnetic Characterization of $MnFe_2O_4$/Polystyrene Core/Shell Nanoparticles," *J. Am. Chem. Soc.,* 124, 14312-14313.
[154] Fan, Q. -L., Neoh, K. G., Kang, E. -T., Shuter, B., and Wang, S. -C., 2007, "Solvent-Free Atom Transfer Radical Polymerization for the Preparation of Poly(Poly(Ethyleneglycol) Monomethacrylate)-Grafted $Fe_3O_4$ Nanoparticles: Synthesis, Characterization and Cellular Uptake," *Biomaterials*, 28, 5426-5436.
[155] Wang, Y., Teng, X., Wang, J. -S., and Yang, H., 2003, "Solvent-Free Atom Transfer Radical Polymerization in the Synthesis of $Fe_2O_3$@Polystyrene Core−Shell Nanoparticles," *Nano Lett.*, 3, 789-793.
[156] Li, G., Fan, J., Jiang, R., and Gao, Y., 2004, "Cross-linking the Linear Polymeric Chains in the ATRP Synthesis of Iron Oxide/Polystyrene Core/Shell Nanoparticles, *Chem. Mater.*, 16, 1835-1837.
[157] Garcia, I., Zafeiropoulos, N. E., Janke, A., Tercjak, A., Eceiza, A., Stamm, M., and Mondragon, I., 2007, "Functionalization of Iron Oxide Magnetic Nanoparticles with Poly(Methyl Methacrylate) Brushes via Grafting-From Atom Transfer Radical Polymerization," *J. Polym. Sci. Part A: Polym. Chem.*, 45, 925-932.
[158] Czaun, M., Hevesi, L., Takafuji, M., Ihara, H., 2008, "A Novel Approach to Magneto-Responsive Polymeric Gels Assisted by Iron Nanoparticles as Nano Cross-Linkers," *Chem. Commun.*, 2124-2126.
[159] Hu, F., Neoh, K. G., Cen, L., Kang, and E. -T., 2006, "Cellular Response to Magnetic Nanoparticles "PEGylated" via Surface-Initiated Atom Transfer Radical Polymerization," *BioMacromolecules*, 7, 809-816.
[160] Saoud, F. M., Tonge, M. P., Weber, W. G., and Sanderson, R. D., 2008, "Magnetite Nanoparticles for the Preparation of Ultrapure RAFT Polymers," *Macromolecules*, 41, 1598-1600.
[161] Goff, J. D., Huffstetler, P. P., Miles, W. C., Pothayee, N., Reinholz, C. M., Ball, S., Davis, R. M., and Riffle, J. S., 2009, "Novel Phosphonate-Functional Poly(ethylene oxide)-Magnetite Nanoparticles form Stable Colloidal Dispersions in Phosphate-Buffered Saline," *Chem. Mater.*, 21, 4784-4795.
[162] Qin J., Laurent, S., Jo, Y. S., Roch, A., Mikhaylova, M., Bhujwalla, Z. M., Muller, R. N., and Muhammed, M., 2007, "A High-Performance Magnetic Resonance Imaging T2 Contrast Agent," *Adv. Mater.*, 19, 1874-1878.
[163] Jain, T. K., Richey, J., Strand, M., Leslie-Pelecky, D. L., Flask, C. A., and Labhasetwar, V., 2008, "Magnetic Nanoparticles with Dual Functional Properties: Drug Delivery and Magnetic Resonance Imaging," Biomaterials, 29, 4012-4021.
[164] Jain, T. K., Foy, S. P., Erokwu, B., Dimitrijevic, S., Flask, C. A., and Labhasetwar, V., 2009, "Magnetic Resonance Imaging of Multifunctional Pluronic Stabilized Iron-Oxide Nanoparticles in Tumor-Bearing Mice," *Biomaterials*, 30, 6748-6756.
[165] Qin, J., Jo, Y. S., and Muhammed, M., 2009, "Coating Nanocrystals with Amphiphilic Thermosensitive Copolymers," *Angew. Chem. Int. Ed.*, 48, 7845-7849.
[166] Hu, F., Neoh, K. G., and Kang, E. -T., 2009, "Synthesis of Folic Acid Functionalized PLLA-b-PPEGMA Nanoparticles for Cancer Cell Targeting," *Macromol. Rapid Commun.*, 30, 609-614.
[167] Nasongkla, N., Bey, E., Ren, J., Ai, H., Khemtong, C., Guthi, J. S., Chin, S. -F., Sherry, A. D., Boothman, D. A., and Gao, J., 2006, "Multifunctional Polymeric Micelles as

Cancer-Targeted, MRI-Ultrasensitive Drug Delivery Systems," *Nano Lett.,* 6, 2427-2430.

[168] Yang, X., Chen, Y., Yuan, R., Chen, G., Blanco, E., Gao, J., Shuai, X., 2008, "Folate-Encoded and $Fe_3O_4$-loaded Polymeric Micelles for Dual Targeting of Cancer Cells," *Polymer*, 49, 3477-3485.

[169] Kim, B. -S., Qiu, J. -M., Wang, J. -P., and Taton, T. A., 2005, "Magnetomicelles: Composite Nanostructures from Magnetic Nanoparticles and Cross-Linked Amphiphilic Block Copolymers," *Nano Lett.,* 5, 1987-1991.

[170] Shi, D., Cho, H. S., Chen, Y., Xu, H., Gu, H., Lian, J., Wang, W., Liu, G., Huth, C., Wang, L., Ewing, R. C., Budko, S., Pauletti, G. M., and Dong, Z., 2009, "Fluorescent Polystyrene-$Fe_3O_4$ Composite Nanospheres for In Vivo Imaging and Hyperthermia, *Adv. Mater.*, 21, 2170-2173.

[171] Lee, H. J., Jang, K. -S., Jang, S., Kim, J. W., Yang, H. -M., Jeong, Y. Y., and Kim, J. D., 2010, "Poly(Amino Acid)S Micelle-Mediated Assembly of Magnetite Nanoparticles for Ultra-Sensitive Long-Term MR Imaging of Tumors," *Chem. Commun.*, 46, 3559-3561.

[172] Frankamp, B. L., Boal, A. K., Tuominen, M. T., and Rotello, V. M., 2005, "Direct Control of the Magnetic Interaction between Iron Oxide Nanoparticles through Dendrimer-Mediated Self-Assembly," *J. Am. Chem. Soc.*, 127, 9731-9735.

[173] Fresnais, J., Berret, J. F., Frka-Petesic, B., Sandre, O., and Perzynski, R., 2008, "Electrostatic Co-Assembly of Iron Oxide Nanoparticles and Polymers: Towards the Generation of Highly Persistent Superparamagnetic Nanorods," *Adv. Mater.*, 20, 3877-3881.

[174] Liem, K. P., Mart, R. J., and Webb, S. J., 2007, "Magnetic Assembly and Patterning of Vesicle/Nanoparticle Aggregates," *J. Am. Chem. Soc.*, 129, 12080-12081.

[175] Toprak, M. S., McKenna, B. J., Mikhaylova, M., Waite, J. H., and Stucky, G. D., 2007, "Spontaneous Assembly of Magnetic Microspheres," *Adv. Mater.*, 19, 1362-1368.

[176] Babic, M., Horak, D., Trchova, M., Jendelova, P., Glogarova, K., Lesny, P., Herynek, V., Hajek, M., and Sykova, E., 2008, "Poly(l-lysine)-Modified Iron Oxide Nanoparticles for Stem Cell Labeling," *Bioconjugate Chem.*, 19, 740-750.

[177] Hou, Y., Kondoh, H., Shimojo, M., Sako, E. O., Ozaki, N., Kogure, T., and Ohta, T., 2005, "Inorganic Nanocrystal Self-Assembly via the Inclusion Interaction of β-Cyclodextrins: Toward 3D Spherical Magnetite," *J. Phys. Chem. B*, 109, 4845-4852.

[178] Okassa, L. N., Marchais, H., Douziech-Eyrolles, L., Herve, K., Cohen-J., S., Munnier, E., Souce, M., Linassier, C., Dubois, P., and Chourpa, I., 2007, "Optimization of Iron Oxide Nanoparticles Encapsulation within Poly(D,L-Lactide-Co-Glycolide) Sub-micron Particles," *Eur. J. Pharm. Biopharm.*, 67, 31-38.

[179] Hamoudeh, M., and Fessi, H., 2006, "Preparation, Characterization and Surface Study of Poly-Epsilon Caprolactone Magnetic Microparticles," *J. Colloid Interface Sci.*, 300, 584-590.

[180] Urban, M., Musyanovych, A., and Landfester, K., 2009, "Fluorescent Superparamagnetic Polylactide Nanoparticles by Combination of Miniemulsion and Emulsion/Solvent Evaporation Techniques," *Macromol. Chem. Phys.*, 210, 961-970.

[181] Cheng, F. -Y., Su, C. -H., Wu, P. -C., and Yeh, C. -S., 2010, "Multifunctional Polymeric Nanoparticles for Combined Chemotherapeutic and Near-Infrared Photothermal Cancer Therapy In vitro and In vivo," *Chem. Commun.*, 46, 3167-3169.

[182] Xu, X., Friedman, G., Humfeld, K. D., Majetich, S. A., and Asher, S. A., 2002, "Synthesis and Utilization of Monodisperse Superparamagnetic Colloidal Particles for Magnetically Controllable Photonic Crystals," *Chem. Mater.*, 14, 1249-1256.
[183] Yang, S., Liu, H., and Zhang, Z., 2008, A Facile Route to Hollow Superparamagnetic Magnetite/Polystyrene Nanocomposite Microspheres via Inverse Miniemulsion Polymerization," *J. Polym. Sci. Part A: Polym. Chem.*, 46, 3900-3910.
[184] Qian, Z., Zhang, Z., and Chen, Y., 2008, "A Novel Preparation of Surface-Modified Paramagnetic Magnetite/Polystyrene Nanocomposite Microspheres by Radiation-Induced Miniemulsion Polymerization," *J. Colloid. Interface. Sci.*, 327, 354-361.
[185] Lu, S., Ramos, J., and Forcada, J., 2007, "Self-Stabilized Magnetic Polymeric Composite Nanoparticles by Emulsifier-Free Miniemulsion Polymerization," *Langmuir,* 23, 12893-12900.
[186] Luo, Y. -D., Dai, C. -A., and Chiu, W. -Y., 2008, "Polystyrene/$Fe_3O_4$ Composite Latex via Miniemulsion Polymerization-Nucleation Mechanism and Morphology," *J. Polym. Sci. Part A: Polym. Chem.*, 46, 1014-1024.
[187] Cui, L., Xu, H., He, P., Sumitomo, K., Yamaguchi, Y., and Gu, H., 2007, "Developing a Hybrid Emulsion Polymerization System to Synthesize $Fe_3O_4$/Polystyrene Latexes with Narrow Size Distribution and High Magnetite Content," J. Polym. Sci. A: Polym. Chem., 45, 5285-5295.
[188] Joumaa, N., Toussay, P., Lansalot, M., and Elaissari, A., 2007, "Surface Modification of Iron Oxide Nanoparticles by a Phosphate-Based Macromonomer and Further Encapsulation into Submicrometer Polystyrene Particles by Miniemulsion Polymerization," *J. Polym. Sci. Part A: Polym. Chem.*, 46, 327-340.
[189] Yang, C., Guan, Y., Xing, J., and Liu, H., 2008, "Surface Functionalization and Characterization of Magnetic Polystyrene Microbeads," *Langmuir,* 24, 9006-9010.
[190] Antonietti, M., and Landfester, K., 2002, "Polyreactions in Miniemulsions," *Prog. Polym. Sci.*, 27, 689-757.
[191] Oh, J. K., Drumright, R., Siegwart, D. J., and Matyjaszewski, K., 2008, "The Development of Microgels/Nanogels for Drug Delivery Applications," *Prog. Polym. Sci.*, 33, 448-477.
[192] Kriwet, B., Walter, E., and Kissel, T., 1998, "Synthesis Of Bioadhesive Poly(Acrylic Acid) Nano- and Microparticles using an Inverse Emulsion Polymerization Method for The Entrapment of Hydrophilic Drug Candidates," *J. Control Release*, 56, 149-158.
[193] Landfester, K., Willert, M., and Antonietti, M., 2000, "Preparation of Polymer Particles in Nonaqueous Direct and Inverse Miniemulsions," *Macromolecules*, 33, 2370-2376.
[194] Sun, Q., and Deng, Y., 2005, "In Situ Synthesis of Temperature-Sensitive Hollow Microspheres via Interfacial Polymerization," *J. Am. Chem. Soc.*, 127, 8274-8275.
[195] Owens, D. E. E., Jian, Y., Fang, J. E., Slaughter, B. V., Chen, Y. -H., and Peppas, N. A., 2007, "Thermally Responsive Swelling Properties of Polyacrylamide/Poly(acrylic acid) Interpenetrating Polymer Network Nanoparticles," *Macromolecules*, 40, 7306-7310.
[196] Missirlis, D., Tirelli, N., and Hubbell, J. A., 2005, "Amphiphilic Hydrogel Nanoparticles. Preparation, Characterization, and Preliminary Assessment as New Colloidal Drug Carriers," *Langmuir,* 21, 2605-2613.
[197] Gao, D., Xu, H., and Philbert, M. A., and Kopelman, R., 2008, "Bioeliminable Nanohydrogels for Drug Delivery," *Nano Lett.*, 8, 3320-3324.

[198] Gaur, U., Sahoo, S. K., De, T. K., Ghosh, P. C., Maitra, A., and Ghosh, P. K., 2000, "Biodistribution of Fluoresceinated Dextran using Novel Nanoparticles Evading Reticuloendothelial System," *Int. J. Pharm.*, 202, 1-10.

[199] McAllister, K., Sazani, P., Adam, M., Cho, M. J., Rubinstein, M., Samulski, R. J., and DeSimone, J. M., 2002, "Polymeric Nanogels Produced via Inverse Microemulsion Polymerization as Potential Gene and Antisense Delivery Agents," *J. Am. Chem. Soc.*, 124, 15198-15207.

[200] Craparo, E. F., Cavallaro, G., Bondi, M. L., Mandracchia, D., and Giammona, G., 2006, "PEGylated Nanoparticles Based on a Polyaspartamide, Preparation, Physico-Chemical Characterization, and Intracellular Uptake," *BioMacromolecules*, 7, 3083-3092.

[201] Oh, J. K., Tang, C., Gao, H., Tsarevsky, N. V., and Matyjaszewski, K., 2006, "Inverse Miniemulsion ATRP: A New Method for Synthesis and Functionalization of Well-Defined Water-Soluble/Cross-Linked Polymeric Particles," *J. Am. Chem. Soc.*, 128, 5578-5584.

[202] Oh, J. K., Siegwart, D. J., Lee, H. -I., Sherwood, G., Peteanu, L., Hollinger, J. O., Kataoka, K., and Matyjaszewski, K., 2007, "Biodegradable Nanogels Prepared by Atom Transfer Radical Polymerization as Potential Drug Delivery Carriers: Synthesis, Biodegradation, in Vitro Release, and Bioconjugation," *J. Am. Chem. Soc.*, 129, 5939-5945.

[203] Oh, J. K., Siegwart, D. J., and Matyjaszewski, K., 2007, "Synthesis and Biodegradation of Nanogels as Delivery Carriers for Carbohydrate Drugs," *BioMacromolecules*, 8, 3326-3331.

[204] Siegwart, D. J., Oh, J. K., Gao, H., Bencherif, S. A., Perineau, F., Bohaty, A. K., Hollinger, J. O., and Matyjaszewski, K., 2008, "Biotin-, Pyrene-, and GRGDS-Functionalized Polymers and Nanogels via ATRP and End Group Modification," *Macromol. Chem. Phys.*, 209, 2179-2193.

[205] Siegwart, D. J., Srinivasan, A., Bencherif, S. A., Karunanidhi, A., Oh, J. K., Vaidya, S., Jin, R., Hollinger, J. O., and Matyjaszewski, K., 2009, "Cellular Uptake of Functional Nanogels Prepared by Inverse Miniemulsion ATRP with Encapsulated Proteins, Carbohydrates, and Gold Nanoparticles," *BioMacromolecules*, 10, 2300-2309.

[206] Bencherif, S. A., Siegwart. D. J., Srinivasan, A., Horkay, F., Hollinger, J. O., Washburn, N. R., Matyjaszewski, K., 2009, "Nanostructured Hybrid Hydrogels Prepared by a Combination of Atom Transfer Radical Polymerization and Free Radical Polymerization," *Biomaterials*, 30, 5270-5278.

[207] Bencherif, S. A., Washburn, N. R., and Matyjaszewski, K., 2009, "Synthesis by AGET ATRP of Degradable Nanogel Precursors for In Situ Formation of Nanostructured Hyaluronic Acid Hydrogel," *BioMacromolecules*, 10, 2499-2507.

[208] Oh, J. K., Perineau, F., and Matyjaszewski, K., 2006, "Preparation of Nanoparticles of Well-Controlled Water-Soluble Homopolymers and Block Copolymers Using an Inverse Miniemulsion ATRP," *Macromolecules*, 39, 8003-8010.

[209] Oh, J. K., Bencherif, S. A., and Matyjaszewski, K., 2009, "Atom Transfer Radical Polymerization in Inverse Miniemulsion: A Versatile Route toward Preparation and Functionalization of Microgels/Nanogels for Targeted Drug Delivery Applications," *Polymer*, 50, 4407-4423.

[210] Landfester, K., and Ramirez, L. P., 2003, "Encapsulated Magnetite Particles for Biomedical Application," *J. Phys. Condens. Matter.*, 15, S1345-61.

[211] Xu, Z. Z., Wang, C. C., Yang, W. L., Deng, Y. H., and Fu, S. K., 2004, "Encapsulation of Nanosized Magnetic Iron Oxide by Polyacrylamide via Inverse Mini-emulsion Polymerization," *J. Magn. & Magn. Mater.*, 277, 136-143.

[212] Tan, L., Pu, H., Ming, J., Chang, Z., Wan, D., and Yin, J., 2010, "Iron Nanoparticles Encapsulated in Poly(AAm-co-MAA) Microgels for Magnetorheological Fluids," *Colloids Surf. A*, 360, 137-141.

[213] Dresco, P. A., Zaitsev, V. S., Gambino, R. J., and Chu, B., 1999, "Preparation and Properties of Magnetite and Polymer Magnetite Nanoparticles," *Langmuir,* 15, 1945-1951.

[214] Wormuth, K., 2001, "Superparamagnetic Latex via Inverse Emulsion Polymerization," J. Colloid. Interface Sci., 241, 366-77.

[215] Arshady, R., 1992, "Suspension, Emulsion, and Dispersion Polymerization: A Methodological Survey," *Colloid Polym. Sci.*, 270, 717-732.

[216] Cawse, J. L., 1997, "Dispersion Polymerization," Editors: El-Aasser, M. S., and Lovell, P., *Emulsion Polymerization and Emulsion Polymers*, John Wiley & Sons Ltd., West Sussex UK,. 743-762.

[217] Song, J. -S., Tronc, F., Winnik, and M. A., 2004, "Two-Stage Dispersion Polymerization toward Monodisperse, Controlled Micrometer-Sized Copolymer Particles," *J. Am. Chem. Soc.*, 126, 6562-63.

[218] Horak, D., Bohacek, J., and Subrt, M., 2000, "Magnetic Poly(2-hydroxyethyl methacrylate-co-ethylene dimethacrylate) Microspheres by Dispersion Polymerization," *J. Polym. Sci. Part A: Polym. Chem.*, 38, 1161–71.

[219] Horak, D., 2001, "Magnetic Polyglycidylmethacrylate Microspheres by Dispersion Polymerization," *J. Polym. Sci. Part A: Polym. Chem.*, 39, 3707-15.

[220] Horak, D., Petrovsky, E., Kapicka, A., and Frederichs, T., 2007, "Synthesis and Characterization of Magnetic Poly(glycidyl methacrylate) Microspheres," *J. Magn. & Magn. Mater.*, 311, 500-506.

[221] Horak, D., Lednicky, F., Petrovsky, E., and Kapicka, A., 2004, "Magnetic Characteristics of Ferrimagnetic Microspheres Prepared by Dispersion Polymerization," *Macromol. Mater. Eng.*, 289, 341-348.

[222] Mackova, H., Kralova, D., and Horak, D., 2007, "Magnetic Poly(N-isopropylacrylamide) Microspheres by Dispersion and Inverse Emulsion Polymerization," *J. Polym. Sci. Part A: Polym. Chem.*, 45, 5884-5898.

[223] Zhang, J., Yu, D., Chen, W., Xie, Y., Wan., W., Liang, H., and Min, C., 2009, "Preparation of Poly(styrene–glucidylmethacrylate)/$Fe_3O_4$ Composite Microspheres with High Magnetite Contents," *J. Magn. & Magn. Mater.*, 321, 572-5727.

[224] Purushotham, S., Chang, P. E. J., Rumpel, H., Kee, I. H. C., Ng, R. T. H., Chow, P. K. H., Tan, C. K., and Ramanujan, R. V., 2009, "Thermoresponsive Core-Shell Magnetic Nanoparticles for Combined Modalities of Cancer Therapy," *Nanotechnology,* 20, 305101-305105.

[225] Li, G., Zeng, D. L., Wang, L., Zong, B., Neoh, K. G., and Kang, E. T., 2009, "Hairy Hybrid Nanoparticles of Magnetic Core, Fluorescent Silica Shell, and Functional Polymer Brushes," *Macromolecules*, 42, 8561-8565.

[226] Zhou, J., Meng, L., Lu, Q., Fu, J., and Huang, X., 2009, "Superparamagnetic submicro-megranates: $Fe_3O_4$ Nanoparticles Coated with Highly Cross-linked Organic/Inorganic Hybrids," *Chem. Commun.*, 6370-6372.

[227] Li, G. L., Liu, G., Kang, E. T., Neoh, K. G., and Yang, X., 2008, "pH-Responsive Hollow Polymeric Microspheres and Concentric Hollow Silica Microspheres from Silica-Polymer Core-Shell Microspheres," *Langmuir,* 24, 9050-9055.

[228] Lee, H., Lee, E., Kim, D. K., Jang, N. K., Jeong, Y. Y., and Jon, S., 2006, "Antibiofouling Polymer-coated Superparamagnetic Iron Oxide Nanoparticles as Potential Magnetic Resonance Contrast Agents for in Vivo Cancer Imaging," *J. Am. Chem. Soc.*, 128, 7383-7389.

[229] Lee, H., Yu, M. K., Park, S., Moon, S., Min, J. J., Jeong, Y. Y., Kang, H. -W., and Jon, S., 2007, "Thermally Cross-Linked Superparamagnetic Iron Oxide Nanoparticles: Synthesis and Application as a Dual Imaging Probe for Cancer in Vivo," *J. Am. Chem. Soc.*, 129, 12739-12745.

[230] Zhang, F., Wang, C. -C., 2009, "Preparation of P(NIPAM-co-AA) Microcontainers Surface-Anchored with Magnetic Nanoparticles," *Langmuir,* 25, 8255-8262.

[231] Oha, J. K., Park, J. M., 2011, "Iron Oxide-based Superparamagnetic Polymeric Nanomaterials:Design, Preparation, and Biomedical Application," Prog. Poly. Sci., 36, 168-189.

[232] Charles, S. W., and Popplewell, J., 1980, "Ferromagnetic Liquids," Editor: Wohfarth E. P., *Ferromagnetic Materials*, North-Holland Publishing Company, Amsterdam, Vol. 2, Chapter 8, 509-524

[233] Taketomi, S., Takahashi, H., Inaba, N., and Miyajima, H., 1991, "Experimental and Theoretical Investigations on Agglomeration of Magnetic Colloidal Particles in Magnetic Fluids," *J. Phys. Soc. Jap.*, 60, 1689-1707.

[234] Hong, C. Y., Horng, H. E., Jang, I. J., Hsu, C. J., Yao, Y. D., and Yang, H. C., 1997, "Ordered Structures in $Fe_3O_4$ Kerosene-based Ferrofluids," *J. Appl. Phys.*, 81, 4275-4277.

[235] Hong, C. -Y., 1999, "Field-induced Structural Anisotropy in Magnetic Fluids," *J. Appl. Phys.*, 85, 5962-5964.

[236] Sano, K., and Doi, M., 1983, "Theory of Agglomeration of Ferromagnetic Particles in Magnetic Fluids," *J. Phys. Soc. Jpn.*, 52, 2810-2815.

[237] Wang, H., Zhu, Y., Boyd, C., Luo, W., Cebers, A., and Rosensweig, R. E., 1994, "Periodic Branched Structures in a Phase-separated Magnetic Colloid," *Phys. Rev. Lett.*, 72, 1929-1932.

[238] Hong, C. Y., Horng, H. E., Jang, I. J., Hsu, C. J., Yao, Y. D., and Yang, H. C., 1998, "Magneto-chromatic Effects of Tunable Magnetic Fluid Grating," *J. Appl. Phys.*, 83, 6771-6773.

[239] Horng, H. -E., Hong, C. -Y., Lee, S. L., Ho, C. H., Yang, S. Y., and Yang. H. C., 2000, "Magnetochromatics Resulted From Optical Gratings of Magnetic Fluid Films Subjected to Perpendicular Magnetic Fields," *J. Appl. Phys.*, 88, 5904-5908.

[240] Scherer, C., and Figueiredo Neto, A. M., 2005, "Ferrofluids: Properties and Applications, *Brazilian Journal of Physics*, 35, 718- 727.

[241] Kim, D. K., Zhang,Y., Voit, W., Rao, K. V., and Muhammed, M., 2001, "Synthesis and Characterization of Surfactant-coated Cuperparamagnetic Monodispersed Iron Oxide Nanoparticles," *J. Magn. & Magn. Mater.*, 225, 30-36.

[242] Gass, J., Poddar, P., Almand, J., Srinath, S., and Srikanth, H., 2006, "Superparamagnetic Polymer Nanocomposites with Uniform $Fe_3O_4$ Nanoparticle Dispersions," *Adv. Funct. Mater.*, 16, 71-75.

[243] Zhang, J., Boyd, C., and Luo, W., 1996, "Two Mechanisms and a Scaling Relation for Dynamics in Ferrofluids," *Phys. Rev. Lett.*, 77, 390-393.

[244] Morales, M. B., Phan, M. H., Pal, S., Frey, N. A., and Srikanth, H., 2009, "Particle Blocking and Carrier Fluid Freezing Effects on the Magnetic Properties of $Fe_3O_4$-Based Ferrofluids," *J. Appl. Phys.,* 105, 07b511-1-3.

[245] Rosensweig, R. E., Kaiser, R., and Miskolczy, G., 1969, "Viscosity of Magnetic Fluid in a Magnetic Field," *J. Colloid Interface Sci.*, 29, 680-686.

[246] McTaguem, J. P., 1969, "Magnetoviscosity of Magnetic Colloids," *J. Chem. Phys.*, 51, 133-136.

[247] Odenbach, S., 2004, "Recent Progress in Magnetic Fluid Research," *J. Phys.: Condens. Matter.*, 16, R1135–R1150.

[248] Kim, J. E. L. J., Lee, S. H., Yu, J. H., Lee, J. H., Park, T. G., and Hyeon, T., 2008, "Designed Fabrication of a Multifunctional Polymer Nanomedical Platform for Simultaneous Cancer- Targeted Imaging and Magnetically Guided Drug Delivery," *Adv. Mater.*, 20, 478-483.

[249] Joshi, H. M., Lin, Y. P., Aslam, M., Prasad, P. V., Schultz-Sikma, E. A., and Edelman, R., 2009, "Effects of Shape and Size of Cobalt Ferrite Nanostructures on Their MRI Contrast and Thermal Activation," *J. Phys. Chem. C*, 113, 17761-17767.

[250] Apostolova, I., and Wesselinowa, J. M., 2009, "Magnetic Control of Ferroelectric Properties for Magnetic Hyperthermia," *Phys. Stat. Sol. B*, 246, 1925-1930.

[251] Li, S., Guo, Z., Liu, Y., Yang, Z., and Hui, H. K., 2009, "Integration of Microfiltration and Anion-Exchange Nanoparticles-based Magnetic Meparation with MALDI Mass Spectrometry for Bacterial Analysis," *Talanta*, 80, 313-320.

[252] Hung, C. W., Holoman, T. R. P., Kofinas, P., and Bentley, W. E., 2008, "Towards Oriented Assembly of Proteins onto Magnetic Nanoparticles," *Biochem. Eng. J.*, 38, 164-170.

[253] Cao, S.-W., Zhu, Y.-J., Ma, M.-Y., Li, L., and Zhang, L., 2008, "Hierarchically Nanostructured Magnetic Hollow Spheres of $Fe_3O_4$ and $\gamma$-$Fe_2O_3$: Preparation and Potential Application in Drug Delivery," *J. Phys. Chem. C*, 112, 1851-1856.

[254] Fernando, P., Yossi, W., Eugenii, K., and Itamar, W., 2003, "Magnetically Amplified DNA Assays (MADA): Sensing of Viral DNA and Single-Base Mismatches by Using Nucleic Acid Modified Magnetic Particles," *Angew. Chem. Int. Ed.,* 42, 2372-2376.

[255] Medeirosa, S. F., Santosa, A. M., Fessi, H., Elaissari, A., 2011, "Stimuli-Responsive Magnetic Particles for Biomedical Applications," *Int. J. Pharm.,* 403, 139-161.

[256] Ugelstad, J., Merk, P. C., Schmidt, R., Ellingsen, T., Berge, A., 1993, "Preparation and Biochemical and Biomedical Applications of New Monosized_Polymer Particles," *Polym. Int.*, 30, 157-168.

[257] Arbab, A., Yocum, G., Kalish, H., Jordan, E., Anderson, S., and Khakoo, A. 2004, "Efficient Magnetic Cell Labeling with Protamine Sulfate Complexed to Ferumoxides for Cellular MRI," *Blood*, 104, 1217-1223.

[258] Safaoık, I., and Safaoıkova, M., 1999, "Use of Magnetic Techniques for the Isolation of Cells," *J. Chromatogr. B*, 722, 33-53.

[259] Pankhurst, Q., Connolly, J., Jones, S., and Dobson, J., 2003, "Applications of Magnetic Nanoparticles in Biomedicine," *J. Phys. D: Applied Physics*, 36, R167-R181.

[260] Jin, H., Lin, J. -M., Wang, X., Xin, T. -B., Liang, S. -X., Li, Z. -J., Hu, and G. -M., 2009, "Magnetic Particle-based Chemiluminescence Enzyme Immunoassay for Free Thyroxine in Human Serum," *J. Pharm. Biomed. Anal.*, 50, 891-896.

[261] Franzreb, M., Siemann-Herzberg, M., Hobley, T., and Thomas, O., 2006, "Protein Purification using Magnetic Adsorbent Particles," *Appl. Microbiol. Biotechnol.*, 70, 505-516.

[262] Berensmeier, S., 2006, "Magnetic Particles for the Separation and Purification of Nucleic Acids," *Appl. Microbiol. Biotechnol.*, 73, 495-504.

[263] Safarik. I., and Safarikova, M., 2004, "Magnetic Techniques for the Isolation and Purification of Proteins and Peptides," *Biomagn. Res. Technol.*, 2, 7.

[264] Shao, D., Xu, K., Song, X., Hu, J., Yang, W., and Wang, C., 2009, "Effective Adsorption and Separation of Lysozyme with PAA-modified $Fe_3O_4$@silica Core/Shell Microspheres," *J. Colloid Interface Sci.*, 336, 526-532.

[265] Lin, J., Lin, Y. S., Kuo, S. T., Jiang, C. M., and Wu, M. C., 2009, "Purification of Soybean Amylase by Superparamagnetic Particles," *Food Chem.*, 117, 94-98.

[266] Abdelhamid, E., 2009, "Magnetic Latex Particles in Nanobiotechnologies for Biomedical Diagnostic Applications: State of the Art," *Macromol. Symp.*, 281, 14-19.

[267] Veyret, R., Elaissari, A., Marianneau, P., Sall, A. A., and Delair, T., 2005, "Magnetic Colloids for the Generic Capture of Viruses," *Anal. Biochem.*, 346, 59-68.

[268] Smith. S., Finzi, L., and Bustamante. C., 1992, "Direct Mechanical Measurement of the Elasticity of Single DNA Molecules by using Magnetic Beads," *Science*, 258, 1122-1126.

[269] Archer, M. J., Lin, B., Wang, Z., and Stenger, D. A., 2006, "Magnetic Bead-Based Solid Phase for Selective Extraction of Genomic DNA," *Anal. Biochem.*, 355, 285-297.

[270] Palecek, E., and Fojta, M., 2007, "Magnetic Beads as Versatile Tools for Electrochemical DNA and Protein Biosensing," *Talanta*, 74, 276-290.

[271] Miltenyi, S., Muller. W., Weichel, W., and Radbruch, A., 1990, "High Gradient Magnetic Cell Separation with MACS," *Cytometry A*, 11, 231-238.

[272] Xia, N., Hunt, T. P., Mayers, B. T., Alsberg, E., Whitesides, G. M., Westervelt, R. M., and Ingber D. E., 2006, "Combined Microfluidic-micromagnetic Separation of Living Cells in Continuous Flow," *Biomed. Microdevices*, 8, 299-308.

[273] Gupta, P. K., and Hung, C. T., 1989, Minireview Magnetically Controlled Targeted Micro-carrier Systems," *Life Sci.*, 44, 175-186.

[274] Jain, R.K., 1991, "Haemodynamic and Transport Barriers to the Treatment of Solid Tumours," *Int. J. Biol.*, 60, 85-100.

[275] Siepmann, J., and Peppas, N. A., 2001, "Modeling of Drug Release from Delivery Systems Based on Hydroxypropyl Methylcellulose (HPMC)," *Adv. Drug Deliv. Rev.*, 48, 139-157.

[276] Langer, R. S., and Peppas, N. A., 1981, "Present and Future Applications of Biomaterials in Controlled Drug Delivery Systems," *Biomaterials*, 2, 201-214.

[277] Langer, R., 1993, "Polymer-controlled Drug Delivery Systems," *Acc. Chem. Res.*, 26, 537-542.

[278] Richardson, T., Peters, M., Ennett. A., and Mooney, D., 2001, "Polymeric System for Dual Growth Factor Delivery," *Nature Biotechnol.*, 19, 1029-1034.

[279] Dobson, J., 2006, "Magnetic Nanoparticles for Drug Delivery," *Drug Develop. Res.*, 67, 55-60.

[280] Zhang, L., Qiao, S. Z., Jin, Y. G., Chen, Z. G., Gu, H. C., and Lu, G. Q., 2008, "Magnetic Hollow Spheres of Periodic Mesoporous Organosilica and $Fe_3O_4$ Nanocrystals: Fabrication and Structure Control," *Adv. Mater.*, 20, 805-809.

[281] Banerjee, S. S., and Chen, D. -H., 2009, "A Multifunctional Magnetic Nanocarrier Bearing Fluorescent Dye for Targeted Drug Delivery by Enhanced Two-Photon Triggered Release," *Nanotechnology*, 20, 185103-185110.

[282] Hirota, Y., Akiyama, Y., Izumi, Y., and Nishijima, S., 2009, "Fundamental Study for Development Magnetic Drug Delivery System," *Physica C*, 469, 1853-1856.

[283] Zhang, R. Y., Wu, C. H., Wang, X. M., Sun, Q., Chen, B. A., and Li, X. M., 2009, "Enhancement Effect of Nano $Fe_3O_4$ to the Drug Accumulation of Doxorubicin in Cancer Cells," *Mater. Sci. Eng. C*, 29, 1697-1701.

[284] Kim, D., Zhang, Y., Voit, W., Rao, K., Kehr, J., and Bjelke, B., 2001, "Superparamagnetic Iron Oxide Nanoparticles for Bio-medical Applications," *Scripta Mater.*, 44, 1713-1717.

[285] Chang, F. Y., Yang, Y. S., Huang, W. H., Yeh, C. S., Tsai, C. Y., Wu, C. L., and Shieh, D. B., 2005, "Haracterization of Aqueous Dispersions of $Fe_3O_4$ Nanoparticles and their Biomedical Applications," *Biomaterials*, 26, 729-738.

[286] Larson, T. A., Bankson, J., Aaron, J., and Sokolov, K., 2007, "Hybrid Plasmonic Magnetic Nanoparticles as Molecular Specific Agents for MRI/Optical Imaging and Photothermal Therapy of Cancer Cells," *Nanotechnology,* 18, 325101-325106.

[287] Hinds, K. A., Hill, J. M., Shapiro, E. M., Laukkanen, M. O., Silva, A._C., Combs, C. A., Varney, T. R., Balaban, R. S., Koretsky, A. P., and Dunbar, C. E., 2003, "Highly Efficient Endosomal Labeling of Progenitor and Stem Cells with Large Magnetic Particles Allows Magnetic Resonance Imaging of Single Cells," *Blood*, 102, 867-872.

[288] Neuwelt, E.., Abbott, N. J., Abrey, L., Banks, W. A., Davis, T., Engelhardt, B., Grammas, P., Nedergaard, M., Nutt, J., Pardridge, W., Rosenberg, G. A., Smith, Q., and Drewes, L. R., 2008, "Strategies to Advance Translational Research into Brain Barriers," *The Lancet. Neurol.*, 7, 84-96.

[289] Basti, H., Ben, T. L., Smiri, L. S., Herbst, F., Vaulay, M. -J., and Chau, F., 2010, "Catechol Derivatives-coated $Fe_3O_4$ and $\gamma$-$Fe_2O_3$ Nanoparticles as Potential MRI Contrast Agents," *J. Colloid Interface Sci.*, 341, 248-254.

[290] Shi, X., Wang, S., Swanson, S., Ge, S., Cao, Z., Van Antwerp, M. E., Landmark, K. J., and Baker Jr, J. R., 2008, "Dendrimer-Functionalized Shell-crosslinked Iron Oxide Nanoparticles for In-Vivo Magnetic Resonance Imaging of Tumors," *Adv. Mater.*, 20, 1671-1678.

[291] Hong, R. Y., Feng, B., Chen, L. L., Liu, G. H., Li, H. Z., Zheng, Y., Wei, D. G., and 2008, "Synthesis, Characterization and MRI Application of Dextran-coated $Fe_3O_4$ Magnetic Nanoparticles," *Biochem. Eng. J.*, 42, 290-300.

[292] Qiao, R., Yang, C., and Gao, M., 2009, "Superparamagnetic Iron Oxide Nanoparticles: from Preparations to In Vivo MRI Applications," *J. Mater. Chem.*, 19, 6274-6293.

[293] Lin, C., Wedeen, V., Chen, J., Yao, C., and Tseng, W., 2003, "Validation of Diffusion Spectrum Magnetic Resonance Imaging with Manganese-enhanced Rat Optic Tracts and Ex-vivo Phantoms," *Neuroimage*, 19, 482-495.

[294] McAteer, M., Sibson, N., Von M. C., Schneider, J., Lowe, A., and Warrick, N., 2007, I*n vivo* Magnetic Resonance Imaging of Acute Brain Inflammation Using Microparticles of Iron Oxide," *Nature Med.*, 13, 1253-1258.

[295] Caravan, P., 2009, "Protein-Targeted Gadolinium-Based Magnetic Resonance Imaging (MRI) Contrast Agents: Design and Mechanism of Action," *Acc. Chem. Res.*, 42, 851-862.
[296] Yathindranath. V., Hegmann, T., Lierop, J., Potter, K., Fowler, C., and Moore, D., 2009, "Simultaneous Magnetically Directed Drug Convection and MR Imaging, *Nanotechnology,* 20, 405101.
[297] Patri, A., Majoros, I., and Baker, J., 2002, "Dendritic Polymer Macromolecular Carriers for Drug Delivery," *Curr. Opin. Chem. Biol.*, 6, 466-471.
[298] Louie, A. Y., Huber, M. M., Ahrens, E. T., Rothbacher, U., Moats, R., and Jacobs, R. E., 2000, "In vivo Visualization of Gene Expression Using Magnetic Resonance Imaging," *Nature Biotechnol.*, 18, 321-325.
[299] Modo, M., Mellodew, K., Cash, D., Fraser, S., Meade, T., Price, J., and Williams, S. C. R., 2004, "Mapping Transplanted Stem Cell Migration after a Stroke: A Serial, In vivo Magnetic Resonance Imaging Study," *J. Neuroimage*, 21, 311-317.
[300] Alexiou, C., Arnold, W., Hulin, P., Klein, R., Renz, H., and Parak, F., 2001, "Magnetic Mitoxantrone Nanoparticle Detection by Histology, X-ray and MRI after Magnetic Tumor Targeting," *J. Magn. & Magn. Mater.*, 225, 187-193.
[301] Strafella, A., Paus, T., Barrett, J., and Dagher, A., 2001, "Repetitive Transcranial Magnetic Stimulation of the Human Prefrontal Cortex Induces Dopamine Release in the Caudate Nucleus," *J. Neurosci.*, 21, RC157-160.
[302] Ma, Z., and Liu, H., 2007, "Synthesis and Surface Modification of Magnetic Particles for Application in Biotechnology and Biomedicine," *China Particuology*, 5, 1-10.
[303] Corchero, J., and Villaverde, A., 2009, "Biomedical Applications of Distally Controlled Magnetic Nanoparticles," *Trends Biotechnol.*, 27, 468-476.
[304] Chan, D. C. F., Kirpotin, D. B., and Bunn Jr., P. A., 1993, "Synthesis and Evaluation of Colloidal Magnetic Iron Oxides for the Site-Specific Radiofrequency-Induced Hyperthermia of Cancer," *J. Magn & Magn. Mater.*, 122, 374-378.
[305] Johannsen, M., Gneveckoow, U., Thiesen, B., Taymoorian, K., Cho, C. H., Waldöfer, N., Scholz, R., Jordan, A., Loening, S. A., and Wust, P., 2007, "Thermotherapy of Prostate Cancer Using Magnetic Nanoparticles: Feasibility, Imaging, and Three-dimensional Temperature Distribution," *Eur. Urol.*, 52, 1653-1662.
[306] Wust, P., Gneveckow, U., Johannsen, M., Böhmer, D., Henkel, T., Kahmann, F., Sehouli, J., Felix, R., Ricke, J., and Jordan, A., 2006, "Magnetic Nanoparticles for Interstitial Thermotherapy: Feasibility, Tolerance and Achieved Temperatures," *Int. J. Hyperthermia*, 22, 673-685.
[307] An, K., and Hyeon, T., 2009, "Synthesis and Biomedical Applications of Hollow Nanostructures," *Nano Today*, 4, 359-373.
[308] Sonoda, A., Nitta, N., and Nitta-Seko, A., Shinich, O., Shigeyuki, T., Yoshio, I., Isamu, N., Jun-ichiro, J., Yasuhiko, T., Masashi, T., Osamu, M., and Kiyoshi, M., 2010, "Complex Comprised of Dextran Magnetite and Conjugated Cisplatin Exhibiting Selective Hyperthermic and Controlled-release Potential," *Int. J. Nanomed.*, 5, 499-504.
[309] Singh, R., Srinivasan, A., and Kothiyal, G., 2009, "Evaluation of $CaO$-$SiO_2$-$P_2O_5$-$Na_2O$-$Fe_2O_3$ Bioglass-ceramics for Hyperthermia Application," *J. Mater Sci. Mater. Med.*, 20, 147-151.

[310] Hamaguchi, S., Tohnai, I., Ito, A., Mitsudo, K., Shigetomi, T., and Ito, M., 2003, "Magnetite Liposomes as Potential Theranostic Agents for MRI and Magnetic Hyperthermia of Vascular Inflammation," *Cancer Sci.,* 94, 834-839.
[311] Fortin, J. -P., Wilhelm, C., Servais, J., Menager, C., Bacri, J. -C., and Gazeau, F., 2007, "Size-Sorted Anionic Iron Oxide Nanomagnets as Colloidal Mediators for Magnetic Hyperthermia," *J. Am. Chem. Soc.*, 129, 2628-2635.
[312] Tyson, R. K., 1998, *Principles of Adaptive Optics*, Academic Press, San Diego, USA.
[313] Grosso, R. P., and Yellin, M., 1977, "The Membrane Mirror as an Adaptive Optical Element," *J. Opt. Soc. Am.*, 67, 399-406.
[314] Ninane, N. M., and Jamar, C. A., 1996, "Parabolic Liquid Mirrors in Optical Shop Testing," *Appl. Opt.*, 35, 6131-6139.
[315] Cabanac, R., and Borra, E. F., 1998, "A Search for Peculiar Objects with the NASA Orbital Debris Observatory 3 Meter Liquid Mirror Telescope," *Astrophys. J.*, 509, 309-323.s
[316] Wuerker, R., 1997, "Ballastic Liquid Mirror Telescope Lidar Alignment," *Opt. Eng.*, 36, 1421-1424.
[317] Yogev, D., and Efrima, S., 1988, "Novel Silver Metal Liquid Like Films," *J. Phys. Chem.*, 92, 5754-5760.
[318] Gordon,K. C., McGarvey, J. J., and Taylor, K.P., 1989, "Enhanced Raman Scattering from Liquid Metal Films Formed From Silver Sols," *J. Phys. Chem.*, 93, 6814-6817.
[319] Dery, J. -P., Ermanno, F. B., and Anna, M. R., 2008, "Ethylene Glycol Based Ferrofluid for the Fabrication of Magnetically Deformable Liquid Mirrors," *Chem. Mater.*, 20, 6420-6426.
[320] Gingras, J., Dery, J. P., Yockell-Lelievre, H., Borra, E. F., and Ritcey, A. M., 2006, "Surface Films of Silver Nanoparticles for New Liquid Mirrors," *Colloids and Surfaces A: Physicochem. Eng. Aspects*, 279, 79-86.
[321] Dubois, E., Cabuil, V., Boue, F., and Perzynski, R., 1999, "Structural Analogy between Aqueous and Oily Magnetic Fluids," *J. Chem. Phys.*, 111, 7147-7160.
[322] Tourinho, F. A., Franck, R., and Massart, R., 1990, "Aqueous Ferrofluids Based on Manganese and Cobalt Ferrites," *J. Mater. Sci.*, 25, 3249-3254.
[323] Maity, D., and Agrawal, D. C., 2007, "Synthesis of Iron Oxide Nanoparticles under Oxidizing Environment and their Stabilization in Aqueous and Non-aqueous Media," *J. Magn. & Magn. Mater.*, 308, 46-55.
[324] Shulman, G. D., 1977, "Magnetic Fluid Sealing for Rotating Shafts has Winding and Supplementary Magnet on Sealed Shaft," USSR-544 808.
[325] Raj, K. and Moskowitz, R., 1980, "A Review of Damping Applications of Ferrofluids," *IEEE Tram. Magnetics*, 16, 358-363.
[326] Gardineer, B. A., and Sambucetti, C. J., 1974, "Obtaining $Fe_3O_4$ of High Magnetic Moment for Magnetic Inks," *IBM Tech. Disclosure Bull.*, 17, 1820-1830.
[327] Rex, D. K., 1978, "Magnetic Ink Mist Bar Printer," *IBM Tech. Disclosure Bull.*, 21, 450-451.

In: Nanofluids
Editor: Yuwen Zhang
ISBN: 978-1-62618-165-6


*Chapter 6*

# APPLICATION OF OXIDE-WATER NANOFLUIDS IN FORCED CONVECTION LOOPS: PERFORMANCE EVALUATION

***Shijo Thomas and C. B. Sobhan****
School of Nano Science and Technology,
National Institute of Technology Calicut, India

## ABSTRACT

Nanofluids, as alternate heat transfer fluids, have gained enormous attention from research laboratories and industries in recent times. The thermophysical properties and the overall convective heat transfer performance of nanofluids have been investigated extensively. This chapter aims to review the methodologies adopted for quantification of the forced convection behavior of oxide-based nanofluids. The factors that lead to the enhancement of heat transfer in nanofluids are also discussed. The influence of the two major thermophysical properties, the thermal conductivity and the dynamic viscosity are discussed. The effects of parameters like inertia, Brownian diffusion, thermophoresis, diffusiophoresis, Magnus effect, fluid drainage and gravity are also brought out. Apart from the literature review, results from experiments at the authors' laboratory on forced convection experiments at high pressure loops are also presented, which show a moderate increase in heat transfer coefficient at pressures as high as 5 bar.

## 1. INTRODUCTION

The modern trends in process intensification and miniaturization have resulted in the quest for effective heat dissipation methods in a wide spectrum of engineering applications, such as microelectronic packages, manufacturing and materials processing, energy conversion devices and bio-medical systems. Conventional methods of heat removal have been found rather inadequate to deal with high intensities of heat fluxes. One of the effective methods for

* Corresponding author. Email: csobhan@nitc.ac.in.

enhancing convective heat dissipation from surfaces is by the applications of cooling fluids with exceptional heat removal capabilities. Hence, introduction of methodologies to enhance the thermal transport behavior of engineering fluids has become an key area of interest. A number of investigations have been undertaken in the past, on the heat transfer behavior of suspensions of particulate solids in liquids, which are expected to be cooling fluids of enhanced capabilities, due to the much higher thermal conductivities of the suspended solid particles, compared to the base liquids. However, most of the earlier studies were focused on suspensions of millimeter or micron sized particles, which, although showed some enhancement in the cooling performance, also exhibited problems such as sedimentation and clogging. The gravity of these problems has been more significant in systems using passages of small dimensions.

A recent introduction into the domain of enhanced-property cooling fluids has been that of nanoparticle suspensions, termed nanofluids by Choi [1]. Advances in nanotechnology have made it possible to synthesize particles in the size range of a few nanometers, which, when suspended and stabilized in base fluids, produce nanofluids. The observed advantages of nanofluids over suspensions of micron sized particles include better stability and lower penalty on pumping power due to pressure drop, along with reduced pipe wall abrasion, on top of higher effective thermal conductivity. One of the earliest reported investigations on the enhancement of thermal conductivity in nanofluids is by Masuda et al. [2]. It has been observed by various investigators that the suspension of nanoparticles produce anomalous enhancements in various thermophysical properties of base fluids, which effect is increasingly helpful in making their use as cooling fluids very effective [3, 4, 5, 6].

## 2. Forced Convection Heat Transfer

Some of the basic quantities and correlations in forced convection heat transfer are presented below [7], as these will be helpful in understanding convective heat transfer in nanofluids. Heat transfer by forced convection depends strongly on the flow regime (whether the flow is laminar, transition or turbulent), the flow structure (whether the flow is fully developed or it is developing flow) and the flow geometry, such as whether the flow is internal or external flow, as well as the shape of the conduit or surface.

The flow regime is decided based on the Reynolds number:

$$\mathrm{Re}=\frac{\overline{U}.D_h.\rho}{\mu} \tag{1}$$

where, $\overline{U}_{inlet}$ is the mean inlet fluid velocity(m/s), $D_h$ is the hydraulic diameter(m), ρ is the density of the fluid (kg/m$^3$) and μ, the absolute viscosity of the fluid (kg/m s).

The hydrodynamic entry length, which decides whether the flow in a closed conduit is fully developed or not, in its velocity distribution, is given by

$$x \approx 0.05\,\mathrm{Re}_D.D_i \text{ for laminar flow} \tag{2}$$

$$10 \leq (x/D) \leq 60 \text{ for turbulent flow} \tag{3}$$

where Pr is the Prandtl number of the fluid, defined as $\Pr = \frac{\mu C_p}{k}$.

Similarly, the thermal entry lengths in a closed conduit are given by:

$$x \approx 0.05 \,\mathrm{Re}_D . \Pr D_i \text{ for laminar flow} \tag{4}$$

$$10 \leq (x/D) \leq 60 \text{ for turbulent flow} \tag{5}$$

The heat flux at a dissipating surface in convective heat transfer is given by the Newton's law of cooling, as follows:

$$q'' = h_x.(T_s - T_\infty). \tag{6}$$

where q" is the heat flux ($W/m^2$), $h_x$ is the local convection coefficient ($W/m^2.K$), $T_s$ is the heater surface temperature (K) and $T_\infty$ is the free stream fluid temperature (K).

By applying Fourier's law to the fluid at the wall (y = 0) in the thermal boundary layer at a distance 'x' from the leading edge, the surface heat flux can also be written as:

$$q'' = k.\left.\frac{dT}{dy}\right|_{y=0} \tag{7}$$

where k is thermal conductivity of the fluid (W/m K).

Combining equations (6) and (7), the local convection coefficient is given by,

$$h_x = \frac{k.\left.\frac{dT}{dy}\right|_{y=0}}{(T_s - T_\infty)} \tag{8}$$

Integrating the local heat transfer rate over the entire surface will yield the total heat transfer rate, Q (W).

$$Q = \bar{h}.A.(T_s - T_\infty) \tag{9}$$

where average convection coefficient $\bar{h} = \frac{1}{A_s} \int_{A_s} h_x.dA_s$

For fully developed laminar fluid flow through a circular pipe, the Nusselt number (defined as Nu = h D/k) is given by:

$$Nu_D = 4.36 \quad (10)$$

For fully developed turbulent flow through a circular pipe the Nusselt number can be calculated using the Dittus-Boelter correlation:

$$Nu = 0.023.\mathrm{Re}^{0.8}\,\mathrm{Pr}^{0.4} \quad (11)$$

In the thermally developing flow, the Nusselt number in the entry region for laminar flow can be obtained using the Shah equation (Eq. (12)) and, for turbulent flow, using the Gnielinski correlation ( Eq. (13)).

$$\left.\begin{array}{ll} Nu(x) = 1.953\left(\mathrm{Re}\,.\,\mathrm{Pr}\,\dfrac{D}{x}\right)^{1/3} & for\ \mathrm{Re}\,.\,\mathrm{Pr}\,\dfrac{D}{x} \geq 33.3 \\ Nu(x) = 4.364 + 0.0722\left(\mathrm{Re}\,.\,\mathrm{Pr}\,\dfrac{D}{x}\right) & for\ \mathrm{Re}\,.\,\mathrm{Pr}\,\dfrac{D}{x} < 33.3 \end{array}\right\} \quad (12)$$

$$Nu(x) = \frac{f/8(\mathrm{Re}-1000)\,\mathrm{Pr}}{1+12.7\sqrt{f/8}(\mathrm{Pr}^{2/3}-1)}\left[1+\left(\frac{d}{L}\right)^{2/3}\right] \quad for\ f = \frac{1}{(1.82\log_{10}\mathrm{Re}-1.64)^2} \quad (13)$$

## 3. Convective Heat Transfer in Nanofluids

Maxwell [8] investigated analytically the effect of suspensions of particles in base fluids on the thermal conductivity of the fluid. Maxwell considered dilute suspensions of spherical particles with thermal conductivity $k_p$ in a matrix with thermal conductivity $k_m$. The effective thermal conductivity $k_e$ of the suspension with volume concentration $v_p$ is given by the Maxwell relation:

$$k_e = k_m + 3v_p\frac{k_p - k_m}{2k_m + k_p - v_p(k_p - k_m)}k_m \quad (14)$$

The predictions by Maxwell's equation have been observed to agree reasonably well with suspensions of low volume fraction of particles of large sizes, but not with the effective thermal conductivity of nanofluids [9]. While the reasons for the anomalous enhancements observed in the effective properties of nanoparticle suspensions have been under investigation using fundamental theoretical models such as molecular dynamics simulations [10, 11], the practical application of nanofluids as a cooling option, especially in miniature domains, has already been undertaken extensively and effectively [12, 13]. Quantitative analysis of the heat transfer capabilities of nanofluids based on experimental measurements has been a topic of current interest. Invariably, enhancements in the thermal conductivity, viscosity and forced convection heat transfer coefficient, over the base fluids, have been observed in the investigations.

Nanofluids have significant potentials as coolants for various industrial applications such as those in space, nuclear, automotive and process industries. Their enhanced heat transfer capabilities make them good replacements for conventional cooling fluids, without requiring much of modifications in the cooling systems. The enhancement in heat transfer, on a first level, can be inferred as due to the increased thermal conductivity in nanofluids. However, the convective heat transfer coefficient depends not only on the thermal conductivity, but on the viscosity, specific heat and density of the fluid medium. All of these basic properties which get modified by the inclusion of nanoparticles are expected to alter the fundamental flow and heat transport mechanisms in the base fluid, so that the influence on the heat transfer coeffieints isessentially a synergic effect.

A number of investigations have been reported on the heat transfer performance of different types of nanofluids, under various flow regimes and different configurations of the heat transfer equipment. Convective heat transfer studies have been carried out in the developing region [16, 24] as well as under fully developed conditions [14]. Studies have been reported pertaining to laminar [14, 15, 18, 19, 20], transition [22, 25] and turbulent [19, 23] regimes of flow. Single phase and two phase flows have been analyzed with axial and radial flow directions [18]. Constant heat flux [16, 19] and constant temperature [15, 20] boundary conditions have been investigated. Studies have also been reported on flow and heat transfer in compact passages such as microchannels [17, 21]. The major parameters of interest in convective heat transfer in nanofuids are the concentration, nanoparticle material, particle size, particle shape, the base fluid, fluid temperature, presene and concentration of additives, the pH value, and the flow structure and regime. A compilation of the significant investigations, and a comparison of the convective heat transfer coefficients in different nanofluids at various flow and heat transfer conditions reported in the literature [14-36] are shown in Table 1.

## 3.1. Convection in Regular Liquid Loops

Some interesting investigations presented in the literature, on forced convection heat transfer in liquid loops under ordinary working conditions, while using nanofluids are discussed here.

Hwang et al. [14] investigated the flow and convective heat transfer characteristics of Aluminum Oxide - Water nanofluids flowing under fully developed laminar flow, through a straight circular stainless tube of 1.812 mm diameter and 2500 mm length, with constant heat flux applied on the surface. The experimental setup is shown reproduced from [14], in Figure 1. Nanofluids of volume fractions in the range of 0.01-0.3% were prepared by a two-step method, by dispersing $Al_2O_3$ nanoparticles of diameter 30±5 nm in the base fluid. The density, specific heat, viscosity and thermal conductivity of the nanofluid were measured separately. The heat transfer coefficient was calculated as given by Eq. (15), using the measured fluid temperature, surface temperature and fluid flow rate.

**Table 1. Experimental investigation of convective heat transfer in nanofluids**

| Sl.No. | Reference | Nano-particle | Base fluid | Flow regime | Wall Boundary condition | Concentration | Enhancement in heat transfer coefficient | Pressure drop/friction factor |
|---|---|---|---|---|---|---|---|---|
| 1 | Hwang et al. [14] | Al2O3(30 ±5 nm) | Water | Fully developed laminar flow with | Constant Heat Flux | 0.01-0.3 volume % | @ Re = 700 for 0.3%, heat transfer coeff., h increases by 8% | Friction factor follows $f = 64/Re_D$ |
| 2 | Heris et al. [15] | Al2O3 | Water | Laminar, Re:700-2050 | Constant wall temp. | 0.2, 0.5, 1.0, 1.5, 2.0, 2.5% volume | @ Peclet no:, Pe=6000 for 2.5%, h increases by 41% | ΔP = 200 Pa/m @ Re=700<br>ΔP = 700 Pa/m @ Re=2000 |
| 3 | Anoop et al. [16] | Al2O3(45 nm and 150 nm) | Water | Laminar thermally developing flow | Constant Heat Flux | 1 wt%, 2 wt%, 4 wt% and 6 wt% | @ x/D = 147, Re = 1550 and 4%, for 45 nm h increases by 25% and for 150 nm h increases by 11%. | - |
| 4 | Lee et al. [17] | Al2O3(36 nm) | Water | Laminar flow in microchannels, $Re_{Dh}$=140-941 | Constant heat flux | 1%, 2% by vol | @ Q=300W, Re=800 for 2%, h increases by 17%. | @ Re=800<br>ΔP = 21000 Pa for 2vol%.<br>ΔP = 15000 Pa for water. |
| 5 | Gherasim et al. [18] | Al2O3(47 nm) | Water | Laminar radial flow | Constant heat flux | 2, 4 and 6% by volume | @q"=3900 W/m2, disk spacing of 2mm and Re=500 for 4 %, heat transfer is doubled | - |

| Sl.No. | Reference | Nano-particle | Base fluid | Flow regime | Wall Boundary condition | Concentration | Enhancement in heat transfer coefficient | Pressure drop/friction factor |
|---|---|---|---|---|---|---|---|---|
| 6 | Kim et al. [19] | $Al_2O_3$(20-50 nm), amorphous carbonic nanofluids (20 nm) | Water | Laminar and turbulent flows | Constant heat flux | Amorphous carbonic nanofluids @3.5 vol%, $Al_2O_3$ nanofluids @3 vol%. | @x/D=50, Re=1460 for 3% $Al_2O_3$, h increases by 25%. @x/D=50, Re=6020 for 3% $Al_2O_3$, h increases by 15%. | - |
| 7 | Heris et al. [20] | CuO(50-60 nm), $Al_2O_3$(20 nm) | Water | Laminar flows | Constant wall temp. | 0.2-3 vol % | @Pe=6500 for 3% $Al_2O_3$ Nu=8.5. @Pe=6500 for 3% CuO Nu=8. | - |
| 8 | Jung et al. [21] | $Al_2O_3$(170 nm) | Water, Water-Ethylene glycol 50:50 | Laminar flow in rectangular microchannel | Constant heat flux | 0.6%, 1.2%, 1.8% by vol. | @x/D=0, Re=284 for 1.8% in water, h increases by 40%. @x/D=0, Re=32 for 1.8% in water-EG, h increases by 14%. | Friction factors comparable with that of water. |
| 9 | Sharma et al. [22] | $Al_2O_3$(47 nm) | Water | Hydrodynamically and thermally developed Transition flow. | Constant heat flux | 0.02%, 0.1% by vol. | for 0.1% in the range of Re=3500-8000 heat transfer enhanced by 14-24% | - |

**Table 1. (Continued)**

| Sl.No. | Reference | Nano-particle | Base fluid | Flow regime | Wall Boundary condition | Concentration | Enhancement in heat transfer coefficient | Pressure drop/friction factor |
|---|---|---|---|---|---|---|---|---|
| 10 | Duangthongsuk et al. [23] | $TiO_2$(21 nm) | Water | Turbulent flow, Re-4000-17000 | double pipe counter flow heat exchanger | 0.2 vol% | h increases by 6-11% for the flow range of Re=4000-17000 | pressure drop and friction factor of the nanofluid are close to those of water |
| 11 | Ding et al. [24] | MWCNT | Water | Laminar flow | Cosntant heat flux | 0.1%, 0.25% & 0.5% by vol. | @x/D=150, Re=1200 for 0.1% h increases by 150% | - |
| 12 | Yu et al. [25] | SiC(170 nm) | Water | Re=3300-13000 | Constant heat flux | 3.7 vol% | @Re=10000 h is enhanced by 60% | The pumping power penalty for SiC-water is lesser than for $Al_2O_3$-water. |
| 13 | Ding et al. [26] | Titanate(20 nm), CNT, titanate nanotubes(d=10nm & l=100nm), nano diamond(2-50 nm) | Water | Thermally developing laminar and turbulent flow | Constant heat flux | 0-4 vol% | Heat transfer deteriorates for ethylene glycol-based titania and aqueous-based nano-diamond nanofluids. Water-CNT nanofluids give max enhancement. | |
| 14 | Nguyen et al. [27] | $Al_2O_3$(36 nm) | Water | Confined and submerged impinging jet on a flat, horizontal circular heated surface. Re= 3800-88000. | | 0-6 vol% | for the nozzle-to-heated-surface distance of 5 mm h increases by 26% over base fluid. 6% nanofluid showed heat transfer | |

| Sl.No. | Reference | Nano-particle | Base fluid | Flow regime | Wall Boundary condition | Concentration | Enhancement in heat transfer coefficient | Pressure drop/friction factor |
|---|---|---|---|---|---|---|---|---|
| | | | | | | | performance inferior to that of base fluid | |
| 15 | Pantzali et al. [28] | CuO(30nm) | Water | Plate heat exchanger with cold water flow rate in the range of Re=1-400 | | 4 vol% | Heat removal rates for nanofluids almost same as base fluid for same flow rate. | Pumping power increased by 40% for nanofluid. |
| 16 | Ho et al. [29] | Al2O3(33 nm) | Water | Microchannel, Re:226-1676 | Uniform wall temperature condition | 1, 2 vol% | For 1 vol% nanofluid Nu increases by 40 and 53% for Re=332 and 1641. | friction factor for nanofluids and water are almost same. |
| 17 | Li and Xuan [30] | Fe3O4(10 nm) | Water and 3 wt% sodium dodecyl benzene sulfonate | Rectangular channel | Platinum wire with constant current flow. | 1 vol% | Nu for magnetic fluid decreases in presence of magnetic field parallel or perpendicular to flow direction. Nu increases by 40.6% in presence of a magnetic field gradient along the flow direction. | |
| 18 | He et al. [31] | TiO2(95 nm) | Water | Vertical circular pipe. Re=900, | Constant heat flux | 1, 2.5, 4.9 wt% | @Re=1500 and 1.1 vol% 'h' enhances by 12%.@Re=5900 | Pressure drop in laminar regime matches with Hagen-Poiseuille equation at Re<1500. Pressure drop in turbulent regime matches |

**Table 1. (Continued)**

| Sl.No. | Reference | Nano-particle | Base fluid | Flow regime | Wall Boundary condition | Concentration | Enhancement in heat transfer coefficient | Pressure drop/friction factor |
|---|---|---|---|---|---|---|---|---|
| | | | | 1500, 4800 | | | and 1.1 vol% 'h' enhances by 40%. | with Blasius equation at Re=2300-4000. |
| 19 | Chen et al. [32] | Titanate nanotubes( 10nm diameter and 100nm length) | Water | Re=1100-2300 | Constant heat flux | 0.5, 1, 2.5 wt% | @Re=1700, x/D=50.4, and 0.5, 1.0 and 2.5 wt% 'h' enhances by 5.6%, 13.2%, and 13.5%. | |
| 20 | Mansour et al. [33] | Al2O3(36 nm) | Water | Mixed convection in a horizontal and vertical copper tube | Constant heat flux | 0-4 vol% | As concn. increases from 0-4% h decreases by 10% in thermally developing region. @Re=500 for a horizontal tube for 4 vol% h decreases by 14% in fully developed flow. In vertical pipe h does not vary with concentration. | |
| 21 | Fotukian and Esfahany [34] | CuO(30-50 nm) | Water | Turbulent flow, Re=6000-31,000 | Constant surface temperature | 0.015- 0.236 vol% | @0.015%: for Re=6620 h increases by 32% and Re=29990 h increases by 22% @0.236 vol%: for | For 0.031 vol% pressure drop increased by 20%. |

| Sl.No. | Reference | Nano-particle | Base fluid | Flow regime | Wall Boundary condition | Concentration | Enhancement in heat transfer coefficient | Pressure drop/friction factor |
|---|---|---|---|---|---|---|---|---|
| | | | | | | | Re=8100 and Re=31,110, h increases by 23% | |
| 22 | Suresh et al. [35] | CuO(~15.7 nm) | Water | Helically dimpled tube under turbulent flow, Re=2500-6000 | Constant heat flux | 0.1, 0.2, 0.3 vol% | For dimpled tube Nu increases by 19.49%, 27.06%, 39.53% for 0.1, 0.2 and 0.3 vol%. | Friction factor of nanofluids almost same as that of water. |
| 23 | Fotukian and Esfahany [36] | γ-Al2O3(20 nm) | Water | Turbulent flow, Re=6000-31,000 | Constant surface temperature | 0.03-0.135 vol% | @Re=8300 and 0.135% h increases by 45%. | @Re=20000 and 0.135 vol% pressure drop increases by 30%. |

$$h(x)=\frac{q''}{T_s(x)-T_m(x)} \tag{15}$$

where $T_m(x)$ and $h(x)$ are the mean temperature of the nanofluid and the local heat transfer coefficient, respectively. The mean temperature of the fluid at any axial location was calculated as,

$$T_m(x)=T_{m,i}+\frac{q''P}{\dot{m}C_p}x \tag{16}$$

where, P, $\dot{m}$ and $C_p$ are the surface perimeter, the mass flow rate and the heat capacity, respectively. The experimental setup was calibrated using the Shah equation (Eq. (17)) for laminar flow under constant heat flux condition and the experimental values were found to have good agreement within 3% error.

$$\left.\begin{array}{l} Nu(x)=3.302\left(\left(\frac{x}{D}\right)/(\mathrm{Re.Pr})\right)^{1/3}-1,\quad for\left(\frac{x}{D}\right)/(\mathrm{Re.Pr})\le 0.00005 \\ Nu(x)=1.302\left(\left(\frac{x}{D}\right)/(\mathrm{Re.Pr})\right)^{1/3}-1,\quad for\, 0.00005<\left(\frac{x}{D}\right)/(\mathrm{Re.Pr})\le 0.0015 \\ Nu(x)=4.364+8.68\left(10^3\left(\frac{x}{D}\right)/(\mathrm{Re.Pr})\right)^{0.506}\exp\left(-41\left(\frac{x}{D}\right)/(\mathrm{Re.Pr})\right),\quad for\left(\frac{x}{D}\right)/(\mathrm{Re.Pr})\ge 0.001 \end{array}\right\} \tag{17}$$

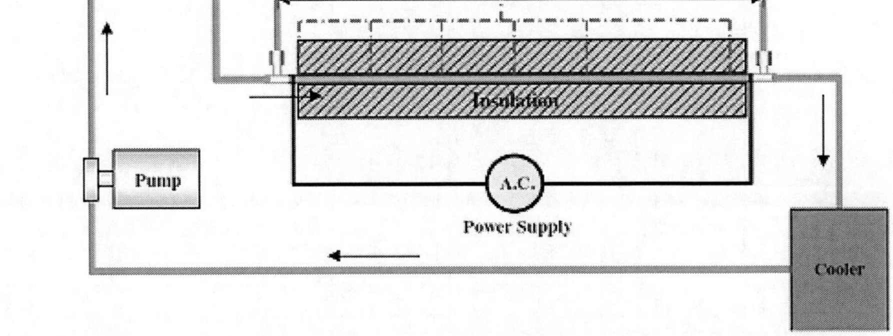


Figure 1. Experimental set up of Hwang et al. reproduced from [14].

Experimental results showed that at 0.3% of nanoparticles by volume, the convection coefficient is enhanced by 8% when compared to that of the base fluid, water. It is to be noted that this enhancement is much more than the 1.44% enhancement in thermal conductivity obtained for the same case. The Shah equation was found to be inadequate to predict the convective heat transfer coefficient in the nanofluid.

Heris et al. [15] performed experimental investigations on convective heat transfer in $Al_2O_3$-water nanofluids flowing through a circular tube with a constant wall temperature, under laminar flow. Figure 2 shows the experimental setup used by them, reproduced from the publication.

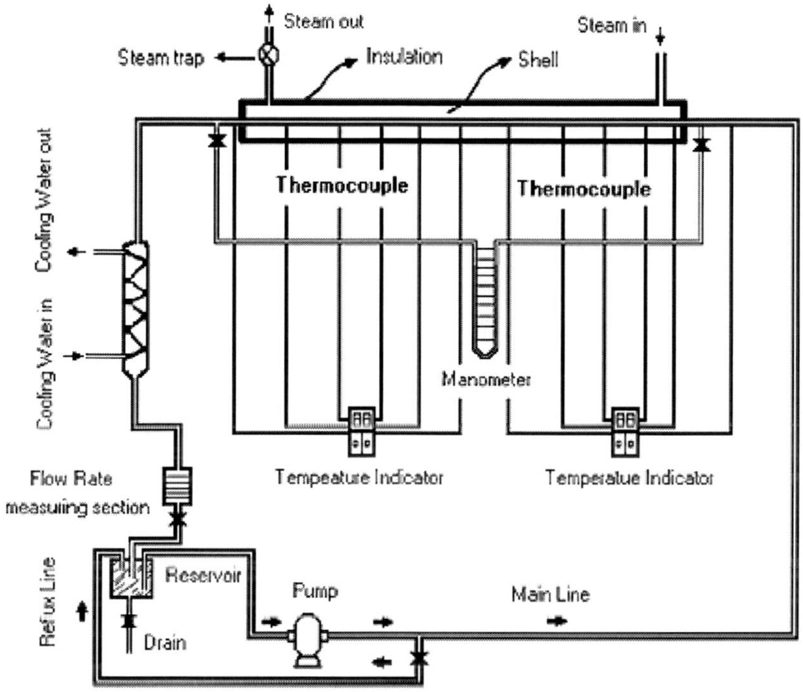


Figure 2. Experimental set up of Heris et al. reproduced from[15].

The Nusselt number was estimated for nanoparticle concentrations of 0.2%, 0.5%, 1.0%, 1.5%, 2% and 2.5%, for a range of Reynolds number (700-2050), and various values of the Peclet number. The flow and temperature measurements had an uncertainty of 1% and 2% respectively. The experimental setup was conditioned by comparing the Nusselt number values for pure distilled water with the Sieder-Tate equation (Eq. (18)).

$$\overline{Nu_{nf}}(th) = 1.86\left(\mathrm{Re}_{nf}.\mathrm{Pr}_{nf}.\frac{D}{L}\right)^{1/3}\left(\frac{\mu_{nf}}{\mu_{w}}\right)^{0.14} \tag{18}$$

where the nanofluid Reynolds number and the Prandtl number for the nanofluid are given by:

$$\mathrm{Re}_{nf} = \frac{\rho_{nf}.\overline{U}.D}{\mu_{nf}} \tag{19}$$

$$\mathrm{Pr}_{nf} = \frac{Cp_{nf}.\mu_{nf}}{k_{nf}} \tag{20}$$

The physical properties of the nanofluid, such as the density, viscosity, specific heat and thermal conductivity were calculated at the average bulk temperature. These were calculated depending on the corresponding properties of nanoparticles ($\rho_{np}$, $\mu_{np}$, $Cp_{np}$, $k_{np}$) and those of water ($\rho_w$, $\mu_w$, $Cp_w$ $k_w$).

The density of the nanofluid is given by Eq. (21), where $\varphi$ is the percentage volume of nanoparticles added to water.

$$\rho_{nf} = \phi.\rho_{np} + (1-\phi).\rho_w \tag{21}$$

The viscosity of the nanofluid is given by Eq. (22), the Einstein equation [38]:

$$\mu_{nf} = \mu_w.(1 + 2.5.\phi) \tag{22}$$

The specific heat of nanofluids is calculated from the mass averaged expression given by Eq. (23):

$$Cp_{nf} = \frac{\phi(Cp_{np}.\rho_{np}) + (1-\phi).Cp_w.\rho_w}{\rho_{nf}} \tag{23}$$

The thermal conductivity of the nanofluid is calculated using the correlation proposed by Yu and Choi [36] given in Eq. (24), where $\beta$ is the ratio of the nanolayer thickness to the particle radius and was taken as 0.1.

$$k_{nf} = \left[\frac{k_{np} + 2k_w + 2(k_{np} - k_w)(1+\beta)^3\phi}{k_{np} + 2k_w - (k_{np} - k_w)(1+\beta)^3\phi}\right]k_w \tag{24}$$

The experimental convective heat transfer coefficient is calculated using Eq. (25), where $(T_w - T_b)_{LM}$ is the logarithmic mean temperature difference:

$$\overline{h_{nf}} = \frac{Cp_{nf}.\rho_{nf}.\overline{U}.A.(T_{b2} - T_{b1})}{\pi.D.L.(T_w - T_b)_{LM}} \tag{25}$$

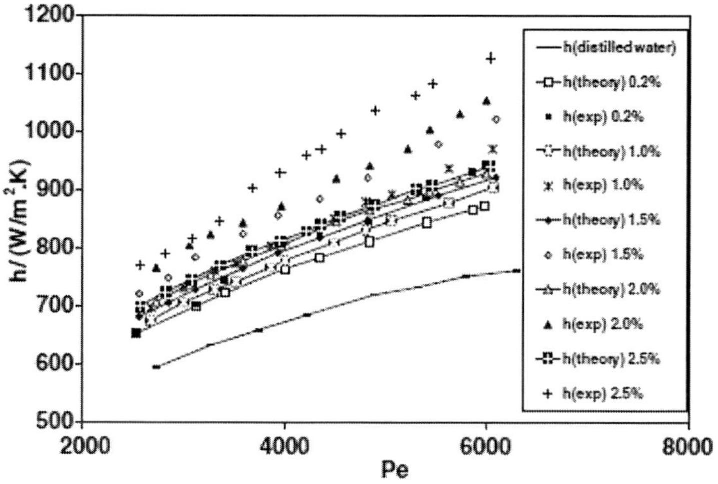


Figure 3. Variation of experimental heat transfer coefficient with Peclet number at different particle concentrations, compared with the heat transfer coefficient obtained using Sieder-Tate equation, reproduced from [15]

The uncertainty in the heat transfer coefficient, pressure drop, Peclet number, Nusselt number and Reynolds number were estimated to be 3%, 3%, 3%, 4%, and 2.5%, respectively. The variation of the heat transfer coefficient with the Peclet number is reproduced in Figure (3).

The heat transfer coefficient was found to increase with the particle concentration and the Peclet number. The enhancement was found to be more significant at higher values of the Peclet number. The heat transfer coefficients obtained from experimental measurements were found to exceed the predictions using the Sieder-Tate equation.

Anoop et al. [16] investigated the effect of the particle size on convective heat transfer in developing flow in a tube with a constant surface heat flux. The convective heat transfer coefficient was measured while using nanofluids containing $Al_2O_3$ nanoparticles of average sizes 45nm and 150 nm. In the developing flow region and for a Reynolds number of 1500, the 45 nm sized particles gave 25% enhancement in heat transfer compared with 11% by the 150 nm particles, for a concentration of 4 % by weight. The enhancement in the local heat transfer coefficient was also found to decrease, from the developing to fully developed region. For a concentration of 4 % (by weight) of 45 nm particles and an approximate Reynolds number of 1500, the enhancement in heat transfer coefficient was 31% at x/D = 63, while it was 10% at x/D = 244. The uncertainty in the measurement of the thermal conductivity was found to be less than 2%, and that for viscosity measurement was estimated to be 0.5%. A systematic uncertainty analysis yielded the maximum error in the Reynolds number and the Nusselt number to be around 3.24% and 2.45% respectively. A correlation for the local Nusselt number proposed by Anoop et al., applicable for alumina nanofluids in the developing region $50<x/D<200$ for laminar flow with $500 < Re < 2000$ is given as:

$$Nu_x = 4.36 + \left[a.x_+^{-b}(1+\phi^c).\exp^{(-dx_+)}\right]\left[1+e.\left(\frac{d_p}{d_{ref}}\right)^{-f}\right] \tag{26}$$

where $a = 6.219 \times 10^{-3}$, $b = 1.1522$, $c = 0.1533$, $d=2.5228$, $e= 0.57825$, $f=0.2183$, $d_{ref} = 100$ nm, $d_p$ = diameter of the nanoparticles in nm and $\Phi$ = volume fraction of nanoparticles.

Gherasim et al. [18] experimentally investigated the heat transfer enhancement capabilities of coolants with $Al_2O_3$ nanoparticles suspended in water, flowing inside a radial flow cooling system. From the time required to accumulate a certain quantity of fluid, the mass flow rate was calculated. The heat flux was varied electrically by the heating elements. The Reynolds number was calculated as in Eq. (19), where $D$ is given by $2\delta$, and $\delta$ being the distance separating the disks. The local heat transfer coefficient $h_r$ was obtained as:

$$h(r)=\frac{q''}{T_{w,r}-T_{b,r}} \tag{27}$$

The bulk temperature at a given radial section ($T_{b,r}$) was calculated as:

$$T_{b,r}-T_{b,i}=\frac{q''\pi r^2}{\dot{m}C_p} \tag{28}$$

where $T_{b,r}$ and $T_{b,i}$ are the bulk temperatures at a given radius and at the inlet respectively.

Considering all the uncertainties on experimental measurements, the average relative errors in Nusselt number calculations were estimated as 12.1%, 11.5% and 11% for cases with particle volume fractions of 2%, 4% and 6% respectively. The experiments were aimed at investigating the effect of nanofluids in a steady laminar flow between the disk and a flat plate, with axial entry and radial exit, as encountered in turbomachinery. The heat transfer coefficient was found to increase with the particle concentration and the flow rate and, decrease with an increasing gap between the disks.

Kim et al. [19] have reported convective heat transfer experiments in a straight circular tube with constant heat flux on the surface, for the flow of water-based nanofluids of alumina (20-50 nm) and amorphous carbon (~20 nm). Experiments were conducted using nanofluids containing 3.5% by volume of $Al_2O_3$ nanoparticles and 3% by volume of amorphous carbonic nanoparticles, with no surfactants added. The pipe surface temperatures along the heating section and the bulk fluid temperatures at the inlet and exit of the heating section were measured using thermocouples. All measurements were done at a constant inlet temperature of 22°C for the nanofluid. The heat flux was kept constant at 60 and 295 W for laminar and turbulent measurements respectively. With the measured temperatures, heat flux, and flow rate, the local heat transfer coefficient was calculated using Eq. (15) and Eq. (16). The thermal conductivity of alumina nanofluids showed an enhancement of 8-12% in the temperature range of 22-52°C, whereas the carbonic nanofluids gave an enhancement of only 2.6% at 52°C. The viscosity of the carbonic and alumina nanoparticles showed 11-14% enhancement in the measured temperature range. The experimental setup was calibrated using de-ionized water and the results obtained were compared with the Shah equation (Eq. (13)) for laminar flow and the Gnielinski equation (Eq. (14)) for turbulent flow under constant heat flux condition. The experiments at Re = 1460 showed an enhancement of 15% in the early entrance region and 12% at the exit region for alumina nanofluids. Corresponding values for carbonic nanofluids were 8 and 2%

respectively. In the turbulent regime (Re=6020), the convection coefficient got enhanced by 20% for alumina nanofluids, whereas the carbonic nanofluid showed a behavior similar to that of de-ionized water.

Heris et al. [20] experimentally measured the heat transfer coefficients in copper oxide (50-60 nm)-water and aluminum oxide (20nm)-water nanofluids for a constant temperature boundary condition and compared them with values obtained from the Sieder-Tate equation (Eq. (18)) utilizing the properties of nanofluids. The experiments were conducted on nanofluids with concentrations in the range of 0.2-3.0 % by volume, and flow corresponding to the Reynolds number in the range 650-2050. The experimental results showed that the Sieder-Tate equation under-predicts the heat transfer coefficients, as the enhancement in the case of nanofluids is attributed to the particle type, particle size, base fluid, flow regime and boundary conditions. The ratio of the experimentally measured heat transfer coefficient to that predicted by Sieder-Tate equation increased with the Peclet number and nanoparticle concentration. $Al_2O_3$ nanoparticles produced better enhancement in the heat transfer coefficient than CuO nanoparticles at same Peclet number. Both $Al_2O_3$ and CuO based nanofluids showed a maximum heat transfer for 2.5% concentration and a lower value at higher concentrations. The changes in the flow structure due to the presence of nanoparticles were understood to influence the heat transfer coefficient, along with the variations in the thermophysical properties of the nanofluids.

Ding et al. [24] experimentally investigated the effect of flow conditions, particle concentration and pH value on the convective heat transfer characteristics of carbon nanotube (CNT)-water nanofluids. A copper pipe of 4.35mm internal diameter and 970mm length, with constant surface heat flux applied by a 300W heater was used as the heating section. The CNT nanofluids were stabilized using surfactants like sodium dodecyl benzene sulfonate or gum arabic (0.25% by weight). The experimental setup was initially conditioned using distilled water and the results were found to agree with the Shah equation. The viscosity of CNT nanofluids were found to be decreasing with an increasing shear rate. Since the shear rate inside the pipe would be higher near the wall than the core, the viscosity of the nanofluid also would not be a constant at any cross section. Hence the Reynolds number calculations were done considering the viscosity of nanofluids at zero shear. The heat transfer coefficient was found to get enhanced in the presence of CNTs and was noticed to increase with the CNT concentration. The percentage enhancement in the heat transfer coefficient of CNT- nanofluids, with respect to distilled water, was found to increase with x/D until a limiting value, and decrease after reaching a maximum. The x/D ratio at which the enhancement becomes a maximum, in turn, was found to increase with the CNT concentration as shown in Figure 4.

The enhancement in heat transfer was not much influenced by the Reynolds number at Re<1100, whereas the system exhibited a distinct improvement at Re>1100. A Reynolds number of 1100 corresponds to a shear rate of approximately 500 $s^{-1}$. This value corresponds to the point at which the CNT nanofluid ceases to have the shear thinning behavior and the viscosity reaches a constant value. The enhancement in convective heat transfer was attributed to the features such as the possible increase in static thermal conductivity, increase in dynamic thermal conductivity at varying shear rates existing at the pipe surface and the core, high aspect ratio of CNTs, the delay in development of the boundary layer, shorter boundary layer thickness and non uniform shear rate resulting in particle migration across the flow area.

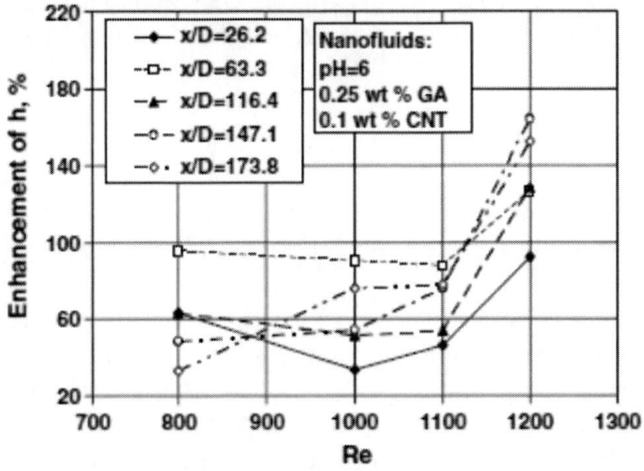


Figure 4. Enhancement of the heat transfer coefficient as a function of the Reynolds number at different axial positions, reproduced from [24].

Yu et al. [25] conducted experiments on convective heat transfer using water-silicon carbide (170 nm) nanofluids with a concentration of 3.7% by volume, in the range of Reynodls number 3300-13000. The test section consisted of a stainless steel tube of 2.27 mm inside diameter, 4.76mm outside diameter and 58 cm heated length. The test section was heated using an external resistance heater. The outside wall temperatures and fluid temperatures at the inlet and the outlet were measured. The inlet wall temperature was calculated using Eq. (29):

$$T_{win}(x) = T_{wout}(x) + \frac{q'(x)}{4\pi k_w(x)} \frac{1+\ln(d/D)^2 - (d/D)^2}{1-(d/D)^2} \tag{29}$$

where q'(x) is the local heat generated in the test section per unit length, d is the inside diameter, D is the outside diameter and $k_w(x)$ is the thermal conductivity of stainless steel corresponding to the outside wall temperature.

The comparison of heat transfer calculated at different values of the Reynolds number assumes that the average fluid velocity remains a constant. But the higher viscosity of nanofluids results in a higher fluid velocity requirement for the same Reynolds number, than those for the base fluid. The heat transfer coefficient of SiC nanofluids were found to be 7% lower than that for the base fluid for a constant velocity comparison, even though the thermal conductivities are higher than those of the base fluid.

The enhanced thermal conductivity of nanofluids reduces the thermal resistance in the boundary layer, but the enhanced viscosity increases the thermal resistance in the boundary layer. The net effect of these on the heat transfer depends on the competing effects. The fluid properties appearing in the Dittus-Boelter correlation can be expressed by the Mouromtseff number, Mo, given by Eq. (30):

$$Mo = \frac{k^{0.6}\rho^{0.8}Cp^{0.4}}{\mu^{0.4}} \tag{30}$$

A higher value of Mo represents better enhancement. The value of Mo for a typical experiment using $Al_2O_3$ nanofluid is 0.75 and SiC nanofluid is 0.89 [46]. A plot of the Nusselt number against $Re^{0.8}Pr^{0.4}$ shows that the heat transfer in SiC nanofluid is enhanced over that in water. The heat transfer enhancement ratio with respect to the Gneilinski correlation shows a 50-60% enhancement in the case of nanofluids. To estimate the practical utility of nanofluids, the pumping power requirement also need to be estimated. A nanofluid merit parameter expressed as the ratio of heat transfer enhancement to the pumping power increase has been given as given as, $(h_{nf}/h_{bf})/(PumpingPower_{nf}/PumpingPower_{bf})$.

The merit parameter value was obtained as 0.8 for SiC nanofluid as compared to 0.6 for $Al_2O_3$ nanofluid, indicating that the former is a better cooling fluid. A comparison of the heat transfer enhancement ratio with respect to the Gneilinki correlation when compared with the modified Gneilinski correlation (Eq. (31)) [47] showed that the modified Gneilinski correlation under-predicts the enhancement in heat transfer coefficient.

$$Nu_b = \frac{\frac{f}{8}\left(Re_b - 1000\right)Pr_b}{1 + \delta_v^+ \sqrt{\frac{f}{8}}\left(Pr_v^{2/3} - 1\right)} \tag{31}$$

In the above, subscripts ‘b’ and ‘v’ represent properties corresponding to the concentration of nanofluids in the bulk fluid and laminar sublayer respectively. The laminar sublayer thickness $\delta^v$ is given by

$$\delta_v \approx \delta_v^+ \left(\mu_v / \rho / \sqrt{\tau_w \rho}\right) \tag{32}$$

where $\tau_w$ is the shear stress in the fluid at the wall and $\delta_v^+$ is a constant determined empirically.

The enhancement noticed over the predictions suggests a mechanism of heat transfer not consistent with pure fluids. The Brownian diffusion and thermophoresis are two mechanisms which could explain the peculiar behavior of nanofluids and are being accounted for in the modified Gneilinski correlation. But these mechanisms predict a higher concentration of nanoparticles near the boundary during heating and a lower concentration during cooling, resulting in enhanced heat transfer over base fluids during heating and deterioration during cooling. The experimentally measured data-points shown in Figure 5 are consistent with the predictions during heating. But at the cooling section while the prediction is that of reduction of heat transfer, the experimental results show an enhancement. Even though Kim et al. [48] predicted particles getting coated on surfaces as a reason for enhancement in pool boiling, experiments by Yu et al. [25] showed that this has no effect on convective heat transfer. Even on surfaces where particles were deposited, the heat transfer values obtained for base fluid as well as SiC nanofluids were unchanged in their experiments.

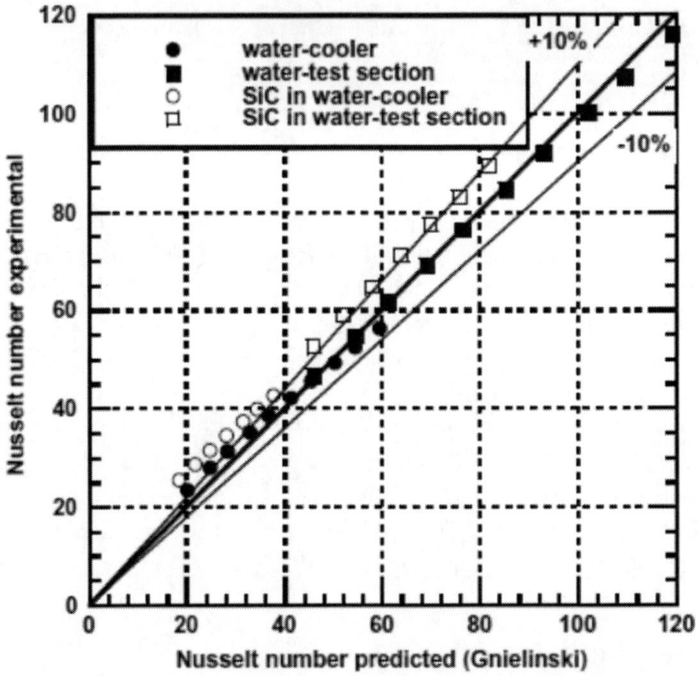


Figure 5. Nusselt number for cooled and heated conditions, reproduced from [25].

Ding et al. [26] experimentally investigated forced convection heat transfer in nanofluids with water and ethylene glycol as base fluids. The nanoparticles used were spherical titania, titanate nanotubes, carbon nanotubes and diamond. All the nanofluids were found to have thermal conductivities larger than the corresponding base fluids. An enhancement was observed in the convective heat transfer in the case of water-based suspensions of carbon nanotubes, titanate nanotubes and titania in the decreasing order, which was found to be higher than the enhancement in thermal conductivity. At the same time, in the case of ethylene glycol-titania nanofluids at low Reynolds numbers, and water-nano-diamond nanofluids, the convective heat transfer was found to be lower than those in the corresponding base fluids, even though an enhancement in thermal conductivity was noticed. In the case of water-nano-diamond nanofluids, the use of SDBS (Sodium Dodecyl Benzene Sulfonate) surfactant was found to produce a foam, which, while migrating to the inner pipe surface, could possibly hinder effective heat transfer. Further, in the case of nano diamond nanofluids, the viscosity behavior under various shear rates was found to deviate from that of other nanofluids. At low shear rates it exhibited shear thinning; a weak shear thickening was produced at medium shear rates, and shear thickening resulted at high shear rates. The high viscosity of ethylene glycol was pointed out as one reason for the deterioration in convective heat transfer in the case of ethylene glycol-titania nanofluids. The uncertainties in the measurement of convective heat transfer were much larger than the reported effective enhancement. The higher value of convective heat transfer coefficient for water-titanate nanofluids over water-titania nanofluids was proposed to be due to the higher aspect ratio of titanate nanotubes.

Nguyen et al. [27] investigated the heat transfer performance of alumina (~36nm) nanofluids when impinged on a flat horizontal circular heated surface. The impingement was made using a nozzle of 3mm diameter, at nozzle-to-heated surface distances of 2, 5 and 10mm. Nanofluids of concentrations 0-6 % by volume, having the Prandtl number in the range of 5-10 and flowing at Reynolds number in the range of 3800-88000 were used. An aluminum block of 30mm diameter was used as the heater surface, to resemble the microprocessor, onto which the nanofluid was impinged as coolant. The entire arrangement was surrounded by a plexiglass cylindrical cover to create a confined environment for the jet. The experimental results showed that for water as well as nanofluids, the heated surface temperature decreases with an increase in the mass flow rate, owing to increased convective heat transfer. The average surface heat transfer coefficient in the case of confined impingement was found to be a function of both the nanofluid concentration and the nozzle-to-heated surface distance. For a nozzle-to-heated surface distance of the 2 mm, distilled water had the highest convection coefficient, above the nanofluids. For 10 mm distance, distilled water and the nanofluid with 2.8% concentration had similar convection coefficients at all mass flow rates, while the nanofluid with 6% showed deterioration in the convection coefficient. At 5 mm distance alone, the nanofluid with s concentration of 2.8% showed a performance superior to that of distilled water, while the nanofluid with 6% concentration still showed a performance inferior to that of water. For a fixed mass flow rate, it was observed that the heat transfer decreases as the nozzle-to-heated surface distance increases, in the case of distilled water. This was attributed to the decrease in fluid velocity at the jet impinging stagnation zone as the nozzle-to-heated surface distance increased. But in the case of nanofluids with 2.8% and 6%, the heat transfer was a maximum for the 5 mm distance. The inferior performance of nanofluids as compared to water was attributed to the recirculation flows existing due to the confined arrangement. The jet after impinging on the heated surface would move out radially, encounter the plexiglass surface, and the heated fluid would be flowing back. The higher thermal conductivity and lower specific heat of nanofluids would result in a decrease in heat transfer due to the presence of this recirculation zone.

He et al. [31] investigated the effect of the size and concentration of spherical $TiO_2$ nanoparticles (~20nm) in water, on the convective heat transfer coefficient during laminar and turbulent flow, while flowing upward through a vertical pipe. Nanofluid concentrations of 1.0%, 2.5% and 4% by weight (0.24%, 0.6% and 1.18% by volume) were used with average particle sizes of 95 nm, 145 nm and 210 nm respectively. The test section used was a vertical copper tube of 3.97mm inner diameter and 1834 mm length with a constant heat flux applied at the surface (at a maximum power supply of 300W). The increase in the average particle size, along with the concentration, resulted in a relative reduction in percentage thermal conductivity enhancement and an increase in the percentage enhancement of viscosity of nanofluids. The convective heat transfer was found to increase in both the laminar and the turbulent regimes. The enhancement was found to be more in the turbulent regime (Re=5900) where there was a 40% enhancement for 1.1% by weight concentration, as against a 12% enhancement in the laminar regime (Re=1500). The particle size was found to have only a marginal effect on the convective heat transfer in $TiO_2$ nanofluids. The increase in the convective heat transfer with an increase in the flow rate was attributed to the reduced boundary layer thickness at higher Reynolds number values.

Bergman [41] investigated the effect of increased thermal conductivity and reduced specific heat of nanofluids relative to the base fluid, on convective heat transfer in microchannels. The effect of an increase in the thermal conductivity is to reduce the temperature difference between the wall and the fluid. A decrease in the specific heat results in an increase in the average temperature of the nanofluid, compared to the base fluid. The effectiveness of nanofluids on device cooling by maintaining reduced device wall temperature will depend a lot on these competing effects. The prediction of the specific heat of nanofluids based on the mass-averaged expression given by Eq. (23) showed good agreement with the experimental measurement [42] and hence was used in the analysis. The thermal conductivity of nanofluids was predicted using the Hamilton-Crosser model given by Eq. (33),

$$k_{nf} = \left[ \frac{k_{np} + (n-1)k_{bf} - (n-1)\phi(k_{bf} - k_{np})}{k_{np} + (n-1)k_{bf} + \phi(k_{bf} - k_{np})} \right] k_{bf} \tag{33}$$

where n = 3 for the nominally spherical 36 nm average diameter nanoparticles.

For a circular tube subjected to constant surface heat flux, with the liquid flowing under fully developed laminar conditions, the analytical value of Nusselt number is 4.36 from classical fluid dynamics. Using this value, The heat transfer coefficient for nanofluids could be calculated using Eq. (15) and Eq. (16).

For an arbitrary case of water-$Al_2O_3$ nanofluid of 0.05 % by volume flowing at a Reynolds number of 798 through a pipe of diameter 0.2 mm and length 100 mm, subjected to a constant surface heat flux of 20 $kW/m^2$, the wall surface temperature and mean fluid temperature were calculated considering a fluid inlet temperature of 30°C. The mean fluid temperature and surface temperature were calculated for both the base fluid and the nanofluid. The effect of the increase in thermal conductivity of nanofluids reflects in a reduced temperature difference between fluid and the surface. But, downstream along the length of the pipe, the reduced specific heat of nanofluids could result in the wall temperature reaching values above the corresponding value while using pure water. This would entail in a degradation in the performance of nanofluids. An energy balance on the fluid in the tube, and application of Newton's law of cooling gives the ratio of the difference between surface temperature at exit and the fluid inlet temperature to the heat flux rate. This figure of merit, which is useful in analyzing cooling systems, was obtained as given in Eq. (34).

$$\frac{T_w(x = L) - T_{m,i}}{q} = \frac{1}{\pi} \left[ \frac{1}{Nu_D k_{nf} L} + \frac{\pi}{\dot{m} c_p n_f} \right] \tag{34}$$

As obvious from the equation, it is necessary to reduce the value of the term within the brackets, to have the device surface temperature lower. A dimensionless "nanofluidic effectiveness" for single phase laminar constant heat flux condition has been presented as in Eq. (35), for which a value greater than zero is preferable [41].

$$\varepsilon_{CF} = 1 - \frac{\left(T_s(x = L)\big|_{nf} - T_{m,i}\right)}{\left(T_s(x = L)\big|_{bf} - T_{m,i}\right)} = 1 - \frac{(1/k_{nf}) + \left(L/\dot{m}\right)(\pi Nu_D / c_{p,nf})}{(1/k_{bf}) + \left(L/\dot{m}\right)(\pi Nu_D / c_{p,bf})} \tag{35}$$

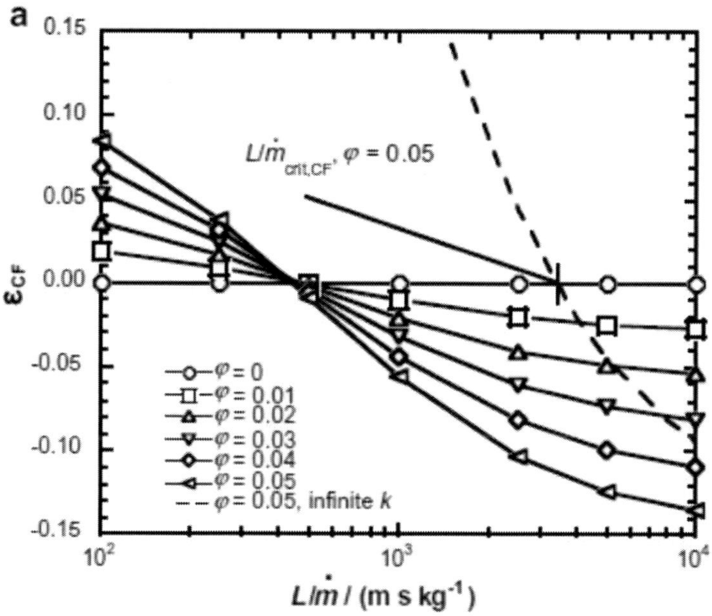


Figure 6. Variation of dimensionless effectiveness with $L/\dot{m}$ for different nanofluid concentrations, reproduced from [41].

It is seen in Figure 6 that at all nanoparticle concentrations, for values of $L/\dot{m} < 500$ ms/kg, gave an enhancement in the thermal performance of nanofluids compared to the base fluid.

Mansour et al. [43] investigated the effect of uncertainties in the properties of nanofluids at different concentrations on the overall heat transfer. They identified two different models for predicting the thermophysical properties of nanofluids at different concentrations. For an elementary heat transfer device the pumping power and length requirement increase with the volume concentration of nanofluids. The first model predicts an 8% increase in length for 10% volume fraction of nanofluid while the second model predicts 33% increase in length, compared to the case with water flow. The analysis established that the operational conditions or design parameters vary significantly with the thermophysical properties of nanofluids.

Xuan and Li [44] conducted experimental investigations into the heat transfer behavior of Copper-Water nanofluids in the turbulent flow regime (Re~10,000-25,000). The test section consisted of a straight brass tube of 10mm inner diameter and 800mm length, with a constant heat flux applied to the pipe surface. The nanofluids contained Cu nanoparticles (<100nm) at concentrations of 0.3%, 0.5%, 0.8%, 1.0%, 1.2%, 1.5% and 2% by volume. In the turbulent flow regime, the Dittus-Boelter correlation given by Eq. 11 predicted well the heat transfer coefficient in conventional fluids. The heat transfer coefficient was found to increase with the flow velocity and particle volume fraction. An increase in the particle volume fraction from 0.3-2% showed an enhancement ratio in the Nusselt number to the tune of 1.06-1.39.

Wen and Ding [45] investigated the convective heat transfer in water-γ- $Al_2O_3$ (27-56 nm) nanofluids flowing through a circular tube with a constant heat flux applied on the tube surface at the test section. The focus of the work was on estimating heat transfer coefficient at the entrance region in the laminar flow regime (Re:500-2100). The test section was made of a copper tube of 970 mm length, 4.5mm inner diameter and 6.4mm outer diameter. The viscosity of the nanofluid was estimated based on the calculation of Prandtl number, and utilized the expression given by the Einstein equation. A surfactant, sodium dodecyl benzene sulfonate (10% of the mass of nanoparticles) was added to the nanofluid containing 0.6, 1.0 or 1.6% by volume of nanoparticles. The heat transfer coefficient was found to be enhanced by 41% and 46% at flow rates corresponding to Re = 1050 and Re = 1600 respectively, close to the entrance region, for a nanofluid concentration of 1.6% by volume. However, farther away from the entrance region, the percentage enhancement in convective heat transfer at Re = 1600 for 1.6% concentration was only 14%. The enhancement was also found to increase with the concentration of the nanoparticles.

Yu et al. [49] has presented a review of the literature, and compared the influence of particle size, shape, material, volume concentration, base fluid material, temperature, additives and acidity individually on the thermal conductivity and heat transfer in nanofluids. The commonly used nanoparticles for preparation of nanofluids are oxides ($Al_2O_3$, CuO), semiconductors(TiO2), Nitride ceramics(AlN, SiN), Carbide ceramics (SiC, TiC), metals (Ag,Au, Cu, Fe), single, double and multi-walled carbon nanotubes (CNT), fullerenes, and composite nanomaterials like nanoparticle core-polymer shell composites. Yu et al. [46] tabulated the most important experimental results on thermal conductivity measurement in nanofluids, classified based on the type of the base fluid, nanoparticle material, particle size, concentration, temperature and preparation method. The heat transfer enhancement in nanofluids were studied for laminar flow, turbulent flow and pool boiling situations. The comparisons showed that for $Al_2O_3$, CuO and MWCNT nanofluids the enhacement ratio is not affected much by the Reynolds number. Experiments by Faulkner et al. [47] using Multi – walled carbon nanotube (MWCNT) nanofluids showed a decrease in heat transfer enhancement ratio with an increase in the particle volume concentration at very low flow rates (Re=3-17). In the turbulent regime, the heat transfer enhancement ratio was found to be unaffected by the Reynolds number, but this was found to increase with the particle volume concentration. For the same concentration metal nanoparticles (Cu) showed more enhancement than oxide nanoparticles. A comparison of the results for single phase heat transfer enhancement of nanofluids has been tabulated for different nanofluids, particle concentration, particle size and flow range [49].

Pool boiling characteristics of nanofluids have been widely investigated by many researchers. The nucleate boiling CHF (Critical Heat Flux) in nanofluids have shown enhancement over pure water [52-61]. Ahn et al. [51] conducted experiments using water-$Al_2O_3$ (47nm) nanofluids with a concentration of 0.01% by volume to quantify the forced convective flow boiling characteristics. The flow channel was rectangular in shape (10mm x 5mm), and was having a copper disk (10mm diameter) heated by conduction at the bottom of the channel. The experiments were conducted with flow velocities in the range of 0-4 m/s and the CHF enhancement was found to increase with the flow velocity. An increase in the flow velocity from 1-4 m/s showed an increase in enhancement of the CHF from 24 to 40%. The changes in surface morphology and contact angles, owing to the deposition of nanoparticles, were found to be producing the enhancement in CHF. The flow velocities and surface wettabilities were found to separately affect the CHF.

### 3.2. Convection in High Pressure Loops

Extensive research has been reported on the superior heat transfer characteristics of nanofluids compared to their constituent base fluids. Investigations usually focus on the measurement of the thermal conductivity, and characterization of the convective heat transfer and boiling heat transfer behavior as compared to conventional fluids. An observation of the results points towards promising and attractive use of the nanofluid option, in a wide range of applications.

However, it appears that almost all investigations that deal with the heat transfer characteristics of nanofluids have been performed at the atmospheric pressure. There are a number of industrial applications which utilize coolants at higher pressures. For instance, Nanofluids at high pressures can be applied in rapid cooling heat treatment processes such as quenching. The objectives of carrying out cooling at high pressures may vary with the application, based on the rate of cooling, material properties, crystal structure of quenched material and so on. For making use of nanofluids in such applications it is necessary to investigate on their heat transfer characteristics at high pressures, as it would be vital to understand the effect of pressure on the heat transfer enhancement and stability of the nanofluids.

Experiments on convection in high pressure loops have been conducted in the authors' laboratory (Nanotechnology Research Laboratory, National Institute of Technology, Calicut, India) to characterize convective heat transfer in titanium oxide $TiO_2$ nanofluids at high pressures, which is an option suitable for heat treatment applications. A method was developed, which makes use of compressed air for increasing the pressure of the fluid circulated in a convective heat transfer, for carrying for performing the investigation. The study has been conducted for different volume fractions (0.01 to 0.3%) of $TiO_2$ nanoparticles (30-50nm), at different flow rates, for three different pressures (1, 3 and 5 bars).

Sodium dodecylbenzene sulfonate (SDBS, supplied by Sigma Aldrich Corporation) was used as the dispersant to improve the stability of nanofluids. Nanofluids with five different $TiO_2$ nanoparticle concentrations (0.01%, 0.03%, 0.1%, 0.2%, and 0.3% volume fraction) were prepared and used to study the enhancement in heat transfer. Approximately 2000 ml of the nanofluid were needed for filling the entire loop. The pre-calculated amount of nanoparticles was dispersed into 500 ml of de-ionized water and the mixture was sonicated continuously for 3 hours using a probe type sonicator (VIBRA-CELL-VCX 750), following which the suspension was diluted gradually to 2000 ml. The suspension was further sonicated in an ultrasonic bath (Hwashin technologies, Korea) for one hour after each step of dilution.

The experimental set-up for measuring the convective heat transfer coefficient used in the laboratory is shown schematically in Figure (7). The set up mainly consists of a flow loop, a test section, a pump, a cooler, a flow rate measurement unit and a pressurizing unit.

A straight stainless steel tube with 12.56 mm inner diameter and 76.5 mm length is used as the test section. It is surrounded by two flexible band heaters, each with 1000 W rating, and each of these heaters is individually controlled by a variable transformer. The heaters are insulated thermally with ceramic wool blankets of thickness around 100 mm, to minimize the heat loss from the tube to the ambient and also to ensure a constant heat flux condition along the test section. Ten T-type (copper-constantan) thermocouples are mounted on the test section at equidistant axial positions from the inlet of the test section, to measure the wall temperature distribution. Two T-type thermocouples are inserted into the flow at the inlet and exit of the test section to measure the bulk temperatures of the flowing nanofluid. The temperature readings from the 12 thermocouples are acquired by a data acquisition system (Agilent Technologies, UK). Two Teflon rod fittings at the inlet and outlet of the test section provide thermal barrier to axial heat conduction. The nanofluid is heated uniformly under a constant heat flux condition. The nanofluid which flows out of the text section is cooled with low temperature water (produced using a constant temperature bath) in a double-pipe counter flow heat exchanger, so that it can re-enter the test section at an inlet temperature around 21°C, maintained constant. The flow rate is controlled with a valve and measured using a rotameter, within an accuracy limit of ± 2.5lph. The pressurizing unit consists of a pressurizing chamber and an air compressor, with the flow accomplished using a centrifugal pump and the pressure monitored using a pressure gauge, as shown in the diagram.

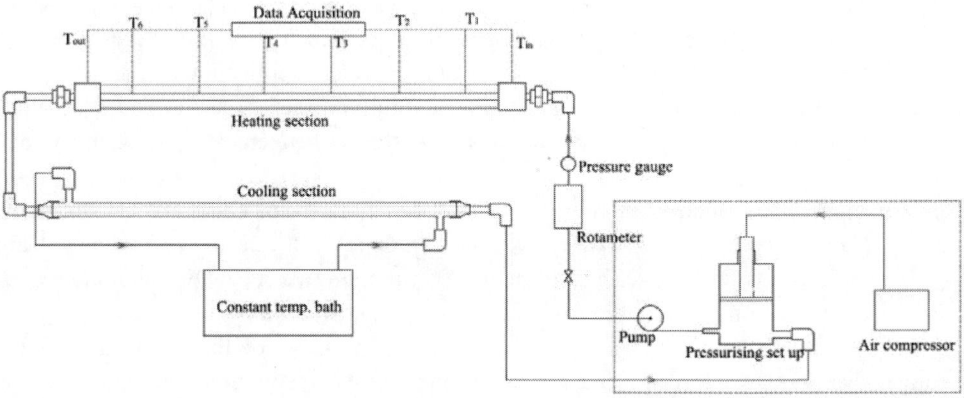


Figure 7. High Pressure Convection Loop.

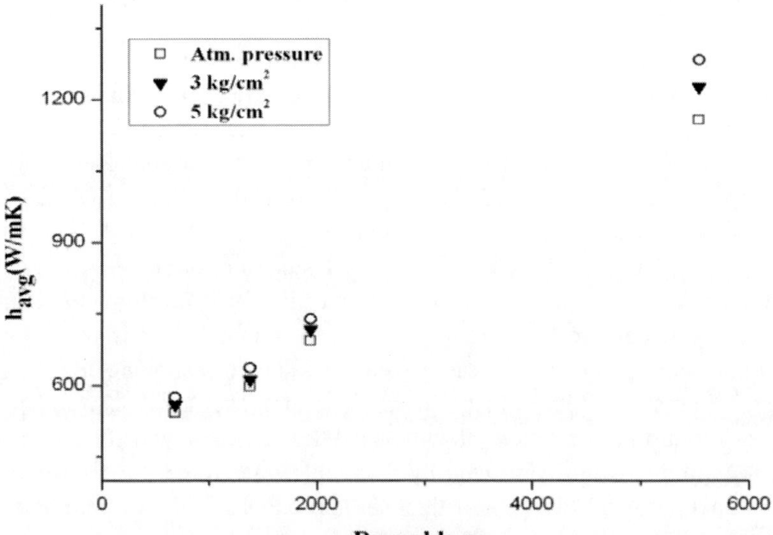


Figure 8. Variation in average heat transfer coefficient with Reynolds number at different pressures (0.03 vol.%).

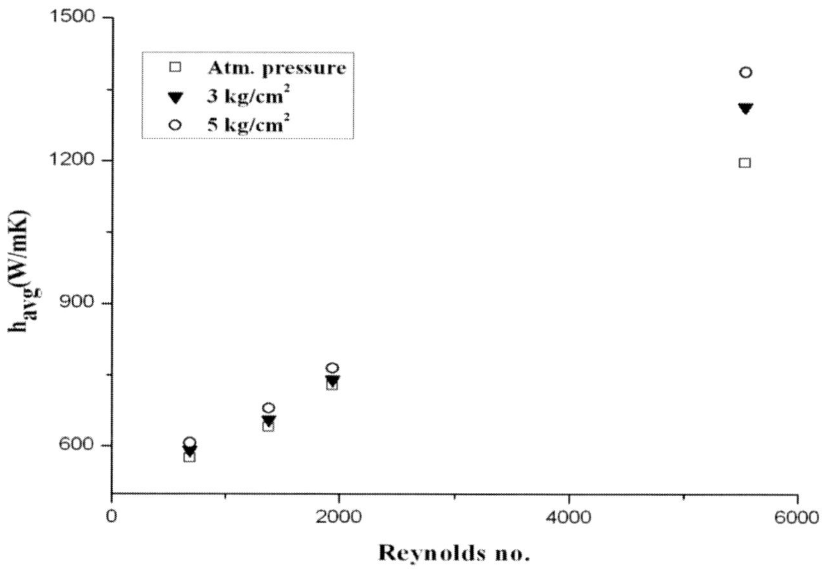


Figure 9. Variation in average heat transfer coefficient with Reynolds number at different pressures (0.1 vol.%).

In the experimental study, the heat transfer coefficient was calculated using Eq. (15) and Eq. (16) and the required thermophysical properties were obtained using Eq. (21), Eq. (22) and Eq. (23). In order to verify the accuracy and reliability of the experimental system, a series of experiments with de-ionized water were performed. The experimental results were compared with the predictions of the well-known Shah equation (12), and were found to agree well. The average heat transfer coefficient at different flow rates and pressures for 0.03% and 0.1% volume concentrations are shown in figures (8) and (9) respectively. For nanofluids containing 0.1% of nanoparticles by volume, the increase in average heat transfer coefficient was found to be 5.6%, 6.3%, 7.2% and 16% respectively for Reynolds number values of 693, 1385, 1940 and 5540. For nanofluid with 0.03% by volume of nanopartices, the improvement percentages were 1%, 2.6%, 3% and 18 % respectively at the above mentioned flow rates. Thus, it could be concluded that better enhancement of heat transfer is effected due to the presence of nanopartices, in the turbulent flow regime than in laminar flow.

### 3.3. Convection in Magnetic Nanofluids

Li et al. [30] have reported investigations on the effect of an applied uniform magnetic field and a magnetic field gradient on convective heat transfer in $Fe_3O_4$-water magnetic nanofluid. The test section used was a square cross section channel (13mm x 13 mm) of 1000 mm length. A thin platinum wire (0.01mmdiameter) with a thin glue layer coated over it for electrical insulation, mounted perpendicular to the main flow direction, acted both as the heat source and the temperature sensor. The steady state wall temperature, which is the platinum wire surface temperature, depends upon the magnetic fluid properties and flow patterns. A solenoid electromagnet provided the uniform magnetic field parallel to the flow direction and an electromagnet provided a uniform magnetic field perpendicular to the flow direction. A permanent magnet kept below the flow channel provided the magnetic field gradient perpendicular to the flow direction. The magnetic nanofluid consisted of 1% by volume of $Fe_3O_4$ (10 nm) nanoparticles dispersed in water along with 3% by weight of sodium dodecylbenzene sulfonate. In the absence of a magnetic field, the heat transfer coefficient for magnetic nanofluids were approximately 8% higher than deionized water at all Reynolds number, owing to the increased energy transport by the magnetic nanoparticles. A deterioration was observed in the heat transfer coefficient under the influence of a uniform magnetic field parallel and perpendicular to the flow direction, compared to the case without an external magnetic field applied. This could be attributed to the increased viscosity of the magnetic nanofluid under the influence of an external magnetic field, as a higher viscosity decreases the ability to flow, which would result in reduced heat transfer. The heat transfer coefficient in the case of a magnetic field applied perpendicular to the flow direction was found to be lower than that when the field is parallel to the flow direction. This was attributed to the possibility that in the case of the perpendicular magnetic field, the magnetic particles will be vertically chain-aligned, blocking the flow to a larger extent than in the case of a parallel magnetic field. Application of a magnetic field-gradient introduces a magnetic volume force (Kelvin force) on the magnetic fluid. A magnetic field-gradient in the direction of flow gives better enhancement in heat transfer than a gradient opposite to the direction of flow. The magnetic field gradient introduces a static pressure gradient as a result of the Kelvin force. If the magnetic field gradient is along the flow direction, the flow is accelerated, enhancing the heat transfer.

## 3.4. Mechanism of Convective Heat Transfer Enhancement in Nanofluids

The heat transfer in nanofluids could get affected by the thermal conductivity, the boundary layer thickness and the delay in boundary layer development [19]. The boundary layer thickness and the rate of its development, in turn could be affected by the particle size, brownian motion of the particles, particle migration and mixing effects near the wall. Experiments by Kim et al. [19] using carbonic nanofluids showed that the enhancement in convective heat transfer is much more than the enhancement in the thermal conductivity, in the laminar regime. This suggests that the disturbances in the thermal boundary layer are the main cause of heat transfer enhancement in carbonic nanofluids in the laminar regime. In the turbulent regime, the enhancement in the thermal conductivity of the nanofluid was found to be the major reason for the enhancement in heat transfer.

The enhancement in the convective heat transfer coefficient which exceeds the enhancement in thermal conductivity, and the failure of the Shah equation to predict the same is a clear indication of the fact that the convective heat transfer in nanofluids are affected by other phenomena also [14]. Experiments by Shon and Chen [37] have shown that the dynamic thermal conductivity of fluids with millimeter sized particles are much higher than the value under static conditions at a Peclet number over 300. In the experiments by Hwang et al. [14], since the nanofluids used were having the Peclet number only of $10^{-5}$, enhancement in thermal conductivity alone could not be stated as the reason for the enhancement in the convection coefficient. The enhancement in the convection coefficient could also be due to the random motion of nanoparticles in the boundary layer, as result of Brownian motion or thermophoresis, to some extent. The contributions due to these are also worth analyzing. A scale-analysis of the energy equation given by Eq. (36) shows that while the thermal conductivity being of order 1, the second and third terms on the right side indicating Brownian diffusion and thermophoresis are of order of $10^{-6}$. Hence, random particle motion cannot be expected to be a major contributor to the enhancement in the convection coefficient.

$$\left[\frac{\partial T}{\partial t}+\vec{v}.\nabla T\right]=\alpha\nabla^{2}T+D_{B}\nabla\phi.\nabla T+D_{T}\frac{\nabla T.\nabla T}{T},\tag{36}$$

In the equation, $\alpha, D_B = \dfrac{k_B T}{3\pi\mu_f d_p}, D_T = K\dfrac{\mu_f}{\rho}\phi$, are the thermal diffusivity, Brownian diffusion coefficient and thermophoretic coefficient respectively.

The convection coefficient in nanofluids might be affected by particle migration resulting from Brownian motion, viscosity gradient, non-uniform shear rate and thermophoresis. An order of magnitude analysis of the corresponding terms in the continuity equation under fully developed laminar flow regime showed that Brownian motion and thermophoresis affects particle migration, while viscosity gradient and non-uniform shear rate have no effect. Particle migration increases the viscosity and thermal conductivity of nanofluids near the centre line. This results in a flattening of the velocity profile at the centre line which decreases the difference between the tube wall temperature and the bulk fluid temperature, hence producing an enhancement in convective heat transfer coefficient.

Experiments by Heris et al. [15] indicated influence of the Peclet number on convective heat transfer coefficient, and led to the observation that the Sieder-Tate equation failed to predict heat transfer in nanofluids. This suggested that the enhancement in heat transfer is not due to the changes in fluid properties alone. The heat transfer at the fluid-surface interface is expected to get enhanced in nanofluids due to an increase in the local temperature gradient, induced by the presence of nanoparticles. Particle migration, clustering due to non-uniform shear rate across the pipe cross section, stochastic effects and dispersion effects could lead to enhancement in energy exchange rates causing the enhancement in heat transfer while using nanofluids. At higher flow rates, dispersion effects and random movement of particles intensify the mixing near the walls, thus flattening the temperature profile and leading to an enhancement in heat transfer. The increased chance of agglomeration is suggested as a possible reason for lesser enhancement at lower flow rates [40].

It can be inferred from the macroscopic definition for convective heat transfer coefficient given by h = $k/\delta_t$ that an increase in thermal conductivity k or/and a decrease in thermal boundary layer thickness $\delta_t$ could lead to an enhancement in the coefficient h. This partially explains the enhancement in heat transfer coefficient observed in experimental results on nanofluids by Ding et al. [26], in relation to the increased thermal conductivity. If nanofluids can be considered as homogeneous fluids, for a fully developed pipe flow with constant heat flux boundary condition, the convective heat transfer coefficient is given by, $h = 4.26.k / D$.

For a given hydraulic diameter of a conduit, this gives the impression that the heat transfer coefficient *h* is only dependent on the thermal conductivity *k*, whereas it was found that in many cases, enhancement in *h* is not proportional to that in *k*. For fully developed turbulent flow conditions at constant wall heat flux, *h* is given by,

$$h = 0.023 \frac{\rho^{0.8} U^{0.8} C_p^{\ 0.33}}{D^{0.2}} \frac{k^{0.67}}{\mu^{0.47}} \tag{37}$$

Since at low concentrations, the density ρ and the heat capacity Cp are approximately the same as that of the base fluid, *h* will be proportional to $k^{0.67}/\mu^{0.47}$ and the addition of nanoparticles should produce a lower percentage enhancement in heat transfer coefficient in the turbulent regime compared to the laminar flow regime. But experimental observations have shown larger enhancement in turbulent regime than in laminar flow [28, 52]. Also at the thermal entrance region, for nanofluids, the entry-length is not found to follow the expression for single phase flows [17, 24, 31, 56].

The considerations presented above suggest that nanofluids cannot be assumed as homogeneous fluids. The non-homogeneous nature in nanofluids could be due to agglomeration occurring, either by sintering during the physical synthesis of nanoparticles at high temperatures or due to solution effects causing the surface charges to decrease. The presence of agglomerates results in particle migration due to velocity and viscosity gradients and enhanced thermal conductivity. The particle migration results in lower particle concentration at the wall region and a reduced thermal boundary layer thickness ($\delta_t$) due to disturbances produced by moving particles. So the heat transfer coefficient is enhanced, if the decrease in $\delta_t$ exceeds a decrease in k or, if $\delta_t$ decreases and k increases. If the decrease in $\delta_t$ is of the same order as the decrease in k, h remains relatively unchanged whereas, if the decrease in $\delta_t$ is lower than the decrease in k, h deteriorates.

Ding et al. [62] investigated the effect of particle migration in laminar pipe-flows of low volume concentration nanofluids. The variations in the particle concentrations at any particular cross section of the pipe were found to be due to shear and viscosity induced particle migration as well as self-diffusion due to Brownian motion. In non-uniform shear flows, particles move from regions of higher shear rate to lower shear rate, from regions of higher viscosity to lower viscosity, and from higher to lower concentration due to Brownian motion. The flux causing particle migration is given by,

$$J = J_\mu + J_b + J_c \tag{38}$$

The particle migration flux due to viscosity gradient is,

$$J_\mu = -K_\mu \dot{\gamma} \phi^2 \left( \frac{d_p^2}{\mu} \right) \left( \frac{d\mu}{d\phi} \right) \nabla \phi \tag{39}$$

The particle migration flux due to non-uniform shear rate is,

$$J_c = -K_c d_p^2 \left( \phi^2 \nabla \dot{\gamma} + \phi \dot{\gamma} \nabla \phi \right) \tag{40}$$

In the above, $K_\mu$ and $K_c$ are constants, μ is viscosity, $\dot{\gamma}$ is shear rate, $d_p$ is particle size and Φ is concentration.

$J_b$ is the migration flux due to Brownian motion given by

$$J_b = -D_b \nabla\phi \tag{41}$$

The Brownian diffusion coefficient $D_b$ is given by,

$$D_b = \frac{k_b T}{3\pi\mu d_p} \tag{42}$$

where $k_b$ is the Boltzman constant ($k_b = 1.38 \times 10^{-23}$ J/K) and T is the temperature.

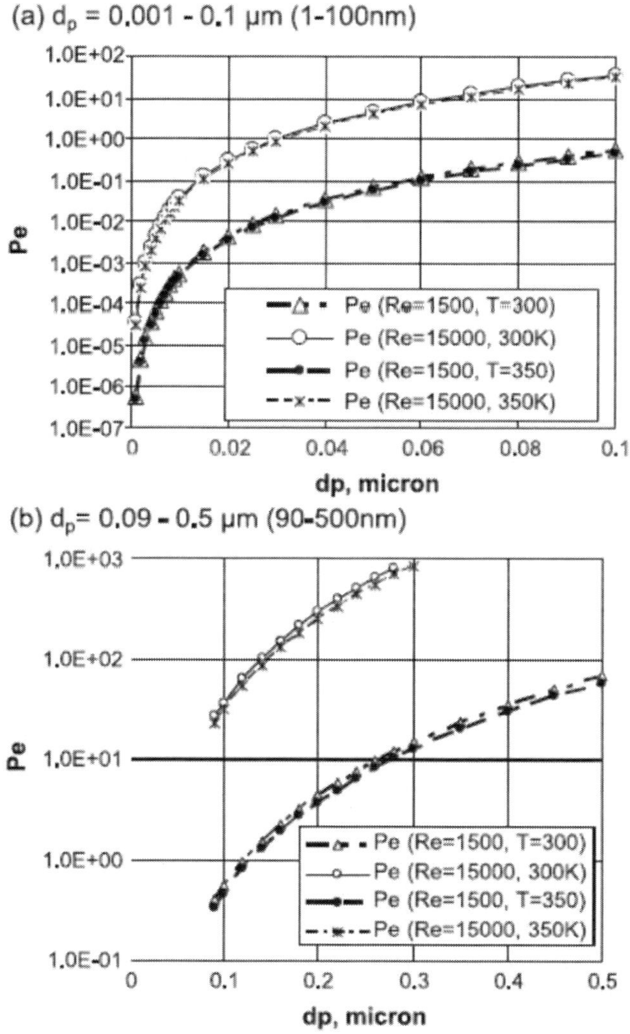


Figure 10. Effect of particle size, temperature and flow on Peclet number, reproduced from [62].

In Newtonian fluids, the shear stress is related to the shear rate by the Eq. (43).

$$\tau = -\mu\dot{\gamma} \tag{43}$$

The non-newtonian models which relate the shear stress and the shear rate are the Power law model, Bingham plastic model, Casson model, Herschel-Bulkley model and Sisko model [63, 64]. Experimental results by various researchers have shown that at low volume fraction, nanofluids exhibit Newtonian behavior. For micron sized particles, Kc and Kμ are of the order of 1/10.

Figure 10 (a) shows that the Peclet number increases with particle size and flow and is not much affected by the temperature. The Brownian diffusion will become significant only at Peclet numbers less than 10. Figure 10 (b) shows that, for a flow corresponding to Re=15,000, the Peclet number is less than 10 for dp< 60nm. Similarly, at Re=1500 for dp<300nm the Peclet number is less than 10. At Peclet number values greater than 10, the particle distribution is non-uniform at the transverse plain, particularly due to the viscosity and shear induced particle migration. For a given flow condition, smaller the particle size, more uniform will be the particle distribution. The higher concentration of particles, lower shear rate and lower particle-particle distance will enhance chances of particle agglomeration at the core of the flow and hence affect fluid stability. The viscosity is higher at the pipe centre at Peclet numbers greater than 10, and the shear thinning/shear-thickening effects need not be considered at low volume fractions. A lower particle concentration and hence lower viscosity at the wall offers an advantage from the point of view of lower pressure drop. For $K_c/K_\mu=1$ and $K_\mu.Pe=10$, at higher particle concentrations, the particle distribution is highly non-uniform for micron sized particles. For a particle concentration of 0.05 and $K_\mu.Pe=10$, the non-uniformity in particle concentration and viscosity increases with $K_c/K_\mu$ and at $K_c/K_\mu< 0.2$, the particle distribution is uniform. Since the effective thermal conductivity of nanofluids depends on particle concentration, the particle migration across the pipe cross section could lead to non-uniform thermal conductivity, the thermal conductivity being lower at the wall. The presence of particles may also affect the flow field resulting in a higher temperature gradient at the wall. The increased temperature gradient and thermal conductivity result in an enhanced convective heat transfer.

The measurement of thermal conductivity in a rotating Couette flow apparatus by Sohn and Chen [65] at low Reynolds number and high Peclet number indicated an enhancement in thermal conductivity at Pe>300 for micrometer sized particles. In the case of nanofluids, the Peclet number was of the order of 1 and hence the enhancement in thermal conductivity could not be attributed as a reason for the enhancement in heat transfer. The decrease in boundary layer thickness due to particle migration was suggested as a possible reason for enhancement in heat transfer, which was much more in percentage than the increase in thermal conductivity. The effect of particle migration was found to be particularly significant in the entrance region due to the larger velocity gradients present there [56]. It was also observed that the thermal developing length for nanofluids are larger than that for conventional fluids, given by x~(0.05Re.Pr)D.

The nanofluid being a solid-liquid mixture, the slip velocity may not be zero, which results in chaotic movement of the particles. In addition to an increase in the thermal conductivity, the random particle movement in the fluid leading to a steep temperature gradient at the wall may also augment heat transfer. In the experiments by Xuan and Li [55], the Dittus-Boelter correlation failed to predict the Nusselt number in the case of nanofluids, particularly at higher particle concentration. The correlation under-predicted the Nusselt number by over 30% at Re=16,000 and particle concentration of 2% by volume. This emphasized the need for a new correlation which considers parameters like volume fraction, in the form given in Eq. 44.

$$Nu_{nf} = f\left( \text{Re}, \text{Pr}, \frac{k_d}{k_f}, \frac{(\rho C_P)_d}{(\rho C_P)_f}, \phi, \textit{size and shape of the particles} \right) \tag{44}$$

Xuan and Li [52] proposed a correlation for the Nusselt number suitable for nanofluids as given by Eq. 45.

$$Nu_{nf} = c_1\left(1.0 + c_2\phi^{m1} Pe_d^{m2}\right)\text{Re}_{nf}^{m3}\,\text{Pr}_{nf}^{0.4} \tag{45}$$

where $\Phi$ is the volume fraction of particles in the base fluid, and $c_1$, $c_2$, $m_1$, $m_2$ and $m_3$ are parameters obtained by data reduction. For the experiments conducted by Xuan and Li, the parameters were obtained as $c_1=0.0059$, $c_2=7.6286$, $m_1=0.6886$, $m_2=0.001$ and $m_3=0.9238$. The particle Peclet number is given by Eq. 46.

$$Pe_d = \frac{u_m d_p}{\alpha_{nf}} \tag{46}$$

The anomalous increase in heat transfer coefficient observed in nanofluids, which is much more than the percentage enhancement in thermal conductivity, can also be attributed to the thermal dispersion and increased turbulence due to presence of particles. A theoretical understanding of the heat transfer in nanofluids has been sought by researchers using homogeneous models and dispersion models. Homogeneous models are extensions of the conventional flow models, with the fluid properties corresponding to those of nanofluids used, and hence they fail to predict the heat transfer in nanofluids. The dispersion models include the effect of heat transfer enhancement due to modified properties, as well as due to the dispersion of particles in the nanofluid.

Buongiorno [58] investigated the effect of inertia, Brownian diffusion, thermophoresis, diffusiophoresis, Magnus effect, fluid drainage and gravity on the slip mechanisms that cause relative movement between particles and base fluid. The salient points pertaining to these investigations are summarized below:

***Inertia***

The slip velocity of nanoparticles in the fluid due to inertia in the presence of turbulent eddies is given by Eq. (47).

$$V_e = V_{eo} e^{-t/\tau_p} \tag{47}$$

where $V_{eo}$ is velocity of turbulent eddies, and the relaxation time is given by,

$$\tau_p = \frac{\rho_p d_p^2}{18\mu} \quad (48)$$

For a typical water-alumina nanofluid ($\mu$~1m Pa.S, $d_p$<100nm and $\rho_p$~4 g/cm$^3$), the relaxation time is less than 2ns which is negligibly small than the turbulent eddy timescales. The distance $S_p$ (stopping distance) which the nanoparticle travels once the turbulent eddy carrying it comes to a stop, given by Eq. (49), is less than 1 nm for $V_{eo}$~0.3m/s.

$$S_p = \frac{\rho_p d_p^2}{18\mu} V_{eo} \quad (49)$$

Hence it can be assumed that the nanoparticles move without any slip along with the turbulent eddies and are not affected by the particle inertia. The laminar sublayer being approximately equal to 20 μm for the given flow conditions and, since the nanoparticle stopping distance is much less (<1nm), the chances of nanoparticle crossing the laminar sublayer under the influence of inertia is remote.

***Brownian Diffusion***

The Brownian motion of the particles is expressed through the Brownian diffusion coefficient $D_B$ given by the Einstein-Stoke's equation (Eq. (50)).

$$D_B = \frac{k_B T}{3\pi\mu d_p} \quad (50)$$

where $k_B$ is the Boltzman constant. For nanoparticles in the size range of 1-100 nm, the Brownian coefficient ranges from $4 \times 10^{-10}$ to $4 \times 10^{-12}$ m$^2$/s. The Brownian diffusion coefficient becomes a significant parameter in slip mechanism under such conditions when the turbulent eddies are absent.

***Thermophoresis***

Thermophoresis is the diffusion of particles in a fluid under the influence of a temperature gradient, which results in the particles moving from a hot to a cold region. The thermophoretic velocity is given by Eq. (51).

$$V_T = -\beta \frac{\mu}{\rho} \cdot \frac{\nabla T}{T} \quad (51)$$

where the proportionality factor, $\beta = 0.26 \dfrac{k}{2k + k_p}$

The expression for β was developed for large particles and was used in the absence of suitable expressions for nanoparticles. For water-alumina nanoparticles (k~1 W/m.K, $k_p$~40 W/m.K) under the influence of a temperature gradient of $10^5$ K/m, it was found that $V_T \sim 2 \times 10^{-6}$

m/s. Thermophoresis is understood to become a significant phenomenon only under the absence of turbulent eddies.

***Diffusiophoresis***

If a particle is suspended in a medium with a concentration gradient, a force tries to pull the material in the direction opposite to the gradient. This phenomenon which results in the particle colliding with diffusing species is called diffusiophoresis. In the case of nanofluids, the base fluid being devoid of particles, the chances of diffusiophoresis taking place in nanofluids are almost zero.

***Magnus Effect***

The shear forces due to fluid flow induced on particles causes them to rotate with an axis perpendicular to the flow direction. This introduces a lift force perpendicular to the flow direction on the particle due to the pressure gradient existing around the particle owing to the relative axial velocity, which is known as Magnus effect. In large particles, the relative axial velocity is large in the laminar sublayer of a turbulent flow. But since in nanofluids the chances of nanoparticles forcing themselves into the laminar sublayer is almost nil, the Magnus effect is not likely in them.

***Fluid Drainage***

As a particle approaches the wall and the distance between the particle and the wall becomes almost equal to the particle diameter, it feels a resistance due to the pressure difference between the two surfaces. In the case of nanoparticles this effect can be relevant only on a small region in the laminar sublayer and hence can safely be neglected.

***Gravity***

The settling velocity of nanoparticles(<100nm) due to gravity is obtained by balancing the buoyancy and viscous forces, as given in Eq. (52). The value is normally less than 1.6 x $10^{-8}$ m/s

$$V_g = \frac{d_p^2(\rho_p - \rho)g}{18\mu} \tag{52}$$

If the time a particle takes to diffuse through a length equal to the particle diameter is considerable, then that mechanism can be considered as significant in the turbulent eddy transport. For a nanoparticle of 100nm diameter, for turbulent transport, $d_p/V_{eo}$~3 x $10^{-7}$ s; for Brownian diffusion, $d_p^2/D_B$~0.002 s; for thermophoresis, $d_p/V_T$~0.05s; and for transport under gravity, $d_p/V_g$~6 s. This shows that in turbulent flow, the particle is carried by the turbulent eddies without any slip, but in the absence of turbulent transport, effects like thermophoresis and Brownian diffusion are significant in the laminar sublayer. The nanoparticles and base fluid can safely be assumed to be under thermal equilibrium by calculating the heat transfer time constants for heat conduction within nanoparticles and in the fluid immediately surrounding the particle. The time constants obtained by calculating $d_p^2/\alpha_p$ and $d_p^2/\alpha_{bf}$ for water-alumina (100 nm) nanofluids are 1ns and 100ns respectively, which are smaller than the time constants associated with Brownian diffusion and thermophoresis, thus supporting the assumption.

Buongiorno [58] concludes from an observation of the continuity, momentum and energy equations that the dispersion effects in nanofluids do not influence heat transfer as much as the heat conduction and convection. Possibly, the variations in thermophysical properties owing to the varying concentration in turbulent flow, Buongiorno [58] assumes that the wall layer consists of a laminar sublayer and a turbulent sublayer. In the laminar sublayer, the Brownian diffusion coefficient, viscosity and thermal conductivity are much larger than particle diffusivity, momentum diffusivity and energy diffusivity respectively, whereas the opposite happens in the turbulent sublayer. In the turbulent sublayer, since the particles are carried along with eddies, the particle concentration is the same as in the bulk. In the laminar sublayer, owing to thermophoresis, for large nanoparticles and higher heat fluxes, the nanoparticle concentration near the wall will be much lower than in the bulk medium.

Savithiri et al. [66] also investigated the effect of slip mechanisms, namely drag, gravity, Brownian motion, thermophoresis, lift, and rotational and magnus effects in nanofluids through a scaling analysis. The effects of the particle shape, particle size, concentration and temperature on various nanofluids were studied. The base fluids used were water and ethylene glycol, and the nanoaprticles were of gold, copper, alumina, titania, silica, carbon nanotubes (CNT) and graphene. The scaling analysis showed that all the slip mechanisms are of significance for the cylindrical parameters, as compared to the spherical and sheet particles, the drag force having the maximum effect and Brownian force the least. Hence the cylindrical particles would have better thermal transport compared to other shapes. As the size (diameter) of cylindrical particles increased, the drag force and the Brownian force decreased, while all other forces increased. For a cylindrical particle of 10 nm diameter, with an increase in concentration, the rotational and Magnus forces decrease, the thermophoresis, drag and lift forces increase, and the gravity force remains unaffected. The Brownian force increases, reaches a maximum at a concentration of 1% by volume and then decreases. With an increase in the nanofluid bulk temperature, the Brownian and Magnus forces increase, thermophoresis, rotational and lift force decrease and, gravity is unaffected. For a given temperature gradient, thermophoresis is found to be inversely proportional to the nanofluid bulk temperature. The slip mechanisms showed similar trends in nanofluids with water and ethylene glycol as base fluids, the only difference being in the magnitude of forces in the two fluids. The water based nanofluids exhibit Brownian force almost 100 times higher, and gravity and Magnus forces an order of magnitude higher, than the corresponding values in ethylene glycol based nanofluids. In ethylene glycol based nanofluids, thermophoresis was 10 orders of magnitude higher and, rotational and drag forces were an order of magnitude higher than in water based nanofluids. The gravity, lift, Magnus, rotational and Brownian forces are all found to be higher in copper and gold nanofluids, among the nanofluids studied; the drag force is higher in silica and alumina nanofluids, and the thermophoresis is stronger in alumina nanofluids. The effect of the slip mechanisms on the overall heat transfer is that the drag and gravity forces tend to reduce the Nusselt number while the Brownian, thermophoresis, lift, rotational and magnus effects tend to enhance the Nusselt number of nanofluids.

Overall, it can be concluded that the enhancement in convective heat transfer coefficient in nanofluids could be due to:

a) increase in thermal conductivity
b) increase in viscosity
c) reduction in thermal boundary layer

d) particle migration due to slip, induced by drag, gravity, Brownian motion, thermophoresis, lift, and rotational and Magnus effects.

The phenomenon that is dominant in any particular nanofluid would depend upon the particle shape, size, concentration, fluid temperature and the flow conditions.

## 4. PRESSURE DROP IN NANOFLUID FLOW

A large number of investigations, both experimental and theoretical, aiming at quantifying convective heat transfer in nanofluids, suggesting the use of nanofluids as an effective cooling option. However, one important issue that raises concern is the enhanced viscosity of nanofluids as compared to the basefluids, which will results in higher pressure losses in passages and hence entails higher pumping power requirements when used in conduits. Various investigations have reported quantative meaurements of the pressure drop and the friction factor in different nanofluids [14, 15, 17, 21, 25, 28, 29, 31, 34-36].

Hwang et al. [14] investigated the flow and pressure drop in $Al_2O_3$ (30±5 nm) - water nanofluid flowing through a straight circular stainless tube of 1.812mm diameter and 2500 mm length, in the fully developed laminar flow regime, with constant heat flux applied on the tube surface. From the measured pressure drop, the Darcy friction was estimated as,

$$f = \frac{-(dp/dx)d_{tube}}{\rho u_m^2/2} \tag{53}$$

Theoretically, for fully developed laminar flow in a circular pipe,

$$f = \frac{64}{\mathrm{Re}_d} \tag{54}$$

The friction factor for $Al_2O_3$ nanofluids was found to agree reasonably well with equation (50) within an error band of 2.4%.

He et al. [31] investigated the effect of particle size and concentration of spherical $TiO_2$ nanoparticles (~20nm) in water-based nanofluids, underlaminar and turbulent flow regimes while flowing upward through a vertical pipe. The nanofluid concentrations of 1.0%, 2.5% and 4% by weight (0.24%, 0.6% and 1.18% by volume) had average particle sizes of 95nm, 145nm and 210nm respectively. The measured pressure drop values in $TiO_2$ nanofluids were compared with the theoretical predictions using Eq. (55).

$$\Delta p = \frac{2 f \rho_f u^2 L}{D} \tag{55}$$

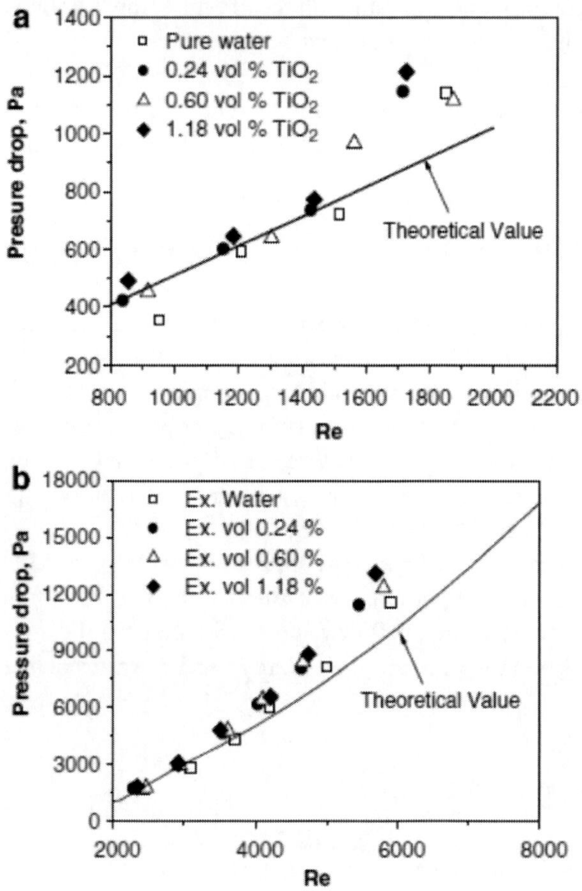


Figure 11. Pressure drop as a function of the Reynolds number in laminar and turbulent flows, reproduced from [31].

The friction factor in the laminar flow regime is given by the appropriate form of the Hagen-Poiseuille equation ($f = 16/Re$), and that in the turbulent regime is given by the Blasius equation ($f = 0.078Re^{-1/4}$).

The experimental results shown in Figure 11 are found to agree well with the theoretical predictions, for most of the Reynolds number range. The slight deviations at higher flow rates could be attributed to the entrance effects, the entrance lengths being higher at higher Reynolds numbers.

Xuan and Li [55] conducted experimental measurements of the pressure drop in turbulent flow of Copper-water nanofluids containing nanoparticles (<100nm) at concentrations of 1.0%, 1.2%, 1.5% and 2% by volume. The friction factor and the pressure drop in the pipe were found to be almost same as those obtained for pure water at the corresponding Reynolds number values.

The reported investigations on nanofluids have indicated that the pressure drop is not greatly affected, and remains almost same as that for the base fluids in the laminar regime. However, in the turbulent regime, a slight increase has been noticed in the pressure drop while using nanofluids.

## CONCLUSION

The option of using nanoparticle suspensions as a heat removal medium has been projected as a method which offers high heat transfer peformance in free convection systems. The enhanced and often anomalous behavior of nanofluids in such applications can be better understood in the light of the special effects imparted to the basic convection mechanisms, due to the presence of the suspended particles, apart from the modifications in the thermophysical properties of the medium. The discussions presented here focused on the state of the art of investigations on the application of nanofluids in forced convection loops. Important studies have been reviewed, and the findings revisited, with a view to understand the observations and the suggested reasons for the improved thermal behavior of nanofluids, as presented by various investigators. A few findings on heat transfer in nanfluids flowing through high pressure loops, investigatd in the authors' laboratory were also presented and discussed. A summary of the investigations and major findings available in the literature have been tabulated and presented in Table 1 for easy reference.

## REFERENCES


[1] Choi, S., 1995, Enhancing Thermal Conductivity of Fluids With Nanoparticles, In: Developments and Applications of Non-Newtonian Flows, D.A. Siginer, and H.P. Wang, eds., ASME, FED 231/MD 66, 99–105.

[2] Masuda, H., Ebata, A., Teramae, K., Hishinuma, N., 1993, "Alteration of Thermal Conductivity and Viscosity of Liquid by Dispersing Ultra-Fine Particles," *Netsu Bussei* **7**(4), 227–233.

[3] Lee, S., Choi, S.U.S., Li, S., Eastman, J.A., 1999, "Measuring Thermal Conductivity of Fluids containing Oxide Nanoparticles," ASME *J. Heat Transfer*, 121(2), 280-289.

[4] Xie, H., Wang, J., Xi, T., Liu, Y., Ai, F., 2002, "Thermal Conductivity Enhancement of Suspensions Containing Nanosized Alumina Particles," *J. Applied Physics*, 91(7), 4568-4572.

[5] Nguyen, C.T., Desgranges, F., Roy, G., Galanis, N., Mare, T., Boucher, S., and Angue Mintsa, H., 2007, "Temperature and Particle-Size Dependent Viscosity Data for Water-Based Nanofluids - Hysteresis Phenomenon," *International Journal of Heat and Fluid Flow*, 28, 1492-1506.

[6] Duangthongsuk, W., and Wongwises, S., 2009, "Measurement of Temperature-Dependent Thermal Conductivity and Viscosity of $TiO_2$-Water Nanofluids," *Experimental Thermal and Fluid Science,* 33, 706–714.

[7] Incropera, F., and DeWitt, D.P., 2007, *Fundamentals of Heat and Mass Transfer*, Wiley India Pvt. Ltd. Ed. 5.

[8] Maxwell, J.C., 1892, *A Treatise on Electricity and Magnetism*, 3rd ed., Newyork Oxford University Press.

[9] Sobhan, C.B., and Peterson, G.P., 2008, *Microscale and Nanoscale Heat Transfer Fundamentals and Engineering Applications*, Taylor and Francis/CRC Press.

[10] Sobhan, C.B., Sankar, N., Mathew, N., and Ratnapal, R., 2006, Molecular Dynamics Modeling of Thermal Conductivity of Engineering Fluids and its Enhancement Using

Nanoparticles, CANEUS 2006 Micro-Nano Technology for Aerospace Applications, Toulouse, France.

[11] Sankar, N., Mathew, N., Sobhan, C.B., 2008, "Molecular Dynamics Modeling of Thermal Conductivity Enhancement in Metal Nanoparticle Suspensions," *International Communications in Heat and Mass Transfer*, 35(7), 867-872.

[12] Pantzali, M.N., Kanaris, A.G., Antoniadis, K.D., Mouza, A.A., and Paras, S.V., 2009, "Effect of Nanofluids on the Performance of a Miniature Plate Heat Exchanger with Modulated Surface," *International Journal of Heat and Fluid Flow*, 30(4), 691-699.

[13] Nguyen, C.T., Roy, G., Gauthier, C., and Galanis, N., 2007, "Heat Transfer Enhancement Using $Al_2O_3$ - Water Nanofluid for an Electronic Liquid Cooling System," *Applied Thermal Engineering*, 27, 1501–1506.

[14] Hwang, K.S., Jang, S.P., and Choi, S.U.S., 2009, "Flow and Convective Heat Transfer Characteristics of Water-Based $Al_2O_3$ Nanofluids in Fully Developed Laminar Flow Regime," *International Journal of Heat and Mass Transfer*, 52, 193–199.

[15] Heris, S.Z., Esfahany, M.N., Etemad, S.Gh., 2007, "Experimental Investigation of Convective Heat Transfer of $Al_2O_3$/water Nanofluid in Circular Tube," *International Journal of Heat and Fluid Flow,* 28, 203–210.

[16] Anoop, K.B., Sundararajan, T., and Das, S.K., 2009, "Effect of Particle Size on the Convective Heat Transfer in Nanofluid in the Developing Region," *International Journal of Heat and Mass Transfer*, 52, 2189–2195.

[17] Lee, J., and Mudawar, I., 2007, "Assessment of the Effectiveness of Nanofluids for Single-Phase and Two-Phase Heat Transfer in Micro-Channels," *International Journal of Heat and Mass Transfer*, 50, 452–463.

[18] Gherasim, I., Roy, G., Nguyen, C.T., Vo-Ngoc, D., 2009, "Experimental Investigation of Nanofluids in Confined Laminar Radial Flows," *International Journal of Thermal Sciences*, 48, 1486–1493.

[19] Kim, D., Kwon, Y., Cho, Y., Li, C., Cheong, S., Hwang, Y., Lee, J., Hong, D., and Moon, S., 2009, "Convective Heat Transfer Characteristics of Nanofluids under Laminar and Turbulent Flow Conditions," *Current Applied Physics*, 9, 119–123.

[20] Heris, S.Z., Etemad, S.Gh., and Esfahany, M.N., 2006, "Experimental Investigation of Oxide Nanofluids Laminar Flow Convective Heat Transfer," *International Communications in Heat and Mass Transfer*, 33, 529–535.

[21] Jung, J.Y., Oh, H.S., and Kwak, H.Y., 2009, "Forced Convective Heat Transfer of Nanofluids in Microchannels," *International Journal of Heat and Mass Transfer*, 52, 466–472.

[22] Sharma, K.V., Sundar, L.S., Sarma, P.K., 2009, "Estimation of Heat Transfer Coefficient and Friction Factor in the Transition Flow with Low Volume Concentration of $Al_2O_3$ Nanofluid Flowing in a Circular Tube and with Twisted Tape Insert," *International Communications in Heat and Mass Transfer*, 36, 503–507.

[23] Duangthongsuk, W., and Wongwises, S., 2009, "Heat Transfer Enhancement and Pressure Drop Characteristics of $TiO_2$–Water Nanofluid in a Double-Tube Counter Flow Heat Exchanger," *International Journal of Heat and Mass Transfer*, 52, 2059–2067.

[24] Ding, Y., Alias, H., Wen, D., and Williams, R.A., 2006, "Heat Transfer of Aqueous Suspensions of Carbon Nanotubes (CNT Nanofluids)," *International Journal of Heat and Mass Transfer*, 49, 240–250.

[25] Yu, W., France, D.M., Smith, D.S., Singh, D., Timofeeva, E.V., and Routbort, J.L., 2009, "Heat Transfer to a Silicon Carbide/Water Nanofluid," *International Journal of Heat and Mass Transfer*, 52, 3606–3612.

[26] Ding, Y., Chen, H., He, Y., Lapkin, A., Yeganeh, M., Siller, L., Butenko, Y.V., 2007, "Forced Convective Heat Transfer of Nanofluids," *Advanced Powder Technology*, 18(6), 813–824.

[27] Nguyen, C.T., Galanis, N., Polidori, G., Fohanno, S., Popa, C.V., and Bechec, A.L., 2009, "An Experimental Study of a Confined and Submerged Impinging Jet Heat Transfer Using Al2O3-Water Nanofluid," *International Journal of Thermal Sciences*, 48, 401–411.

[28] Pantzali, M.N., Mouza, A.A., and Paras, S.V., 2009, "Investigating the Efficacy of Nanofluids as Coolants in Plate Heat Exchangers (PHE)," *Chemical Engineering Science*, 64, 3290 – 3300.

[29] Ho, C.J., Wei, L.C., and Li, Z.W., 2010, "An Experimental Investigation of Forced Convective Cooling Performance of a Microchannel Heat Sink with Al2O3/Water Nanofluid," *Applied Thermal Engineering*, 30, 96–103.

[30] Li, Q., and Xuan, Y., 2009, "Experimental Investigation on Heat Transfer Characteristics of Magnetic Fluid Flow around a Fine Wire under the Influence of an External Magnetic Field," *Experimental Thermal and Fluid Science*, 33, 591–596.

[31] He, Y., Jin, Y., Chen, Ding, Y., Cang, D., and Lu, H., 2007, "Heat Transfer and Flow Behavior of Aqueous Suspensions of TiO2 Nanoparticles (Nanofluids) Flowing Upward through a Vertical Pipe," *International Journal of Heat and Mass Transfer*, 50, 2272–2281.

[32] Chen, H., Yang, W., He, Y., Ding, Y., Zhang, L., Tan, C., Lapkin, A.A., and Bavykin, D.V., 2008, "Heat Transfer and Flow Behavior of Aqueous Suspensions of Titanate Nanotubes (Nanofluids)," *Powder Technology*, 183, 63–72.

[33] Mansour, R.B., Galanis, N., and Nguyen, C.T., 2011, "Experimental Study of Mixed Convection with WatereAl2O3 Nanofluid in Inclined Tube with Uniform Wall Heat Flux," *International Journal of Thermal Sciences*, 50, 403-410.

[34] Fotukian, S.M., and Esfahany, M.N., 2010, "Experimental Study of Turbulent Convective Heat Transfer and Pressure Drop of Dilute CuO/Water Nanofluid inside a Circular Tube," *International Communications in Heat and Mass Transfer*, 37, 214–219.

[35] Suresh, S., Chandrasekar, M., and Chandra Sekhar, S., 2011, "Experimental Studies on Heat Transfer and Friction Factor Characteristics of CuO/Water Nanofluid under Turbulent Flow in a Helically Dimpled Tube," *Experimental Thermal and Fluid Science,* 35, 542–549.

[36] Fotukian, S.M., and Esfahany, M.N., 2010, "Experimental Investigation of Turbulent Convective Heat Transfer of Dilute γ-Al2O3/Water Nanofluid inside a Circular Tube," *International Journal of Heat and Fluid Flow*, 31, 606–612.

[37] Shon, C.W., and Chen, M.M., 1981, "Microconvective Thermal Conductivity in Disperse Two-Phase Mixture as Observed in a Low Velocity Couette Flow Experiment," ASME *J. Heat Transfer*, 103, 47–51.

[38] Drew, D.A., and Passman, S.L., 1999, *Theory of Multi-Component Fluids*, Springer.

[39] Yu, W., and Choi, S.U.S., 2003, "The Role of International Layers in the Enhanced Thermal Conductivity of Nanofluids: A Renovated Maxwell Model," *Journal on Nanoparticle Research*, 5, 167–171.

[40] Yang, Y., Zhang, Z.G., Grulke, E.A., Andersen, W.B., and Wu, G., 2005, "Heat Transfer Properties of Nanoparticle-in-Fluid Dispersions (Nanofluids) in Laminar Flow," *International Journal of Heat and Mass Transfer*, 48(6), 1107–1116.

[41] Bergman, T.L., 2009, "Effect of Reduced Specific Heats of Nanofluids on Single Phase, Laminar Internal Forced Convection," *International Journal of Heat and Mass Transfer*, 52, 1240–1244.

[42] Zhou, S.Q., and Ni, R., 2008, "Measurement of the Specific Heat Capacity of Water-Based Al2O3 Nanofluid," *Appl. Phys. Lett.*, 92, 093123.

[43] Mansour, R.B. Galanis, N., and Nguyen, C.T., 2007, Effect of Uncertainties in Physical Properties on Forced Convection Heat Transfer with Nanofluids," *Applied Thermal Engineering*, 27, 240–249.

[44] Xuan, Y.M., and Li, Q., 2003, "Investigation on Convective Heat Transfer and Flow Features of Nanofluids," ASME *J. Heat Transfer*, 125, 151–155.

[45] Wen, D.S., and Ding, Y.L., 2004, "Experiment Investigation into Convective Heat Transfer of Nanofluids at the Entrance Region under Laminar Flow Conditions," *Int. J. Heat Mass Transfer*, 47, 5181–5188.

[46] Pak, B.C., Cho, Y.I., 1998, "Hydrodynamic and Heat Transfer Study of Dispersed Fluids with Submicron Metallic Oxide Particles," *Exp. Heat Transfer*, 11, 151–170.

[47] Buongiorno, J., 2006, "Convective Transport in Nanofluids," ASME *J. Heat Transfer,* 128, 240–250.

[48] Kim, S.J., Bang, I.C., Buongiorno, J., and Hu, L.W., 2007, "Surface Wettability Change during Pool Boiling of Nanofluids and Its Effect on Critical Heat Flux," *Int. J. Heat Mass Transfer*, 50, 4105–4116.

[49] Yu, W., France, D.M., Choi, S.U.S., and Routbort, J.L., 2008, "Review and Comparison of Nanofluid Thermal Conductivity and Heat Transfer Enhancements," *Heat Transfer Eng.*, 29(5), 432–460.

[50] Faulkner, D.J., Rector, D.R., Davidson, J., Shekarriz, R., 2004, "Enhanced Heat Transfer through the Use of Nanofluids in Forced Convection," Proceedings of IMECE 2004, Anaheim, California, USA.

[51] Ahn, H.S., Kim, H., Jo, H.J., Kang, S.H., Chang, W.P., and Kim, M.H., 2010, "Experimental Study of Critical Heat Flux Enhancement During Forced Convective Flow Boiling of Nanofluid on a Short Heated Surface," *International Journal of Multiphase Flow*, 36, 375–384.

[52] Vassallo, P., Kumar, R., and D'Amico, S., 2004, "Pool Boiling Heat Transfer Experiments in Silica–Water Nano-Fluids," *Int. J. Heat Mass Transfer*, 47, 407–411.

[53] Liter, S.G., and Kaviany, M., 2001, "Pool-Boiling CHF Enhancement by Modulated Porouslayer Coating: Theory and Experiment," *Int. J Heat Mass Transfer*, 44, 4287–4311.

[54] Golubovic, M., Hettiarachchi, M.H.D., and Worek, W.M., 2008, "Nanofluids and Critical Heat Flux," Proceedings of 1st ASME Micro/Nanoscale Heat Transfer International Conference, Tainan, Taiwan, Paper No. MNHT2008-52204.

[55] Park, H.S., Shiferaw, D., Sehgal, B.R., Kim, D.K., and Muhammed, M., 2004, "Film Boiling Heat Transfer on a High Temperature Sphere in Nanofluid," Proceedings of ASME HT/FED 4, 469–476.

[56] Milanova, D., and Kumar, R., 2005, "Role of Ions in Pool Boiling Heat Transfer of Pure and Silica Nanofluids," *Appl. Phys. Lett.,* 87, 233107.

[57] Moreno Jr., G., Oldenburg, S.J., You, S.M., and Kim, J.H., 2005, "Pool Boiling Heat Transfer of Alumina–Water, Zinc Oxide–Water and Alumina–Water + Ethylene Glycol Nanofluids," Proceedings of HT2005 2005 ASME Summer Heat Transfer Conference, San Francisco, California, USA, July 17–22, 2005.

[58] Wen, D., and Ding, Y., 2005, "Experimental Investigation into the Pool Boiling Heat Transfer of Aqueous Based Alumina Nanofluids," *J. Nanoparticle Res.*, 7, 265–274.

[59] Bang, I.C., and Chang, S.H., 2005, "Boiling Heat Transfer Performance and Phenomena of Al2O3–Water Nano-Fluids from a Plain Surface in a Pool," *Int. J. Heat Mass Transfer*, 48, 2407–2419.

[60] Kim, H., Kim, J., and Kim, M.H., 2006, "Effect of Nanoparticles on CHF Enhancement in Pool Boiling of Nano-Fluids," *Int. J. Heat Mass Transfer*, 42, 2003–2013.

[61] Kim, H., Kim, J., and Kim, M.H., 2006, "Experimental Study on the Characteristics and Mechanism of Pool Boiling CHF Enhancement Using Nano-Fluids," ECI International Conference on Boiling Heat Transfer, Spoleto, May 7–12, 2006.

[62] Ding, Y., and Wen, D., 2005, "Particle Migration in a Flow of Nanoparticle Suspensions," *Powder Technology*, 149(2), 84–92.

[63] Turian, R.M., Ma, T.W., Hse, F.L., and Sung, D.J., 1998, "Flow of Concentrated Non-Newtonian Slurries: 1. Friction Loss in Laminar, Turbulent and Transition Flow through Straight Pipe," *Int. J. Multiph. Flow*, 24, 225– 242.

[64] Bournonville, B., and Nzihou, A., 2002, "Rheology of Non-Newtonian Suspensions of Fly Ash: Effect of Concentration, Yield Stress and Hydrodynamic Interactions," *Powder Technology*, 128, 148– 158.

[65] Sohn, C.W., and Chen, M.M., 1981, "Microconvective Thermal Conductivity in Disperse Two Phase Mixture as Observed in a Low Velocity Couette Flow Experiment," *J. Heat Transfer*, 103, 47–51.

[66] Savithiri, S., Pattamatta, A., and Das, S.K., 2011, "Scaling Analysis for the Investigation of Slip Mechanisms in Nanofluids," *Nanoscale Research Letters*, 6, 471.

In: Nanofluids
Editor: Yuwen Zhang

ISBN: 978-1-62618-165-6


*Chapter 7*

# HEAT TRANSFER ENHANCEMENT OF JET IMPINGEMENT COOLING WITH NANOFLUIDS

***Yimin Xuan**, *Qiang Li, Peng Tie and Feng Yu***
School of Power Engineering, Nanjing University of Science and Technology, Nanjing, Jiangsu, P.R. China

## ABSTRACT

Jet impingement cooling is a vital technique for thermal management of electronic devices of high heat flux by impinging fluid on a heater surface due to its high local heat transfer rates. One may find plenty of publications on the heat transfer characteristics of jet impingement cooling technique such as the distance between the jet outlet and the heated surface, the magnitude of the jet velocities, turbulence intensity, the angle of impingement, the array of jets and the roughness of the heater surface. But little attention is paid to the effect of the thermal properties of the impingement fluid, which may be another way to further improve the heat transfer performance of the jet impingement. With development of enhanced heat transfer techniques in conventional systems, the conventional working fluid with low thermal conductivity turns into a bottleneck factor for promoting the performance effectiveness of thermal systems. Improvement of the thermal properties of impinging fluids may become a promising trick of augmenting heat transfer process in the jet impingement cooling system.

Recent advances in the nanotechnology have permitted the creation of a new and special class of fluids, named 'nanofluids'. The thermal conductivity and heat transfer performance of nanofluids are superior to those of the original pure carrier fluids because the suspended nanoparticles remarkably improve energy transport capability of the suspensions. Nanofluid offers the exciting possibility of increased heat transferability of jet impingement.

In this current study, experiments were conducted to investigate the heat transfer enhancement of the nanofluid in the impinging jet cooling system. Two types of nanofluids were used as the working fluid. Cu-water nanofluid was studied in submerged single jet impingement cooling while Cu-ethylene glycol/water nanofluid was introduced into free surface jet array impingement cooling, respectively. In submerged single jet impingement,

* To whom correspondence should be addressed. E-mail: ymxuan@mail.njust.edu.cn.

nanofluids with volume fraction ranged from 1.5% to 3.0% were studied for different jet-to-target spans. Experiments have revealed that the suspended nanoparticles increase the heat transfer performance of the base liquid in the jet impingement cooling system. It had been found that an enhancement of 18.5% compared with the base fluid at the Reynolds number of 8000 when nanofluids are used. In addition, the experiments have shown that the suspended nanoparticles almost result in no extra addition of pressure drop in both submerged single jet and free surface jet array impingement.

**Keywords** Jet impingement; Nanofluids; Enhanced Heat Transfer

## 1. Introduction

With the prevalence of high-density electronic components, energy dissipation becomes a momentous embarrassment and limits the development of electronic industry such as high performance microprocessors, laser diode array, radars and X-ray anodes. The requirement of super-speed electronics and the shrinking size of electronic devices lead to higher and higher heat flux dissipated from the electronic devices, i.e., high-power electronics provides severe challenges for scientists and engineers. It is urgent to find more effective heat removal techniques. If the superfluous heat cannot be removed immediately, electronic components will be heated up to very high temperature, which will adversely depress the performance and reliability of the electronic components. In order to improve the reliability of the electronics at ever-increasing heat flux, more and more efforts have been devoted to exploit new and effective thermal control methods, especially for the cases that conventional forced air cooling techniques fail to satisfy the ever-increasing cooling requirement.

Currently, many electronic devices always use forced convection heat transfer method that a heat sink and fan combination to remove heat flux. The heat is conducted from the heated surface to the heat sink which has a high thermal conductivity. Finally, the waste heat is dissipated to the environment through the fins via forced convection using a fan. Although the heat sink is widely used, it still has its limitations. When the heat sink is used, its surface area should be several times larger than the surface to be cooled. On the other hand, the pins of enhanced surface are far away from the heated surface, which reduces the heat transfer efficiency and may make the effects of local hot-spots even worse. Moreover, the conventional method is difficult to meet heat transfer and cooling requirements of high heat loads.

Although a lot of convective heat transfer techniques are widely applied to cooling electronic devices, jet impingement is becoming an attractive selection for dealing with thermal management issues. Because of the high velocity of the impingement jets, the turbulence around the stagnation area is very strong, which results in the high heat transferability. Conventionally, either air or liquid may be used as the working fluid of jet impingement. One may find plenty of publications on the heat transfer behaviors of jet impingement cooling [1-11]. Since liquid jet impingement may exert several orders of magnitude of heat transfer ability times that of air jet impingement, liquid jet impingement cooling technique invokes more efforts and is one of the most powerful cooling solutions for high-heat-flux removal. According to the distribution and flow of working fluid on the surface to be cooled, there are two kinds of liquid jet impingement cooling: One is submerged jets and another is free surface jets.

In 1965, Bardbury [12] studied heat transfer features for the single jet impingement. The results revealed that the jet impingement has a great potential for thermal management at the high heat flux. Hereafter, more investigations were carried out for jet impingement cooling. For example, Viskanta [13] established an experimental system to study heat transfer performance of jet impingement and he found that the distance of jet-to-target is one of main factors of affecting on the heat transfer characteristics for jet impingement. Webb and Ma [14] studied the influence of Reynolds number on the single jet impingement and their results indicated that the heat transfer coefficient of jet impingement increased with the increasing of the Reynolds number. When both free surface jet impingement and submerged jet were included, a correction could be expressed as $Nu = C\,\mathrm{Re}^m\,\mathrm{Pr}^n$. Hoogendoom [15] and Lytlel [16] studied the influence of the distance of jet-to-target on the heat transfer of jet impingement. In their experiments, a secondary enhancement of heat transfer was observed when a small distance of jet-to-target is used. Ma and Zheng [17] studied the influence of the angle of the jet and they found that the heat transfer performance of jet impingement decreased with the increasing of the impingement angle. Goldstein and Seol [18] investigated the influence of the figure of the jet on heat transfer intensity and revealed found that the circular nozzle has the higher heat transfer coefficient than the narrow nozzle. Stevens [19] analyzed the influence of the jet diameter on the impingement performance.

Some researchers also carried out investigations on the heat transfer characteristics of jet array impingement cooling technique. Jiji et al. [20] investigated that the temperature uniformity of surface to be cooled is improved with the decreasing of the spacing between the adjacent jets. The ratio of jet-to-target distance to the nozzle diameter had no effect on the heat transfer coefficient when it was in the range from 3 to 5. Fabbri and Dhir [21] studied the micro jet array impingement cooling technique. They found that the jet array impingement would experience better heat transferability than the array containing the large jets with the same flow rate. But the micro jet array would cost a bigger pressure drop than the larger jets. Whelan [22] studied both free surface jet array impingement and submerged single jet and pointed out that the submerged jet has a higher heat transfer coefficient than the free surface jet impingement. Sung and Mudawar [23, 24] combined the micro-channel and the micro-jet to enhance the heat transfer of jet array impingement. Goodro, et al. [25] experimentally investigated the effects of hole spacing on spatially-resolved incompressible and compressible jet array impingement heat transfer for a series of Reynolds number and Mach number and they found that Mach number has a significant impact on overall heat transfer. Yan et al. [26] conducted an experimental study on impingement spray cooling and analyzed effects of mass flow rate, nozzle inlet pressure and chamber pressure were investigated in this study. They found that increasing the coolant mass flow rate, nozzle inlet pressure and chamber pressure will have positive effects on the heat transfer effectiveness of impingement spray cooling. Since R134a was used as working fluid in their experimental runs, the heat flux input was limited and heat transfer coefficients up to 5596 $W/m^2\,K$ was achieved with heat flux input around 50,000 $W/m^2$.

It is noted that the previous studies were mainly focused on optimizing the transport parameters such as the distance between the jet outlet and the impingement surface, the magnitude of the jet velocities, turbulence intensity, the angle of impingement and the array of jets et al. To date the literature concerning the heat transfer enhancement of jet working fluid is relatively sparse. In the existing investigations, the thermal conductivity of some traditional natural and organic working fluids such as water and ethylene glycol is too low to satisfy the requirement of heat removal in electronic components with high heat flux. Comparison between the cooling performances of water and air impingement reveals that the properties of working fluid will greatly affect the cooling features and improvement of the thermal properties of jet working fluid may become a trick for improving heat transfer performances of the impinging jet system.

To meet the needs of enhanced heat transfer, an innovative category of heat transfer fluid, nanofluid, has been proposed with development of nanomaterial techniques [27]. In 1995, Dr. Choi of Argonne National Laboratory of USA proposed the concept of nanofluids to exploit the unique properties of nanoparticles in developing advanced heat transfer fluids. Nanofluids are solid-liquid composite materials consisting of solid nanoparticles or nanofibers with sizes typically of 1-100 nm suspended in liquid. Plenty of investigations [28-36] on the nanofluid are conducted and have shown that the nanofluid has superior heat transfer performance than those of conventional pure fluids. For example, compared with deionized water, the thermal conductivity of the nanofluid consisting of deionized water and 5% volume fraction of 100 nm passivated Cu nanoparticles is enhanced by 45% [28]. The experimental results also show that one of the important factors affecting the thermal conductivity of nanofluids is the volume fraction of nanoparticles. The thermal conductivity of the nanofluids remarkably increases with the volume fraction of nanoparticles [28-30]. The experiments for the flow and convective heat transfer of nanofluids showed that the suspended nanoparticles remarkably increase heat transfer performance of the base fluid and the nanofluid has larger heat transfer coefficient than pure water under the same Reynolds number. Compared with water, for example, the convective heat transfer coefficient of nanofluid increased about 60% for the nannofluid with 2.0 vol% Cu nanoparticles under the same Reynolds number [31, 32]. Since the nanoparticles are so small that the nanofluid behaves like a pure fluid, the nanofluids do not cause extra penalty in pump power.

Although most of the existing investigation efforts of nanofluid were devoted to the convective heat transfer performance, it is an intuitive idea that as a new type of heat carrier, the nanofluid may have great potential for enhanced heat transfer of jet impingement cooling. The present chapter is aiming at possible applications of nanofluids to the jet impingement in order to develop a more effective approach for cooling high power electronics. By considering water-based $Al_2O_3$ nanfluid as a single mixture, Manca et al. [37] numerically studied heat transfer performance of a confined slot impinging jet and showed the dependence of the intensity and size of flow vortex structures on the ratio of height $H$ to the jet orifice width $W$, particle volume fraction and Reynolds number. Their numerical results showed that the local Nusselt number profiles exhibited the highest values at the stagnation point and the lowest at the end of the heated plate and the averaged Nusselt number increased for increasing particle concentrations and Reynolds numbers. As mentioned before, the assumption of single-phase model in this study neglected the two-phase nature of nanofluids and just substituted the properties of nanofluids for those of conventional fluids such as water or air, which failed to take the bombardment effect of nanoparticles on the heated surface into account. Nguyen et al. [38] carried out an experimental investigation was performed to study the heat transfer performance of a 36 nm-$Al_2O_3$-particle–water nanofluid in a confined and submerged impinging jet on a flat, horizontal and circular heated surface. They found that the use of a nanofluid can provide a heat transfer enhancement in some cases; conversely, for other combinations, an adverse effect on the convective heat transfer coefficient may occur.

It is well-known that nanofluids consisting of Cu nanoparticles manifest better performance of enhanced heat transfer than that of nanofluids consisting of nonmetallic nanoparticles under the other same conditions. Now we introduce experimental investigations on the impingement heat transfer features of nanofluids consisting of Cu nanopartilces. Several types of Cu nanofluids are used as the working fluid. Cu-water nanofluids (Cu nanoparticles with 25 nm and 100 nm diameter) are involved in the submerged single jet impingement cooling system and Cu-ethylene glycol/water nanofluids (Cu nanoparticles with diameter of 25 nm) are introduced into the free surface jet array impingement cooling system. Through the experimental results, The effects of the structural and operational parameters such as the nanoparticle concentration, Reynolds number, nozzle-to-plate distance, fluid temperature, and nanoparticle diameter as well as the surface dispersant used in preparation of nanofluids on the heat transfer and flow performances of the jet impingement of nanofluids are investigated.

## 2. Submerged Single Jet Impingement Cooling

### 2.1. Experimental System

As showed schematically in Figure 1, an experimental setup is built to study the heat transfer performance and flow features of Cu-water nanofluids in a submerged single jet impingement cooling system. It mainly includes a reservoir tank, a pump, a pipe line, a cooler, a test section and a fluid collection tank. The reservoir tank is used to reserve the nanofluid and monitor the dispersion behavior and stability of the nanofluid. The cooler is used to keep a constant temperature of the nanofluid at the inlet of the test section. The flow rate is controlled by means of two adjusting valves, one at the main flow loop and the other at the by-bass line. A three-way valve is installed at the end of the main flow loop and it provides an access for the nanofluid from the reservoir tank into the fluid collection tank to precisely measure the volume flow rate of the nanofluid. The flowmeter is only used to monitor the flow rate because of its lower precision.

Figure 2 illustrates the construction of the test section. The test section includes upper plate, polyfluortetraethylene plate, jet room, static pressure chamber and heating system. A circular copper piece with 16 mm-diameter and 4 mm-height heated by a ceramic package resistance is used as the heat transfer surface. Six type-K thermocouples are installed upside of the copper piece to measure the average temperature of the heated surface. Two K-type thermocouples are arranged at the entrance and exit port of test section to measure the inlet and outlet temperature of the fluid. Two pressure transducers are installed in both ends of test section to measure system pressure drop of the fluid. Three jet-to-target distances (2, 4 and 6 mm) are studied, namely, $H/D$=1, 2, 3, where $H/D$ is the ratio of jet-to-target distance $H$ to the nozzle diameter $D$.

In order to fabricate a horizontal circular heated surface, a heater block has been designed. This block, a cylindrical all-copper body of diameter 16 mm and thickness of 4 mm, is electrically heated by means of one standard 100 Ω ceramic package resistance heater. The working fluid passes through a vertical nozzle of 2 mm in diameter and forms impinging jet to cool the heated surface. In order to minimize the heat loss from the heater to the ambient, thermal insulation materials are filled and enwrapped around the heater block and the top of the upper plate.

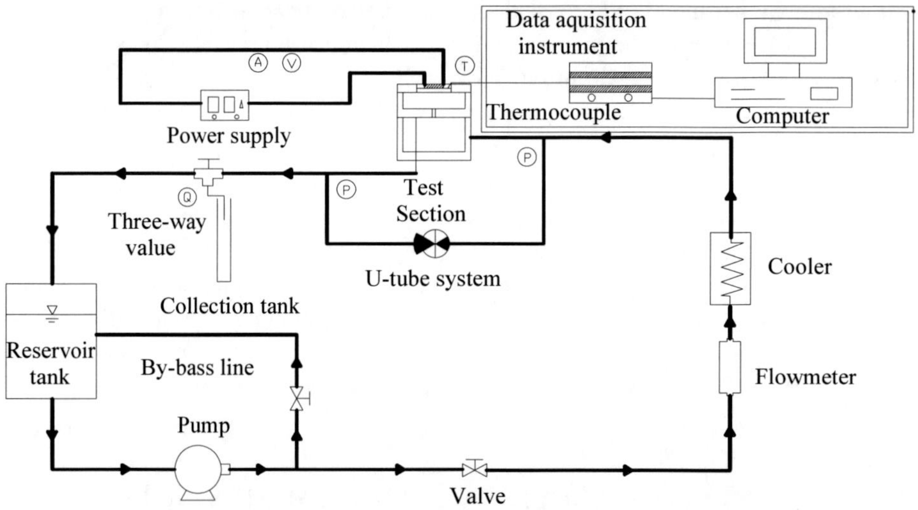


Figure1. Schematic illustration of submerged single jet impingement cooling system.

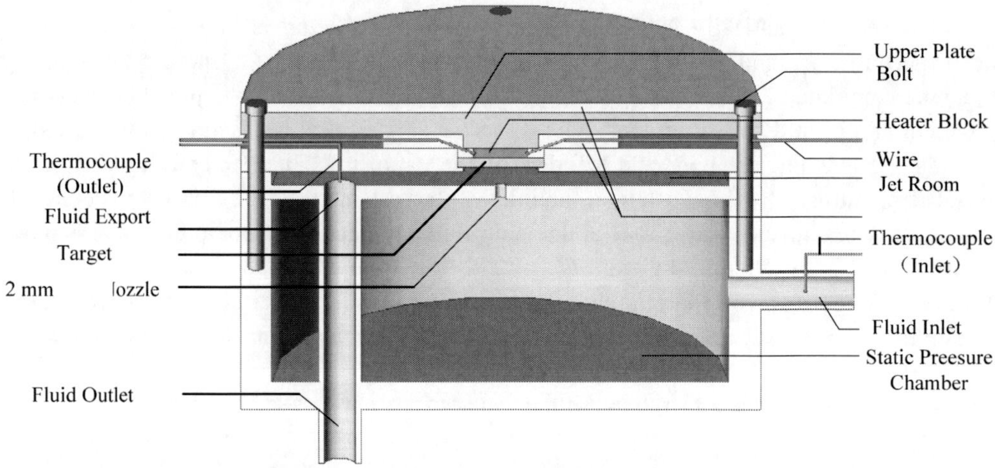


Figure 2. Scheme of test module for submerged single jet impingement cooling system.

## 2.2. Preparation and Thermal Properties of Cu-Water Nanofluid

Two types of Cu-water Nanofluids are prepared by two-phase method, in which Cu nanoparticles of 25 nm in diameter or 100 nm in diameter are directly added into distilled water, respectively. Consequently, the suspension is vibrated for 4 h in an ultrasonic vibrator to obtain stable suspensions. The sample suspensions kept stable in the stationary state with little apparent sediment during the whole experiment process.

The thermal conductivities of the sample Cu-water nanofluid used are measured by the transient double-hot-wires instrument. The experimental results are shown in Figure 3. The experimental results have shown that the suspended nanoparticles dramatically increase thermal conductivity of nanofluids and the thermal conductivity of the nanofluid samples remarkably increases with the volume fraction of suspended nanoparticles.

The viscosity of the above-mentioned sample nanofluids is measured by using NXE-1 viscometer. Figure 4 indicates variation of the viscosity of the nanofluids with the different volume fractions of the nanoparticles.

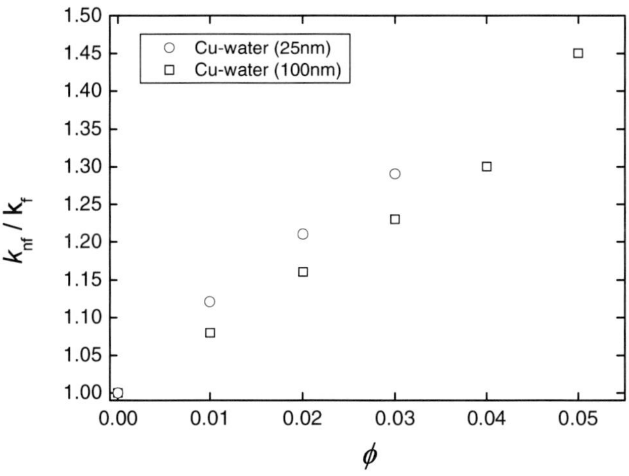


Figure 3. Thermal conductivity of Cu-water nanofluid.

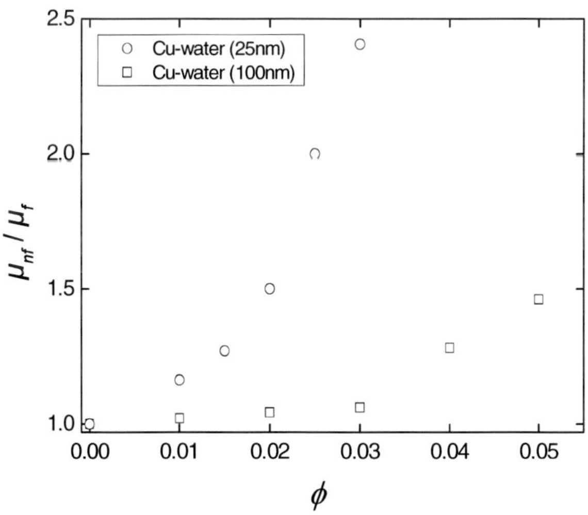


Figure 4. Viscosity of Cu-water nanofluid.

The following equations are used for computing the density, specific heat, and thermal diffusivity of the nanofluid:

$$\rho_{nf} = (1-\phi)\rho_f + \phi\rho_p \tag{1}$$

$$(\rho c_p)_{nf} = (1-\phi)(\rho c_p)_f + \phi(\rho c_p)_p \quad (2)$$

$$\alpha_{nf} = \frac{k_{nf}}{(\rho c_p)_{nf}} = \frac{k_{nf}}{(1-\phi)(\rho c_p)_f + \phi(\rho c_p)_d} \quad (3)$$

## 2.3. Experimental Procedures and Data Processing

During a series of experimental runs the nanofluids with different volume fractions of suspended nanoparticles were used to investigate the effect of the nanoparticle concentration on the enhanced heat transfer performances, in which 1.5%, 2.0%, 2.5%, and 3.0% volume fraction Cu-

water nanofluids were tested. Here the volume fraction of nanoparticles is determined from the mass fraction of nanoparticles. For each of these concentrations, three different cases of jet-to-target distance are studied in turn, i.e., $H$=2, 4 and 6 mm, respectively.

With the collected data of the temperature and the flow rate, the average convective heat transfer coefficient $h$ ($W/m^2 \cdot K$) under the submerged impinging jet can be calculated from the following Newton's cooling law:

$$h_{nf} = \frac{Q}{A(T_s - T_f)} = \frac{UI}{A\left(T_s - \frac{T_{in} + T_{out}}{2}\right)} \quad (4)$$

where $Q$ is the input electric power. $A$ is the area of the heater circular surface. $T_s$, $T_{in}$ and $T_{out}$ are respectively the average temperature of the heated surface, the inlet and outlet temperatures of the fluid.

The Reynolds number and the averaged Nusselt number are defined as follows:

$$\mathrm{Re}_{nf} = \frac{u_m D}{\upsilon_{nf}} \quad (5)$$

$$Nu_{nf} = \frac{h_{nf} D}{k_{nf}} \quad (6)$$

where $u_m$ is the mean impinging velocity. $D$ is the inner diameter of the tube. $\upsilon_{nf}$ is the viscosity of the nanofluid. $k_{nf}$ is the thermal conductivity of the nanofluid.

Uncertainty of the experimental data can be assessed from the measuring errors of parameters such as the heat flux, the temperature, the flow rate and the length. Table 1 and 2 list the experimental error and uncertainty, respectively. It is shown that the maximum uncertainty of the heat transfer coefficient and the friction factor is 1.29% and 2.4%, respectively.

**Table 1. Estimation of experimental errors**

| Sources of the error | Explanation |
|---|---|
| Voltage Meter | The smallest scale is 0.1 V |
| Amperemeter | The smallest scale is 0.01 A |
| Vernier Caliper | The smallest scale is 0.02 mm |
| Thermocouples | The minimum scale of thermometer used for calibration is 0.1 °C |

**Table 2. Uncertainty of experimental data**

| Parameters | Calculation expressions | Results |
|---|---|---|
| $E_h$ | $\sqrt{(\frac{\delta U}{U})^2+(\frac{\delta I}{I})^2+4(\frac{\delta D}{D})^2+(\frac{\delta T}{T})^2}$ | 1.44% |
| $E_{Nu}$ | $\sqrt{E_h^2+(\frac{\delta D}{D})^2+(\frac{\delta K}{K})^2}$ | 3.47% |
| $E_{\mathrm{Re}}$ | $\sqrt{(\frac{\delta v}{v})^2+(\frac{\delta \upsilon}{\upsilon})^2+(\frac{\delta D}{D})^2}$ | 2.93% |
| $E_{\Delta P}$ | $u_{\Delta P} / \Delta P$ | 2.34% |

## 2.4. Results and Discussion

### *2.4.1. Heat Transfer Features of the 25 Nm Cu-Water Nanofluid*

Figure 5 illustrates the convective heat transfer coefficient of the 25 nm Cu-water nanofluid in the submerged single jet impingement cooling system for the different cases of jet-to-target distances and the volume fractions of suspended nanoparticles. The results clearly indicate that application of nanofluids significantly improves the heat transfer features of all the submerged single jet impingement cooling system with three different jet-to-target distances. With respect to the same Reynolds number, Cu-water nanofluid has larger the heat transfer coefficient of jet impingement than pure water. As a whole, the enhanced impingement performance of the nanofluid increases with the volume fraction of nanoparticles. For the nanofluid of containing 3.0% nanoparticles in volume fraction, the impingement heat transfer coefficient is 52% higher in comparison with the case of pure water.

Figure 6 further clearly exhibits the tendency that the heat transfer coefficient of nanofluid impingement remarkably increases with the volume fraction of nanoparticles. While the volume fraction of 25 nm Cu nanoparticles increases from 0.015% to 3.0%, for example, the ratio of the heat transfer coefficient of the Cu-water nanofluid to that of pure water varies from 1.17 to 1.52 under the same Reynolds number $Re$=7500.

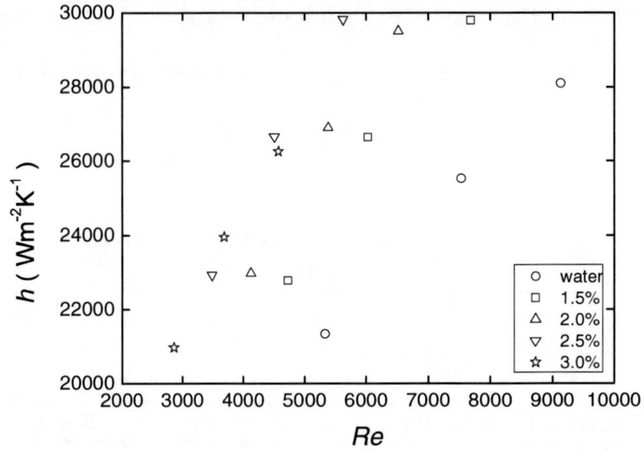


**(a) *H*/*D*=1**

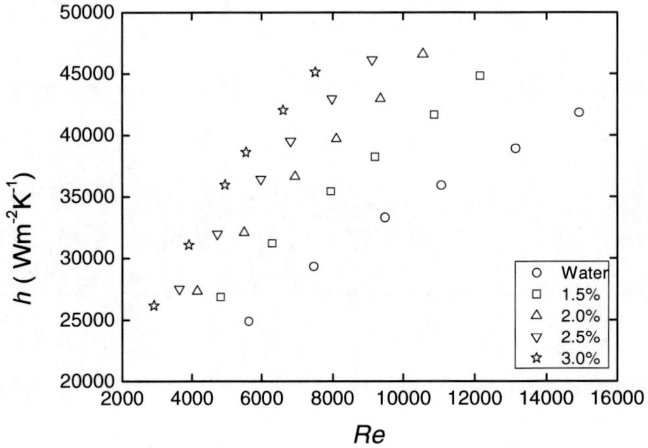


**(b) *H*/*D*=2**

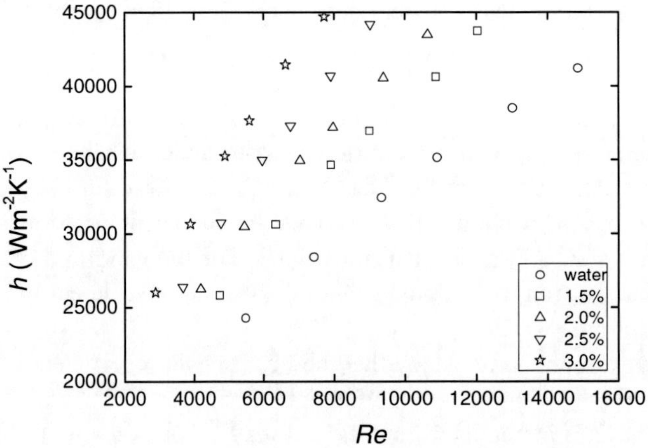


**(c) *H*/*D*=3**

Figure 5. Impingement heat transfer coefficient of Cu-water nanofluid ( $d_p = 25$ nm).

The suspended nanoparticles enhanced energy transport inside the submerged single jet impingement cooling system. The reason may be explained from the following three aspects: The first is that the suspended particles increase the thermal conductivity of the two-phase mixture; the second is that nanoparticle accelerates energy exchange process among the fluid and the heated surface, the nanoparticle and the nanoparticle as well as the nanoparticles and the fluid; the third is that intense bombardment of individual nanoparticles on the surface to be cooled sharply enhances turbulence of the working fluid close to the surface and greatly reduces the thickness of fluid sub-layer nearby the solid surface. All these three factors alter the status of heat exchange between the working fluid and the heated surface exposed to the jet impingement. Therefore, the heat exchange between the nanofluid and the heated surface is enhanced. Obviously, the larger the particle volume fraction is, the higher augment of the impingement heat transfer coefficient is during the present experimental runs.

As shown in Figure 7, the heat transfer coefficient of the 25 nm Cu-water nanofluid increases with the fluid impinging velocity. Furthermore, it is larger than that of the base liquid (water) under the same impinging velocity. But it can be seen that the heat transfer coefficient of the jet impingement of the suspensions does not monotonously increases with the volume fraction of nanoparticles for the different jet-to-target distances. The heat transfer coefficient of the Cu-water nanofluid with a volume concentration of 2.0% is higher than that of the nanofluids with other volume concentration under the condition of constant average velocity. The reasons are not so clear and one may be that the nanofluids with volume fraction of 2.5% and 3.0% have the high viscosities, which adversely suppresses the impinging strength of the nanofluids. From Figure 4, it is found that the viscosities of the nanofluids with volume fraction of 2.5% and 3.0% increase sharply compared with that of the nanofluids with volume fraction of 1.0%, 1.5% and 2.0%. While the volume fraction of the Cu nanoparticles increases from 2.0% to 2.5%, for example, the ratio of the viscosity of the nanofluid to that of water varies from 1.5 to 2.0. The higher viscosity suspensions suppress the heat exchange between the fluid and the heated surface. It is instructive to point out that proper selection of the nanoparticle volume fraction are important for applying nanoparticles to heat transfer enhancement of jet impingement cooling.

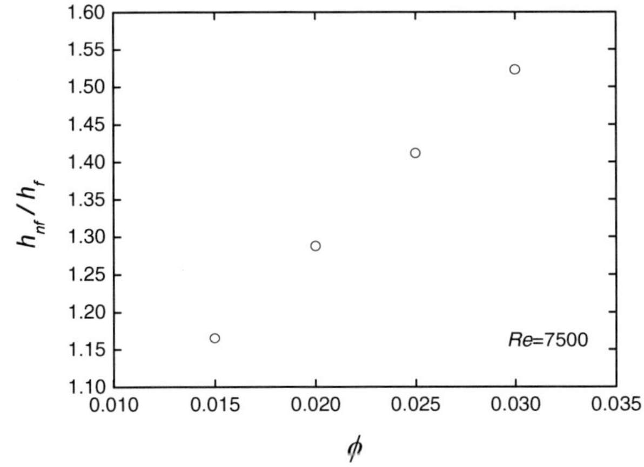


Figure 6. Variation of dimensionless heat transfer coefficients of the 25 nm Cu-water nanofluid with the volume fraction of nanoparticles (*H*/*D*=2).

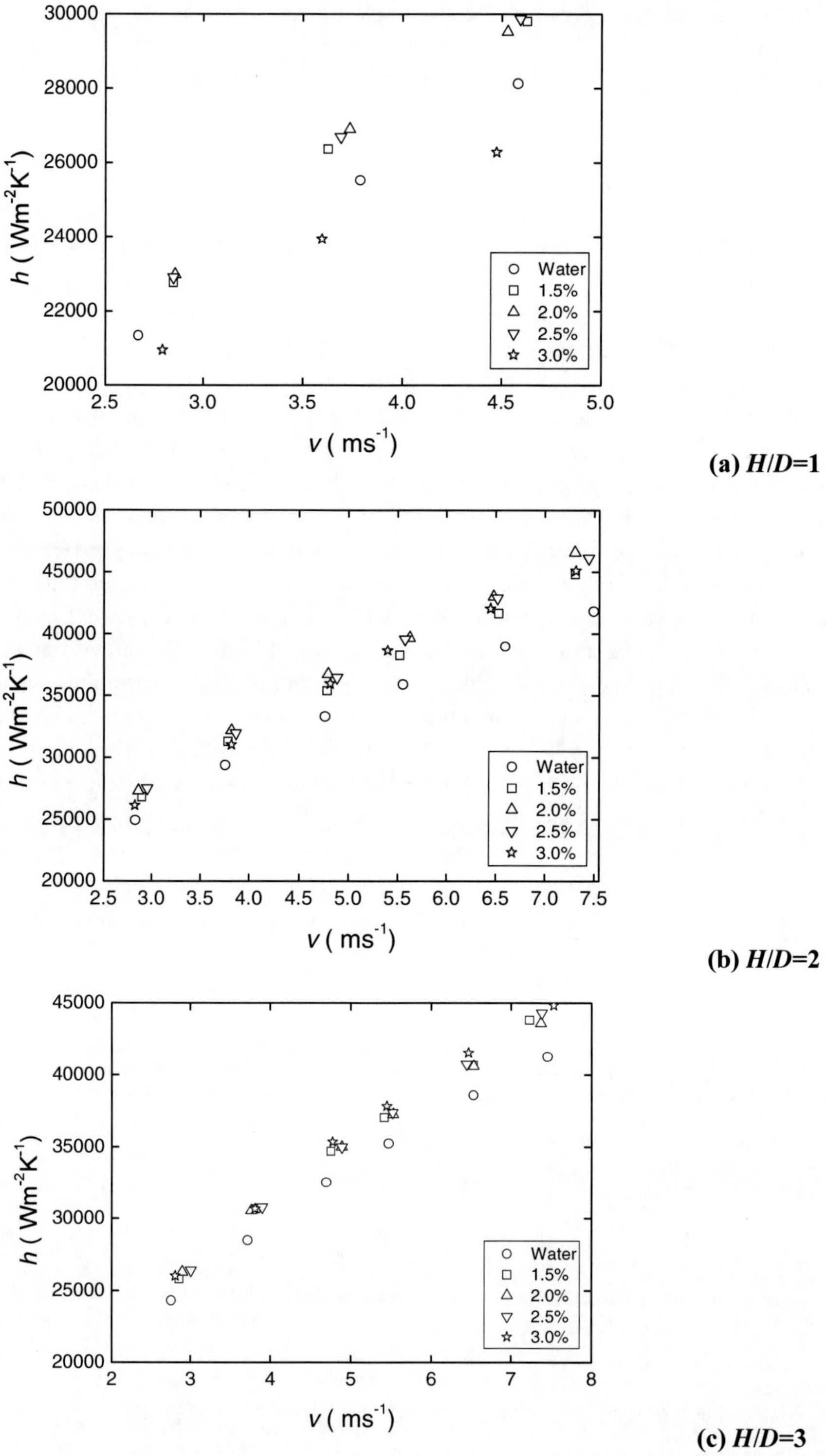


Figure 7. Variation of heat transfer coefficients of the 25 nm Cu-water nanofluid with the impinging velocity.

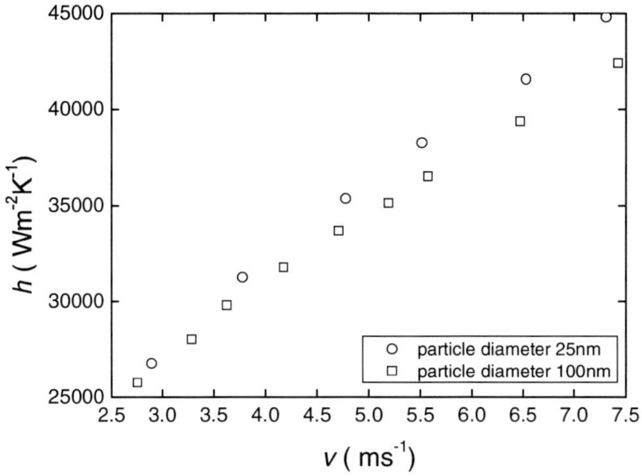


**(a) nanoparticles volume fraction of 1.5%**

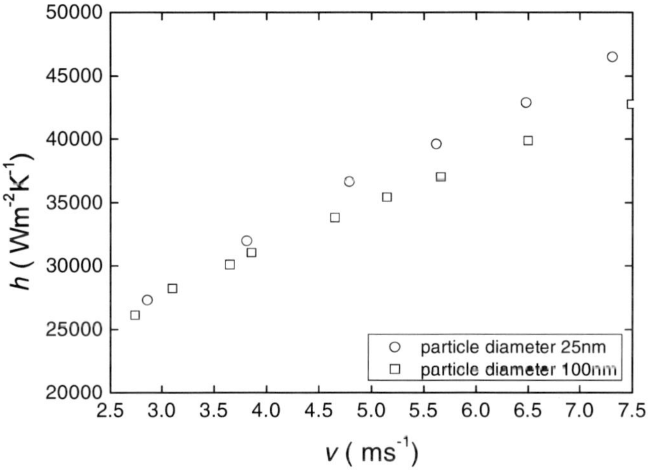


**(b) nanoparticles volume fraction of 2.0%**

Figure 8. Heat transfer coefficients of two types of Cu-water nanofluid (*H*/*D*=2).

### *2.4.2. Size Effect of Nanoparticles on the Heat Transfer Feature of Nanofluids*

It is expected that the nanoparticle size will also exert some effect on the impingement heat transfer performance. The experimental data shown in Figure 8 indicate the size effect of nanoparticles on the heat transfer coefficient of nanofluids in the submerged single jet impingement cooling system, which include the impingement heat transfer coefficients of the nanofluids containing 25 nm and 100 nm Cu nanoparticles with the nanoparticles volume fraction of 1.5% and 2.0%, respectively. These experimental results reveal that the heat transfer coefficient of the nanofluid with Cu nanoparticles of 25 nm in diameter is larger than that of Cu nanoparticles of 100 nm for the same impinging velocity. For the impinging velocity of $5.5\,\mathrm{m/s}$, for example, the heat transfer coefficient of the nanofluids increases from 36000 $\mathrm{W/m^2 \cdot K}$ to 38200 $\mathrm{W/m^2 \cdot K}$ when the diameter of nanoparticles varies from 100 nm to 25 nm. It is obvious that energy transport enhancement induced by the suspended nanoparticles increases with decrease in the size of nanoparticles. The reason may be explained from the following two aspects: One is that the thermal conductivity of the nanofluid with Cu nanoparticles of 25 nm diameter is larger than that of Cu nanoparticles of 100 nm. As shown in Figure 4, the ratio of the thermal conductivity of the nanofluid with the volume fraction of 2.0% to that of the pure base fluid varies from 1.16 to 1.21 with the diameter of nanoparticles varies from 100 nm to 25 nm. Another is that the impinging action of the nanoparticles accelerates energy exchange process between the nanofluid and the heated surface. With respect to the fact that the number of Cu nanoparticle of 25 nm diameter is 4 times larger than that for Cu nanoparticle of 100 nm diameter, heat exchange between the nanofluids and the heated surface is improved as the nanoparticles sizes decrease, which results in enhancement of the convective heat transfer coefficient of the nanofluid. It reminds us of the fact that selection of proper sizes of nanoparticles for preparation of nanofluids is important for achieving the desired impingement cooling.

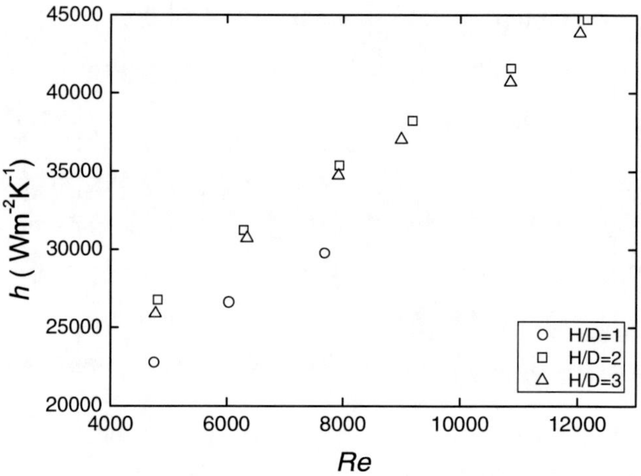


**(a) 25 nm Cu nanoparticles volume fraction of 1.5%**

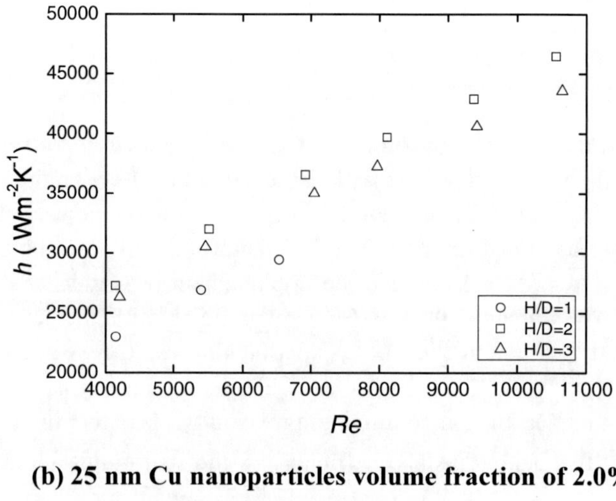


**(b) 25 nm Cu nanoparticles volume fraction of 2.0%**

Figure 9. Variation of Nu with *H*/*D* for Cu-water nanofluid.

### 2.4.3. *Effect of Dimension Ratio H/D on the Heat Transfer Feature of Nanofluids*

Figure 9 illustrates the variation tendency of the Nusselt numbers of two types of 25 nm Cu-water nanofluid containing the nanoparticle volume fraction of 1.5% and 2.0% with the different ratio of jet-to-target distance to the jet hole, $H/D$. For the same Reynolds number, it is found that the Nusselt number of the nanofluid at $H/D$=2 is higher than that of the other ratio values. The reason may be that the phenomena of the reflected fluid from the heated surface is easily occurred at the case of $H/D$=1. The reflected fluid suppresses impinging strength of the impinging suspensions. As for the case of $H/D$=3, the turbulence intensity of the impinging suspensions is weakened because of the larger jet-to-target distance. Therefore, the proper selection of the jet-to-target distance is important for the jet impingement cooling of nanofluids.

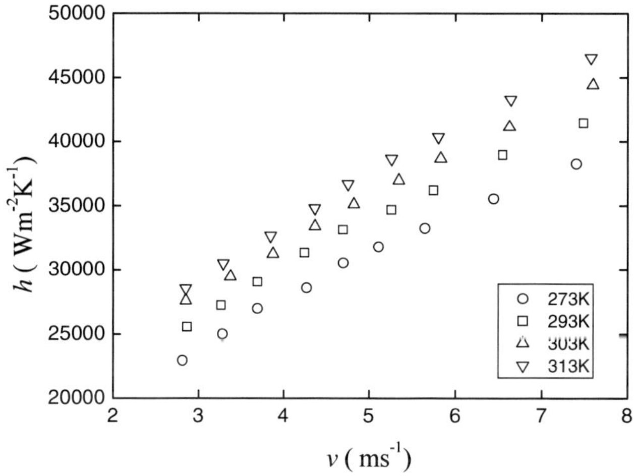


**(a) volume fraction 2.5% of 25 nm Cu nanoparticles**

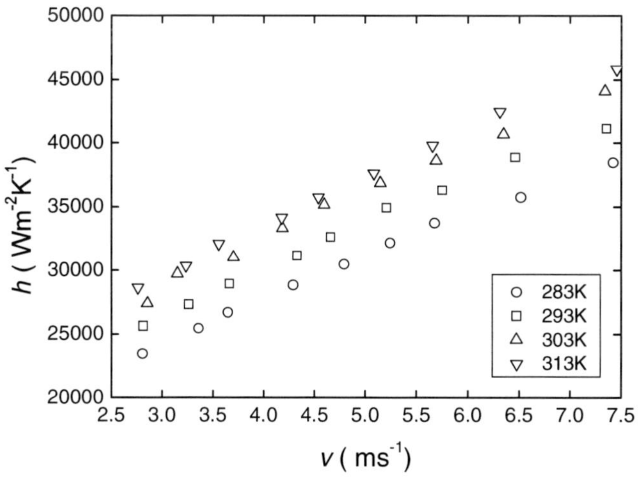


**(b) volume fraction 3.0% of 25 nm Cu nanoparticles**

Figure 10. Variation of the heat transfer coefficients of nanofluid with the temperature.

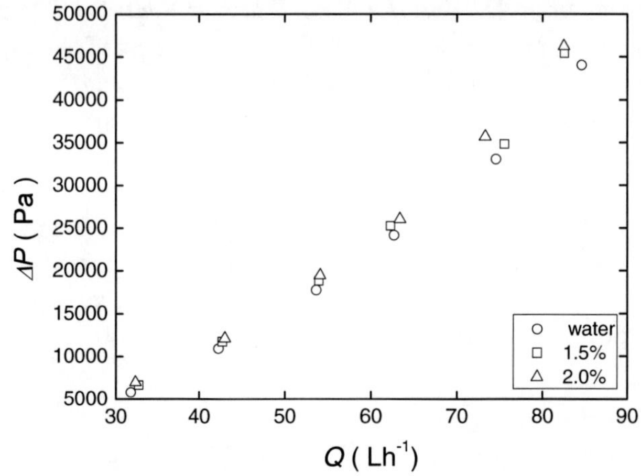


Figure11. Pressure drop of nannofluids in the submerged single impinging jet system (*H*/*D*=2).

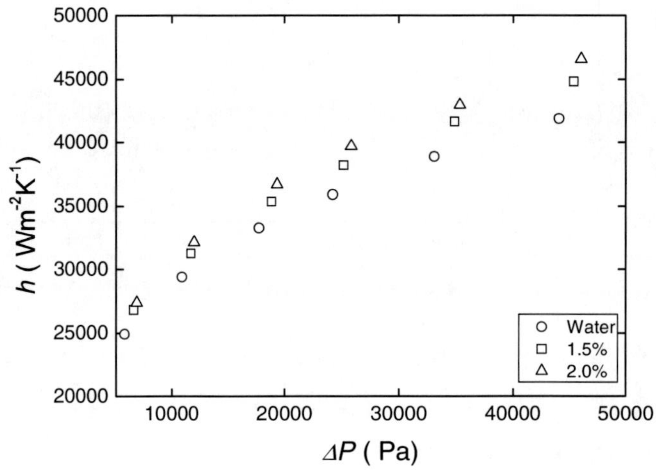


Figure12. Variation of the impingement heat transfer coefficient with the pressure drop in the submersed single impinging jet system.

### *2.4.4. Effect of the Fluid Temperature on the Heat Transfer Feature of Nanofluids*

Figure 10 shows the heat transfer coefficients of two types of 25 nm Cu-water nanofluids with the nanoparticles volume fraction of 2.5% and 3.0% for the different cases of inlet temperature of nanofluids, respectively. The experimental results indicate that the heat transfer coefficient of nanofluid increases with the nanofluid temperature. For the impinging velocity of $4.7\,\mathrm{m/s}$, for example, the heat transfer coefficient of the nanofluids increases from 30500 $\mathrm{W/m^2 \cdot K}$ to 36600 $\mathrm{W/m^2 \cdot K}$ if the inlet temperature of the nanofluids varies from 283 K to 313 K. It is obvious that energy transport enhancement of the nanofluid increases with the nanofluid temperature. The reason is that the thermal conductivity of a nanofluid increases with the nanofluid temperature and the viscosity of base liquid decreases with increase in the fluid temperature, which will improve the jet impingement process.

### *2.4.5. Pressure Drop Experiments of Nanofluids*

It is necessary to learn the flow features of nanofluids besides the heat transfer enhancement feature in order to apply the to the impinging jet cooling technique of nanofluids to any practical cases. In the present paper, the pressure drop of the submersed single impinging jet system using nanofluids and distilled water is experimentally investigated. Figure 11 illustrates the variation of pressure drop $\Delta P$ with the flow rate of nanofluid, in which Cu-water nanofluids of 1.5 % and 2.0% volume fraction of 25 nm Cu nanoparticles are respectively involved. The experimental results indicate that no significant augmentation in pressure drop for the nanofluid was found in all runs of the experiments. Compared with water, the pressure drop of Cu-water nanofluid of 1.5% and 2.0% volume fraction respectively increases 4.03% and 4.73% at the flow rate of 60 L/h. It reveals that the dilute nanofluid will almost cause no extra penalty in pump power.

Figure 12 gives variation of the impingement heat transfer coefficient with the pressure drop for pure water and Cu-water nanofluids of two types with the volume fraction 1.5% and 2.0% of 25 nm Cu nanoparticles. The experimental results indicate that the nanofluid has larger heat transfer coefficient than pure water under the same pressure drop. Obviously, the enhanced heat transfer of nanofluids for the submerged single jet impingement is achieved.

## 2.5. Heat Transfer Correlation of Nanofluids for the Submerged Single Jet Impingement

According to the previous investigation [14], the Nusselt number of the single-phase pure fluid for the submerged single impinging jet can be expressed as follows:

$$Nu = c\,\mathrm{Re}^{m}\,\mathrm{Pr}^{n}[1.0 + a_1(H/D) + a_2(H/D)^2 + a_3(H/D)^3] \qquad (7)$$

Considering the effect of nanoparticles on the enhanced heat transfer of nanofluids, Xuan and Roetzel presented a heat transfer correlation of nanofluids [39]:

$$Nu_{nf} = c_1\left(1.0 + c_2\phi^{m_1}\,\mathrm{Pe}_d^{m_2}\right)\mathrm{Re}_{nf}^{m_3}\,\mathrm{Pr}_{nf}^{0.4} \qquad (8)$$

This expression took the two-phase feature of nanofluids into account. As an extension, therefore, the heat transfer correlation of nanofluids for the submerged single jet impingement can be expressed as follows:

$$Nu_{nf} = c_1(1.0 + c_2\phi^{m_1} Pe_d^{m_2})\mathrm{Re}_{nf}^{m_3}\,\mathrm{Pr}_{nf}^{1/3}[1.0 + a_1(H/D) + a_2(H/D)^2 + a_3(H/D)^3] \qquad (9)$$

With an ensemble of the above experimental data of $\mathrm{Nu}_{nf}$ vs $\varphi$, $\mathrm{Pe}_d$, $\mathrm{Re}_{nf}$, and $\mathrm{Pr}_{nf}$, the coefficient $c_1$ and $c_2$ as well as the exponent $m_1$, $m_2$, and $m_3$ are found from data reduction. The heat transfer correlation of nanofluids for submerged single jet impingement is expressed as follows:

$$Nu_{nf} = 0.2464(1.0 + 2.2061\phi^{0.8464} Pe_d^{0.2715}) \mathrm{Re}_{nf}^{0.5375} \mathrm{Pr}_{nf}^{1/3} \\ [1.0 + 0.3923(H/D) + 0.0086(H/D)^2 - 0.0259(H/D)^3] \quad (10)$$

Figure 13 illustrates both the measured data and the curves of the theoretical predictions of impinging jet heat transfer coefficient of nanofluids from correlation (10). Obviously, there exists a satisfactory coincidence between the results calculated from this reduced correlation and the experimental ones. The discrepancy between both of them falls below 2%. Therefore, correlation (10) can be used to predict the heat transfer coefficient of nanofluids for the submersed single jet impingement.

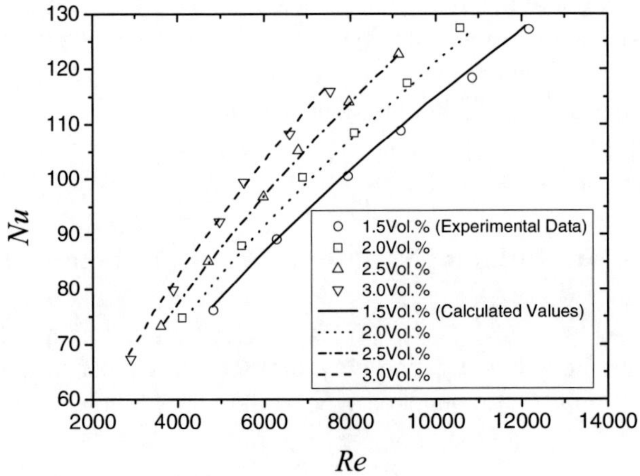


**(a) *H/D*=2**

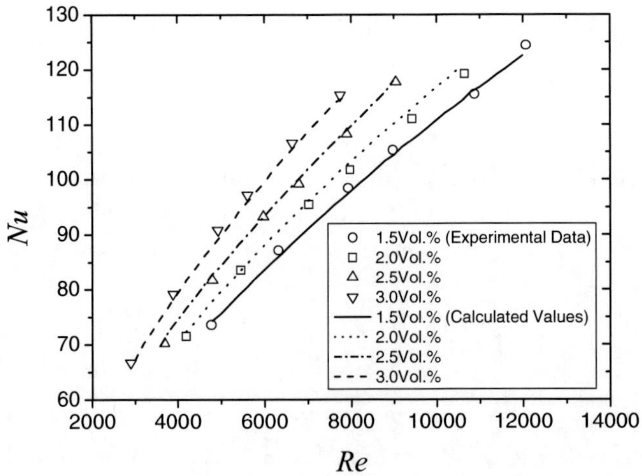


**(b) *H/D*=3**

Figure13. Comparison between the measured data and the calculated values from correlation (10).

# 3. Free Surface Jet Array Impingement Cooling

## 3.1. Experimental Methods

An experimental rig was established to study the heat transfer features of free surface jet array impingement. Figure 14 shows the schematic diagram of a flow system, which is mainly composed of a reservoir, a test module, a cooler and a magnetically-driven gear pump that forces a continuous circulation of liquid. The Cu-ethylene glycol/water nanofluids containing the copper nanoparticles of 25 nm in diameter was used as the working fluid. The cooler is used to keep the constant temperature at the inlet of the test section. An electromagnetic flowmeter with the uncertainty of 1% was used to measure the flow rate of the nanofluid.

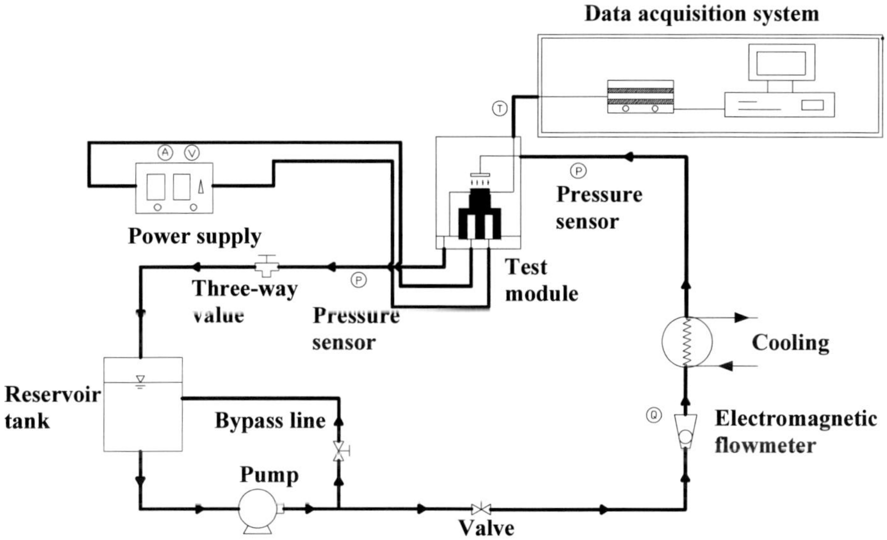


Figure 14. Schematic of flow system for free surface jet array impingement cooling.

Figure 15 illustrates the construction and assembly of the test module, which consists of a copper heating block, a jet plate, support plate, and 6 cartridge heaters. The top of a copper block, which is a flat circular surface with a diameter of 20 mm, constitutes the heated surface onto which the liquid jet from 1.5 mm diameter nozzle is impinged. Heat is supplied to the test surface from the cartridge heaters, which are inserted into bores machined in the underside of the copper block. The cartridge heaters are powered by a DC power, and their total power input is measured by a voltage meter and an amperemeter.

The jet plate is also machined from oxygen-free copper. Five nozzles are equidistantly drilled into the jet plate (as shown in Figure 16). The distance between the nozzles and the target is fixed to be 15 mm. The upper support plate and lower support plate are fabricated from Teflon which provides thermal insulation. A protrudent step which is 10 mm higher than the outlet of the working fluid was made to prevent the test surface being submerged by the fluid. Before the experiment, the test surface was cleaned by acetone. Four outlets were used to remove the working fluid to ensure the free surface jet impingement. During the experiment it was found that the fluid was effectively removed and did not cover the whole test surface. Test surface temperature and heat transfer coefficient distributions were determined with aid of four thermocouples (K-type) inserted a small distance below the test surface. Also inserted into the inlet and the outlet of the impinging room are two type-K thermocouples, which respectively measure the inlet and the outlet temperatures of the working fluid. The pressure transducers were connected to the inlet and the outlet of the impinging room to respectively measure the pressures at the inlet and outlet of the test module. To minimize heat loss from the experimental system to the ambient, the heating block rests on a solid insulating ceramic plate and is laterally encased with a thick layer of ceramic fiber.

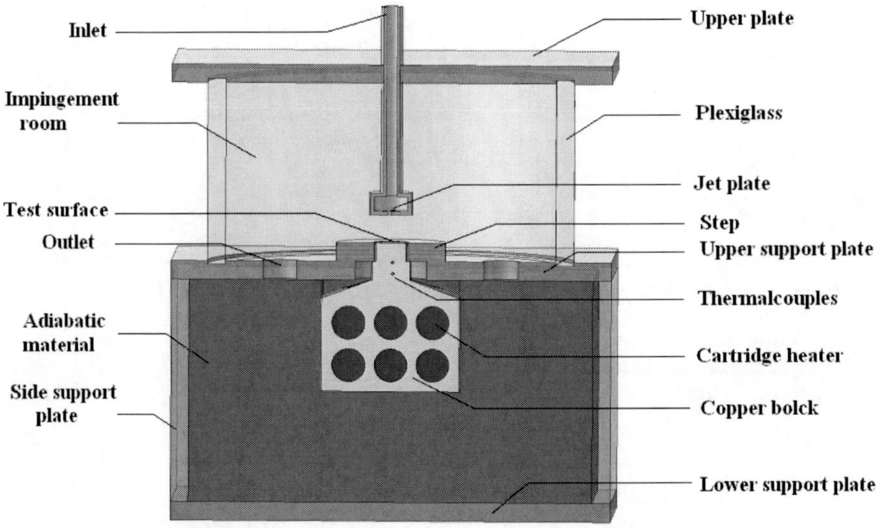


Figure 15. Cross-section of test module for free surface jet array impingement cooling system.

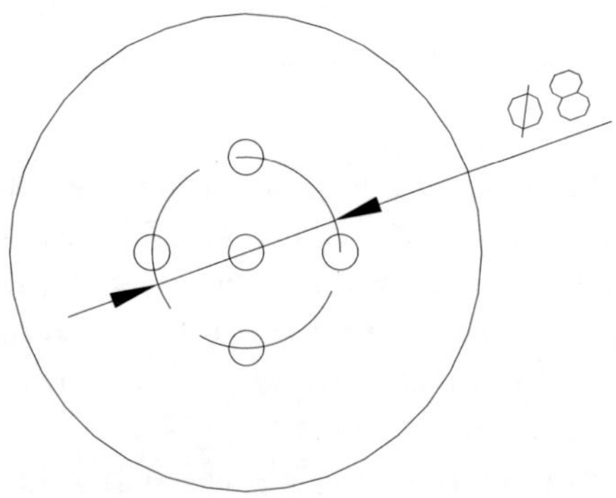


Figure 16. Location of five nozzles at the jet plate.

As mentioned before, Cu-ethylene glycol/water nanofluids containing Cu nanoparticles of 25 nm in diameter was used as the working fluid for the free surface jet array impingement cooling system. The sample nanofluids are prepared by direct mixing of the Cu particles and 50:50 (by weight) ethylene glycol/water mixture. The nanofluids with different particle volume fractions are used in the experiment, in which Cu-ethylene glycol/water nanofluids with a series of nanoparticle volume fractions of 0.187%, 0.374%, 0.56%, 0.746%, 0.933%, 1.12% and 1.343% are involved.

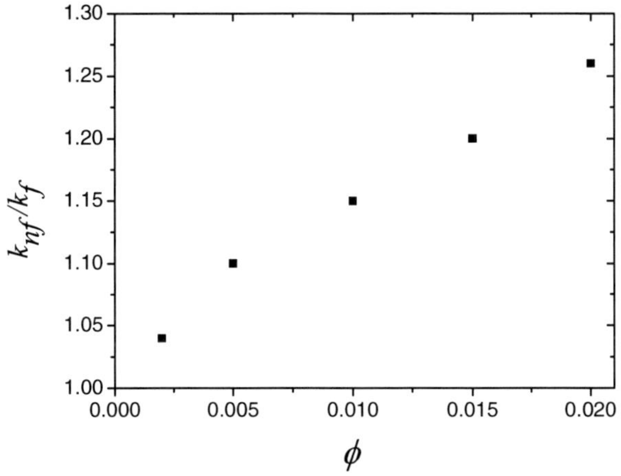


Figure 17. Thermal conductivity of Cu-ethylene glycol/water nanofluids.

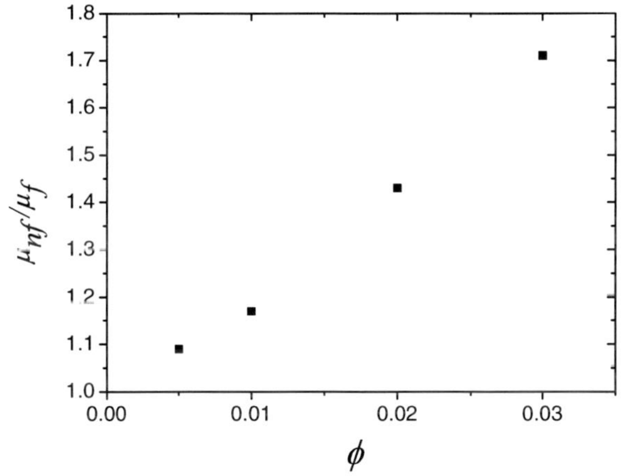


Figure 18. Viscosity of Cu-ethylene glycol/water nanofluids.

**Table 3. Uncertainty of free surface jet array impingement cooling system**

| Parameters | Calculation expressions | Results |
|---|---|---|
| $E_h$ | $\sqrt{(\frac{\delta U}{U})^2+(\frac{\delta I}{I})^2+4(\frac{\delta D}{D})^2+(\frac{\delta T}{T})^2}$ | 1.45% |
| $E_{Nu}$ | $\sqrt{E_h^2+(\frac{\delta D}{D})^2+(\frac{\delta K}{K})^2}$ | 4.19% |
| $E_{Re}$ | $\sqrt{(\frac{\delta v}{v})^2+(\frac{\delta \upsilon}{\upsilon})^2+(\frac{\delta D}{D})^2}$ | 3.21% |
| $E_{\Delta P}$ | $u_{\Delta P}/\Delta P$ | 2.4% |

Figures 17 and 18 respectively shows the experimental data of thermal conductivity and the viscosity of Cu-ethylene glycol/water nanofluids obtained from the previous paper [40].

Being similar to the above-mentioned analysis procedure of submerged single jet impingement, experimental data processing and experimental uncertainties for the free surface jet array impingement cooling system can be obtained. Table 3 lists the experimental uncertainty.

## 3.2. Results and Discussions

### *3.2.1. Heat Transfer Features of the Cu-Ethylene Glycol/Water Nanofluids for Free Surface Jet Array Impingement*

Figure 19 shows the temperature difference $\Delta T = T_s - T_{in}$ between the heated surface and the corresponding fluid inlet as a function of the flow rate of the working fluid by respectively using Cu-ethylene glycol/water nanofluid of 0.56% particle volume fraction and just ethylene glycol/water mixture. During this experiment, the temperature of the fluid inlet was set to be 294.2 K and the heat flux was fixed to be 50 W/cm$^2$. It can be observed that the temperature $\Delta T$ considerably decreases with increase in the impinging velocity of the fluid.

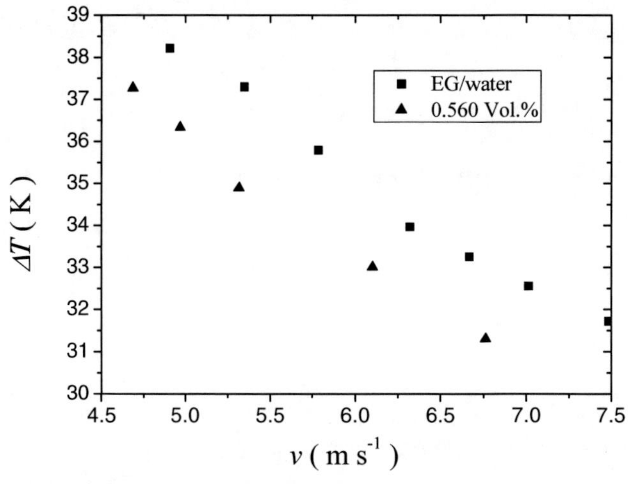


Figure 19. Variation of the temperature difference between the heated surface and the inlet of the working fluid with the impinging velocity.

Such behavior is due to increase in the forced convection heat transfer coefficient with increase in the fluid velocity. On the other hand, it has been found that for a given impinging velocity, the temperature difference $\Delta T$ when using ethylene glycol/water mixture as the working fluid was clearly higher than that for using the nanofluid of Cu-ethylene glycol/water. For the impinging velocity of 5.3 m/s, for example, the temperature difference $\Delta T$ is 37.3 K by using ethylene glycol/water mixture and 34.8 K by using the nanofluid with the nanoparticle volume fraction of 0.56%. It is shown that the Cu-ethylene glycol/water nanofluid has a higher heat transfer coefficient than that of ethylene glycol/water mixture. As illustrated in Figure 20, for example, the impingement heat transfer coefficient is 13332 $W/m^2 \cdot K$ corresponding to ethylene glycol/water mixture and 14247.98 $W/m^2 \cdot K$ corresponding to the Cu-ethylene glycol/water nanofluid for the impinging velocity of 5.3 $m/s$. Compared with the ethylene glycol/water mixture, the impingement heat transfer coefficient of Cu-ethylene glycol/water nanofluid with the nanoparticle volume fraction of 0.56% increased about 6.87%. Therefore, it can be concluded that the suspended nanoparticles increase heat transfer performance of the base fluid for free surface jet array impingement.

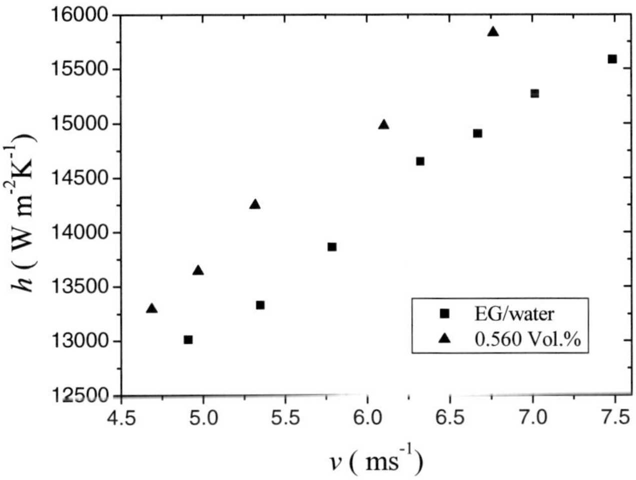


Figure 20. Variation of the heat transfer coefficient with impinging velocity.

### *3.2.2. Effect of Particle Volume Fraction on the Heat Transfer Feature of Nanofluids*

Figure 21 shows the heat transfer coefficient of the sample Cu-ethylene glycol/water nanofluids with the different particle volume fractions for free surface jet array impingement. The experimental results indicate that the particle volume fraction is one of the vital factors affecting the heat transfer feature of nanofluids. Heat transfer of the nanofluids enhances with increase in the volume fraction of the suspended nanoparticles within the range between 0.187%-1.12%. For a given Reynolds number, for example, the ratio of the heat transfer coefficient of the nanofluids to that of the base fluid varies from 1.05 to 1.18 when the volume fraction of the nanoparticles increases from 0.187% to 1.12%. The suspended nanoparticles enhanced energy exchange between the heated surface and the nanofluid. But it is noted that the heat transfer coefficient of the nanofluid with nanoparticles volume fraction of 1.343% no longer increases with increasing the nanoparticle volume fraction. In such case the heat transfer coefficient is reversely smaller than that of the nanofluid with nanoparticles volume fraction of 1.12%. Such phenomenon implies a fact that there exists an optimal particle volume fraction that would maximize the surface heat transfer for the free surface jet array impingement of nanofluids. More efforts are needed to get deeper insight into the mechanism.

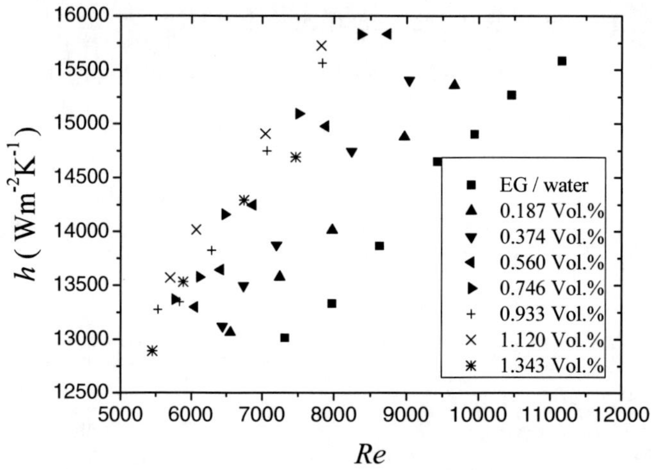


Figure 21. Variation of the heat transfer coefficient with the nanoparticle volume fraction.

### *3.2.3. Pressure Drop Experiments for Free Surface JetArray Impingement*

The pressure drops of the sample Cu-ethylene glycol/water nanofluids in the free surface jet array impingement cooling system are experimentally measured. Figure 22 illustrates the pressure drop between the inlet and outlet pressure of jet impingement as a function of the fluid flow rate. Compared with the base liquid, the experimental results show that no significant augmentation in pressure drop for the nanofluid is found in all runs of the experiment. The maximal pressure drop of the nanofluid with nanoparticles volume fraction of 1.343% are about 6% higher than that of ethylene glycol/water mixture.

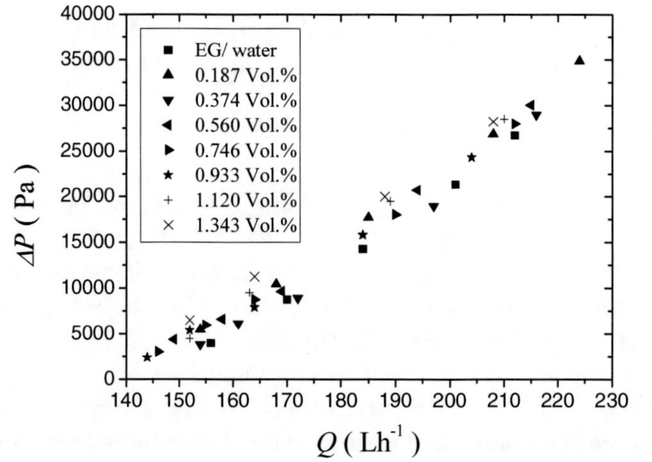


Figure 22. Pressure drops of the free surface jet array impingement.

### *3.2.4. Effects of the Dispersant on Heat Transfer Featuresof Nanofluids*

It is well known that some auxiliary activators or dispersants are added into the nanofluids to prevent the aggregation among the nanoparticles while preparing nanofluids. The addition of dispersant can effectively improve the stability and evenness of the nanofluids, which may simultaneously alter the transport properties and heat transfer characteristics of nanofluids. Here it is emphasized that no dispersant was added into either Cu-water nanofluids or Cu-ethylene glycol/water nanofluids in the above-mentioned experiments. In order to investigate the effects of the dispersant on heat transfer features of nanofluids, a dispersant is added into the Cu-ethylene glycol/water nanofluid during preparation of the sample and the heat transfer features of the sample nanofluid for free surface jet array impingement are measured.

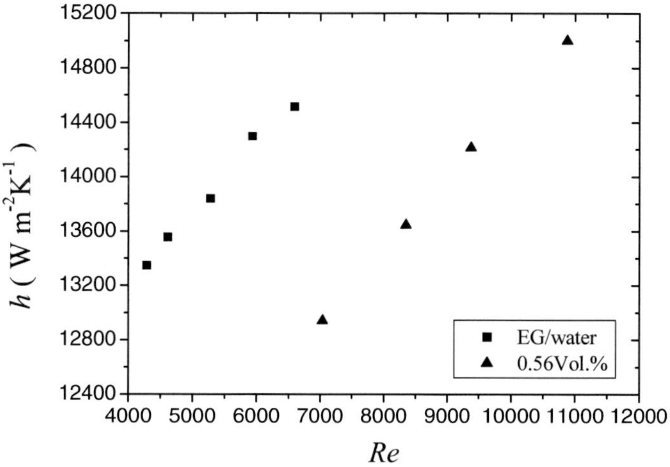


Figure 23. Effects of the dispersant on heat transfer feature of nanofluid.

In the experiment, sodium dodecylbenzenesulfonate (SDBS) was selected as the dispersant to be added into Cu-ethylene glycol/water nanofluid with particle volume fraction of 0.56%. The amount of SDBS is 7%, which refers to the weight percentage of the Cu nanoparticles. Figure 23 illustrates the experimental results of the heat transfer coefficient of the sample nanofluid with the dispersant as a function of the Reynolds number. An unexpected result occurred! Obviously, the results reveal that addition of dispersant into the nanofluid makes heat transfer feature of free surface jet array impingement more complicated and the nanofluid with the dispersant does not guaranty an enhancement of heat transfer. The experimental data shown in Figure 23 have clearly indicated the possible reverse effect of nanofluids with dispersant. Compared with the case of using nanofluids without any addition of dispersant, the nanofluid with dispersant led to adverse results, i.e., a decrease of impingement heat transfer coefficient which is even smaller than that for the base liquid. For the Reynolds number of 7000, for example, the heat transfer coefficient of the nanofluid with the dispersant is about 10% lower than that of the base liquid of ethylene glycol/water mixture. The dispersant SDBS plays a hindering role on the heat transfer of nanofluid. This phenomenon may be explained from the following two aspects: One is that the addition of dispersant increases the viscosity of the nanofluid and results in a negative effect on the impingement of the sample nanofluid, which suppresses impinging strength of the suspensions; another is that the dispersants SDBS may be adhered on the heated surface, which contaminates the heated surface and changes the surface characteristics. In fact, addition of dispersant or surfactants during preparation of nanofluids is one of the main factors of leading to discrepancy among the experimental data from different research groups. One must pay much more attention on utilization of adding dispersant into nanofluids. Of course, understanding on such effect of the dispersant on heat transfer features of nanofluid is still faint and more investigation efforts are needed.

## CONCLUSION

In the present chapter Cu-nanofluids were used as the working fluid for the jet impingement to develop effective cooling technique for high-power electronics. The heat transfer features of the nanofluids were experimentally investigated. The effects of the nanoparticle concentration, Reynolds number, nozzle-to-plate distance, fluid temperature, and nanoparticle diameter as well as dispersant on the heat transfer and flow performances of nanofluids for the jet impingement have been discussed. The main conclusions are as follows:

1) In the submerged single jet impingement cooling system, the suspended nanoparticles remarkably increase heat transfer performance of the base fluid. Cu-water nanofluid exhibits larger impingement heat transfer coefficient than pure water under the same Reynolds number. The heat transfer enhancement of Cu-water nanofluid increases with the volume fraction of Cu nanoparticles and the nanofluid temperature. The heat transfer coefficient of the nanofluid with Cu nanoparticles of 25 nm diameter is larger than that of Cu nanoparticles of 100 nm.
2) By considering the effects of the suspended nanoparticles as well as the condition of impinging jet, a new heat transfer correlation of nanofluids for the submerged single jet impingement has been proposed.
3) In the free surface jet array impingement cooling system, heat transfer of Cu-ethylene glycol/water nanofluid enhances with increase in the volume fraction of the suspended nanoparticles within the range of 0.187%-1.12%. But such feature does not monotonously occur. The impingement heat transfer coefficient of the nanofluid with nanoparticles volume fraction of 1.343% is smaller than that of the nanofluid with nanoparticles volume fraction of 1.12%.
4) The pressure drop of impinging jet system using nanofluids with different nanoparticle volume fraction incurs little or almost no augmentation compared with the base fluid.
5) Dispersant and/or surfactant for preparing nanofluids may greatly affect the impingement heat transfer of nanofluids. In some cases, the addition of dispersant suppresses the impingement heat transfer process between the heated surface and nanofluids. For application of nanofluids to heat transfer enhancement, one should pay attention to the negative effect of surfactants or dispersants used in preparation of nanofluids.

## REFERENCES

[1] Martin, H., 1977, "Heat and Mass Transfer between Impinging Gas Jets and Solid Surface," *Advances in Heat Transfer*, 13, 1-60.

[2] Womac, D.J., Ramadhuani, S., Incropera, F.P., 1993, "Correlating Equations for Impingement Cooling of Small Heat Sources with Single Circular Liquid Jets," *Journal of Heat Transfer*, 115, 106-115.

[3] Fangjun, S., Plesniak, M.W., and Sojka, P.E., 2007, "Experimental Investigation of Flow Physics in Indeterminate Origin Nozzle Jet," *Journal of Experimental Mechanics*, 22, 276-284.

[4] Pavlova, A.A., Otani, K., and Amitay, M., 2008, "Active Control of Sprays Using a Single Synthetic Jet Actuator," *International Journal of Heat and Fluid Flow*, 29, 131-148.

[5] Katti, V., and Prabhu, S.V., 2008, "Influence of Spanwise Pitch on Local Heat Transfer Distribution for In-Line Array of Circular Jets with Spent Air Flow in Two Opposite Directions," *Experimental Thermal and Fluid Science*, 33, 84-95.

[6] Geers, L.F.G., Tummers, M.J., Bueninck, T.J., and Hanjalic, K., 2008, "Heat Transfer Correlation for Hexagonal and In-Line Array of Impinging Jets," *International Journal of Heat and Mass Transfer*, 51, 5389-5399.

[7] Pan, Y., and Webb, B.W., 1995, "Heat Transfer Characteristics of Array of Free-Surface Liquid Jets," *Journal of Heat Transfer*, 117, 878-883.

[8] Zhou, J.W., and Herwig, H., 2007, "Heat Transfer Characteristics of Precessing Jets Impinging on a Flat Plate: Further Investigations," *International Journal of Heat and Mass Transfer*, 50, 4488-4492.

[9] Shu, F., 2005, Flow Physics in Indeterminate Origin Nozzle Jets and Passive Control of Sprays, Doctoral dissertation, Purdue University.

[10] Jeng, T.M., and Tzeng, S.C., 2007, "Experimental Study of Forced Convection in the Metallic Porous Block Subject to a Confined Slot Jet," *International Journal of Thermal Sciences*, 46, 242-1250.

[11] O'Donovan, T.S., and Murray, D.B., 2007, "Jet Impingement Heat Transfer- Part I: Mean and Root-Meat-Square Heat Transfer and Velocity Distributions," *International Journal of Heat and Mass Transfer*, 50, 3291-3301.

[12] Bradbury, L.J.S., 1965, "The Structure of a Self-Preserving Turbulent Plane Jet," *Journal of Fluid Mechanics*, 23, 31-64.

[13] Viskanta, R., 1980, "Heat Transfer to Impinging isothermal Gas and Flame Jets," *Experimental Thermal and Fluid Science*, 6, 111-134.

[14] Webb, B.W., and Ma, C.F., 1995, "Single Phase Liquid Jet Impingenment Heat Transfer," *Advances in Heat Transfer*, Hartnett, J. P., ed., New York: Academic Press, 26, 105-217.

[15] Hoogendoorn, C.J., 1977, "The Effect of Turbulence on Heat Transfer at a Stagnation Point," *International Journal of Heat and Mass Transfer*, 20, 1333-1338.

[16] Lytle, D., and Webb, B.W., 1994, "Air Jet Impingement Heat Transfer at Low Nozzle-Plate Spacing," *International Journal of Heat and Mass Transfer*, 37, 1687-1697.

[17] Ma, C.F., and Zheng, Q., 1997, "Local Convective Heat Transfer from a Vertical Flat Surface to Oblique Submerged Impinging Jets of Large Prandtl Number Liquid," *Experimental Thermal and Fluid Science*, 10, 1-10.

[18] Xie, H.Q. and Chen, L.F., 2009, "Mechanism of Enhanced Convective Heat Transfer Coefficient of Nanofluids," *Acta Physica Sinica*, 58, 2513-2517.

[19] Stevens, J., Pan, Y., and Webb, B.W., 1992, "Effect of Nozzle Configuration on Transport in the Stagation Zone of Axisymmetric, Impinging Free-Surface Liquid Jets-I, Turbulent Flow Structure," *Journal of Heat Transfer*, 114, 874-879.

[20] Jiji, L.M., and Dagan, Z., 1988, "Experimental Investigation of Single-Phase Multijet Impingement Cooling of an Array of Microelectronic Heat Source," Proceedings of the International Symposium on Cooling Technology for Electronic Equipment, W. Aung, ed., Hemisphere Publishing Corporation, Washington, D.C., 1988, 333-351.

[21] Fabbri, M., and Dhir, V.K., 2005, "Optimized Heat Transfer for High Power Electronic Cooling Using Array of Microjets," *Journal of Heat Transfer*, 127, 760-769.

[22] Whelan, B.P., and Robinson, A.J., 2009, "Nozzle Geometry Effects in Liquid Jet Array Impingement," *Applied Thermal Engineering*, 29, 2211-2221.

[23] Sung, M.K., and Mudawar, I., 2008, "Single-Phase and Two-Phase Heat Transfer Characteristics of Low Temperature Hybrid Micro-Channel/Micro-Jet Impingement Cooling Module," *International Journal of Heat and Mass Transfer*, 51, 3882-3895.

[24] Sung, M.K. and Mudawar, I., 2008, "Single-Phase and Two-Phase Hybrid Cooling Schemes for High-Heat-Flux Thermal Management of Defense Electronics," *International Journal of Heat and Mass Transfer*, 51, 4342-4352.

[25] Goodro, M., Park, J., Ligrani, P., Fox, M., and Moon, H. K., 2008, "Effects of Hole Spacing on Spatially-Resolved Jet Array Impingement Heat Transfer," *International Journal of Heat and Mass Transfer*, 51,6243–6253.

[26] Yan, Z.B., Toh, K.C., Duan, F., Wong, T.N., Choo, K.F., Chan, P.K., and Chua, Y.S., 2010, "Experimental Study of Impingement Spray Cooling for High Power Devices," *Applied Thermal Engineering*, 30, 1225–1230.

[27] Choi, S.U.S., 1995, "Enhancing Thermal Conductivity of Fluids with Nanoparticles," Developments and Applications of Non-Newtonian Flows, Siginer D A, Wang H P, eds., New York: ASME, FED-Vol. 231/MD-Vol. 66, 99-103

[28] Xuan, Y.M., and Li, Q., 2000, "Heat Transfer Enhancement of Nanofluids," *International Journal of Heat and Fluid Flow*, 21(1), 58-64

[29] Li, Q., and Xuan, Y.M., 2000, "Experimental Investigation on Transport Properties of Nanofluids," *Heat Transfer Science and Technology 2000*, Wang Buxuan, eds., Higher Education Press, Beijing, 757-762

[30] Xuan, Y.M, Li, Q., Hu, W.F., 2003, "Aggregation Structure and Thermal Conductivity of Nanofluids," *AIChE Journal*, 49(4), 1038-1043

[31] Xuan, Y.M., and Li, Q., 2003, "Investigation on Convective Heat Transfer and Flow Features of Nanofluids," *Journal of Heat Transfer*, 125(1), 151-155

[32] Li, Q., and Xuan, Y.M., 2002, "Convective Heat Transfer and Flow Characteristics of Cu-Water Nanofluid," *Science in China* (Series E), 45(4), 408-416.

[33] Choi, S.U.S., Zhang, Z.G., X Yu, Z.G., Lockwood, F.E., and Grulke, E.A., 2001, "Anomalously Thermal Conductivity Enhancement in Nanotube Suspensions," *Applied Physics Letters*, 79, 2252-2254

[34] Eastman, J.A., Choi, S.U.S., Li, S., Yu, W., and Thompson, L.J., 2001, "Anomalously Increased Effective Thermal Conductivities of Ethylene Glycol-Based Nanofluids Containing Copper Nanoparticles," *Applied Physics Letters*, 78(6), 718-720

[35] Xie, H.Q., Lee, H., Youn, W., and Choi, M., "Nanofluids Containing Multiwalled Carbon Nanotubes and Their Enhanced Thermal Conductivities," *Journal of Applied Physics*, 94, 4967-4971

[36] Wen, D., and Ding, Y., 2004, "Experimental Investigation into Convective Heat Transfer of Nanofluids at the Entrance Region under Laminar Flow Conditions," *International Journal of Heat and Mass Transfer*, 47, 5181-5188.

[37] Manca, O., Mesolella, P., Nardini, S., and Ricci, D., 2011, "Numerical Study of a Confined Slot Impinging Jet with Nanofluids," *Nanoscale Research Letters*, 6, 188.

[38] Nguyen, C.T., Galanis, N., Polidori, G., Fohanno, S., Popa, C.V., and Bechec, A.L., 2009, "An Experimental Study of a Confined and Submerged Impinging Jet Heat Transfer Using $Al_2O_3$-Water Nanofluid," *International Journal of Thermal Sciences*, 48, 401–411.

[39] Xuan, Y.M., and Roetzel, W., 2000, "Conceptions for Heat Transfer Correlation of Nanofluids," *International Journal of Heat and Mass Transfer*, 43, 3701-3707

[40] Chen, B., 2005, Investigation on Heat Transfer Characteristics of the Nanofluids under the Different Temperatures, Master's thesis, Nanjing University of Science and Technology (in Chinese).

In: Nanofluids
Editor: Yuwen Zhang
ISBN: 978-1-62618-165-6


*Chapter 8*

# NANOFLUID BUBBLE FORMATION AND BOILING HEAT TRANSFER

***Saeid Vafaei****
School of Engineering, University of Warwick
Coventry, UK

## ABSTRACT

The purpose of this chapter is to demonstrate the effects of nanoparticles on boiling heat transfer coefficient and heat flux. Effects of nanoparticles on force balance at the triple line, behavior of triple line, wettability, roughness, active nucleation site density, initiation of nucleation, dynamics of bubble growth, bubble departure, advancing and receding contact angles, wetting and rewetting, boiling heat transfer coefficient and critical heat flux will be discussed in this chapter.

Nanoparticles have two main roles on boiling heat transfer phenomenon, a) suspended nanoparticles modify the physical properties of nanofluids; b) deposited nanoparticles modify the characteristics of heated substrate. Behavior of the triple line might be changed by deposited nanoparticles as well as the suspended nanoparticles inside nanofluids. This chapter attempts to explore and discuss the effects of both mechanisms on boiling heat transfer phenomenon in details.

Droplet and bubble formation methods are developed to investigate the effect of nanoparticles on force balance and behavior of triple. In case of droplet method, gravity is in favor of the expansion of triple line and reduction of contact angle which will reduce the accuracy of evaluation of effects of nanoparticles on behavior of triple line. Droplet contact angle changes with volume, so droplet method can not be a unique method to study the effects of nanoparticles on behavior of triple line including surface wettability, thus asymptotic contact angle will be discussed to investigate the behavior of triple line, regardless of gravity.

In case of bubble growth method, gravity affects as buoyancy force and lifts the bubble upward, so triple line can move more freely and effect of nanoparticles on triple line could be noted easier. Formation of bubbles inside nanofluids has a great capacity to express effects of nanoparticles on behavior of triple line, bubble growth, waiting time,

* Email: saeid.vafaei@ymail.com.

bubble formation time, bubble frequency, departure process, receding and advancing dynamic contact angles, wetting and rewetting. Indeed, effects of nanoparticles depend on concentration, and characteristics of nanoparticles such as shape, size, coating and material.

## INTRODUCTION

Nowadays nanofluids are applied in many industrial, automotive, electronic, biomedical, and other applications [1]. Introducing nanoparticles into the base liquid would change the physical properties of the nanofluids such as thermal conductivity [2], viscosity [3], liquid-gas surface tension [4], solid surface tensions [5], and possibly gas-liquid-solid interactions at the triple line. Variation of physical properties of nanofluids depends on concentration and characteristics of nanoparticles as well as the base liquid. Nanoparticles have a great potential to engineer and optimize the physical properties of the working fluids for a specific need and at the same time they create many complexities. One of the biggest difficulties in developing and designing nanofluids is to understand the interactions of nanoparticles-nanoparticles and nanoparticles-molecules. Understanding the mechanisms of interactions of nanoparticles with surrounded molecules and nanoparticles would help to predict the variation of the physical properties and wettability of the nanofluids better.

One of the most interesting applications of nanofluids is to improve the behavior of triple line and surface wettability. The wettability of nanofluids depends on distribution of nanoparticles in triple region, liquid-gas, solid surface tensions and interactions of nanoparticles with gas-liquid-solid at the triple line. Nanofluids have also great opportunities for enhancing the pining behavior of the triple line [6-8]. Variation of the wettability and pinning behavior of the triple line [9] has a significant role on the boiling heat transfer of nanofluids. As results of evaporation of nanofluids inside microlayer region beneath the vapor bubble, concentration of nanoparticles inside the microlayer increases, so nanoparticles are deposited on the heated substrate and form a tiny porous layer of nanoparticles. The porous layer of deposited nanoparticles changes the characteristics of the heated substrate such as surface roughness [10-11], equilibrium, receding and advancing contact angles [12]. The surface roughness has a key impact on equilibrium, receding and advancing contact angles [13]. The combinations of effects of nanoparticles on surface wettability, receding, advancing contact angles and surface roughness affect the performance of boiling heat transfer coefficient and critical heat flux. Understanding of nexus between surface wettability, roughness and nucleation site density has a significant impact on the modification of nanofluid boiling heat transfer. Several investigations of the boiling heat transfer of nanofluids have been conducted, which considered the effects of nanoparticles on wettability, roughness, nucleation site density, receding, and advancing contact angles. However, the responsible mechanisms of enhancement of the nanofluid boiling heat transfer coefficient and critical heat flux are still unknown. Nanofluids have a significant effect on convective heat transfer in micro [14] and macro [15] channels as well. In this chapter, the effects of nanoparticles on the physical properties of working fluids, force balance at triple line, wettability, roughness, initiation of nucleation, bubble growth, waiting time, bubble formation time, bubble frequency, receding and advancing contact angles, departure process, critical heat flux, and boiling heat transfer coefficient will be investigated.

## NANOFLUID SURFACE WETTABILITY

The affinity of liquid for the solid substrate is referred to the wettability of a fluid. Liquids with weak affinities for a solid substrate collect themselves into beads while those with high affinities for the solid surface maximize the liquid-solid contact area. The affinity of liquids for substrate depends on force balance on triple line. Triple line could expand more with reduction of resistance forces. For a given droplet volume, as the affinity of liquids increases, the liquid spreads more over the surface, thus the contact angle decreases [16]. In fact, there are relations between gravity, wetting area and contact angle. Gravity has important role in favor of spreading of droplet liquid over solid substrate [5]. The liquid spreads over the substrate until pushing and resisting forces are balanced at the triple line. The Young equation describes the surface tension force balance at the triple line, regardless of gravity. Effect of gravity and drop size [17-18] on behavior of droplet triple line has been investigated experimentally but there is no accurate theoretical modelling yet. Homogeneity and surface roughness are effective parameters on variation of contact angle [13, 19-22]. Effects of surface roughness on wettability and droplet contact angle modelled theoretically [13] and observed experimentally [23]. Water droplet contact angle on stainless steel and copper substrates is reported to decrease, as roughness increases. As the roughness of copper surface increases further, water droplet contact angle increases again [23]. Indeed, the surface roughness increases the actual contact area and total liquid-solid interactions [24]. There have been many attempts to understand and formulate the effective forces at the triple line such as liquid-gas, solid-gas, and solid-liquid surface tensions. The nature and mechanisms of the forces at the triple line is not clear yet.

Adding nanoparticles inside the base liquids modifies the physical properties of the base liquid and as results behavior of the triple line will be more complicated. Suspended nanoparticles inside the base liquid change the liquid-gas [4], solid surface tensions [5] and eventually force balance at the triple line. The surface tensions have a significant role on behavior of triple line such as radius of triple line and contact angle. Presence of nanoparticles at the triple line can introduce many complexities and affect distribution of nanoparticles in triple region, interactions with gas, liquid and solid molecules at the triple line. Existence of nanoparticles at the triple line changes the gas-liquid-solid interactions and might vary the solid-gas surface tension. Information on distribution of nanoparticles in triple region is very limited. Gathering and spreading of macro scale latex particles have been observed inside the triple region. Position and oscillations of the particles in triple region were monitored. Period of oscillation of particles was found to be around 2 Hz [25]. Different spreading and thinning (layering) behavior of nanoparticles have been observed at the triple region. The nanoparticles inside triple region were observed to spread stepwise with the layering phenomenon through film thinning at the triple line on a smooth hydrophilic glass surface. Number of layer of nanoparticles decreases in a stepwise pattern in triple region, as nanoparticles get closer to the edge of triple line. Apparently the layering phenomenon of the nanoparticles at the triple region depends on many factors such as concentration and characteristics of nanoparticles, nanoparticle charge, solid-liquid-gas materials, and film size [26]. Knowing distribution and concentration of naoparticles at the triple line region is essential. Presence of nanoparticles at the triple region has a significant role on behavior of the triple line.

Study of the particles ring from dried coffee droplet was demonstrated that the dispersed solids in a drying droplet were migrated to the edge of the droplet and formed a solid ring. In fact, evaporation of liquid from the edge of a droplet can be refilled by liquid from interior therefore resulting outward flow could carry dispersed particles to the edge by capillary flow. Shape and thickness of deposited particles might be controlled by speed of evaporation [27-28]. Spreading of aluminium-ethanol nanofluid on hydrophobic Teflon-AF coated substrate has been revealed that the nanoparticles could enhance the wetting dynamics in the vicinity of the triple line, for concentrations up to 1 w%. The nanoparticles were changed the equilibrium, receding, and advancing contact angles while no change of liquid-gas surface tension was observed [29]. That might suggest, the solid surface tensions at the triple line have a significant role on behavior of triple line. The experimental evidence for bismuth telluride-deionized water nanofluid has been showed equilibrium droplet contact angle, liquid-gas and solid surface tensions changed as the concentration of nanoparticles increased. As bismuth telluride nanoparticle concentration increased, the equilibrium contact angle increased to a maximum and then decreased. Liquid-gas surface tension of nanofluids was showed opposite trend with variation of concentration [4, 30].

The effects of the nanoparticle on the behavior of triple line were monitored by injecting air into the gold-deionized water and alumina-deionized water nanofluids from a stainless steel nozzle. Dynamics of triple line and bubble growth have been captured by a fast camera and attached microscope. Interestingly, suspended nanoparticles were changed waiting time, bubble formation time, bubble frequency, dynamics of bubble growth and bubble departure process without any sign of nanoparticle deposition on top of the substrate [5-8, 46]. Bubble contact angle was found to be higher than that of deionized water, while radius of triple line inside gold nanofluid was noticed to be less than that of deionized water. Practically, suspended gold nanoparticles were enhanced the pinning behavior of the triple line inside the nanofluids by pushing the liquid phase toward gas phase. Similar gold nanofluid and stainless steel substrate were employed to compare the gold nanofluid droplet contact angle with that of deionized water. Gold nanofluid droplet contact angle was observed to have a higher wettability and pushes the liquid toward the gas phase more. Similar phenomenon was observed by bubble formation inside gold nanofluid. Opposite behaviors were found inside alumina nanofluid. Obviously, material of nanoparticles has a key impact on behavior of triple line inside nanofluids [6-8, 46]. It was similarly noticed that alumina nanofluid enhanced the droplet contact angle on top of the stainless steel substrate compared to water [32].

A tiny porous layer of deposited nanoparticles was observed to form on a heated substrate, in nanofluid boiling heat transfer phenomenon. Deposited nanoparticles change the surface roughness, wettability, receding, and advancing contact angles [31-33, 12]. The effects of suspended nanoparticles mostly lie directly on the triple line region and affect the distribution of nanoparticles in triple region, solid surface tensions, interactions of gas-liquid-solid in the triple region, and possibly the liquid-gas surface tension as well. Variation of surface tensions and gas-liquid-solid interactions at the triple line region might be responsible for behavior of the triple line inside nanofluids. In addition, it has been shown theoretically that particles could spread the triple line to a distance of 20-50 times of the particle diameter through a structural disjoining pressure and self-ordering of particles in a confined wedge. However, the structural disjoining force only becomes significant while concentration of particles is high, i.e. over 20 v% [34]. Concentration of nanofluids is below 20 v% in most of the applications.

Exploring of interactions of nanoparticles-nanoparticles and nanoparticles with surrounding molecules of base liquid, solid surface, and gas at the triple line provides a better understanding of concentration and spreading of nanoparticles in the triple region. Nanoparticles have enormous impacts on the force balance and behavior of triple line. Effects of nanoparticles on behavior of triple line have a key impact on dynamics of bubble growth, departure process, wetting and rewetting, advancing and receding contact angles, boiling heat transfer and critical heat flux. Suspended nanoparticles have a great potential to modify nanofluids and enhance boiling heat transfer coefficient and critical heat flux. It is essential to be able to engineer the nanofluids to enhance the boiling heat transfer coefficient and critical heat flux simultaneously. Such a simultaneous enhancement might be provided by nanoparticles [12].

## TRIPLE LINE FORCE BALANCE

Liquid-gas, $\sigma_{\lg}$, solid-liquid, $\sigma_{sl}$, and solid-gas, $\sigma_{sg}$ surface tensions are major effective forces at the triple line. Schematics of surface tension forces for bubbles and droplets are given in Figure 1. The Young equation describes force balance at the triple line as follows

$$\sigma_{\lg} \cos\theta_e = \sigma_{sg} - \sigma_{sl} \tag{1}$$

where $\theta_e$ is the equilibrium contact angle. The Young equation is associated with several restrictions and has never been experimentally verified. Its application is limited to the situations where the substrate is ideal and contact angle is size independent [16, 35]. The Young equation cannot be applied directly except for long droplets [18]. The left side of equation (1) is size dependent while the right side of the equation contains physical properties which make the equation inconsistent. Force balance between liquid-gas and solid surface tensions has a significant role on the interfacial shapes of bubbles and droplets. However, modelling and prediction of dynamics of triple line are challenging and still unclear.

The liquid-gas surface tension can be found for most of materials, but still solid-liquid and solid-gas surface tensions are not easily available. Several independent approaches have been employed to calculate the solid surface tensions [35-36], such as Berthelot's combining rule [37], $\sigma_{sl} = (\sqrt{\sigma_{\lg}} - \sqrt{\sigma_{sg}})^2$, the modified Berthelot's rule [38], $\sigma_{sl} = \sigma_{\lg} + \sigma_{sg} - 2\sqrt{\sigma_{\lg}\sigma_{sg}}\exp-\beta(\sigma_{\lg} - \sigma_{sg})^2$, the alternative formulation [39-40], $\sigma_{sl} = \sigma_{\lg} + \sigma_{sg} - 2\sqrt{\sigma_{sg}\sigma_{\lg}}(1 - \beta_o(\sigma_{\lg} - \sigma_{sg})^2)$, and the equation of state formulation [41-42], $\sigma_{sl} = \dfrac{(\sqrt{\sigma_{\lg}} - \sqrt{\sigma_{sg}})^2}{1 - 0.015\sqrt{\sigma_{\lg}\sigma_{sg}}}$, where $\beta = 0.000115(m^2/mJ)^2$ and $\beta_o = 0.0001057\,(m^2/mJ)^2$.

The given correlations have been compared against of each other for some materials and a relatively good agreement has been observed [36]. The exact prediction of solid surface tensions is not possible yet and requires more investigations. Characteristics of substrate, force balance at the triple line (see Figure 1) and gravity are the main factors on bubble and droplet shapes. Droplet contact angle would change with volume and can not be a unique criterion to measure effect of nanoparticles on triple line.

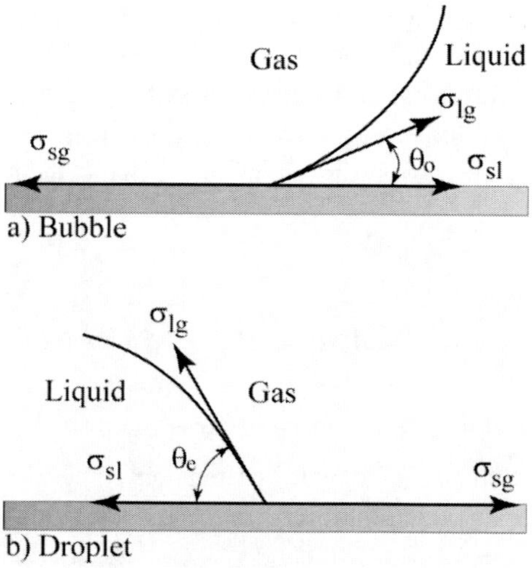


Figure 1. Schematic of effective forces at the triple line.

## NANOFLUID DROPLET SHAPE

Nanoparticles would change force balance and behavior of triple line, consequently they modify boiling heat transfer phenomenon. So, the nanofluid droplet contact angle was employed [32, 43] to predict the effects of nanoparticles on behavior of the triple line. Effects of nanoparticles on droplet contact angle were investigated [29-30] by many researchers in order to predict the behavior of triple line, liquid-gas [4], and solid surface tensions [5].

To form a droplet on a solid substrate, nanofluid has to be injected slowly on top of the substrate by a syringe. A goniometer can be employed to measure parameters of a droplet such as contact angle, $\theta_e$, radius of contact line, $r_d$, and location of apex, $\delta$ (see Figure 2). The Young-Laplace equation normally needs to be solved to predict the profile of droplet, volume, radius of curvature at apex and liquid-gas surface tension.

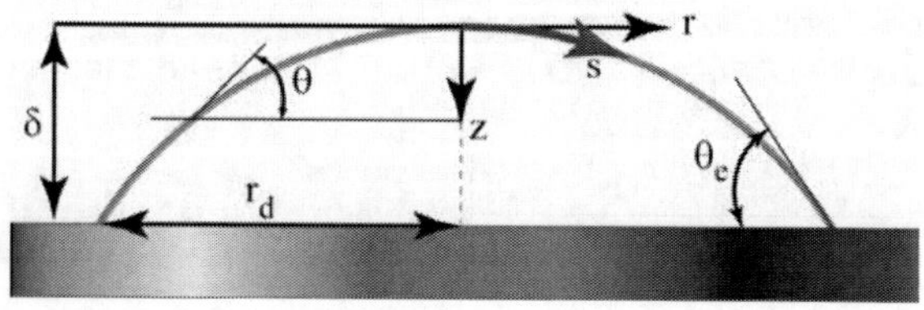


Figure 2. Schematic of a droplet shape.

The Young-Laplace equation demonstrates that the pressure difference across the interface is equal to the product of the curvature by gas-liquid surface tension. Assuming solid substrate is perfectly smooth and homogeneous without any contact angle hysteresis, droplet shape can be predicted by the Young-Laplace equation under equilibrium condition. Theoretical prediction of droplet shape by the Young-Laplace equation has been examined with experimental data, and a good agreement was observed [4, 18, 30, 35, 44].

The Young-Laplace equation can be seen as

$$\Delta p = \left(\frac{1}{R_1} + \frac{1}{R_2}\right)\sigma_{\mathrm{lg}} \tag{2}$$

where $R_1$ and $R_2$ are the radii of curvatures, i.e. $R_1$ is the radius of curvature describing the latitude as it rotates and $R_2$ is the radius of curvature in a vertical section of the droplet describing the longitude as it rotates. The centres of $R_1$ and $R_2$ are on the same line vertical to the interface but in the different locations. $\Delta p$ is pressure difference between gas, $p_g$, and liquid, $p_l$, phases (see Figure 2). The liquid and gas pressures are

$$p_l(z) = \frac{2\sigma_{g\,\mathrm{ln}}}{R_o} + P_o + \rho_l g z \tag{3}$$

$$p_g(z) = P_o + \rho_g g z \tag{4}$$

$$R_1 = ds/d\theta \text{ and } R_2 = r/\sin\theta \tag{5}$$

where $P_o$ is the ambient pressure, and $\theta$ is the running contact angle. Substituting equations (3-5) into equation (2), the Young-Laplace equation is

$$\frac{d\theta}{ds} = \frac{2}{R_o} + \frac{gz}{\sigma_{\mathrm{lg}}}(\rho_l - \rho_g) - \frac{\sin\theta}{r} \tag{6}$$

In case of nanofluids, liquid-gas surface tension of nanofluid, $\sigma_{\mathrm{lg}\,n}$, needs to be used instead of $\sigma_{\mathrm{lg}}$. System of ordinary differential equations (7-9) along with Young-Laplace equation (6) needs to be solved to predict the droplet shape.

$$\frac{dr}{ds} = \cos\theta \tag{7}$$

$$\frac{dz}{ds} = \sin\theta \tag{8}$$

$$\frac{dV}{ds} = \pi r^2 \sin\theta \tag{9}$$

This system of ordinary differential equations avoids the singularity problem at the droplet, since

$$\left.\frac{\sin\theta}{r}\right|_{s=0} = \frac{1}{R_o} \tag{10}$$

where $R_o$ is the radius of curvature at the apex.

Knowing two parameters of a droplet shape, system of ordinary differential equations (6-9) can be solved to obtain the axisymmetric droplet shape and volume, using following boundary conditions [5, 35].

$$r(0) = z(0) = \theta(0) = V(0) = 0 \tag{11}$$

Knowing three parameters of a droplet shape, the ordinary differential equations (6-9) can be solved to obtain the droplet shape and liquid-gas surface tension simultaneously. Figure 3 compares the prediction of the Young-Laplace equation with profile of 2.5 *nm* bismuth telluride nanofluid droplet. A good agreement between the experimental data and numerical prediction of the Young-Laplace equation was observed [4].

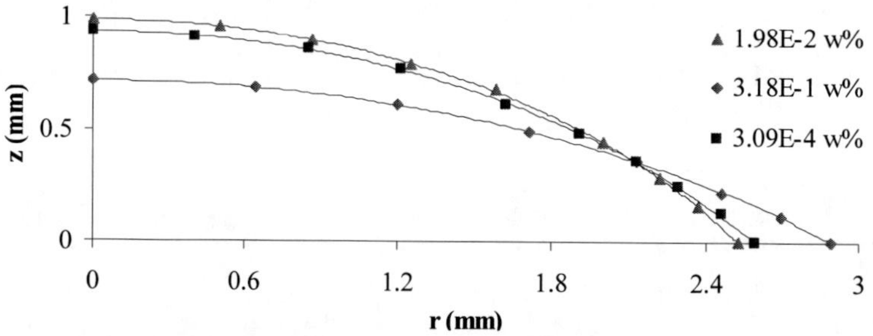


Figure 3. Experimental droplet shape (mixture of deionized water and 2.5 *nm* bismuth telluride nanoparticles on silicon wafer) versus the numerical solution of the Young-Laplace equation [4].

## Asymptotic Contact Angle

Droplet contact angle is a function of volume, so surface wettability can not be measured with droplet contact angle. As effect of gravity decreases, droplet shape changes and eventually becomes a spherical cap under the condition of zero gravity. Droplet contact angle under no gravity is only a function of the physical properties of solid, gas and liquid, which is called asymptotic contact angle, $\theta_S$ [18]. Asymptotic contact angle can be considered as a physical property that explains the interactions between gas, liquid, and solid at the triple line. In case of no gravity, the force balance at the triple line can be expressed precisely as

$$\sigma_{\lg} \cos\theta_S = \sigma_{sg} - \sigma_{sl} \quad (12)$$

Equation (12) is a precise force balance equation at the triple line and can be called the modified form of Young's equation, assuming substrate is smooth and homogenous. The right and left sides of equation (12) are size independent and all elements of the equation are physical properties.

Asymptotic contact angle is a unique criterion to measure surface wettability. Asymptotic contact angle can be used to measure effect of nanoparticles on surface wettability as well. Asymptotic contact angle can be measured experimentally or calculated theoretically. In this chapter, asymptotic contact angle is determined theoretically, knowing parameters of droplet in normal condition ($g = 9.8\ m/s^2$).

As indicated by hypothesis 2 from reference [18], '*For given liquid, gas and solid materials, the interfacial force on a droplet, normal to the solid surface, is uniquely defined; in particular, this force is independent of the acceleration of gravity*'. This means, the normal component of the droplet interfacial force is independent of gravity and can be written as

$$F^i_{n,a} = F^i_{n,s} \quad (13)$$

where $F^i_{n,a}$ is normal component of the interfacial force of an axisymmetric droplet and $F^i_{n,s}$ is normal component of the interfacial force of a spherical droplet. Normal component of interfacial force with and without gravity are

$$F^i_{n,a} = \sigma_{gl} 2\pi r_d \sin\theta_o \quad (14)$$

$$F^i_{n,s} = \sigma_{gl} 2\pi r_{d,S} \sin\theta_S \quad (15)$$

Radius of triple line at zero-gravity is

$$r_{d,s} = \left[\frac{3V}{\pi(2+\cos\theta_S)(1-\cos\theta_S)^2}\right]^{1/3} \sin\theta_S \quad (16)$$

Knowing equations (14-16), equation (13) can be written as follows

$$r_d \sin\theta_o = r_{d,s} \sin\theta_S \quad (17)$$

$$r_d \sin\theta_o = \left[\frac{3V}{\pi(2+\cos\theta_S)(1-\cos\theta_S)^2}\right]^{1/3} \sin^2\theta_S \tag{18}$$

The combination of equations (12) and (17) expresses a new formulation at the triple line as

$$\sigma_{\lg}\sqrt{1-(\frac{r_d \sin\theta_o}{r_{d,s}})^2} = \sigma_{sg} - \sigma_{sl} \tag{19}$$

Equations 18 and 19 can be used for both nanofluid and pure liquids. In general, the Young-Laplace equation can be solved to predict droplet shape, knowing physical properties, and two droplet characteristics such as the radius of triple line and location of apex. Having droplet volume, $V$, contact angle, $\theta_o$, and the radius of the contact line, $r_d$, asymptotic contact angle, $\theta_S$, can be calculated, using equation (18). Asymptotic contact angle introduces an accurate criterion to evaluate and measure effects of important parameters on surface wettability such as characteristics of nanoparticles, liquid, gas, and solid materials. Figures 4 and 5 illustrate the variation of asymptotic contact angle with concentration of bismuth telluride nanofluid on silicon wafer and glass substrates for 2.5 *nm* and 10.4 *nm* nanoparticle sizes. The figures clearly demonstrate that location of the maximum asymptotic contact angle and droplet contact angle shifted to the right side as nanoparticle size increased. The maximum asymptotic contact angle and droplet contact angle were occurred at lower nanofluid concentration, as nanoparticle size increased. In other words, the nanoparticle size has an important role on the triple line behavior. Smaller nanoparticles are more effective to change contact angle and liquid-gas surface tension of nanofluids [4, 30].

Effects of substrate on the asymptotic contact angle can be seen in Figures 4 and 5. Table 1 expresses the conversion between concentration (weight of bismuth telluride nanoparticles per weight of deionized water) and –Ln (concentration). Figures 6 and 7 compare the differences between the measured contact angle and the asymptotic contact angle with concentration of nanofluid on both silicon wafer and glass substrates for 2.5 *nm* and 10.4 *nm* bismuth telluride nanoparticles. These figures exhibit, a) for a given substrate, as nanoparticle size changes the maximum contact angle occurs at different nanoparticle concentration, b) droplet contact angle changes dramatically with adding small amount of nanoparticles into water. Pure water droplet contact angle on glass and silicon wafer is less than 5 degree, c) as concentration of nanoparticle increases, the contact angle increases and after a maximum, the contact angle start decreasing [30]. Similar trends were observed for receding and advancing contact angles [29]. The detail of the variation of contact angle with concentration of bismuth telluride nanofluid for 2.5 *nm* and 10.4 *nm* nanoparticle sizes on glass and silicon wafer substrates can be found in reference [30].

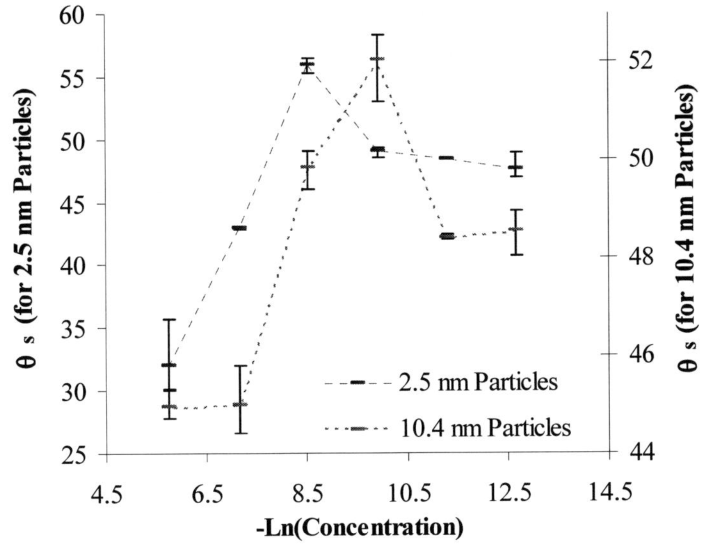


Figure 4. Variation of asymptotic contact angle with concentration of bismuth telluride nanofluid on silicon wafer substrate for nanoparticles in size of 2.5 *nm* and 10.4 *nm* [5].

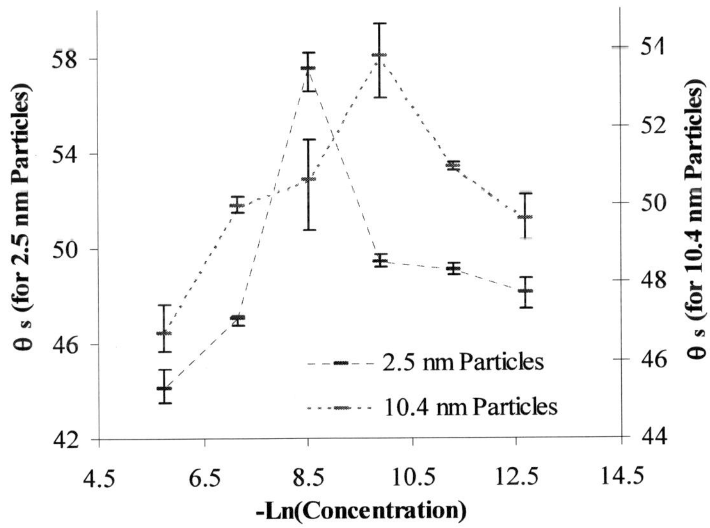


Figure 5. Variation of asymptotic contact angle with concentration of bismuth telluride nanofluid on glass substrate for nanoparticles in size of 2.5 *nm* and 10.4 *nm* [5].

**Table 1. The conversion between concentration (weight of bismuth telluride nanoparticles per weight of deionized water) and -Ln (concentration)**

| Concentration | 3.09E-6 | 1.24E-5 | 4.95E-5 | 1.98E-4 | 7.93E-4 | 3.18E-3 |
|---|---|---|---|---|---|---|
| -Ln (Concentration) | 12.68 | 11.29 | 9.91 | 8.52 | 7.13 | 5.75 |

The repulsion forces between the negatively charged carboxylic groups of the functionalized bismuth telluride nanoparticles and the hydroxyl groups (OH–) on the silicon, and silicon dioxide surfaces has a significant role on the variation of droplet contact angle. The repulsive forces between solid and nanofluids induce a non-wetting behavior that increases with concentration of the nanoparticles therefore the contact angle initially increases with concentration. Moreover, the solid surface tensions of the nanofluids near the triple line seem to be changing because of the cohesive and adhesive interactions between water molecules and nanoparticles at the interfacial surfaces [30]. Since both force balance at the contact line and the corresponding surface tensions may lead to change the contact angle, further investigations, including analysis of the effects of nanoparticles on the solid-liquid, solid-gas and liquid-gas surface tensions and different substrates are required to understand the responsible mechanisms of interactions of nanoparticles with surrounded molecules and nanoparticles inside the nanofluids.

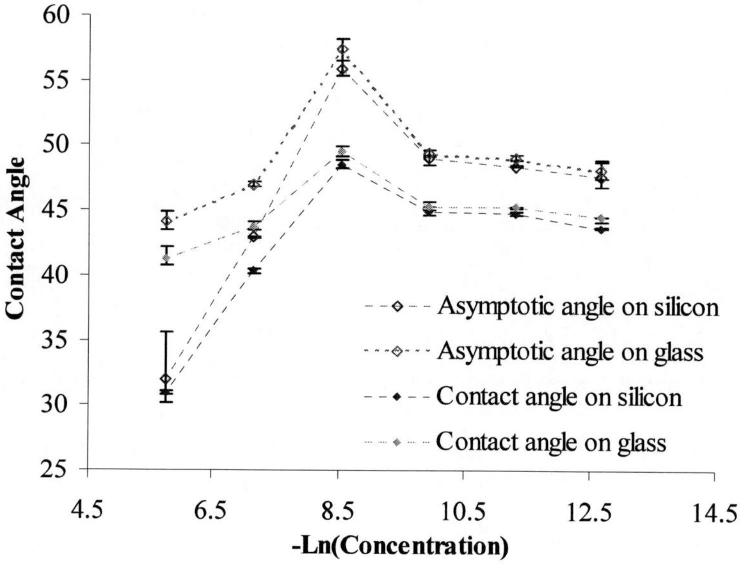


Figure 6. Variation of asymptotic contact angle and droplet contact angle with concentration of nanofluids for 2.5 *nm* bismuth telluride nanoparticles on top of a silicon wafer and glass substrates [5].

The asymptotic contact angle and droplet contact angle follow similar trends, opposite to the trend of liquid-gas surface tension of nanofluids. Gas-liquid surface tension of nanofluids initially decreases as concentration of nanoparticles increases, reaching a minimum, and then increases again (see Figures 8 and 9). Moreover, for a given particle size and concentration, the droplet and asymptotic contact angles depend on substrate which means solid surface tensions have a significant role on force balance at the triple line (see Figure 7). The effects of nanoparticles on liquid-gas and solid surface tensions change the force balance at the triple line and consequently change the droplet profile. For a given substrate, effects of nanoparticles on bubbles and droplets mostly lie on triple line and can be seen through either contact angle or radius of contact line. Solid surface tensions can be calculated from equation (12), knowing asymptotic contact angle and liquid-gas surface tension.

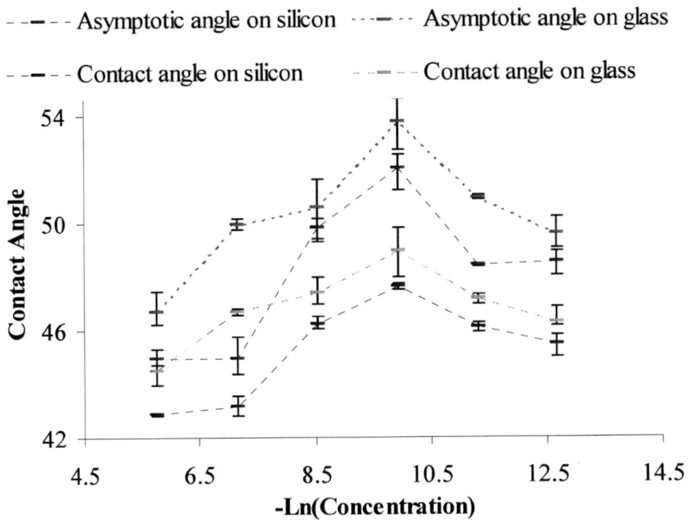


Figure 7. Variation of asymptotic contact angle and droplet contact angle with concentration of nanofluids for 10.4 *nm* bismuth telluride nanoparticles on top of a silicon wafer and glass substrates [5].

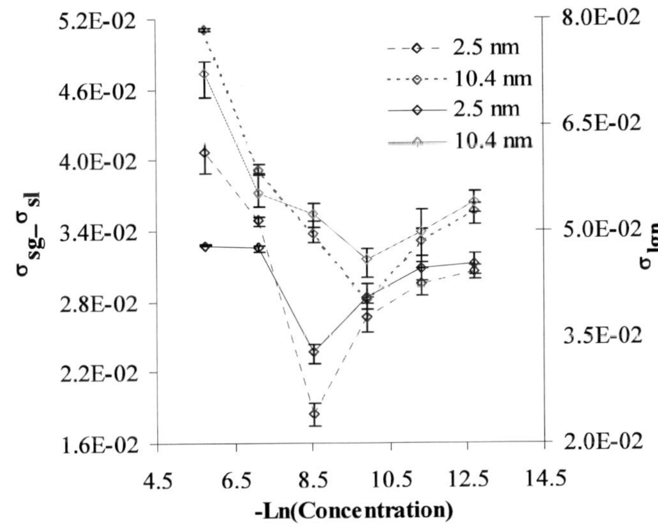


Figure 8. Variation of difference between solid-gas and solid-liquid surface tensions and liquid-gas surface tension with bismuth telluride nanoparticle concentration for 2.5 *nm* and 10.4 *nm* nanoparticle sizes on silicon wafer substrate (Dashed line left side and solid line right side) [5].

Figures 8 and 9 illustrate variation of liquid-gas, $\sigma_{\lg n}$, and solid surface tensions, $\sigma_{sg} - \sigma_{sl}$, of nanofluids with concentration, for 2.5 *nm* and 10.4 *nm* nanoparticle sizes on top of glass and silicon wafer. These figures indicate a) liquid-gas and solid surface tensions of nanofluids is affected more by smaller nanoparticles, b) independency of the liquid-gas surface tension of nanofluids from substrate demonstrates the accuracy of the calculations, c) material of the substrate affects the solid surface tensions. It is well known (see equation 12) that variations of solid surface tensions with nanoparticle size and concentration have important roles on surface wettability. Still there is a debate whether the existence of nanoparticles at the triple line would change the gas-liquid-solid interactions and solid-gas surface tension or not. Further investigations are required to understand the effects of nanoparticles on solid-gas and solid-liquid surface tensions.

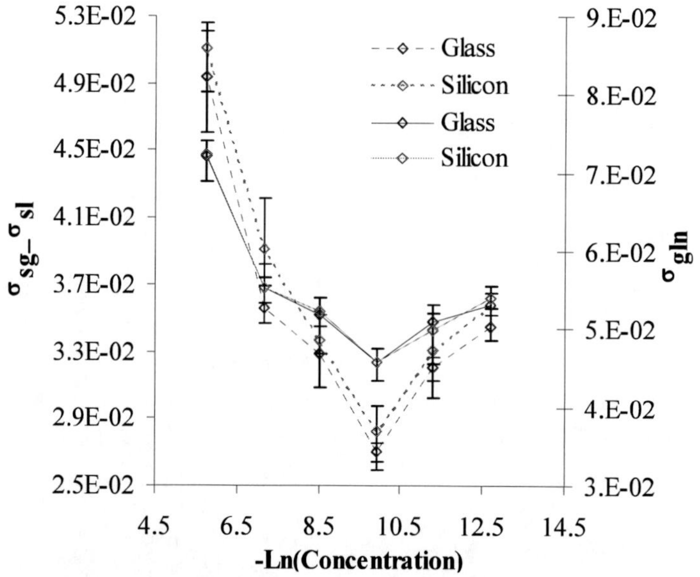


Figure 9. Variation of liquid-gas, $\sigma_{\lg n}$, and solid surface tensions, $\sigma_{sg} - \sigma_{sl}$, with bismuth telluride concentration for 10.4 *nm* nanoparticles on silicon wafer and glass substrates (Dashed line left side and solid line right side) [5].

Comparing the variation of asymptotic contact angle and droplet contact angle with nanofluid concentration clearly shows how force balance at the triple line would change with gravity. For instance, the difference between asymptotic contact angle and droplet contact angle is higher at the lower nanofluid surface tension. Figure 10 demonstrates the variation of contact angle and radius of triple line with concentration of the nanofluid for normal and no gravity conditions. The radius of triple line and contact angle has an opposite trend with variation of concentration. As gravity reduces, the force balance at the triple line changes, thus the droplet contact angle increases and the radius of triple line decreases. Similar behavior observed for 10.4 *nm* nanoparticle bismuth telluride on top of the glass and silicon wafer substrates. The error bars represent the difference between the averaged contact angle for three similar droplets and corresponding maximum and minimum contact angles in each series. The source of error can be due to the measurement, non-homogeneity of the substrate, surface roughness, and local variations of concentration of the nanofluids.

The experimental evidences show how the asymptotic contact angle reflects the effect of variation of nanofluids or substrates regardless of droplet size which leads to the standard and unique way of evaluation and measurement of surface wettability. The asymptotic contact angle depends on gas-liquid-solid materials, characteristics and concentration of nanoparticles. The combinations of effects of nanoparticles on liquid-gas and solid surface tensions affect the force balance at the triple line and surface wettability. The receding and advancing contact angles have an important role in many applications such as boiling heat transfer as well as surface wettability. In fact, the dewetting and wetting phenomenon throughout the formation and detachment of bubbles can be studied respectively by the receding, and advancing contact angles.

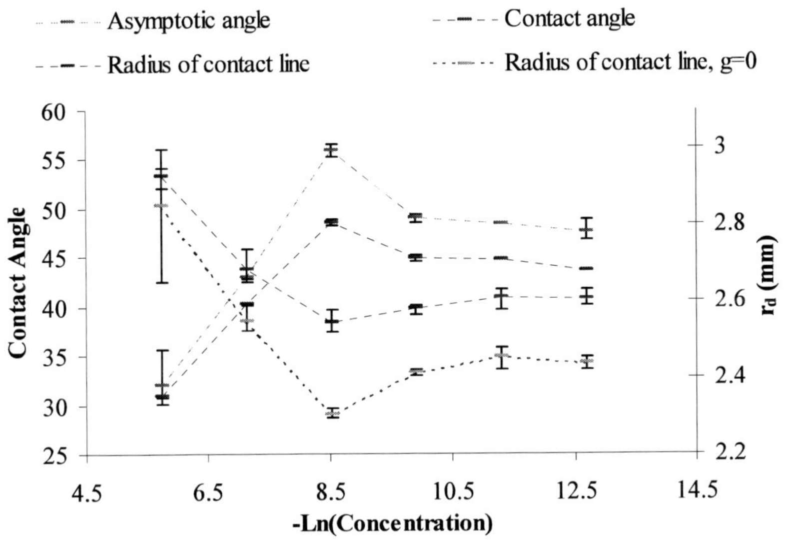


Figure 10. Variation of asymptotic contact angle, droplet contact angle, radius of contact line, with, and without gravity with concentration of nanofluids for 2.5 *nm* bismuth telluride nanoparticles on top of a silicon wafer [5].

# Theoretical Analysis of Axisymmetric Bubble Growth

## Prediction of Bubble Shape

The Young-Laplace equation has been solved to predict the axisymmetric bubble shape under quasi steady state conditions. Prediction of the Young-Laplace equation has been examined by experimental data [6-7, 45-50], theoretical modeling such as volume of fluid method and level-set method [51-52]. A good agreement is observed between prediction of the Young-Laplace equation, experimental and numerical results. It is assumed that solid surface is perfectly smooth and homogeneous, the bubble shape is axisymmetric, the gas flow rate is steady and slow enough to ignore the liquid-gas shear stress, and the variation of momentum on top of the bubbles is negligible. The Young-Laplace equation predicts the bubble shape up to the point of critical equilibrium, where the bubble has the maximum volume. Assuming bubble is growing in a quasi-steady state condition, the Young-Laplace equation on the interfacial surface can be expressed by equation (2), where $\Delta p$ is pressure difference between gas, $p_g$, and liquid, $p_l$, phases (see Figure 11). The gas pressure as a function of $z$ is

$$p_g(z) = \frac{2\sigma_{\mathrm{lg}}}{R_o} + P_o + \rho_g gz + \rho_l gh \tag{20}$$

where the first term on the right hand side is pressure difference at the bubble apex, $P_o$ is ambient pressure, and the last two terms are the hydrostatic gas and liquid pressures. The liquid pressure is

$$p_l(z) = P_o + \rho_l g(h+z) \tag{21}$$

Substituting equations (20, 21, 5) into equation (2), the Young-Laplace equation can be rewritten as

$$\frac{d\theta}{ds} = \frac{2}{R_o} - \frac{gz}{\sigma_{\mathrm{lg}}}(\rho_l - \rho_g) - \frac{\sin\theta}{r} \tag{22}$$

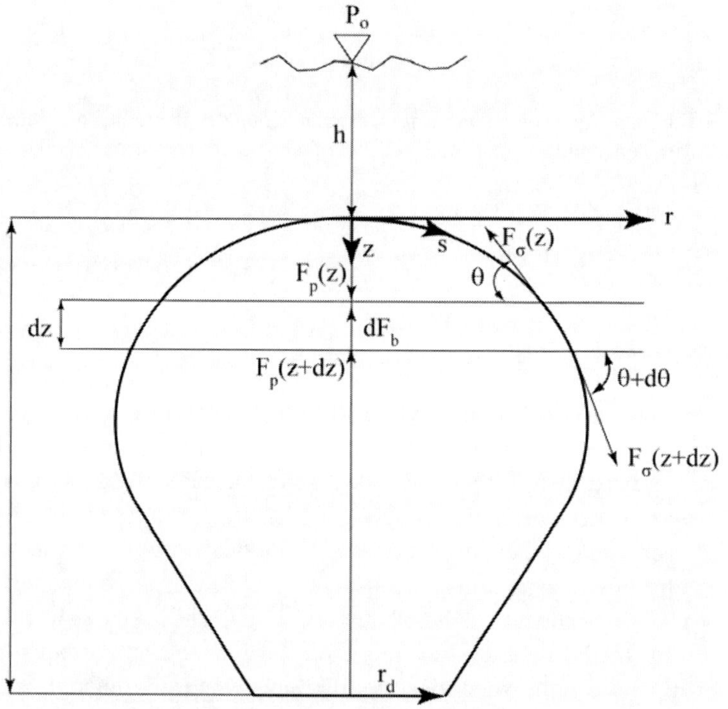


Figure 11. Schematics of an axisymmetric bubble with the forces acting on a slice of bubble, extending between $z$ and $z$+d$z$, along the vertical axis [6, 45].

To predict the bubble shape inside nanofluids, gas-liquid surface tension of nanofluids, $\sigma_{\mathrm{lg}n}$, needs to be employed instead of $\sigma_{\mathrm{lg}}$. The system of ordinary differential equations (7-9, 22) can be solved to predict the axisymmetric bubble shape, knowing the boundary conditions (see equation 11) and two parameters of the bubble shape such as bubble height, and radius of triple line [6-7, 45, 49-50].

## Analytical Expression of Bubble Volume

Using force balance method, an analytical expression is developed to express a relation between bubble volume, $V$ , radius of triple line, $r_d$ , bubble height, $\delta$ , liquid-gas surface tension, $\sigma_{lg}$ , and radius of curvature at apex, $R_o$ . Figure 11 shows the schematic of the forces on a slice of bubble, extending between $z$ and $z$+dz, along the vertical axis of an axisymmetric bubble [6, 45].

Force balance equation along the vertical direction is

$$\sum F_z = -dF_b(z) - F_p(z+dz) + F_p(z) - F_\sigma(z)\sin\theta + F_\sigma(z+dz)\sin(\theta + d\theta) = 0 \qquad (23)$$

where $F_b$ is buoyancy force, $F_p$ is force due to pressure and $F_\sigma$ is force due to surface tension. Equation (23) is simplified as

$$-dF_b(z) - d[F_p(z) - F_\sigma(z)\sin\theta] = 0 \qquad (24)$$

where the individual elements are as follows

The buoyancy force is

$$dF_b = (\rho_l - \rho_g)g\pi r^2 dz \qquad (25)$$

The force due to hydrostatic pressure is

$$F_p(z) = [p_g(z) - p_l(z)]\pi r^2 \qquad (26)$$

The force due to surface tension is

$$F_\sigma(z) = \sigma_{lg} 2\pi r \qquad (27)$$

where the gas and liquid pressure on liquid-gas interfacial surface are given by equations (20-21). By substituting equations (25-27) into equation (24), the resultant equation is

$$-(\rho_l - \rho_g)g\pi r^2 dz - d[(p_g(z) - p_l(z))\pi r^2 - 2\sigma_{lg} r\pi \sin\theta] = 0 \qquad (28)$$

By integrating equation (28) along the $z$ axis, an analytical expression is obtained as

$$(\rho_l - \rho_g)gV + \frac{2\sigma_{lg}}{R_o}\pi r_d^2 - (\rho_l - \rho_g)g\delta\pi r_d^2 - 2\sigma_{lg} r_d \pi \sin\theta = 0 \qquad (29)$$

Equation (29) gives an analytical expression between parameters of an axisymmetric bubble, under the low gas flow rate conditions. This equation also shows that there are various forces acting on the bubble such as buoyancy force, $(\rho_l - \rho_g)gV$, force due to Laplace pressure, $\frac{2\sigma_{lg}}{R_o}\pi r_d^2$, hydrostatic force of the gas and liquid phases, $-(\rho_l - \rho_g)g\delta\pi r_d^2$, and vertical component of surface tension force, $-2\sigma_{lg} r_d \pi \sin\theta_o$. Inside nanofluids, gas-liquid surface tension of nanofluids, $\sigma_{lgn}$, has to be used instead of $\sigma_{lg}$.

## Nanofluid Bubble Formation

Suspended nanoparticles inside liquid change the physical properties of nanofluids such as viscosity, liquid-gas, and solid surface tensions. As a result behavior of bubble triple line and dynamics of formation of bubble vary. Effects of nanoparticles on bubble growth and departure process were investigated experimentally by injecting gas into water and nanofluids through different nozzles. Bubble growth was captured by a high speed camera equipped with an optical microscope head. Bubble height and radius of triple line was measured from the captured pictures during the bubble formation. Knowing bubble height and radius of triple line, the Young-Laplace equation was solved to predict the bubble shape and rest of its characteristics [6-8, 45-46, 49]. Figure 12 compares the prediction of the Young-Laplace equation inside 5 *nm* gold nanofluid and deionized water with experimental data for a nominal gas flow rate of 0.249 ml/min. The Young-Laplace equation predicts the bubble shape relatively accurate from needle nozzle, inside pure liquids and nanofluids, as long as a) gas flow rate is small enough to negate the shear stress, b) equilibrium condition exists between gas and liquid and c) while bubble is not stretched upward. At the necking and detachment periods, bubble starts being stretched, so the effect of viscosity becomes important. Prediction of the Young-Laplace is not accurate enough during the necking and detachment periods. Figure 13 shows a comparison between numerical prediction of the Young-Laplace equation and the analytical expression (29) of bubble volume inside the water and gold nanofluid of 0.01 w%. A good agreement was observed between numerical prediction of the Young-Laplace equation and analytical expression (29). Figure 14 shows the variation of force due to Laplace pressure, vertical component of surface tension, and buoyancy forces with time inside the gold nanofluid (0.01 w%) for a nominal gas flow rate of 0.254 ml/min.

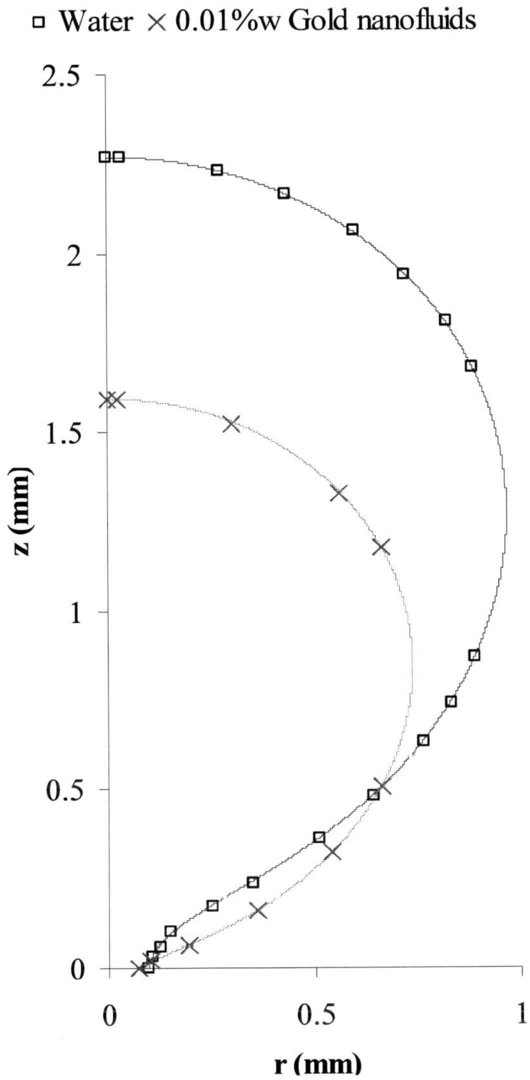


Figure 12. Comparison of bubble shape and prediction of the Young-Laplace equation for a nominal flow rate of 0.249 ml/min (Exp: Dotted points, Prediction: solid lines) [7].

Radius of curvature at apex keeps increasing with time, thus force due to Laplace pressure keeps decreasing. Laplace pressure is virtually the same as vertical component of surface tension as long as buoyancy force is negligible. By rapid expansion of the bubble, the effect of buoyancy force increases, and pushes the bubble upward thus, the bubble contact angle and vertical component of surface tension force increase. Buoyancy force is a responsible reason for the variation of bubble contact angle and the rest of the forces in bubble expansion period.

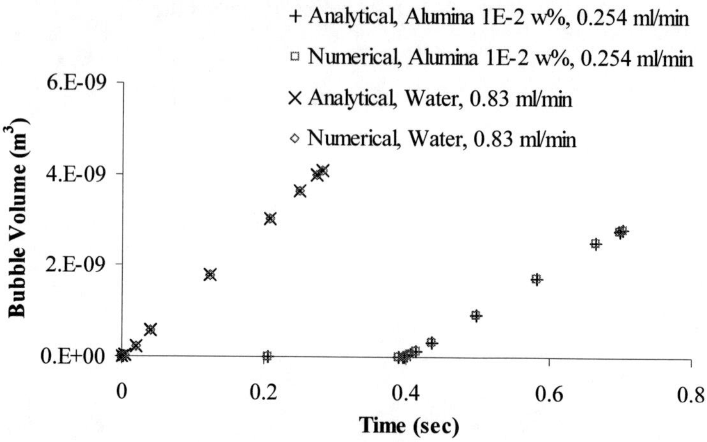


Figure 13. Comparison between bubble volume predicted by numerical solution of the Young-Laplace equation and analytical expression (29), inside deionized water and 5 *nm* gold nanofluid with concentration of 0.01 w% [6].

Evolution of formation of bubble inside gold nanofluid and deionized water are different. Bubbles start forming inside gold nanofluid earlier, and departing with smaller volume and higher frequency. For instance, 25% reduction in bubble departure volume and ~1/3 higher bubble frequency was observed for bubbles forming inside gold nanofluid under a gas flow rate of 0.249 ml/min. Such differences are partly associated with reduction of gold nanofluid surface tension and might be influenced by the dynamics of triple line. The bubble frequency increases with gas flow rate, inside both deionized water and gold nanofluid. Departure bubble volume exhibits a weak dependency on gas flow rate. Increasing of flow rate, contributes to raise the amount of gas that enters into the bubble and at the same time supplements the necking phenomenon, which has a tendency to reduce the departure volume. Both factors are important on bubble departure volume [6].

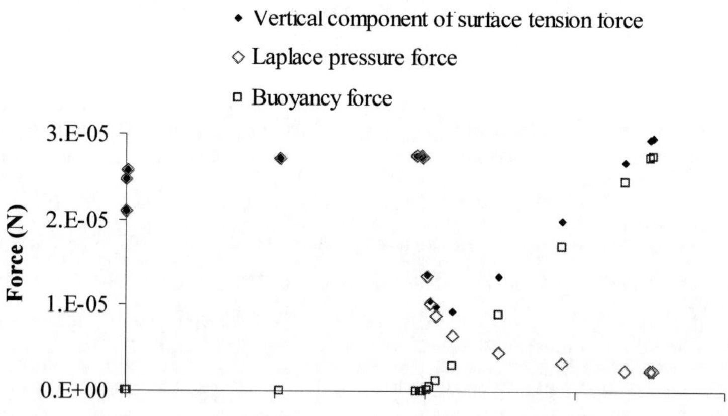


Figure 14. Variation of force due to Laplace pressure, vertical component of surface tension, and buoyancy with time inside 5 *nm* gold nanofluid (0.01 w%) for a nominal gas flow rate of 0.254 ml/min [6].

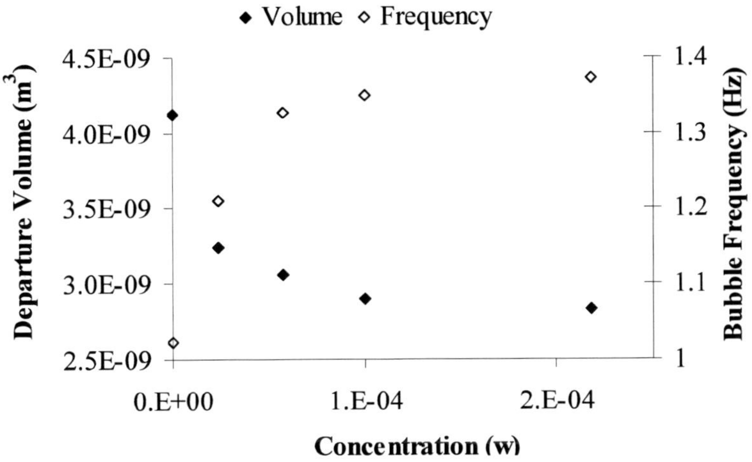


Figure 15. Variation of bubble volume departure and bubble frequency with concentration, inside 5 *nm* gold nanofluid, for a nominal gas flow rate of 0.254 ml/min [6].

Figure 15 shows the variation of bubble volume and bubble frequency with concentration of gold nanofluid. Vertical component of surface tension force is nearly identical with buoyancy force, at the departure point. Reduction of surface tension force reduces the buoyancy force and thus requires less bubble volume to be detached [6, 53-55]. No effect of liquid-gas surface tension on bubble departure volume was reported [56-57] whereas others showed that the bubble volume increases with decreasing surface tension as bubbles were allowed to grow further [58]. There are difficulties to assessing the role of the surface tension alone, as the bubble departure volume is modulated with many other variables such as diameter of nozzle, radius of triple line, wettability, gas-liquid-solid physical properties and gas flow rate.

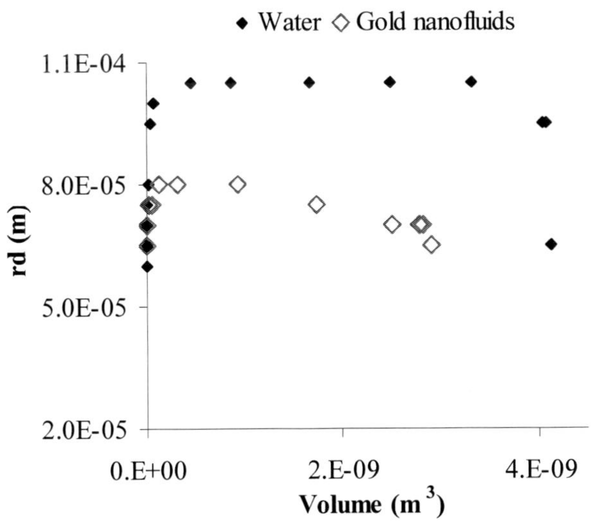


Figure 16. Variation of the radius of contact line of bubble with volume, inside deionized water and 5 *nm* gold nanofluid (0.01 w%), for a nominal gas flow rate of 0.249 ml/min [7].

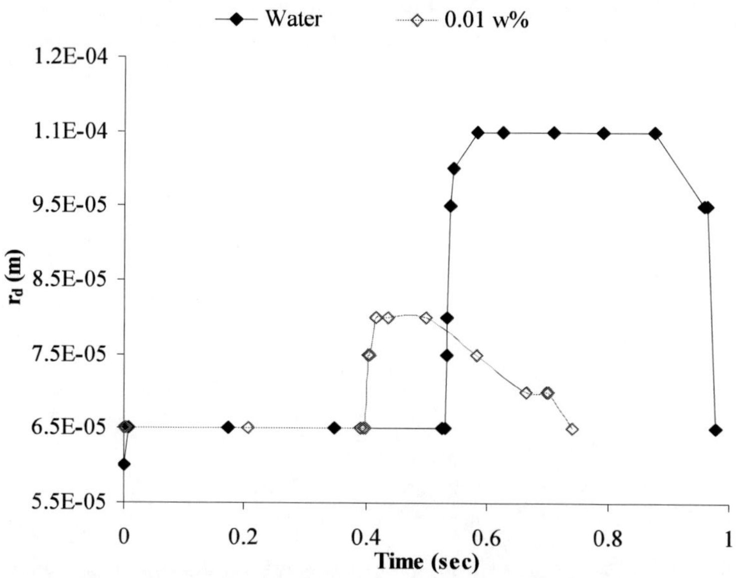


Figure 17. Variation of the radius of contact line with time, inside deionized water and 5 *nm* gold nanofluid (0.01 w%), for a nominal gas flow rate of 0.254 ml/min [6].

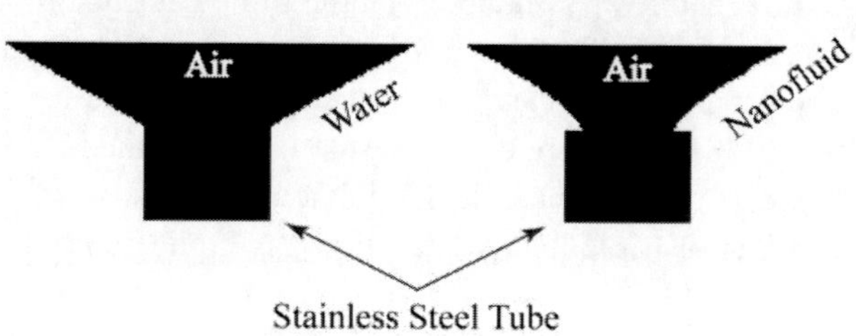


Figure 18. Snapshot of bubble shape and radius of contact lines inside deionized water (left) and 0.01 w% gold nanofluid (right), for a given bubble volumes and a nominal gas flow rate of 0.48 ml/min [7].

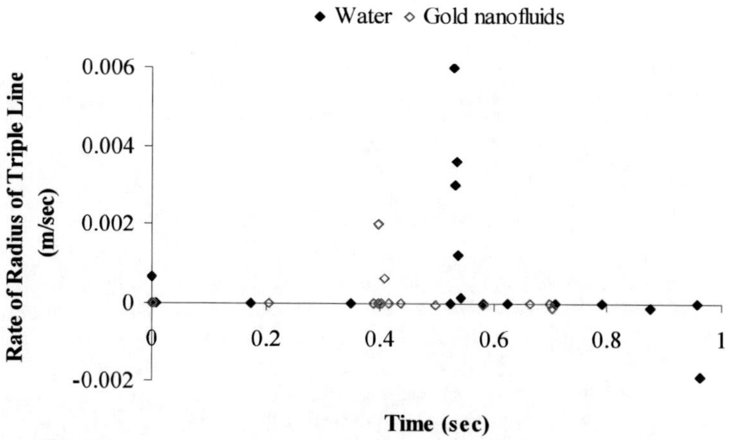


Figure 19. Variation of the rate of radius of triple line with time, inside deionized water and 5 *nm* gold nanofluid (0.01 w%), for a nominal gas flow rate of 0.254 ml/min [6].

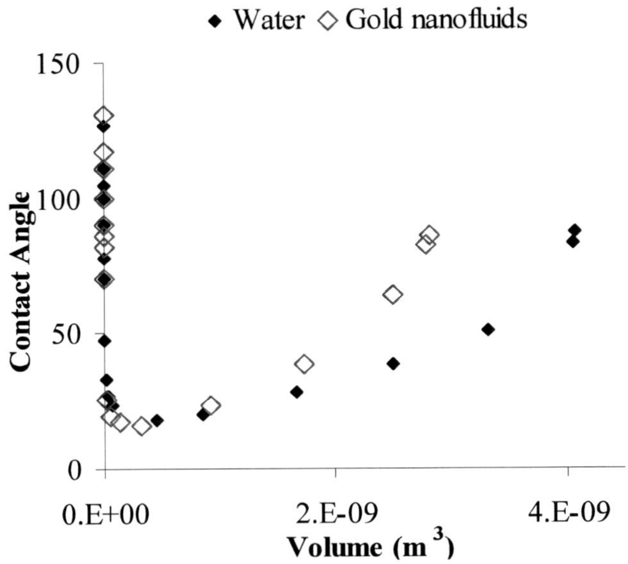


Figure 20. Variation of the contact angle of bubble with volume, inside the deionized water and 5 *nm* gold nanofluid (0.01 w%), for a nominal gas flow rate of 0.249 ml/min [7].

Figures 16 and 17 illustrate the variation of the radius of triple line with bubble volume and time inside deionized water and gold nanofluid. Radius of triple line initially has a period of fixed value for small bubble volumes and is followed by a rapid expansion, constant period and reduction through the necking and departure period when the contact line starts shrinking. Interestingly, the radius of triple line inside deionized water spreads to the outer edge of nozzle wall during the bubble growth period, while the triple line is pinned inside gold nanofluid somewhere in the middle of the tube wall, during the bubble growth. Similar pinning behavior of triple line was observed for higher gold concentrations and different gas flow rates (0.015 ~ 0.83 ml/min).

Figure 18 shows a snapshot of bubble shapes at the vicinity of triple line inside deionized water (left) and 0.01 w% gold nanofluid (right) for a given bubble volume and nominal gas flow rate of 0.48 ml/min. The experimental evidences demonstrate that the gold nanoparticles improve the pining behavior of triple line without existence of any porous layer of deposited nanoparticles. Suspended gold nanoparticles change the liquid-gas and solid surface tensions, enhance pinning behavior of triple line, reduce the radius of triple line and increase the wetting area. Variation of liquid-gas and solid surface tensions with dispersed gold nanoparticles is responsible for variation of force balance (see Figure 1) at the triple line and reduction of radius of triple line. The nanoparticles have a great potential to change the liquid-gas [4] and solid surface tensions [5]. Figure 19 explains that the velocity of the expansion of radius of triple line inside gold nanofluid is lower than that inside deionized water which confirms that the opposition force against of expansion of triple line is higher inside gold nanofluid. Figure 20 explains the variation of bubble contact angle with volume, inside deionized water and 5 *nm* gold nanofluid for given gas flow rate. For small bubble volume, where buoyancy force is negligible, as bubble volume increases, contact angle decreases. As bubble volume increases further, the effect of the buoyancy force becomes effective and pushes the bubble upward therefore contact angle starts increasing. Bubble contact angle inside water and gold nanofluid are virtually the same before reaching to the minimum value. The difference between the bubble contact angle inside deionized water and gold nanofluid increases at relatively large bubble volumes when effect of buoyancy force becomes more effective. In fact, pinning behavior of triple line inside gold nanofluid is more noticeable where buoyancy force is dominant. In case of bubble formation, buoyancy force pushes the bubble upward, so triple line can move more freely and effects of nanoparticles can be noticed easier. For a given bubble volume, bubble contact angle inside gold nanofluid is found to be slightly higher than that of deionized water which might be related to the pinning of triple line and dynamics of bubble growth.

The electron microscopy analysis of the surface morphology revealed that the nanoparticle deposition is negligible, since gold nanofluid is stable and there is no heating and boiling involved. Variation of contact angle and radius of triple line inside water and gold nanofluids has nothing to do with substrate modifications. Suspended nanoparticles inside the base liquid affect distribution of nanoparticles in triple region, liquid-gas, and solid surface tensions. Indeed the existence of nanoparticles changes the force balance at the triple line, and as results varies the behavior of triple line such as contact angle and radius of triple line [5].

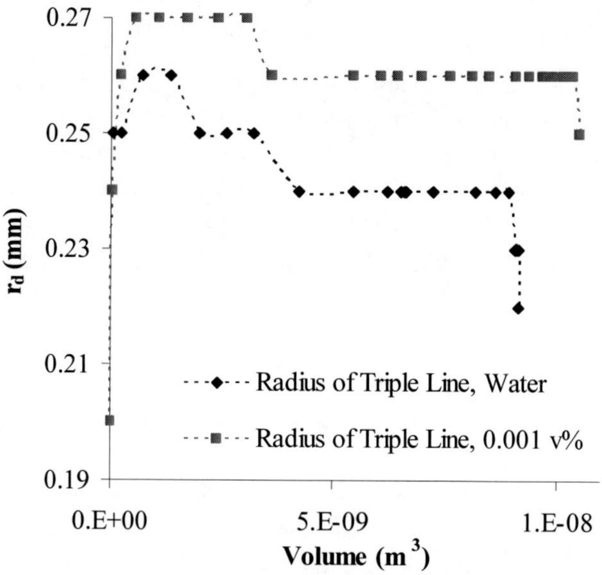


Figure 21. Variation of radius of triple line with volume on top of a stainless steel substrate plate inside water and 25 *nm* alumina nanofluid (0.001 v%). Nominal gas flow rate is 0.5 ml/min and internal diameter of the stainless steel substrate nozzle is 0.4 *mm* [46].

Similarly, air was injected into alumina nanofluid (0.001 v%) through a stainless steel substrate with 0.4 *mm* internal diameter to see the evolution of bubble formation. Figure 21 shows the variation of radius of triple line on top of a stainless steel nozzle inside the deionized water and alumina nanofluid (0.001 v%). At the early stage of bubble expansion, before buoyancy force becomes significant, variation of bubble contact angle inside deionized water and 25 *nm* alumina nanofluid is quite similar. In the early stage, as bubble volume increases, the contact angle decreases. As bubble volume expands further, bubble contact angle starts increasing. During this period, for a given bubble volume, contact angle inside alumina nanofluid is below that inside deionized water while inside gold nanofluid, the contact angle was above that of deionized water. Radius of triple line inside alumina nanofluid is higher than that inside water, while an opposite behavior observed inside gold nanofluid. Since liquid-gas surface tension of alumina nanofluid and water are virtually the same for a low concentration of nanofluids [32], modification of solid surface tensions is responsible for different behavior of triple line inside water and alumina nanofluid. Indeed, modification of the solid surface tensions and dynamics of bubble formation together could contribute to the reduction of the bubble contact angle inside the alumina nanofluid [46]. Since the concentrations of gold and alumina nanofluids are low, the effect of disjoining pressure can be negligible. The structural disjoining force only becomes significant at relatively high particle concentrations, i.e. over 20 v% [34].

The behavior of the triple line has a key impact on the evolution of the formation and detachment of bubbles in boiling heat transfer phenomenon. Effects of suspended nanoparticles on dynamics of triple line, bubble growth and departure process can be studied by bubble formation method and eventually extended into boiling heat transfer phenomenon.

## Nanofluid Boiling Heat Transfer

Nanofluid boiling heat transfer is a complex phenomenon with many applications in macro scale devices. There are enormous numbers of investigations about nanofluid boiling heat transfer that allow us to understand the effects of nanoparticles on the boiling heat transfer coefficient and critical heat flux. Nanofluid boiling heat transfer depends on many factors such as a) physical properties of nanofluids such as liquid-gas and solid surface tensions, viscosity, thermal conductivity, liquid and vapor density, vapor and liquid enthalpies, b) characteristics of heated substrate such as roughness, active nucleation site density, behavior of triple line, wetting and rewetting, receding and advancing contact angles and c) near surface hydrodynamics such as departure bubble volume, bubble frequency, and hot/dry spot dynamics [11].

Physical properties of working fluids can be modified by adding different solid nanoparticles. The concentration, size, shape, coating and material of nanoparticles has a significant role on the modification of the physical properties of nanofluids. The degree of effects of these factors on physical properties of nanofluids is not clear yet. For instance, some of the nanoparticles might enhance the liquid-gas surface tension significantly [4] and some not [32]. As results, nanoparticles have a great potential to engineer a better working fluids for a targeted application. Nanoparticles have promising effects on liquid-gas, and solid surface tensions, behavior of the triple line, wetting and dewetting and active nucleation site density of the heated substrate. Nanofluids can be a prime candidate for boiling heat transfer phenomenon. The aim is to find a suitable combination of base liquid and nanoparticles to enhance the boiling heat transfer coefficient and critical heat flux simultaneously.

Nanofluid boiling heat transfer has started initially in macro scale level, however nowadays nanofluids are employed in micro scale devices as well. One of the applications of boiling heat transfer is the cooling of electronic apparatuses. Electronic devices are being miniaturized and more powerful everyday, consequently, working fluid and heated substrate of the cooling system need to be developed and modified to raise the capacity of the heat removal. Nanofluids can be employed for boiling heat transfer inside the microchannels however the clogging phenomenon of deposited nanoparticles needs to be taken more into the consideration as channel size decreases.

## Nanofluid Boiling Heat Transfer and Nanoparticles Deposition

As temperature of the heated substrate goes above the saturated temperature, the vapor bubbles start initiating at active nucleation sites on the heated substrate, and expand with the evaporation of the surrounded superheated liquid. Eventually the buoyancy force will dominate and push the vapor bubble upward and then bubble detachment takes place. Vapor bubble expands due to the evaporation of the superheated liquid at the triple line, around the bubble and inside the microlayer trapped beneath the bubbles [59] as it can be seen in Figure 22.

Similar mechanisms participate to expand the vapor bubbles inside nanofluids in boiling heat transfer on heated substrate. As results of evaporation of nanofluids inside the microlayer beneath the bubbles, the concentration of nanoparticles inside the microlayer increases and distance between nanoparticles inside the nanofluids decreases, thus the possibility of collision, agglomeration and deposition of the nanoparticles increase. Deposited nanoparticles form a tiny porous layer of nanoparticles on heated substrate [61-65]. Thickness of deposited nanoparticles increases in boiling heat transfer over time and characteristics of the heated substrate are being modified continuously [63, 65]. Deposited nanoparticles on heated substrate modify the surface roughness, active boiling site density, surface wettability, wetting and dewetting, receding and advancing contact angles. As results, nanoparticles have enormous opportunities to enhance boiling heat transfer coefficient and critical heat flux simultaneously. Characteristics, thickness and form of the porous layer of deposited nanoparticles depend on several of factors such as distribution of nanoparticles in triple region, concentration and characteristics of nanoparticles, surface temperature, evaporation rate, and heat flux.

Thickness of the deposited nanoparticles in flow boiling depends on several more factors such as nanofluid mass flow rate, channel size, and subcooling. Stainless steel microchannels were used to run flow boiling heat transfer by water and alumina nanofluid. Microchannel was grooved after running the flow boiling heat transfer by deionized water and alumina nanofluid (see Figure 23) in order to see the porous layer of deposited nanoparticles on heated surface. Scattering electron microscope (SEM) was employed to see the deposited nanoparticles in detail at the outlet and inlet of the heating tube. Figure 24 shows the variation of deposited nanoparticles on heating surface in the vicinity of the outlet of heating tube after flow boiling test with deionized water and 25 *nm* alumina nanofluid with 0.1 v%, 0.01 v%, and 0.001 v% concentrations inside stainless steel microchannels with an internal diameter of 0.51 *mm*. As the concentration of nanofluid increases, more particles were deposited on heated surface. Apparently the porous layer of deposited nanoparticles was thicker and more condensed for nanofluids at a higher concentration. Figure 25 demonstrates the difference between deposited nanoparticles at the inlet and outlet of the heated surface after flow boiling test with 25 *nm* alumina nanofluid of 0.1 v% inside stainless steel microchannels with an internal diameter of 0.51 *mm*. Temperature of the heating surface and consequently the evaporation rate is higher at the outlet of tube, thus deposited nanoparticles are thicker with larger agglomerations. To prevent the clogging of the micro size channels by deposited nanoparticles, it is essential to engineer the naoparticles to be removable with fluid flow. The sticky behavior of some nanoparticles on heated substrate increases the thickness of deposited nanoparticles with time and causes blockage [66] inside the microchannels. The blockage of microchannels by deposited nanoparticles can be prevented with engineering of the channel size, flow specifications, characteristics of nanoparticles and solid surface.

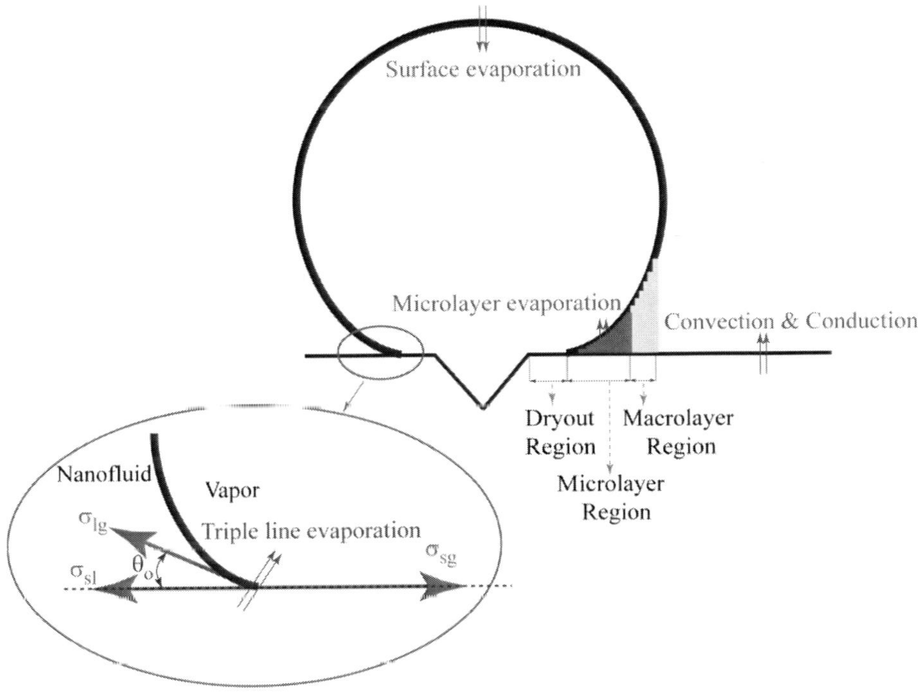


Figure 22. Schematic of bubble heat transfer and mechanisms of bubble expansion at the active nucleation site on a heated substrate.

Material of nanoparticles might affect the pattern of deposition of nanoparticles on heated surface. Formation of porous layer of deposited nanoparticles on heated surface was confirmed by nanofluid pool boiling [61, 67], flow boiling inside macrochannels [43] and microchannels [31, 33]. Formation of porous layer of deposited nanoparticles needs to be engineered in order to eliminate or minimize the negative impacts on performance of nanofluid boiling heat transfer. Further investigation is required to understand the positive and negative impacts of the formation of the porous layer of deposited nanoparticles from thermal fluidic and material science perspectives. As channel size decreases, sensitivity of the flow boiling heat transfer to the deposited nanoparticles increases. Thus, it is necessary to try to create better conditions to minimize or eliminate the deposition of nanoparticles in nanofluid boiling heat transfer.

Figure 23. Test section of a heated stainless steel microchannel with an internal diameter of 0.51 *mm* [33].

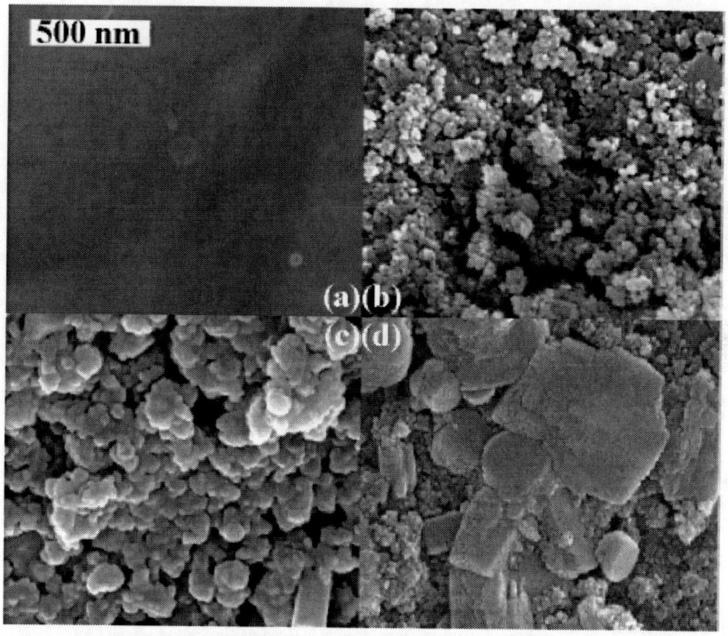


Figure 24. SEM pictures of the test section of a stainless steel microchannel with an internal diameter of 0.51 *mm*, at the vicinity of outlet, after flow boiling test with (a) deionized water, (b) 0.001 v % alumina nanofluid, (c) 0.01 v % alumina nanofluid, and (d) 0.1 v % alumina nanofluid [33].

The porous layer of deposited nanoparticles on heated substrate has two significant effects, a) improvement of surface roughness which affects the nucleation site density [10, 43, 61-64, 68-69] and b) modification of force balance at the triple line, behavior of triple line, surface wettability, wetting and dewetting, advancing and receding contact angles [33, 43, 67]. Variation of surface roughness indirectly affects behavior of triple line, surface wettability, receding and advancing contact angles. Understanding of the nexus between these effects is essential to create a better design which would enhance the boiling heat transfer coefficient, and critical heat flux of nanofluids.

Figure 25. SEM pictures of the test section of stainless steel microchannel with an internal diameter of 0.51 *mm* after flow boiling test with 0.1 v % alumina nanofluid at the (a) inlet and (b) outlet of the microchannel [33].

## EFFECTS OF NANOPARTICLE DEPOSITION

The porous layer of deposited nanoparticles changes the force balance at the triple line, wettability, roughness, active nucleation site density, receding and advancing contact angles. Active nucleation site density relies on the force balance at the triple line roughness, wettability, wall superheat, thermophysical properties of the heated surface and working fluids [70-71]. Naoparticles change most of these factors consequently active nucleation site density could be affected. Variation of roughness by deposited nanoparticles depends on concentration and characteristics of nanoparticle, surface roughness, and thickness of deposited nanoparticles (deposition period, and rate of deposition). Deposited nanoparticles by 20-50 *nm* alumina nanofluid pool boiling were observed to decrease the surface roughness when size of nanoparticles were one or two orders of magnitude smaller than of roughness (0.2-1.2 $\mu m$) of heated surface. Small nanoparticles sit on rough surface and reduce the surface roughness [68]. Obviously, nanoparticles deposit in agglomerated form, so the surface roughness of heated substrate decreases when size of agglomerated form of nanoparticles is smaller than of cavities. Likewise, in 20-25 *nm* $ZrO_2$ nanofluid (0.005 v%) pool boiling, the surface roughness was observed to decrease from $R_a = 0.642\,\mu m$ to $0.604\,\mu m$, the surface roughness was decreased as number of nanofluid boiling run was increased [10]. Similar experiments were conducted with 40-50 *nm* alumina nanofluid flow boiling inside the stainless steel macrochannels and it was realized that the surface roughness inside the macrochannel was $1.82\,\mu m$ as purchased, $2.28\,\mu m$ after critical heat flux test by 0.001 v% alumina nanofluid, $1.86\,\mu m$ after critical heat flux test by 0.01 v% alumina nanofluid, and $1.72\,\mu m$ after critical heat flux test by 0.1 v% alumina nanofluid. Surface roughness was slightly increased by deposited nanoparticles, as concentration of alumina nanofluid increased to 0.001 v%. However as the concentration of nanofluid was increased further, the surface roughness was decreased [43].

In contrary, the machine vision was employed to measure the nucleation site density of heated substrate after alumina nanofluid pool boiling test and it was observed that nucleation site density increased, when size of nanoparticles (47 *nm*) was much smaller than roughness of heated surface ($R_a$ =0.524 $\mu m$). In fact, alumina nanoparticles were sitting in the large cavities and multiplying them by splitting a single nucleation site into multiple ones but if size of nanoparticles (47 *nm*) was in order of magnitude of surface roughness ($R_a$ =48 *nm*), nucleation site density was decreased significantly [62]. In contrast, atomic force microscopy (AFM) was used to measure surface roughness of heated substrate after pool boiling with deionized water and multiple pool boiling running with 40 *nm* ZnO nanofluid (2.3 v%). Surface roughness of heated substrate was observed to increase from 60 *nm* after pool boiling test with deionized water to 440 *nm* following by seven times running of the nanofluid pool boiling and it was realized that the surface roughness increased with the number of nanofluid pool boiling runs increased [64]. The surface roughness of heated substrate was increased by nanofluid pool boiling, while alumina nanoparticles (47 *nm* average) size was bigger than roughness of heated substrate ($R_a$ =37.32 *nm*) [69].

Obviously, existing experimental evidence demonstrates that effects of deposited nanoparticles on a heated substrate are inconsistent and further investigation is required to understand the responsible causes, mechanisms, and effects of the deposited nanoparticles on variation of surface roughness. Effect of roughness of heated substrate, concentration and characteristics of nanoparticles on roughness of tiny porous layer of deposited nanoparticles need to be investigated carefully.

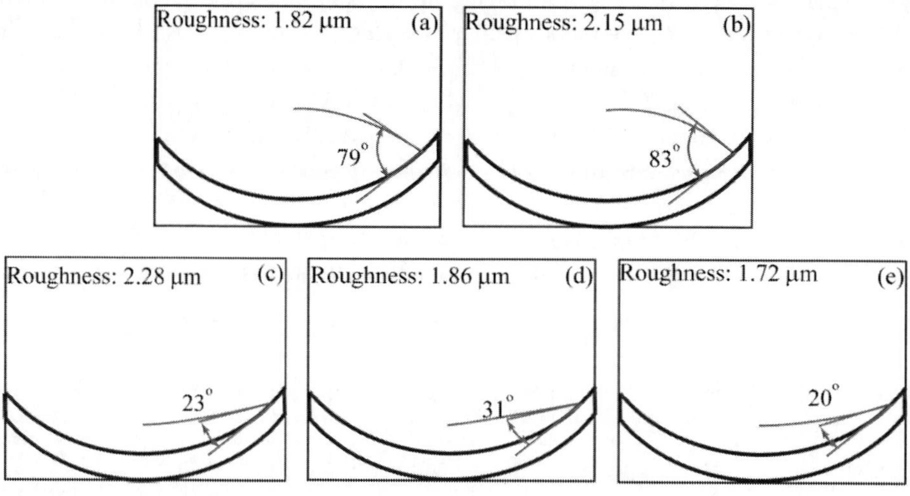


Figure 26. Schematics of variation of water droplet contact angle on heated surfaces a) as-purchased, b) after boiling in water, c) after boiling in 0.001 v% alumina nanofluid, d) after boiling in 0.01 v% alumina nanofluid, e) after boiling in 0.1 v% alumina nanofluid for mass flux of 2500 kg/m$^2$s. The average alumina nanoparticle size is 40-50 *nm*. Real pictures can be seen in reference [43].

Porous layer of deposited nanoparticles not only changes the surface roughness but also affects the surface wettability of the heated substrate. The experimental results have been demonstrated that porous layer of deposited nanoparticles affects the surface wettability more than the surface roughness of heated substrate [12, 43]. Figure 26 illustrates the variation of roughness and wettability of surface of stainless steel channel after running by water and alumina nanofluid flow boiling with different concentration. Droplet contact angle after flow boiling test with water was similar to as-purchased stainless steel tube and was around 79°-83°. However range of droplet contact angle on substrate after flow boiling test by different alumina nanofluid concentrations was around 20°-31°. Obviously deposited nanoparticles make a big difference to the surface wettability of the heated substrate compared to the variation of surface roughness [43]. Likewise, nanofluid pool boiling was conducted on a flat stainless steel substrate and similar variation in surface wettability was observed [11, 61, 72]. Practically, deposited nanoparticles on heated surface modify the solid surface tensions, force balance at the triple line and consequently droplet contact angle would be changed.

There are complicated nexus between active nucleation site density, wettability and roughness of the heated surface. As surface wettability increases, possibility of filling of cavities with liquid increases therefore active nucleation site density decreases. Active nucleation site density has a key role on nanofluid boiling heat transfer. Table 2 demonstrates a brief literature review of nanofluid boiling heat transfer coefficient and critical heat flux.

## NANOFLUID BOILING HEAT TRANSFER COEFFICIENT

A porous layer of deposited nanoparticles was observed to form on top of a heated surface in nanofluid boiling heat transfer. Obviously, deposited nanoparticles would change boiling heat transfer coefficient. This section illustrates the important factors in nanofluid boiling heat transfer coefficient and discusses the reasons for the deterioration or enhancement of the boiling heat transfer coefficient.

Researchers have tried to justify variation of nanofluid boiling heat transfer coefficient by finding a connection between the boiling heat transfer coefficient and variation in roughness of the heated surface caused by deposited nanoparticles. It has been suggested that small nanoparticles sit on relatively rough surfaces, and reduce the roughness of the heated substrate and nucleation site density by increasing the likelihood of filling of the cavities by deposition of nanoparticles. As results, pool boiling heat transfer coefficient was reduced and wall superheat increased [68, 73]. Similar pool boiling experiments were conducted on rough heated surface (0.5-0.7 $\mu m$), inside 20-25 *nm* $ZrO_2$ nanofluid and it was observed that even though the roughness of the heated surface decreased inside nanofluids (0.005 v%), but pool boiling heat transfer coefficient increased [10]. Other investigations were explained how smaller alumina nanoparticles were settled into the relatively large cavities and divided and multiplied the cavities, thus increased the nucleation site density and pool boiling heat transfer coefficient. It has been noticed when alumina nanoparticles were relatively in size of the surface roughness, the nanoparticles were sited inside the cavities and prevented nucleation and consequently reduced the pool boiling heat transfer coefficient. Boiling heat transfer coefficient deteriorated more, as the concentration of nanofluids increased [62]. In contrast, nanofluid pool boiling heat transfer coefficient was reduced, for heated surface with less roughness ($R_a = 37.22$ *nm*) compared to average size of nanoparticles (47 *nm*). Even though, the roughness of the heated surface increased with the concentration of alumina nanoparticles, but the pool boiling heat transfer coefficient decreased [69].

Range of heat flux might change the effects of deposited nanoparticles on pool boiling heat transfer coefficient. Effect of concentration of nanoparticles on pool boiling heat transfer coefficient is observed to be negligible in lower heat flux, since only large cavities are active at low heat fluxes. At higher heat fluxes, smaller cavities are also activated but heat transfer coefficient was noticed to be decreased as concentration of nanoparticles increased. The filling of the smaller cavities by deposited nanoparticles, and reduction of nucleation site density might be the possible reason for deterioration of pool boiling heat transfer coefficient [73].

Boiling heat transfer coefficient inside water and nanofluids was noticed to be affected by material of heated substrate. For instance, boiling heat transfer coefficient inside water and $TiO_2$ nanofluid on copper was higher than that on an aluminium heated substrate. On top of copper heated substrate, boiling heat transfer coefficient of $TiO_2$ nanofluid with concentrations below 0.0001 v% was observed to be higher than that of water. While $TiO_2$ nanofluid boiling heat transfer coefficient on aluminium heated substrate was always less than that of water [74]. Material of solid, vapor and working fluid plays a significant role on effective parameters on boiling heat transfer such as evolution of formation of bubble, bubble frequency, bubble detachment, behavior of interfacial surface and triple line, wetting and dewetting behavior in bubble formation and detachment. Obviously, behavior of triple line has a vital impact on dynamics of bubble growth and consequently boiling heat transfer coefficient and critical heat flux. Behavior of triple line can be expressed by the contact angle and radius of spreading of triple line. It is crucial to measure these two parameters at boiling conditions, while majority of studies have been done in equilibrium and ambient conditions [75].

Poor thermal conductivity of deposited nanoparticles might increase the thermal resistance and possibly deteriorate the nanofluid boiling heat transfer coefficient. Experimental evidence [65] was shown that $TiO_2$ nanofluid pool boiling heat transfer increased with thickness of porous layer of deposited nanoparticles. No significant change of nanofluid boiling heat transfer coefficient occurred throughout the deposition of nanoparticles. This indicates that the effect of the thermal resistance of porous layer of deposited nanoparticles is negligible compared to the boiling convection, even though the thermal conductivity of $TiO_2$ porous layer is poor. Similar experiments were conducted [73] on high thermal conductive nanoparticles such as 10 *nm* copper and it was realized that deposited copper nanoparticles deteriorated the nanofluid boiling heat transfer coefficient. Thermal resistance of deposited nanoparticles might not be the possible reason for deterioration of copper nanofluid boiling heat transfer coefficient. Effects of the thickness of deposited nanoparticles on the boiling heat transfer coefficient were investigated by increasing of thickness of the deposited nanoparticles, layer by layer. Forty-five minutes of nanofluid pool boiling for each run was employed to deposit a layer of nanoparticles on top of the heated surface and it was realized that heat transfer coefficient inside water increased as the number of layers of deposited nanoparticles on heated substrate increased but heat transfer coefficient inside nanofluid increased for the first layer of deposited nanoparticles and then the nanofluid heat transfer coefficient decreased slightly and remained nearly constant, and after that declined for each additional cycle of nanofluid boiling [64]. Pool boiling heat transfer coefficient inside water increased as thickness of deposited nanoparticles increased, thus effect of thermal resistance of deposited nanoparticles on top of the heated surface might be negligible. Different heat transfer coefficient inside water and nanofluids with additional layers of the deposited nanoparticles on top of the heated surface demonstrates that the deterioration of the heat transfer coefficient inside the nanofluids might be related to the combination of variation in roughness and wettability of the heated surface.

Roughness of the heated surface is a key concern to increase the number of cavities in order to initiate bubbles or enhance the nucleation site density. However, not all of the cavities are able to initiate bubbles. Flooded cavities by working fluid are not active to initiate bubbles. Enhancement of roughness of heated surface is not enough to improve boiling heat transfer coefficient. In fact, cavities that are not completely flooded by working fluid able to initiate bubbles and promote bubble nucleation and consequently enhance the boiling heat transfer coefficient. Reduction of the surface wettability prevents the flooding of cavities by working fluid and activates more nucleation sites. To understand the effects of surface wettability on boiling heat transfer coefficient, three coated wires were employed to provide hydrophilic, superhydrophilic, and hydrophobic heated surfaces. The heated substrates were coated with 24-50 *nm* $SiO_2$ nanoparticles with different surface treatments to create variety of surface wettabilities. No change in surface roughness was observed after coating of heated surfaces which indicates that the given nanoparticles coat the microscale deformities conformably. The hydrophobic heated substrate was observed to have a higher active nucleation site density, thus higher water boiling heat transfer coefficient. While the superhydrophilic heated substrate degraded the water boiling heat transfer coefficient and hydrophilic heated substrate made no change compared to the bare surface [12].

The relation between nucleation site density and surface wettability is formulated [76] as follows

$$N_a(sites/cm^2) \propto 5\times10^5(1-\cos\theta)D_c^{-6} \tag{1}$$

$N_a, \theta, D_c$ are respectively active nucleation site density, cavity mouth diameter and contact angle. The expression describes the nexus between activation of nucleation sites and wettability of surface by working fluids. Experimental results confirm that even though the surface roughness increased by deposited nanoparticles, boiling heat transfer coefficient of nanofluid (mixture of 110-210 *nm* alumina nanoparticles with water 0.01 v%) deteriorated [11, 32, 72]. The surface wettability enhancement by deposited nanoparticles [11, 32, 72] might be the possible reason for deterioration of boiling heat transfer coefficient. Enhancement [77-81], deterioration [65] and lack of response [82-83] of nanoparticles on boiling heat transfer coefficient was reported by several others.

Interactions of nanoparticles inside triple region with gas, liquid and solid have significant effects on nanofluid boiling heat transfer coefficient and yet it is not completely understood. Most of existing investigations of effects of nanoparticles on surface wettability and behavior of triple line have been done in equilibrium and ambient conditions which is different from boiling conditions. Thus, extra studies are required to understand the interactions of nanoparticles with solid, gas and liquid at the triple line on boiling conditions. The nanofluid boiling heat transfer suffered a notable lack of consistency of the effects of nanoparticles on boiling heat transfer coefficient. There are enormous discrepancies among the existing experimental results and explanations of the mechanisms of effects of nanoparticles. Therefore further investigation is essential to understand the mechanisms and improve the effects of nanoparticles on boiling heat transfer coefficient. Enhancement of nanofluid boiling heat transfer coefficient is a significant milestone toward industrial applications of nanofluid boiling heat transfer.

## Nanofluid Critical Heat Flux

Variation of wettability and roughness of heated substrate by the porous layer of deposited nanoparticles are discussed in previous sections. Existing evidence was demonstrated that variation of surface roughness was not much compared to the modifications of surface wettability by deposited nanoparticles on heated surface (see Figure 26) [43]. Initially, the enhancement of wettability by deposited nanoparticles [43, 60, 72] was believed to be a possible reason for enhancement of critical heat flux inside nanofluids. Effects of surface wettability on critical heat flux was considered by the macrolayer dryout model [84-85], and the hot/dry spot theory [86] as well. However, further results indicate that all of the hydrophobic, hydrophilic, and superhydrophilic surfaces enhance the pool boiling critical heat flux. The hydrophobic, hydrophilic, and superhydrophilic heated surfaces were made by thin film coating of nanoparticles built up with layer by layer [12]. Since the pool boiling critical heat flux is enhanced by hydrophobic and hydrophilic surfaces, the surface wettability might not be the only possible reason for critical heat flux enhancement. Critical heat flux was also modelled theoretically by considering the receding contact angle [23, 87]. Experimental results were verified the importance of receding contact angle on critical heat flux. Heated surfaces with low receding contact angles were observed to enhance the critical heat flux [12]. Practically, the best candidate has to enhance the boiling heat transfer coefficient and critical heat flux simultaneously. The hydrophobic heated surface (20 Bilayers PAH/ $SiO_2$, calcinated, fluorosilane) enhanced the boiling heat transfer coefficient and critical heat flux simultaneously and had a low surface wettability, high advancing and low receding contact angles. The equilibrium, advancing and receding contact angles of the hydrophobic heated surface respectively were 140°, 160°, and 20° [12].

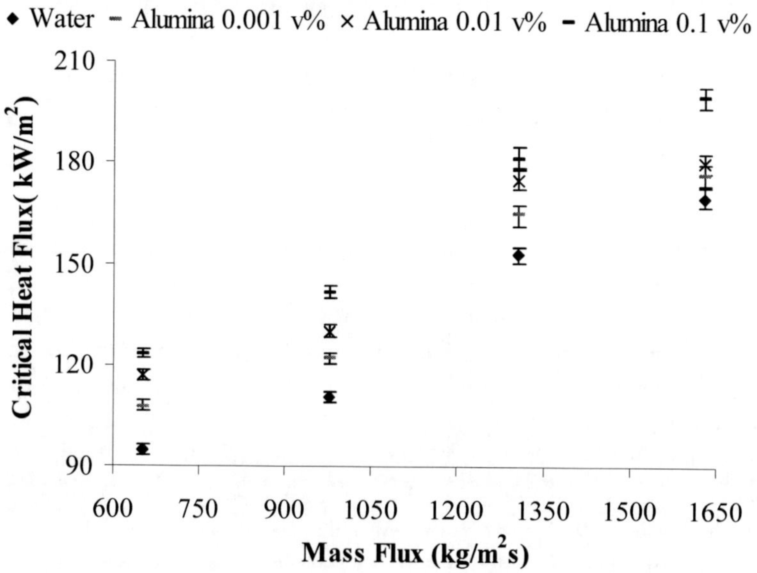


Figure 28. Variation of the critical heat flux with mass flux for deionized water and alumina nanofluid with a concentration of 0.001-0.1 v%. The initial subcooling is ~45 $^oC$ [31].

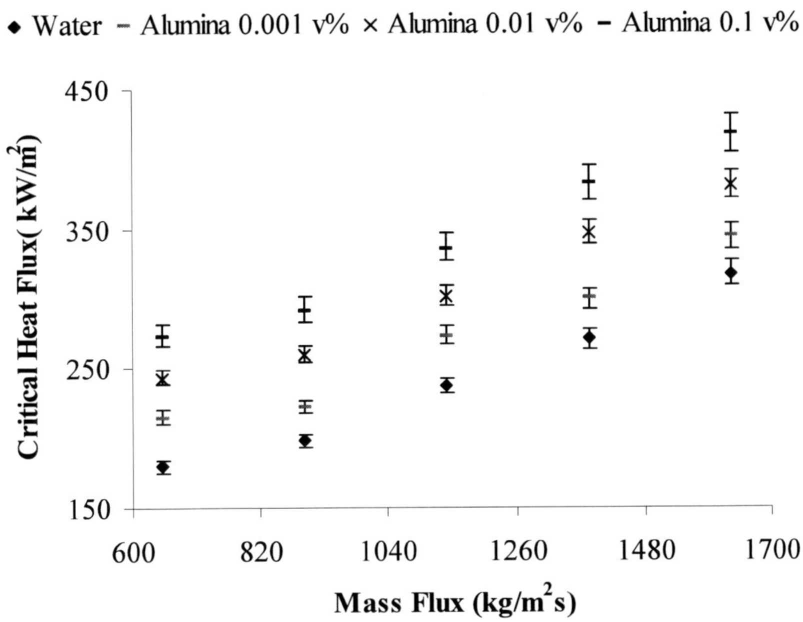


Figure 27. Variation of the critical heat flux with mass flux for deionized water and alumina nanofluid with concentration of 0.001-0.1 v%. The initial subcooling is ~93 $^{o}C$ [33].

The receding contact angle appears when triple line moves towards the liquid phase and the advancing contact angle forms when the contact line moves toward the gas phase. The difference between the receding and advancing contact angle is called hysteresis contact angle. Surface roughness, non homogeneity of working fluids and surface are effective parameters on hysteresis, receding and advancing contact angles. Receding and advancing contact angles in pool boiling was observed when bubble respectively expanded laterally and vertically [88]. Further investigations are required to find out relation between surface roughness, wettability, receding and advancing contact angles and eventually the responsible reasons for the enhancement of critical heat flux and boiling heat transfer coefficient. It is essential to find the particular conditions to enhance the boiling heat transfer coefficient and critical heat flux of nanofluids simultaneously.

Very limited studies have been conducted to investigate the flow boiling critical heat flux of nanofluids. In macro scale size, two stainless steel tubes with internal diameters of 5.54 *mm* [43] and 8.7 *mm* [89] were used to study the nanofluid critical heat flux. Alumina and zinc oxide nanofluid with concentrations of 0.001-0.1 v% were observed to enhance the flow boiling critical heat flux up to 53%. Similarly, critical heat flux enhancement of diamond nanofluid was 38%. The effective parameters of the flow boiling heat transfer of nanofluids (alumina, zinc oxide, and diamond), such as boiling point, surface tension, thermal conductivity and viscosity were found to be virtually identical to those of pure water [32] for given concentrations.

In micro scale size, fresh single stainless steel microchannels with an internal diameter of 510 $\mu m$ were employed to investigate the critical heat flux of subcooled flow boiling of nanofluid, under low mass flow rate conditions (600-1650 kg/m$^2$s). Figure 27 shows the variation of critical heat flux with mass flow rate and concentration of alumina nanoparticles.

A remarkable critical heat flux enhancement was achieved for very dilute alumina nanofluid inside microchannels. For instance, the average critical heat flux enhancements of 10.73%, 28.25% and 40.94% were obtained respectively for concentrations of 0.001 v%, 0.01 v% and 0.1 v% at mass flow rate of 1385.18 kg/m$^2$s. Critical heat flux enhancement within the whole experimental range was observed to be about 8-51% for initial subcooling of ~93 $^oC$ and 4-31% for initial subcooling of ~45 $^oC$. For a same mass flux, critical heat flux increased with concentration of alumina nanofluid. The maximum critical heat flux enhancement accrued at the lowest mass flux 651.84 kg/m$^2$s and maximum concentration of alumina nanofluid, 0.1 v%. Figure 28 shows the variation of critical heat flux of deionized water and alumina nanofluid (0.001-0.1 v%) with mass flux for initial subcooling of ~45 $^oC$. Repeatability of enhancement of flow boiling critical heat flux of alumina nanofluid inside the microchannels was observed to be weak while experimental results for water were very repeatable. Repeatability of critical heat flux enhancement was deteriorated as concentration of nanofluid increased. Further analysis by scanning electron microscopy (SEM) was showed that morphology of heated surface was changed each time after nanofluid flow boiling by deposited nanoparticles.

The degree of surface modification and thickness of deposited nanoparticles depends on concentration and characteristics of nanoparticles, roughness of the heated surface, stability of nanofluid, interactions of nanoparticles with heated surface and boiling period. The effects of deposited nanoparticles on flow boiling heat transfer and critical heat flux increased, as the channel size decreased. Microchannels are more sensitive to the deposited nanoparticles than macrochannels [33].

Figures 29 and 30 show the variation of the inlet/outlet temperatures and pressures, temperature of the hottest point of the channel, and heat power with time for concentration of 0.01 v%, where the mass flux and average heat flux respectively were 1629.62 kg/m$^2$s and 247 kW/m$^2$. Pressure and temperature fluctuations were in phase, but there was a phase shift between power and pressure [33].

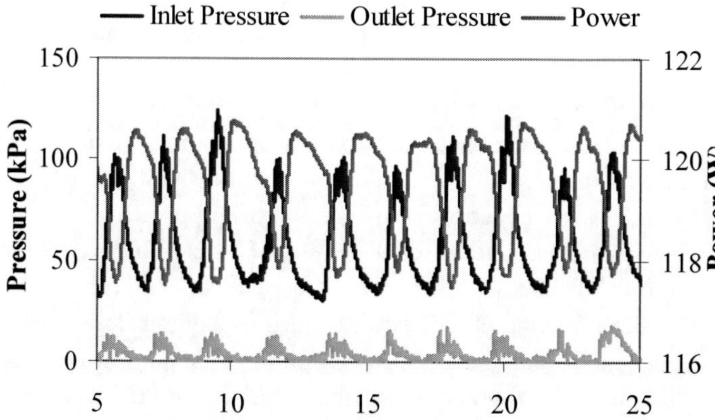


Figure 29. Variation of pressure and heat power with time for mass flux of 1629.62 kg/m$^2$s. The alumina nanofluid concentration and initial subcooling respectively are 0.01 v% and ~93 $^oC$ [33].

**Table 2. The literature review of existing nanofluid heat transfer coefficient and critical heat flux**

| Reference | Nanofluids | Remarks |
|---|---|---|
| Vafaei et al. [31] (2011)<br>Vafaei et al. [33] (2010) | Mixture of 25 *nm* $Al_2O_3$ nanoparticles and water with concentration of 0.001-0.1 v%<br><br>Inside 0.5 *mm* stainless steel tubes. | • The repeatability of the flow boiling heat transfer coefficient and critical heat flux with alumina nanofluid was difficult to establish for the stainless steel microchannels. As nanoparticle concentration increased, the repeatability deteriorated.<br>• The critical heat flux with alumina nanofluid increased as mass flux increased.<br>• For a given mass flux, the critical heat flux increased as concentration of alumina nanofluid increased.<br>• The critical heat flux inside the microchannel increased as initial subcooling increased.<br>• The maximum critical heat flux enhancement happened at the lowest mass flux. The critical heat flux enhancement within whole experimental range was about 8-51% for initial subcooling of $\sim 93\,^{o}C$, and 4-31% for initial subcooling of $\sim 45\,^{o}C$.<br>• As concentration of nanofluids increased, more nanoparticles appeared on heated surface. The porous layer of deposited nanoparticles being thicker and more condensed for higher nanofluids concentrations.<br>• Since temperature of heating surface was higher at the outlet, the deposition rate was higher and the deposited nanoparticles at the outlet were thicker with larger agglomerations.<br>• The mode of breakdown due to extensive burnout for water was a complete azimuthal failure, while a pinhole type of failure happened for alumina nanofluid. The deposited nanoparticles on a heated substrate mitigated the propagation of hot spot and changed the mode of breakdown from complete failure by water to pinhole failure by nanofluids.<br>• Behavior of triple line inside the alumina nanofluid and water was investigated by injecting air into the nanofluids to form bubbles at ambient temperature. No deposited nanoparticles were observed on the substrate. Presence of nanoparticles at the triple line were modified the solid surface tensions and force balance at the triple line thus behavior of triple line such as radius of triple line and bubble contact angle inside |

**Table 2. (Continued)**

| Reference | Nanofluids | Remarks |
|---|---|---|
| | | alumina nanofluid was changed.<br>• Radius of triple line inside the alumina nanofluid was higher than that inside water.<br>• The bubble frequency inside water and alumina nanofluid were virtually the same, however the bubble formation period inside alumina nanofluid was longer than that inside water. (The waiting period inside the alumina nanofluid was shorter).<br>• Presence of nanoparticles were decreased the bubble contact angle inside the nanofluids compared to that inside water. |
| Forrest et al. [12] (2010) | Water on coated heated substrates by 24 *nm* and 50 *nm* $SiO_2$ nanoparticles. | • Using layer-by-layer assembly method, the stainless steel flat plate and nickel wire were coated by $SiO_2$ nanoparticles to change the surface wettability and create hydrophilic, superhydrophilic and hydrophobic heated surfaces.<br>• When 24-50 *nm* $SiO_2$ nanoparticles were deposited on a substrate with roughness of $R_a = 0.290\,\mu m$ through the layer-by-layer assembly method, no change of surface roughness was observed, which indicates that the given nanoparticles coat the microscale deformities conformably.<br>• More bubble nucleation sites were observed with hydrophobic surface compared to superhydrophilic and hydrophobic heated wires.<br>• Boiling heat transfer coefficient was enhanced on the hydrophobic heated wire compared to hydrophilic, and superhydrophilic heated surfaces. The hydrophobic surface showed 127% heat transfer coefficient enhancement over untreated surfaces.<br>• The hydrophobic surface exhibited high static contact angle, 141°, high advancing contact angle, 160°, and low receding contact angle, 20°, so contact angle hysteresis was large.<br>• Critical heat flux enhancement was observed for all coated surfaces regardless of final treatment. But the superhydrophilic wire displayed the greatest critical heat flux enhancement.<br>• As coating thickness increased the critical heat flux was enhanced. |

| Reference | Nanofluids | Remarks |
| --- | --- | --- |
| | | • Superhydrophilic heated surface was displayed the greatest critical heat flux enhancement, but deteriorated the boiling heat transfer coefficient. Thus hydrophobic surface could be a better candidate since it was enhanced the critical heat flux and boiling heat transfer coefficient simultaneously. |
| White et al. [64] (2010) | Mixture of 40 *nm* ZnO nanoparticles and water with concentration of 2.3 v% | • Nanofluid pool boiling heat transfer coefficient was higher than that of water at the first running on top of the fresh and clean heated surface, furthermore, the nanofluid heat transfer coefficient increased with heat flux inside the nanofluids and water. The reason for the nanofluid heat transfer coefficient enhancement was related to enhancement of thermal properties of nanofluid rather than the surface roughness improvement from the porous layer of deposited nanoparticles.<br>• As results of nanofluid boiling, a layer of deposited nanoparticles was observed to build up on heated substrate.<br>• Forty-five minutes of nanofluid pool boiling for each run was employed to deposit nanoparticles on heated surface. It was realized that heat transfer coefficient inside the water increased as the number of the deposited nanoparticles layers on heated substrate increased Variation of roughness by porous layer of deposited nanoparticles on heated substrate was responsible for water heat transfer coefficient enhancement.<br>• Similarly the boiling heat transfer coefficient inside the nanofluids was measured on heated surface after depositing a layer of nanoparticles and it was observed that nanofluid heat transfer coefficient increased for the first layer of deposited nanoparticles and then the nanofluid heat transfer coefficient slightly decreased and remained nearly constant, subsequently declined for each additional cycle of nanofluid boiling. It was suggested that reason for the declination of nanofluid heat transfer coefficient was related to the suppression of bubble nucleation and transport by suspended nanoparticles. |

**Table 2. (Continued)**

| Reference | Nanofluids | Remarks |
|---|---|---|
| Phan et al. [65] (2010) | Mixture of 80-100 *nm* $Al_2O_3$ nanoparticles and water with concentration of 0.01 v%<br><br>Mixture of 15-20 *nm* Au nanoparticles and water with concentration of 0.0003 v%<br><br>Mixture of 20-30 *nm* $TiO_2$ nanoparticles and water with concentration of 0.01 v% | • Deposited nanoparticles in nanofluid pool boiling formed a porous layer of nanoparticles. The thickness of the porous layer increased with duration of boiling and concentration of nanoparticles<br>• The deposited $TiO_2$ nanoparticles significantly reduced the water pool boiling heat transfer coefficient.<br>• The heat transfer coefficient inside water was higher than that of $TiO_2$ nanofluid and the pool boiling heat transfer coefficient inside the $TiO_2$ nanofluid increased with time. Since there was no significant change in $TiO_2$ nanofluid heat transfer coefficient during the nanoparticle deposition, the effect of thermal resistance of $TiO_2$ porous layer was assumed to be negligible thus, thermal resistance of the porous layer of deposited nanoparticles cannot be the possible reason for deterioration of nanofluid heat transfer coefficient, even though the thermal conductivity of $TiO_2$ porous layer is low.<br>• The porous layer of deposited $TiO_2$ nanoparticles enhanced the surface wettability of water, and reduced the active nucleation site density thus deteriorated the pool boiling heat transfer coefficient inside the water. |
| Kwark et al. [63] (2010) | Water based mixture of 139 *nm* ± 100 *nm* $Al_2O_3$, 143 *nm* ± 80 *nm* CuO, 86 *nm* ± 50 *nm* diamond nanoparticles with concentration of 0.001- 1 gr/l | • The effect of given nanoparticles was observed to be negligible on thermal conductivity, surface tension, and viscosity of nanofluids at low concentrations.<br>• The porous layer of deposited nanoparticles was formed due to evaporation of the liquid sublayer underneath the vapor bubbles.<br>• The deposited nanoparticles constantly modified the heated surface in nanofluid pool boiling. The thickness of deposited nanoparticles depends on the boiling period and heat flux. |
| Suriyawong et al. [74] (2010) | Mixture of 21 *nm* $TiO_2$ nanoparticles and water with concentration of | • For a given surface roughness, boiling heat transfer coefficient on top of the aluminium heated substrate was higher than that of copper. More differences appeared at the higher heat fluxes. |

| Reference | Nanofluids | Remarks |
|---|---|---|
| | 0.00005-0.01 v% | • As results of higher nucleation site density, the higher surface roughness gave a higher boiling heat transfer coefficient for both copper and aluminium heated surfaces.<br>• On top of the copper heated substrate with 0.2 $\mu m$ roughness, the nanofluid heat transfer coefficient increased respectively 7 % and 15 % for 0.00005 v% and 0.0001 v% concentrations. The nanofluid heat transfer coefficient decreased for concentrations above 0.0001 v%. For surface roughness of 0.4 $\mu m$ similar trends were observed with the variation of nanofluid heat transfer coefficient but the nanofluid heat transfer coefficient increased respectively 2 % and 4 % for 0.00005 v% and 0.0001 v% concentrations.<br>• Nanofluid heat transfer coefficient decreased for concentration above 0.0001 v%.<br>• The sedimentation of nanoparticles below a 0.0001 v% concentration was prevented by upward flow boiling thus, the nanofluid heat transfer coefficient was enhanced. |
| Park et al. [90] (2009) | Mixture of polyvinyl pyrrolidone polymer (PVP), Carbon nanotubes (CNT) (D=10-20 *nm* and L=10-50 $\mu m$) and water with concentration of 0.0001-0.05 v% | • Pool boiling heat transfer coefficient inside the suspended CNT was lower than that of pure water. As CNT concentration increased, the heat transfer coefficient of aqueous solutions with CNT decreased.<br>• The CNTs were deposited on the heated substrate during bubble formation and formed a thin layer of deposited carbon nanotubes on heated substrate which reduced the active nucleation sites, and thus deteriorated the pool boiling heat transfer coefficient.<br>• Polyvinyl pyrrolidone polymer (PVP) dispersant alone did not play any significant role in heat transfer coefficient in entire heat flux range. |
| Kathiravan et al. [73] (2009) | Mixture of 10 *nm* Cu nanoparticles and water with concentration of 0.25-1 w% | • Effect of concentration of nanoparticles on pool boiling heat transfer coefficient was negligible in lower heat fluxes, since only large cavities were active in low heat flux.<br>• At higher heat fluxes, heat transfer coefficient was observed to decrease as concentration of nanofluids increased. At higher heat fluxes the smaller cavities were |

**Table 2. (Continued)**

| Reference | Nanofluids | Remarks |
|---|---|---|
| | Mixture of 10 *nm* Cu nanoparticles and water plus 9 w% sodium dodecyl sulphate (SDS) | also activated. These cavities might be filled by deposited nanoparticles thus reducing the pool boiling heat transfer coefficient.<br>• Roughness of the heated surface decreased after nanofluid pool boiling. The average roughness of heated surface reduced from 1.09 $\mu m$ to 571.66 *nm* after nanofluid pool boiling.<br>• The nanofluid heat transfer coefficient with surfactant (SDS) was relatively higher than that of nanofluids without surfactant (SDS). The surfactant reduced the surface tension, thus increased the heat transfer coefficient.<br>• The critical heat flux inside the nanofluids was more than that of water. As nanofluid concentration increased, the critical heat flux increased.<br>• Adding surfactant to the nanofluids decreased the critical heat flux. |
| Kim et al. [91] (2008)<br>Kim et al. [89] (2008)<br>Kim et al. [43] (2009) | Mixture of 40-50 *nm* $Al_2O_3$, 50-90 *nm* ZnO, and 4 *nm* diamond nanoparticles and water with concentration of 0.001-0.1 v%.<br><br>Inside 5.54 and 8.7 *mm* stainless steel tubes. | • Nanofluid critical heat flux was higher than that of water. The critical heat flux increased as concentration of nanofluids and mass flux increased.<br>• The boiling point, surface tension, thermal conductivity, and viscosity of given nanofluids were found to be virtually identical to those of pure water.<br>• The mode of breakdown due to extensive burnout by water was azimuthally a complete failure but by alumina nanofluid was localized the burnout with pinhole failure. The deposited nanoparticles on heated surface enhanced the surface wettability, and thus mitigated the propagation of the hot spot and caused only pinhole failure.<br>• The roughness ($R_a$) of the heated surface was 1.82 $\mu m$ as-purchased, 2.15 $\mu m$ after critical heat flux test by water, 2.28 $\mu m$ after critical heat flux test by 0.001 v% alumina nanofluid, 1.86 $\mu m$ after critical heat flux test by 0.01 v% alumina nanofluid, and 1.72 $\mu m$ after critical heat flux test by 0.1 v% alumina nanofluid. The average of surface roughness for all cases was 2 $\mu m$.<br>• The deposited nanoparticles enhanced surface wettability of heated substrate. The |

| Reference | Nanofluids | Remarks |
|---|---|---|
| | | droplet contact angle of heated surface was 79° as-purchased, 83° after critical heat flux test by water, 23° after critical heat flux test by 0.001 v% alumina nanofluid, 31° after critical heat flux test by 0.01 v% alumina nanofluid, and 20° after critical heat flux test by 0.1 v% alumina nanofluid.<br>• The variation of surface roughness compared to surface wettability enhancement by deposited nanoparticles was small.<br>• The increase of surface wettability can play an important role in critical heat flux enhancement.<br>• The maximum critical heat flux enhancement was 53% for alumina and zinc oxide nanofluid and 38% for diamond nanofluid. |
| Chopkar et al. [10] (2008) | Mixture of 20-25 *nm* $ZrO_2$ nanoparticles and water with concentration of 0.005-0.15 w%<br><br>Mixture of $ZrO_2$ and water plus 1 v% of tetramethyl ammonium hydroxide as a surfactant | • The pool boiling heat transfer coefficient inside nanofluids without surfactant was higher compared to pure water at the lowest concentration of 0.005 v%. As nanofluid concentration increased, the heat transfer coefficient decreased and eventually the heat transfer coefficient of nanofluids at 0.15 v% reached below pure water.<br>• The deposited nanoparticles in nanofluid pool boiling decreased the roughness of the heated surface. The surface roughness and heat transfer coefficient decreased, with repetition of pool boiling on the heated substrate. A similar trend was observed, adding surfactant (1 v%). |
| Narayan et al. [62] (2007) | Mixture of $Al_2O_3$ nanoparticles (47 *nm*, 150 *nm*) and water with concentration of 0.5-2 w% | • Nanofluid pool boiling heat transfer coefficient was examined on heated surfaces with three kind of surface roughness of 528, 98, and 45 *nm*.<br>• For the roughest heater ($R_a = 524$ *nm*), the higher pool boiling heat transfer coefficient was consistently observed compared to the other two heaters. A significant heat transfer coefficient enhancement was observed, using nanofluid compared to water. |

**Table 2. (Continued)**

| Reference | Nanofluids | Remarks |
|---|---|---|
| | | • A slight nanofluid heat transfer coefficient enhancement was observed on the heater with surface roughness of $R_a = 98$ *nm* for concentration of 0.5 w%. As nanofluid concentration increased the nanofluid heat transfer coefficient deteriorated.<br>• For the smoothest heater ($R_a = 48$ *nm*), a lower heat transfer coefficient was consistently observed compared to the other two heaters. The heat transfer coefficient was reduced using nanofluids. As nanofluid concentration increased, the nanofluid heat transfer coefficient decreased.<br>• The number of nucleate sites greatly decreased, when the average particle size was in order of roughness of heated surface. The nanoparticles sat on nucleation sites and hindered the nucleation, thus deteriorated the boiling heat transfer coefficient.<br>• The number of nucleate sites greatly increased, when average particle size was much smaller than that of roughness of the heater. The smaller nanoparticles sat in nucleation sites, and divided a single nucleation site into multiple ones, thus the nucleation site density and boiling heat transfer coefficient increased. |
| Kim et al. [92] (2007) | Water based mixture of $Al_2O_3$, $TiO_2$, $SiO_2$ nanoparticles with concentration of 0.00001-0.1 v% | • Deposited nanoparticles on the heated surface by boiling heat transfer enhanced the surface wettability and caused the capillary wicking on the porous surface.<br>• The deposited nanoparticles enhanced the capillarity of the surface. Sufficient increase in capillarity moved the contact line in a lateral direction thus a capillary liquid flow occurred. The lateral liquid flow by capillary wicking supplied fresh liquid to the dry region beneath the vapor bubbles, delayed the irreversible growth of hot spots and critical heat flux. |
| Park et al. [93] (2007) | Mixture of Carbon nanotubes (CNT) (D=10-20 *nm* and L=10-50 $\mu m$) and water with | • Carbon nanotubes enhanced the boiling heat transfer of base fluid such as water and R22 at low heat flux. As heat flux increased the enhancement of boiling heat transfer decreased.<br>• No fouling from CNTs was observed on top of the heated surface. |

| Reference | Nanofluids | Remarks |
|---|---|---|
| | concentration of 1 v%<br><br>Mixture of Carbon nanotubes (CNT) (D=10-20 *nm* and L=10-50 $\mu m$) and R22 with concentration of 1 v% | |
| Kim et al. [72] (2007)<br>Kim et al. [61] (2007)<br>Kim et al. [67] (2006) | Water based mixture of 110-250 *nm* $Al_2O_3$, 110-250 *nm* $ZrO_2$, 20-40 *nm* $SiO_2$ nanoparticles with concentration of 0.001-0.1 v% | • The variation of viscosity, thermal conductivity, and gas-liquid surface tension with nanofluids concentration in a given range was negligible.<br>• The deposited nanoparticles on heated surface were observed soon after nanofluid pool boiling began. The evaporation of the developed thin liquid microlayer beneath the vapor bubble increased the nanofluid concentration in the microlayer with subsequent settlement of the nanoparticles, thus the nanoparticles kept depositing.<br>• The deposited alumina naoparticles increased roughness of heated surface from 0.1 $\mu m$ to 2 $\mu m$ respectively after boiling inside water and 0.1 v% alumina nanofluid. Similar results obtained with $SiO_2$ and $ZrO_2$ nanoparticles.<br>• The 5 μL water droplet contact angle decreased from 71° to 20° respectively after boiling inside water and 0.01 v% alumina nanofluid. Similar results were obtained with $SiO_2$ and $ZrO_2$ nanoparticles.<br>• The deposited nanoparticles increased the roughness of the heated surface, but since the surface wettability was enhanced simultaneously, the active nucleation site density decreased, thus the boiling heat transfer coefficient inside the nanofluid decreased.<br>• The surface wettability enhancement by deposited nanoparticles on the heated surface could lead to the critical heat flux enhancement. The critical heat flux increased up to 52%, 75%, and 80% respectively inside $Al_2O_3$, $ZrO_2$, and $SiO_2$ nanofluid.<br>• As nanofluid concentration increased, pool boiling critical heat flux increased inside |

**Table 2. (Continued)**

| Reference | Nanofluids | Remarks |
|---|---|---|
| | | • $SiO_2$ nanofluid, decreased inside $Al_2O_3$ nanofluid, and increased to a maximum and then decreased again inside the $ZrO_2$ nanofluid.<br>• A review of the prevalent critical heat flux theories demonstrated that higher wettability could enhance the critical heat flux. |
| Bang et al. [69] (2005) | Mixture of 10-100 *nm* (47 *nm* in average) $Al_2O_3$ nanoparticles and water with concentration of 0.5-4 v% | • The pool boiling heat transfer coefficient of nanofluids was lower than that of water, and heat transfer coefficient decreased by increasing nanoparticles concentration.<br>• The surface roughness of the heated surface ($R_a = 37.22$ *nm*) was smaller than of the average size of nanoparticles (47 *nm*). The surface roughness increased on average with increasing nanoparticle concentration.<br>• The boiling heat transfer coefficient decreased as nanofluid concentration increased.<br>• The nanofluid critical heat flux enhanced 32% and 13% respectively on horizontal and vertical flat surfaces due to delayed boiling activity. |
| Kim et al. [94] (2003) | Mixture of $Al_2O_3$ nanoparticles and water with concentration up to 0.05 g/l. | • The critical heat flux inside alumina nanofluid was enhanced.<br>• Below 0.01 g/l concentration, as the concentration of nanofluids increased the critical heat flux increased. Higher than 0.01 g/l concentration, the critical heat flux was not observed to be enhanced.<br>• The departure bubble volume and total bubble formation period inside the alumina nanofluid increased significantly compared to those in pure water. |
| Das et al. [68] (2003) | Mixture of 38 *nm* $Al_2O_3$ nanoparticles and water with concentration of 1-4 v% | • The given nanofluid enhanced the wall superheat and deteriorated the boiling heat transfer coefficient.<br>• For a given heat flux, as nanoparticle concentration increased, the temperature of the heated substrate increased and the heat transfer coefficient decreased.<br>• Since alumina nanoparticles were one or two orders of magnitude smaller than that of roughness of heated surface (0.2 - 1.2 $\mu m$), nanoparticles sat on the relatively rough substrate and reduced the roughness of the heated surface, thus deteriorated the heat transfer coefficient. |

| Reference | Nanofluids | Remarks |
|---|---|---|
| | | • The viscosity of nanofluid increased with nanoparticle concentration but remained Newtonian in nature.<br>• The nanofluid thermal conductivity was enhanced by temperature and nanoparticle concentration.<br>• The variation of alumina nanofluid surface tension with nanoparticle concentration was negligible. |

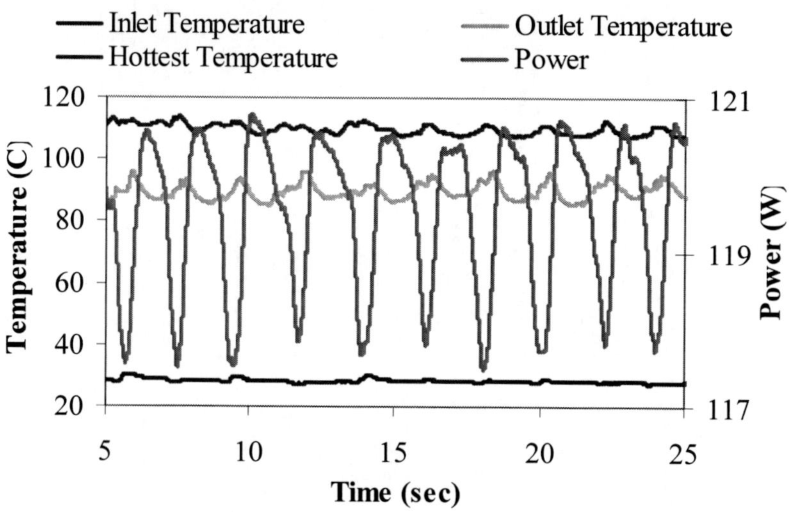


Figure 30. Variation of temperature and heat power with time for mass flux of 1629.62 kg/m$^2$s. The alumina nanofluid concentration and initial subcooling respectively are 0.01 v % and ~93 $^oC$ [33].

## CONCLUSION

Nanoparticles have a great potential to improve the physical properties of nanofluids. However, there are several challenges that need to be addressed such as agglomeration, segregation and deposition of nanoparticles. Deposition of the nanoparticles is one of the major disadvantages of nanofluid flow boiling heat transfer, especially in microchannels. The deposition of nanoparticles will be more problematic for micro scale heating systems. These issues would be resolved as technology of coating and manufacturing of nanoparticles will develop further. The improvement of the technology of manufacturing of nanoparticles will provide more effective and efficient nanofluids. For instance, better coating and smaller nanoparticles respectively hinder the agglomeration and deposition of nanoparticles. Having a variety of nanoparticles with different material, coating, shape, and size provides more possibility for better nanofluids.

Nanoparticles interact with surrounding nanoparticles and molecules of liquid, gas, and solid surface at the triple region. Understanding the interactions of nanoparticles with surrounding nanoparticles and molecules is necessary to be able to predict the physical properties of nanofluids and behavior of triple line such as pinning of triple line, wettability, advancing and receding contact angles. In fact, engineering of nanofluids for specific applications depends on understanding of the behavior of nanoparticles inside the base liquid. The right combination of base liquid and nanoparticles, further development of the technology of coating and manufacturing of nanoparticles, improvement of characteristics of nanoparticles, understanding of the mechanisms of interactions of molecules-nanoparticles and naoparticles-nanoparticles are the key concerns which need to be better understood in future. The effect of nanoparticles on behavior of triple line can be studied by droplet and bubble formation methods.

In case of droplet method, gravity is in favor of expanding of triple line and decreasing of contact angle. As results, droplet method is not accurate when effect of nanoparticles on triple line is weak. The profile and contact angle of droplet depend on gravity and physical properties of gas, liquid and solid such as liquid-gas and solid surface tensions. Droplet contact angle cannot be a unique criterion to measure surface wettability, since droplet contact angle changes with volume in normal gravity. So, asymptotic contact angle was developed to measure surface wettability. The asymptotic contact angle demonstrates the force balance and interactions of gas-liquid-solid at the triple line, regardless of gravity. The asymptotic contact angle is a precise and unique criterion to measure surface wettability. The asymptotic contact angle has a great potential to measure the effects of concentration and characteristics of nanoparticles on surface wettability.

In case of bubble formation method, gravity is in favor of pushing bubble upward as buoyancy force, so triple line can move more freely and effects of nanoparticles on triple line can be seen easier. For instance, modification of pinning behavior of triple line by gold nanofluid is more visible in bubble formation method. Injecting air into the liquid or nanofluid has a great capacity to illustrate the behavior of triple line similar to what is happening in boiling heat transfer. Behavior of triple line has a significant role on boiling heat transfer and critical heat flux.

The effects of nanoparticles on liquid-gas and solid surface tensions, wettability, advancing and receding contact angles, wetting and dewetting, pinning behavior of the triple line and roughness of surface, have significant impacts on the variation of nanofluid boiling heat transfer coefficient and critical heat flux. Nanoparticles have enormous opportunities to enhance boiling heat transfer coefficient and critical heat flux simultaneously which is ultimate goal in nanofluid boiling heat transfer process.

Investigation of deposition of nanoparticles in nanofluid boiling is necessary from both nanoparticle material science and fluid flow heat transfer points of views in order to remove the disadvantages of the deposited nanoparticles in nanofluid boiling, especially inside the microchannels. The minimization or elimination of deposition of nanoparticles in nanofluid boiling heat transfer inside the microchannels is essential. The deposition of nanoparticles in nanofluid boiling heat transfer inside the macrochannels could be challenging for long periods of operation as well.

There are enormous discrepancies among the existing experimental results so, it is essential to understand the effects of nanoparticles on thermal conductivity, viscosity, deposition of nanoparticles, surface roughness, wettability, receding and advancing contact angles, boiling heat transfer coefficient, and critical heat transfer. The behavior of nanofluids depends on several variables such as the material of the base liquid, characteristics of the nanoparticles (size, shape, concentration, coating and material) characteristics of the substrate and physical conditions (temperature, pressure, etc). The degree of our understanding of the role of each one of these factors is different from one to another. Variation of anyone of these factors could change the outcomes and be the reason for the discrepancies among the existing experimental results.

## REFERENCES


[1] Wang, K.V., and Leon, O.D., 2010, "Applications of Nanofluids: Current and Future," *Advances in Mechanical Engineering*, 2010, 519659.

[2] Buongiorno, J., Venerus, D. C., Prabhat, N., McKrell, T., Townsend, J., Christianson, R., Tolmachev, Y. V., Keblinski, P., Hu, L., Alvarado, J. L., Bang, I. C., Bishnoi, S. W., Bonetti, M., Botz, F., Cecere, A., Chang, Y., Chen, G., Chen, H., Chung, S. J., Chyu, M. K., Das, S. K., Di Paola, R., Ding, Y., Dubois, F., Dzido, G., Eapen, J., Escher, W., Funfschilling, D., Galand, Q., Gao, J., Gharagozloo, P. E., Goodson, K. E., Gutierrez, J. G., Hong, H., Horton, M., Hwang, K. S., Iorio, C. S., Jang, S. P., Jarzebski, A. B., Jiang, Y., Jin, L., Kabelac, S., Kamath, A., Kedzierski, M. A., Kieng, L. G., Kim, C., Kim, J.-H., Kim, S., Lee, S. H., Leong, K. C., Manna, I., Michel, B., Ni, R., Patel, H. E., Philip, J., Poulikakos, D., Reynaud, C., Savino, R., Singh, P. K., Song, P., Sundararajan, T., Timofeeva, E., Tritcak, T., Turanov, A. N., Van Vaerenbergh, S., Wen, D., Witharana, S., Yang, C., Yeh, W.-H., Zhao, X.-Z., Zhou, S.-Q., 2009, "A Benchmark Study on the Thermal Conductivity of Nanofluids," *J. Appl. Phys,* 106, 094312.

[3] Godson, L., Raja, B., Mohan Lal, D., Wongwises, S., 2010, "Experimental Investigation on the Thermal Conductivity and Viscosity of Silver-Deionized Water Nanofluid," *Experimental Heat Transfer*, 23, 317-332.

[4] Vafaei, S., Purkayastha, P., Jain, A., Ramanath, G., and Borca-Tasciuc, T., 2009, "The Effect of Nanoparticles on the Liquid-Gas Surface Tension of $Bi_2Te_3$ Nanofluids," *Nanotechnology*, 20, 185702-185708.

[5] Vafaei, S., Wen, D., and Borca-Tasciuc, T., 2011, "Nanofluids Surface Wettability through Asymptotic Contact Angle," *Langmuir*, 27, 2211-2218.

[6] Vafaei, S., and Wen, D., 2010, "Bubble Formation in a Quiescent Pool of Gold Nanoparticle Suspension," *Advances in Colloid and Interface Science*, 159, 72-93.

[7] Vafaei, S., and Wen, D., 2010, "Effect of Gold Nanoparticles on the Dynamics of Gas Bubbles," *Langmuir*, 26, 6902-6907.

[8] Vafaei, S., and Wen, D., 2010, "The Effect of Gold Nanoparticles on the Spreading of Triple Line," *Microfluidics and Nanofluidics*, 8, 843-848.

[9] Sefiane, K., 2006, "On the Role of Structural Disjoining Pressure and Contact Line Pining in Critical Heat Flux Enhancement During Boiling Nanofluids," *Applied Physics Letters*, 89, 0441061-3.

[10] Chopkar, M., Das, A.K., Manna, I., and Das, P.K., 2008, "Pool Boiling Heat Transfer Characteristics of $ZrO_2$-Water Nanofluids from a Flat Surface in a Pool," *Heat Mass Transfer*, 44, 999-1004.

[11] Buongiorno, J., Hu, L.W., and Bang, I.C., "Towards an Explanation of the Mechanism of Boiling Critical Heat Flux Enhancement in Nanofluids," ASME 2007 5th International Conference on Nanochannels, Microchannels, and Minichannels (ICNMM2007), 989-995.

[12] Forrest, E., Williamson, E., Buongiorno, J., Hu, L.W., Rubner, M., and Cohen, R., 2010, "Augmentation of Nucleate Boiling Heat Transfer and Critical Heat Flux Using Nanoparticle Thin-Film Coatings," *International Journal of Heat and Mass Transfer*, 53, 58-67.

[13] Wenzel, R.N., 1936, "Resistance of Solid Surfaces to Wetting by Water," Industrial *and* Engineering *Chemistry*, 28, 988-994.

[14] Vafaei, S. and Wen, D., 2012, "Convective Heat Transfer of Aqueous Alumina Nanosuspensions in a Horizontal Mini-Channel," *Heat Mass Transfer*, 48, 349-357.

[15] Rea, U., McKrell, T., Hu, L.W., and Buongiorno, J., 2009, "Laminar Convective Heat Transfer and Viscous Pressure Loss of Alumina-Water and Zirconia-Water Nanofluids," *International Journal of Heat and Mass Transfer*, 52, 2042–2048.

[16] Carey, V.P., 2008, *Liquid–Vapor Phase-Change Phenomena*, $2^{nd}$ Ed., Hemisphere Publishing Corporation.

[17] Gaydos, J. and Neumann, A.W., 1987, "Dependence of Contact Angles on Drop Size and Line Tension," *Journal of Colloid and Interface Science*, 120, 76-86.

[18] Vafaei, S, and Podowski, M.Z., 2005, "Analysis of the Relationship between Liquid Droplet Size and Contact Angle," *Advances in Colloid and Interface Science*, 113, 133-146.

[19] Cassie, A.B.D., 1948, "Contact Angle," Discuss, *Faraday Soc*, 3, 11-16.

[20] Baxer, S., and Cassie, A.B.D., 1945, "The Water Repellency of Fabrics and a New Water Repellency Test," *Journal of Text. Inst*., 36(4), 67-90.

[21] Cassie, A.B.D., Baxter, S., 1944, "Wettability of porous surfaces"*Trans. Faraday Soc*, 40, 546-551.

[22] Wolansky, G., and Marmur, A., 1999, "Apparent Contact Angles on Rough Surfaces: the Wenzel Equation Revisited," *Colloids and Surfaces A: Physicochemical and Engineering Aspects*, 156, 381-388.

[23] Kandlikar, S.G., and Steinke, M.E., 2002, "Contact Angles and Interface Behavior during Rapid Evaporation of Liquid on a Heated Surface," *International Journal of Heat and Mass Transfer*, 45, 3771-3780.

[24] Yildirim, E., 2006, Surface Chemistry of Solid and Liquid Interfaces, Blackwell.

[25] Wasan, D.T, and Nikolov, A.D., 2003, "Spreading of Nanofluids on Solids," *Nature*, 423, 156-159.

[26] Nikolov, A., Kondiparty, K., Wasan, D., 2010, "Nanoparticle Self-Structuring in a Nanofluid Film Spreading on a Solid Surface," *Langmuir*, 26, 7665-7670.

[27] Deegan, R.D., Bakajin, O., Dupont, T.F., Huber, G., Nagel, S.R., and Witten, T.A., 1997, "Capillary Flow as the Cause of Ring Stains from Dried Liquid Drops," *Nature*, 389, 827-829.

[28] Deegan, R.D., Bakajin, O., Dupont, T.F., Huber, G., Nagel, S.R., and Witten, T.A., 2000, "Contact Line Deposits in an Evaporating Drop," *Physical Review E,* 62, 756-765.

[29] Sefiane, K., Skilling, J., and MacGillivray, J., 2008, "Contact Line Motion and Dynamic Wetting of Nanofluid Solutions," *Advances in Colloid and Interface Science*, 138, 101-120.

[30] Vafaei, S. Borca-Tasciuc, T, Podowski, M.Z., Purkayastha, A., Ramanath, G., and Ajayan, P.M., 2006, "Effect of Nanoparticles on Sessile Droplet Contact Angle," *Nanotechnology*, 17, 2523-2527.

[31] Vafaei, S., and Wen, D., 2011, "Flow Boiling Heat Transfer of Alumina Nanofluids in Single Microchannels and the Roles of Nanoparticles," *Journal of Nanoparticle Research*, 13, 1063-1073.

[32] Kim, S.J., Bang, I.C., Buongiorno, J., and Hu, L.W., 2007, "Surface Wettability Change during Pool Boiling of Nanofluids and Its Effect on Critical Heat Flux," *International Journal of Heat and Mass Transfer*, 50, 4105-4116.
[33] Vafaei, S., and Wen, D., 2010, "Critical Heat Flux (CHF) of Subcooled Flow Boiling of Alumina Nanofluids in a Horizontal Microchannel," *Journal of Heat Transfer*, 132, 102404-4.
[34] Chengara, A., Nikolov, D., Wasan, D.T., Trokhymchuk, A., Henderson, D., 2004, "Spreading of Nanofluids Driven by the Structural Disjoining Pressure Gradient," *Journal of Colloid and Interface Science*, 280, 192-201.
[35] Neumann, A.W., and Spelt, J.K., 1996, *Applied Surface Thermodynamics*, M. Dekker.
[36] Kwok, D.Y., and Neumann, A.W., 1999, Contact Angle Measurement and Contact Angle Interpretation," *Advances in Colloid and Interface Science*, 81, 167-249.
[37] Berthelot, D., 1898. Sur le mélange des gaz. Comptes rendus hebdomadaires des séances de l'Académie des Sciences, 126, 1703–1706, 1857–1858.
[38] Li, D., and Neumann, A.W., 1990, "Reformulation of the Equation of State for Interfacial Tensions," *Journal of Colloid Interface Science*, 137, 304-307.
[39] Kwok, D.Y., and Neumann, A.W., 2000, "Contact Angle Interpretation in Terms of Solid Surface Tension," *Colloids and Surfaces A: Physicochemical and Engineering Aspects*, 161, 31-48.
[40] Kwok, D.Y., 1998, Contact Angles and Surface, Energetics, Ph.D. Thesis, University of Toronto.
[41] Driedger, O., Neumann, A.W., and Sell, P.J., 1965, "An Equation *of* State Approach for Surface Free Energies," *Kolloid-Z. Z. Polym*, 201, 52-57.
[42] Neumann, A.W., Good, R.J., Hope, C.J., and Sejpal, M., 1974, "An Equation-of-State Approach to Determine Surface Tensions of Low-Energy Solids from Contact Angles," *Journal of Colloid Interface Science*, 49, 291-304.
[43] Kim, S.J., McKrell, T., Buongiorno, J., and Hu, L.W., 2009, "Experimental Study of Flow Critical Heat Flux in Alumina-Water, Zinc-Oxide-Water, and Diamond-Water Nanofluids," ASME *Journal of Heat Transfer*, 131, 0432041.
[44] Vafaei, S., and Podowski, M.Z., 2004, "The Modeling of Liquid Droplet Shape on Horizontal and Inclined Surfaces," 5th ICMF, Yokohama, Japan, May 30-June 4, 2004.
[45] Vafaei, S., Borca-Tasciuc, T., and Wen, D., 2010, "Theoretical and Experimental Investigation of Quasi-Steady-State Bubble Growth on Top of Submerged Stainless Steel Nozzles," *Colloids and Surfaces A: Physicochemical and Engineering Aspects*, 369, 11-19.
[46] Vafaei, S., and Wen, D., 2011, "*Spreading of Triple Line and Dynamics of Bubble Growth Inside Nanoparticle Dispersions on Top of A Substrate Plate,*" *Journal of Colloid and Interface Science*, 362, 285-291.
[47] del Rio, O.I., and Neumann, A.W., 1997, "Axisymmetric Drop Shape Analysis: Computational Methods for the Measurement of Interfacial Properties from the Shape and Dimensions of Pendant and Sessile Drops," *Journal of Colloid and Interface Science*, 196, 136-147.
[48] Gerlach, D., Biswas, G., Durst, F., Kolobaric, V., 2005, "Quasi-Static Bubble Formation on Submerged Orifices," *International Journal of Heat and Mass Transfer*, 48, 425-438.

[49] Vafaei, S., and Wen, D., 2010, "Bubble Formation on a Submerged Micronozzle," *Journal of Colloid and Interface Science*, 343, 291-297.

[50] Vafaei, S., Angeli, P., and Wen, D., 2011, "Bubble Growth Rate from a Stainless Steel Substrate and Needle Nozzles," *Colloids and Surfaces A: Physicochemical and Engineering Aspects*, 384, 240-247.

[51] Gerlach, D., Tomar, G., Biswas, G., and Durst, F., 2006, "Comparison of Volume-of-Fluid Methods for Surface Tension-Dominant Two-Phase Flows," *International Journal of Heat and Mass Transfer*, 49, 740-754.

[52] Chakraborty, I., Ray, B., Biswas, G., Durst, F., Sharma, A., Ghoshdastidar, P.S., 2009, "Computational Investigation on Bubble Detachment from Submerged Orifice in Quiescent Liquid under Normal and Reduced Gravity," *Physics of Fluids*, 21, 0621031-17.

[53] Datta, R.L., Napier, D.H., and Newitt, D.M., 1950, "The Properties and Behaviour of Gas Bubbles Formed at a Circular Orifice," *Transactions of the Institution of Chemical Engineers*, 28, 14-26.

[54] Benzing, R.L., and Myers, J.E., 1955, "Low Frequency Bubble Formation at Horizontal Circular Orifices," *Industrial and Engineering Chemistry*, 47, 2087-2090.

[55] Coppock, P.D., and Meiklejohn, G.T., 1951, "The Behavior of Gas Bubbles in Relation to Mass Transfer," *Transactions of the Institution of Chemical Engineers*, 29, 75-86.

[56] Quigley, C.J., Johnson, A.I., Harris, B.L., 1955, "Size and Mass Transfer Studies of Gas Bubbles," *Chemical Engineering Symposium Series*, 16, 31-44.

[57] Davidson, J.F., and Schuler, B.O.G., 1960, "Bubble Formation at an Orifice in an Inviscid Liquid," *Transactions of the Institution of Chemical Engineers*, 38, 335-342.

[58] Martin, M., Montes, F.J., Galan, M.A., 2006, "On the Influence of the Liquid Physical Properties on Bubble Volumes and Generation Times," *Chemical Engineering Science*, 61, 5196-5203.

[59] Collier, J.G., and Thome, J.R, 1996, *Convective Boiling and Condensation*, 3rd Ed., Oxford Science Publications.

[60] Gerardi, C., Buongiorno, J., Hu, L.W., McKrell, T., 2010, "Study of Bubble Growth in Water Pool Boiling through Synchronized, Infrared Thermometry and High-Speed Video," *International Journal of Heat Mass Transfer*, 53, 4185-4192.

[61] Kim, S.J., Bang, I.C., Buongiorno, J., and Hu, L.W., 2007, "Surface Wettability Change during Pool Boiling of Nanofluids and Its Effect on Critical Heat Flux," *International Journal of Heat Mass Transfer*, 50, 4105-4116.

[62] Narayan, G.P., Anoop, K.B., Das, S.K., 2007, "Mechanism of Enhancement/Deterioration of Boiling Beat Transfer Using Stable Nanoparticle Suspensions over Vertical Tubes," *Journal of Applied Physics*, 102, 0743171-7.

[63] Kwark, S.M., Kumar, R., Moreno, G., Yoo, J., and You, S.M., 2010, "Pool Boiling Characteristics of Low Concentration Nanofluids," *International Journal of Heat Mass Transfer*, 53, 972-981.

[64] White, S.B., Shih, A.J., and Pipe, K.P., 2010, "Effects of Nanoparticle Layering on Nanofluid and Base Fluid Pool Boiling Heat Transfer from a Horizontal Surface Under Atmospheric Pressure," *Journal of Applied Physics*, 107, 1143021-6.

[65] Phan, H.T., Caney, N., Marty, P., Colasson, S.P., Gavillet, J., 2010, "Surface Coating with Nanofluids: The Effects on Pool Boiling Heat Transfer," *Nanoscale and Microscale Thermophysical Engineering*, 14, 229-244.

[66] Lee, J., and Mudawar, I., 2007, "Assessment of the Effectiveness of Nanofluids for Single-Phase and Two-Phase Heat Transfer in Micro-Channels," *International Journal of Heat Mass Transfer*, 50, 452-463.

[67] Kim, S.J., Bang, I.C., Buongiorno, J., 2006, "Effects of Nanoparticle Deposition on Surface Wettability Influencing Boiling Heat Transfer in Nanofluids," *Applied Physics Letters*, 89, 1531071-3.

[68] Das, S.K., Putra, N., and Roetzel, W., 2003, "Pool Boiling Characteristics of Nano-Fluids," *International Journal of Heat and Mass Transfer*, 46, 851-862.

[69] Bang, I.C., and Chang, S.H., 2005, "Boiling Heat Transfer Performance and Phenomena of $Al_2O_3$-Water Nanofluids from a Plain Surface in a Pool," *International Journal of Heat Mass Transfer*, 48, 2407-2419.

[70] Benjamin, R.J., and Balakrishnan, A.R., 1997, "Nucleation Site Density in Pool Boiling of Saturated Pure Liquids: Effect of Surface Microroughness and Surface and Liquid Physical Properties," *Experimental Thermal and Fluid Science*, 15, 32-42.

[71] McHale, J.P., and Garimella, S.V., 2010, "Bubble Nucleation Characteristics in Pool Boiling of a Wetting Liquid on Smooth and Rough Surfaces," *International Journal of Multiphase Flow*, 36, 249-260.

[72] Kim, S.J., Bang, I.C., Buongiorno, J., and Hu, L.W., 2007, "Study of Pool Boiling and Critical Heat Flux Enhancement in Nanofluids," *Bulletin of the Polish Academy of Sciences, Technical Sciences*, 55, 211-216.

[73] Kathiravan, R., Kumar, R., Gupta, A. Chandra, R., 2009, "Characterization and Pool Boiling Heat Transfer Studies of Nanofluids," *Journal of Heat Transfer*, 131, 0819021.

[74] Suriyawong, A., and Wongwises, S., 2010, "Nucleate Pool Boiling Heat Transfer Characteristics of $TiO_2$-Water Nanofluids at Very Low Concentrations," *Experimental Thermal and Fluid Science*, 34, 992-999.

[75] Nagai, N., and Carey, V.P., 2002, "Assessment of Surface Wettability and Its Relation to Boiling Phenomena," *Thermal Science and Engineering*, 10, 1-9.

[76] Wang, C.H., and Dhir, V.K., 1993, "Effect of Surface Wettability on Active Nucleation Site Density during Pool Boiling of Water on a Vertical Surface," *Journal of Heat Transfer*, 115, 659-669.

[77] Dinh, T.N., Thu, J.P., and Theofanous, T.G., 2004, "Burnout in High Heat Flux Boiling: the Hydrodynamic and Physico-Chemical Factors," 42$^{nd}$ AIAA Aerospace Sciences Meeting and Exhibit, Reno, Nevada, 5-6 January 2004.

[78] Wen, D., and Ding, Y., 2005, "Experimental Investigation into the Pool Boiling Heat Transfer of Aqueous Based C-Alumina Nanofluids," *Journal of Nanoparticle Research*, 7, 265-274.

[79] Soltani, S., Etemad, S.G., and Thibault, J., 2010, "Pool Boiling Heat Transfer of Non-Newtonian Nanofluids," *International Communications in Heat and Mass Transfer*, 37, 29-33.

[80] Soltani, S., Etemad, S.G., and Thibault, J., 2009, "Pool Boiling Heat Transfer Performance of Newtonian Nanofluids," *Heat and Mass Transfer*, 45, 1555-1560.

[81] Henderson, K. Park, Y.G., Liu, L., and Jacobi, A.M., 2010, "Flow-Boiling Heat Transfer of R-134a-Based Nanofluids in a Horizontal Tube," *International Journal of Heat and Mass Transfer*, 53, 944-951.

[82] You, S.M., Kim, J., and Kim, K.H., 2003, "Effect of Nanoparticles on Critical Heat Flux of Water in Pool Boiling Heat Transfer," *Applied Physics Letter*, 83, 3374-3376.

[83] Vassallo, P., Kumar, R., and D'Amico, S., 2004, "Pool Boiling Heat Transfer Experiments in Silica–Water Nano-Fluids," *International Journal of Heat and Mass Transfer*, 47, 407-411.

[84] Haramura, Y., and Katto, Y., 1983, "A New Hydrodynamic Model of CHF Applicable Widely to both Pool and Forced Convection Boiling on Submerged Bodies in Saturated Liquids," *International Journal of Heat and Mass Transfer*, 26, 389-399.

[85] Sadasivan, P., Chappidi, P.R., Unal, C., and Nelson, R.A., 1982, "Possible Mechanisms of Macrolayer Formation," *International Communications in Heat and Mass Transfer*, 19, 801-815.

[86] Theofanous, T.G., and Dinh, T.N., 2006, "High Heat Flux Boiling and Burnout as Microphysical Phenomena: Mounting Evidence and Opportunities," *Multiphase Science and Technology*, 18, 1-26.

[87] Kandlikar, S.G., 2001, "A Theoretical Model to Predict Pool Boiling CHF Incorporating Effects of Contact Angle and Orientation," *Journal of Heat Transfer*, 123, 1071-1079.

[88] Mukherjee, A., and Kandlikar, S.G., 2007, "Numerical Study of Single Bubbles with Dynamics Contact Angle during Nucleate Pool Boiling," *International Journal of Heat and Mass Transfer*, 50, 127-138.

[89] Kim, S.J., McKrell, T., Buongiorno, J., Hu, L.W., 2008, "Alumina Nanoparticles Enhance the Flow Boiling Critical Heat Flux of Water at Low Pressure," *Journal of Heat Transfer*, 130, 0445011.

[90] Park, K.J., Jung, D., and Shim, S.E., 2009, "Nucleate Boiling Heat Transfer in Aqueous Solutions with Carbon Nanotubes Up To Critical Heat Fluxes," *International Journal of Multiphase Flow*, 35, 525-532.

[91] Kim, S.J., McKrell, T., Buongiorno, J., Hu, L.W., 2008, "Experimental Study of Flow Critical Heat Flux in Low Concentration Water-Based Nanofluids," Proceedings of MNHT2008, Micro/Nanoscale Heat Transfer international Conference, January 6-9, 2008, Tainan, Taiwan, 599-603.

[92] Kim, H.D., and Kim, M.H., 2007, "Effect Of Nanoparticle Deposition on Capillary Wicking that Influences the Critical Heat Flux in Nanofluids," *Applied Physics Letters*, 91, 0141041.

[93] Park, K.J., and Jung, D., 2007, "Enhancement of Nucleate Boiling Heat Transfer Using Carbon Nanotubes," *International Journal of Heat and Mass Transfer*, 50, 4499-4502.

[94] You, S.M., Kim, J.H., and Kim, K.H., 2003, "Effect of Nanoparticles on Critical Heat Flux of Water in Pool Boiling Heat Transfer," *Applied Physics Letters*, 83, 3374-3376.

In: Nanofluids
Editor: Yuwen Zhang

ISBN: 978-1-62618-165-6
2013 Nova Science Publishers, Inc.

*Chapter 9*

# NANOFLUID OSCILLATING HEAT PIPE

***Hongbin Ma, Corey Wilson, Il Yoon and Yuwen Zhang***
Department of Mechanical and Aerospace Engineering, University of Missouri, Columbia, MO, US

## ABSTRACT

An Oscillating Heat Pipe (OHP) is a heat transfer device that can be utilized to transfer a large amount of heat from heating to cooling sections. A typical OHP is fabricated from a long capillary tube bent into many turns with an evaporator and a condenser located at two opposite sides, and an adiabatic region between the evaporator and condenser. In order to form a train of vapor bubbles in an OHP, the liquid and vapor must not be able to stratify in the channel. When nanoparticles are added in the base fluid in an OHP, the OHP effectively utilizes 1) the oscillating motion in the OHP to suspend nanoparticles; 2) the higher heat capacity of nanofluid; 3) the higher thermal conductivity; 4) the nanoparticle effect on the fluid field; and 5) the nanoparticle effect on the surface wetting conditions. The unique feature of the oscillating motion fully utilizes the nanofluid features to significantly enhance the heat transport capability in an OHP.

In the fabrication process of OHPs, preparation, charging the OHP with nanofluids and sealing it are inevitable and requiring careful attention for successful testing. Nanofluid was first used as the working fluid in a 12-turn OHP by the Center for Thermal Management at the University of Missouri in 2006. The results of using nanofluids showed a significant improvement over results which used the traditional base fluid of water as the working fluid. The improvement was especially significant at low power; however, even at high power the improvement was measurable. The parameters affecting heat transfer performance were systematically investigated by researchers around the world. It was reported that the thermal resistance for the $Al_2O_3$-water nanofluid was lower than that of pure water, while the thermal performance of $SiO_2$-water nanofluid was worse than pure water. There is an optimal nanofluid concentration for the OHP. It is also worth noting that the effect of nanoparticle deposition on thin-film evaporation has not been investigated and might play a significant role in the thermal performance of the nanofluid OHP.

## 1. Introduction

Effective thermal management has become one of the most serious challenges in the modern electronic components and systems. Power levels have far exceeded the possible heat transport capacity of commonly used heat rejection methods like forced convection, heat sinks, and traditional two-phase capillary-driven passive heat pipes [1, 2]. Faster and smaller electronic components and systems demand highly efficient thermal management. Moreover, today's electronics also require cooling devices that have high reliability, structural simplicity, and low cost.

An Oscillating Heat Pipe (OHP) [3, 4] is a heat transfer device that can be utilized to transfer a large amount of heat from heating to cooling sections. An OHP, as shown in Figure 1, consists of an evaporating section, adiabatic section, and a condensing section, a typical OHP is fabricated from a long capillary tube bent into many turns with an evaporator and a condenser located at two opposite sides, and an adiabatic region between the evaporator and condenser. As heat is added on the evaporating section, the vaporization in the evaporating section causes the vapor volume expansion, and the heat removal in the condensing section causes the vapor volume contraction. The expansion and contraction of vapor volume plus the thermal energy added on the system generates the oscillating motion of liquid plugs and vapor bubbles in the system. OHP is also referred to as the pulsating heat pipe (PHP) in literature. The oscillating motions in OHP depend on the dimensions, working fluids, operating temperature, surface conditions, heat flux and total heat load, orientation, number of turns, and most importantly the filling ratio, $V_l/V_t$, where $V_l$ is the volume occupied by the liquid in the system and $V_t$ is the total volume. Utilizing phase-change heat transfer and forced convection, the heat is transferred from the evaporating section to the condensing section.

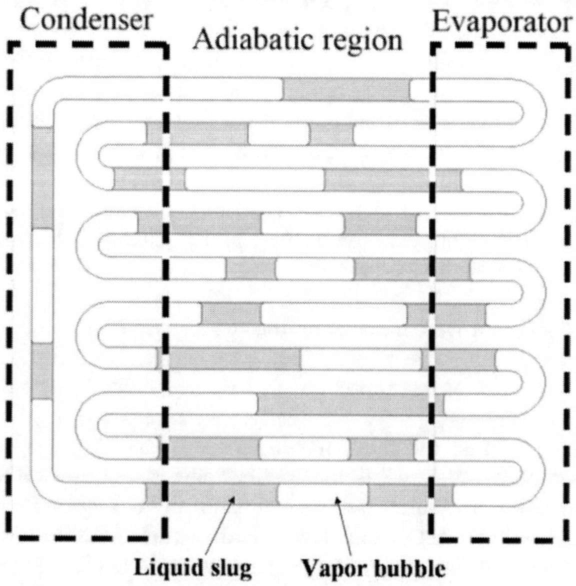


Figure 1. Schematic of a typical OHP.

For a typical OHP, forced convection plays a key role in heat transfer from the evaporator to the condenser sections. In other words, most heat transferred from the evaporator to the condenser is by sensible heat. Thermal conductivity, viscosity, and heat capacity of working fluid play more important roles in an OHP than the surface tension and latent heat in a conventional wicked heat pipe. Many working fluids, such as water [5, 6], acetone [7, 8], 134a [8], and nanofluids [5, 7, 9–11], have been investigated. Among these working fluids, nanofluids significantly affect the heat transport capability in an OHP.

Adding a small amount of nanoparticles into the base fluid results in the creation of a nanofluid. Fabrication technology advanced over the last decade has enabled producing and processing materials at the nanoscale level, making the use of their novel properties in modern devices. Choi [12] reported that effective thermal conductivity can be increased by uniformly suspending a very small quantity, preferably less than 1% by volume, of nanoparticles in conventional coolants. Most recently, an International Nanofluid Property Benchmark Exercise (INPBE) was conducted by over 30 organizations worldwide to measure the thermal conductivity of identical samples of nanofluids [13]. They found that the tested nanofluids' thermal conductivity enhancement could be increased by increasing particle loading, particle aspect ratio and by decreasing the base fluid's thermal conductivity. They also suggested that the effective medium theory was in good agreement with the experimental data and that no anomalous enhancement of thermal conductivity was achieved in nanofluids tested in their exercise. Although they cannot effectively increase the effective thermal conductivity, nanoparticles added to the base fluid can increase the convection heat transfer [14, 15].

Ma et al. [5, 9] demonstrated that when the nanofluid is used in an OHP, the heat transport capability can be significantly increased. When nanoparticles are added to OHP's base fluid, the OHP effectively utilizes 1) the oscillating motion in the OHP to suspend nanoparticles; 2) the higher heat capacity of nanofluid; 3) the higher thermal conductivity; 4) the nanoparticle effect on the fluid field; and 5) the nanoparticle effect on the surface wetting conditions. The unique feature of the oscillating motion fully utilizes the nanofluid features to significantly enhance the heat transport capability in an OHP.

## 2. NANOFLUIDS

### 2.1. Development of Nanofluids

Convection is the key mechanism for convective heat transfer such as flow in pipes or ducts. The convection is divided into two mechanisms: the conduction and bulk advection. This means that it is important to use a high thermal conductivity liquid for enhancing convective heat transfer when the liquid is used as a working fluid. But, the thermal conductivity of a liquid is always lower than that of a solid; especially the thermal conductivity of a metallic solid, which is much higher than a non-metallic solid because of the existence of the free electrons in a metallic solid.

Due to the high thermal conductivity of metallic material, Maxwell (1981) suggested adding metallic particles into the base liquid to enhance the heat transfer rate of liquid. Previous research sought to suspend solids measured as mm or μm in a base liquid to utilize the higher thermal conductivity of solids for enhanced heat transfer. However, the solids do not suspend well in the base liquid because they are relatively dense causing solids dispersed unevenly in the base liquid. In addition, because the mm or μm level sized particles are relatively large, they could cause clogging in the micro fluid channel. Abrasion and pressure drop are also disadvantages to use of mm or μm scale solid particles. Hence, researchers could not successfully use solid particles in the base liquid for enhancing heat transfer.

To overcome these limits, particles smaller than μm scale were needed. Due to the development of nanotechnology to produce nanosized particles, the idea of using particles of nanometer scale was devised. That is, nanometer scale particles were suspended in the base liquid and these liquids containing nanoparticles are nanofluids [12]. In other words, nanofluids are heat transfer fluids where nanoparticles are suspended in liquid. Nanofluids as a heat transfer fluid have significant advantages. Das et al. [16] described benefits of nanofluids as follows: high heat conduction, stability, micro channel cooling without clogging, reduced chances of erosion and reduction in pumping power. It is thought that high heat conduction is achieved by the large surface area of the particles. Because of the large amount of nanoscale sized particles, high specific surface area is presented. This, in turn, causes high heat transfer rate by conduction. The mobility of the nanoparticles might also be accounted for the enhanced heat transfer by micro-convection of the fluid. The stability is achieved by the smaller size and less weight of the particles. This also could have less sedimentation of particles. The less sedimentation of nanoparticles is an important factor to overcome the settling of particles, which gives more stability. Preventing the formation of clogging in the micro channel is another benefit of using nanofluids. The micro channel is usually operated at high heat load, requiring a good heat transfer fluid to cool down. It is definitely better to use a fluid containing high thermal conductivity particles for better heat transport in the channel. However, the microscale particle could cause clogging in the microchannel because its size would be similar to the size of the channel. Thus, it is useful to use the liquid containing nanoscale particles which are a couple of orders smaller than the size of channel. The reduced chances of erosion are achieved by the small size and mass of the nanoparticles. Because it has smaller momentum, erosion of the component is less likely to happen. Reduced pumping power can also be achieved by the enhanced thermal conductivity of nanofluids. It is known that pumping power by a factor of ten should be increased to reach the heat transfer by a factor of two when a conventional heat transfer fluid is used. Thus, using nanofluid, of which thermal conductivity is greatly increased, reduces the pumping power to achieve the desired heat transfer rate.

## 2.2. Mechanisms of Nanofluids

Many experimental studies reported significantly improved effective thermal conductivity of nanofluids. The mechanisms of nanofluid is not fully explained. Wang and Mujumdar [17] suggest two categories of heat transfer mechanisms in the nanofluids in their review paper. The first mechanism consists of conventional static conduction and Brownian motion. The conventional static conduction explains the heat transfer in nanoparticles and base liquids themselves, while the Brownian motion addresses mixing of particles and base liquid. For this mechanism, the particle size, volume fraction of particles to the base liquid, thermal conductivities of particles and base liquid, and their operating temperature are important factors which determine performance. The second mechanism is a composite theory that a nanoparticle is the core and a nanolayer forms a shell around the core. This composite is, then, immersed in the base liquid. However, the conduction mechanism between the nanoparticle and base liquid is not understood yet.

Keblinski et al. [18] and Eastman et al. [19] explained the mechanisms of the increase of the thermal conductivity in more detail. They proposed four possible mechanisms: 1) the nanoparticle motion, 2) liquid layering at liquid-particle interface, 3) the nature of heat transport in nanoparticles, and 4) the effects of nanoparticle clustering. In the case of stationery nanofluids, the only discussed is heat conduction. In the theory of nanoparticle motion, the nanoparticles collide into each other and transfer energy from one particle to another, and, in turn, enhance the thermal conductivity while they are traveling in the base liquid. It was also suggested that the thermal conductivity might be enhanced just by Brownian motion of nanoparticles without collision. To confirm that nanoparticle motion is accounted for in the enhancement of thermal conductivity, they compared the time for a nanoparticle, of 10 nm diameter, to move a distance equal to its diameter by Brownian forces ($\tau_D$), and the time that heat moves the same distance by thermal diffusion in the liquid ($\tau_H$). The result showed that the time ratio ($\tau_D/\tau_H$) was about 500 and is about 25 when the diameter of a nanoparticle is reduced to 0.5 nm. In other words, the thermal diffusion in the liquid is much faster than the Brownian diffusion of a nanoparticle. Thus, the effect of nanoparticle motion for enhancing the thermal conductivity is negligible. However, the nanoparticle motion could indirectly affect the particle clustering, which is one of the possible explanations for the enhancement of thermal conductivity, which will be explained later.

The theory of liquid layering at liquid-particle interface is based on the well known fact that the liquids tend to form structural orderings at the solid-liquid interfaces. A crystalline solid, of which atomic structure is well ordered, has a much higher thermal conductivity than the bulk liquid, which has an atomic structure that is less ordered. Thus, it would be expected that liquid layering at the solid-liquid interface has a higher thermal conductivity than the one of the bulk liquid because its atomic structure is better ordered than the bulk liquid. This theory postulates that the thermal conductivity of the liquid layering is the same as that of the solid. From this postulation, liquid layering around the solid enlarges the effective volume of the composite; in turn, resulting in enhanced thermal conductivity. In Figure 2, it clearly shows that the larger thickness of layered-liquid thickness shows higher thermal conductivity.

Yu and Choi [20] reached the same conclusion. They suggested that a liquid nanolayer structure forms a thermal bridge, which is key to enhancing heat transfer between solid nanoparticles and bulk liquid, because this solid-like nanolayer structure, as shown in Figure 3, is expected to have higher thermal conductivity than a bulk liquid. However, simulations studied by Xue et al. [21] show that there is no effect on thermal transport at liquid layering. Figure 4 shows the density profile and the associated temperature profile through solid-liquid interface. The density profile shows that the two layers of significant layering and about three layers of weaker layering are formed. The temperature gradient in solid is constant, which means the thermal conductivity in the solid is high. However, the temperature profile through the liquid layering at the solid-liquid interface is similar to the one at the bulk liquid located far from the interface. Thus, the thermal conductivity of the interface is similar with that of the bulk liquid. This means that the thermal conductivity of the liquid layering is much lower than the solid and is same as the bulk liquid. However, since this simulation study is done for a simple mono-atomic fluid, it cannot tell that there is no effect on enhanced thermal conductivity at the liquid layering of more complex molecular liquids.

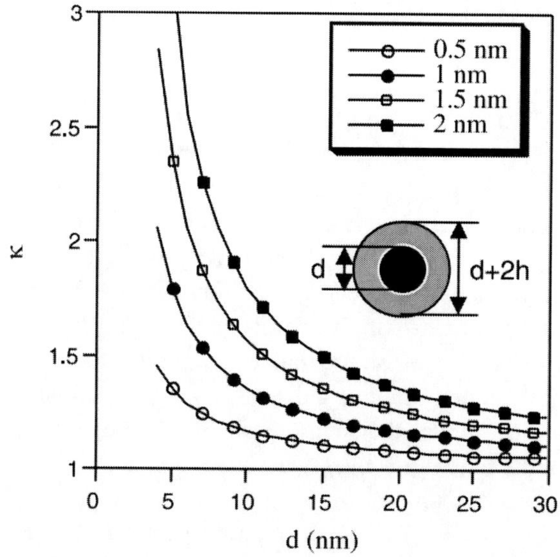


Figure 2. Excessive thermal conductivity enhancement, $\kappa$, as a function of particle diameter, d [18].

Figure 3. Schematic cross section of nanofluid structure consisting of nanoparticles, bulk liquid, and nanolayers at solid-liquid interface [20].

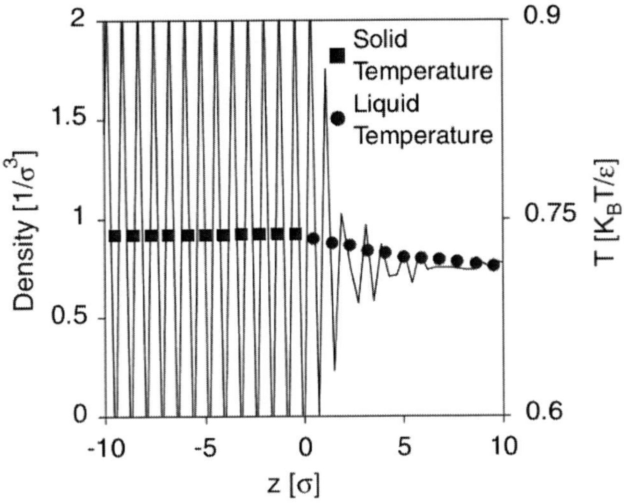


Figure 4. Distributions of number density of atoms (solid line) and average temperatures of the atomic layers (dots) at the liquid–solid interface for the wetting liquid [21].

The enhanced thermal conductivity of nanofluids also might be explained by the nature of the heat transport in nanoparticles. From the macroscopic view, heat is transferred by the phonons in the crystalline solid, which means the energy propagation by lattice vibration. The mean free path of the phonon is expressed as [22]

$$l = \frac{10aT_m}{\gamma T} \tag{1}$$

where $T_m$ is a melting point, $a$ is a lattice constant, $\gamma$ is the Gruneisen parameter [23]. The mean free path of the phonon is too long compared with the size of the nanoparticle to diffuse the energy. For example, the mean free path, $l$, for a typical nanoparticle, such as $Al_2O_3$, is just 35 nm. Thus, the phonon moves in ballistic motion across the nanoparticle. This means that the heat transport by phonon diffusion is not valid in the nanoparticles. As a result, the heat transport theory in the nanoparticle must be based on the ballistic phonon transport. However, in general, it is not expected that the heat transport by ballistic phonon transport can be more effective than the very fast diffusive phonon transport, causing the significantly enhanced thermal conductivity in nanofluids. One possible explanation would be that when the ballistic phonon from one particle passes through the liquid and reaches the other adjacent particle, resulting in transferring its energy. In this case, the distance between the adjacent particles should be very close, about the order of thickness in the layered liquid (~1-2 nm), because the mean free path of a phonon in the liquid is very short. This theory is likely to be supported by two facts: the distance between adjacent particles in the nanofluids is very short, and the distance could be closer due to the constant Brownian motion.

The effect of nanoparticle clustering is another possible explanation of the enhanced thermal conductivity. Heat transfers very fast within clusters, resulting in the high thermal transport. The effective volume of clusters is larger than the physical volume of the particles.

This means, it forms a path of very low thermal resistance in the clusters, capable of transferring a large amount of heat at a given time with high thermal conductivity. Because larger effective volume of the cluster causes higher thermal conductivity, a loosely packed cluster can have higher thermal conductivity. Even, the cluster that has no physical contact between its particles could have dramatically increased thermal conductivity. This may account for the mechanism of the much enhanced thermal conductivity of nanofluids experimentally observed. However, generally, clusters may have negative effect on the heat transport because small particles can be settled out of the base liquid.

## 2.3. Fabrication of Nanofluid

The fabrication of the nanofluids is generally divided into a one-step method and a two-step method. Yu and Xie [24] in a review paper give a brief explanation of these two methods for the preparation of nanofluids wherein the one-step method produces the nanoparticles and disperses them into the base liquid, which is processed simultaneously. By simplifying the procedures, drying, storage, transportation, and dispersion of nanoparticles are avoided. In addition, agglomeration of nanoparticles is minimized, nanoparticles are suspended more stably in the base liquid, and they are distributed uniformly throughout the liquid. However, this method cannot synthesize the nanofluid in a large scale, and the cost for production is high. Akoh et al. [25] first introduced this one-step method, named VEROS (Vacuum Evaporation onto a Running Oil Substrate). Eastman et al. [26] revised the method where the nanocrystalline Cu was evaporated resistively into the pump oil, while the liquid was placed in a rotating cylinder to produce a liquid thin layer above a resistively-heated evaporation source

In the two-step method, two separated procedures are needed to produce the nanofluids. Dry nanopowder is fabricated by chemical or physical method first, and then this powder is dispersed into the base liquid. In the fabrication process, several techniques are used to prevent particles from aggregating and achieve better dispersion because agglomeration of particles might occur, especially in the process of drying, storage, and transportation, which causes settlement of particles, clogging, and possibly lowering of thermal conductivity. These techniques include intensive magnetic force agitation, ultrasonic agitation, high-shear mixing, homogenizing, ball milling and adding surfactants (surface active agents). The two-step method is more economic than the one-step method because the nanopowder synthesis industry is developed on a large scale. Thus, this is the most widely used technique to fabricate nanofluids.

The base fluids for nanofluids include water, ethylene- or tri-ethylene-glycols and other coolants, oil and other lubricants, bio-fluids, polymer solutions and other common fluids. The materials for nanoparticles can be oxide ceramics (e.g., $Al_2O_3$ and CuO), metal carbides (SiC), nitrides (AlN and SiN), metals (Al and Cu), nonmetals (graphite, carbon nanotubes), layered materials (Al + $Al_2O_3$ or Cu + C), and functionalized nanoparticles.

## 2.4. Enhancement of Thermal Conductivity in Nanofluids

The very reason that nanofluids have received considerable attention is the enhancement of thermal conductivity. The significantly increased thermal conductivity of nanofluids compared to the base liquid has been reported by many researchers. Das et al. [16] reviewed the experimental results of the enhanced thermal conductivity. The nanoparticles used so far are roughly divided into three groups, ceramic particles (e.g., $Al_2O_3$, CuO, SiC, AlN and SiN), pure metallic particles (e.g., Al and Cu), and carbon nanotubes (CNTs). The base fluids are divided into four groups, water, ethylene glycol, transformer oil, and toluene.

Nanofluids with $Al_2O_3$ and CuO as nanoparticles and water and ethylene glycol as a base liquid were investigated to measure their thermal conductivity using the transient hot-wire method (THW) by Lee et al. [27]. Their results clearly indicate that nanofluids have higher thermal conductivity than that of base liquids. The highest increase, about 20%, was observed from 4% volume fraction of CuO suspended into ethylene glycol. It also showed that the higher thermal conductivity was achieved when the ethylene glycol was used as a base liquid. Wang et al. [28] also reported the thermal conductivity of CuO-water and $Al_2O_3$-water nanofluids, but their nanoparticle sizes were smaller than the ones used by Lee et al. [25]. The results showed different thermal conductivity from the ones by Lee et al. [25], which means that the size of particle affects thermal conductivity. It was concluded that the smaller nanoparticles in the base liquid have a higher thermal conductivity. The experimental data from Wang et al. [28] also showed that the sample preparation method also affects the thermal conductivity of nanofluids. It is remarkable that the nanofluids with oxide ceramic particles show the enhanced thermal conductivity, although oxide ceramic materials generally do not have high thermal conductivity. Metallic nanofluids showed higher thermal conductivity when compared to the nanofluids with ceramic particles. Xuan and Li [29] measured the thermal conductivity of copper nanoparticles suspended into the transformer oil. The nanoparticle size was about 100 nm, which was much larger than the ceramic nanoparticles used in Ref. [27-28]; the result was 55 % increased thermal conductivity compared to that of the base fluid with 5 % volume fraction. Furthermore, Eastman et al. [30] reported a 40 % increase of thermal conductivity with 0.3 % concentration of 10 nm size copper nanoparticles suspended in ethylene glycol. It clearly showed that the thermal conductivity is largely affected by nanoparticle size. Another experimental result about the size effect of nanoparticles was reported by Patel et al. [31]. They observed relatively lower thermal conductivity of silver nanoparticles suspended in water. Even though silver has high thermal conductivity, the silver-water nanofluid did not have much enhancement. This could be because the size of silver nanoparticle is relatively large, about 60 ~ 80 nm. This result showed that the size of the nanoparticle is more dominant than e particle conductivity and concentration when determining the thermal conductivity of nanofluids.

# 3. Oscillating Heat Pipe

## 3.1. Development of Oscillating Heat Pipe

Since Grover et al. [32] published their test results and coined the term "heat pipe" in 1964, the conventional heat pipe has been widely used in many applications. In the conventional heat pipe, heat is supplied into the evaporator to evaporate the liquid which flows to the condenser. The vapor is condensed when it arrives at the condenser section, and the condensed liquid flows back to the evaporator through the wick structure. During the operation, this working cycle is repeated as long as the heat is supplied to the evaporator and dissipated from the condenser. However, the heat pipe has intrinsic limitations such as capillary, viscous, sonic, entrainment, and boiling limits. Thus, there have been efforts to overcome these limits and achieve better performance as the demand for removing excessive heat from electronic chips grows. An oscillating heat pipe (OHP) was first developed by Akachi [1] in 1988, and was suggested as a promising alternative heat transfer device to the conventional heat pipe. Akachi [1] described in his patent that the newly devised OHP solved the following problems that could be arisen in the application of heat pipes: 1) less amount of heat transfer due to the low limit of heat transport performance by the mutual interference between vapor flow and liquid flow due to the opposite direction flow, 2) inevitable presence of wick structure, causing high thermal resistance at high heat load, 3) the occurrence of abnormalities due to a water hammer action, causing intermittent heat transportation, 4) the presence of limit of the length and diameter of heat pipes due to the mutual actions of a liquid resistance and wick limit, 5) limited mounting orientation of the heat pipe, due to the limit of liquid transportation from the condenser to the evaporator through wick structure, 6) difficulty in mountings on the heated and cooled objects because of no flexibility, and 7) occurrence of mutual interference in the working fluid stream. Since the invention of the OHP, extensive studies have been carried out by many researchers showing that OHPs have better performance than conventional heat pipes under certain conditions.

From the above description about conventional and oscillating heat pipes, the advantages of the OHP compared to the conventional heat pipe can be described as follows: 1) lower pressure drop of liquid flow due to the absence of the wick structure, enabling the liquid to flow through easily from the condenser to the evaporator, and vice versa, 2) higher heat transfer performance by convection of the liquid and vapor in the OHP, 3) additional heat transfer by thin film evaporation and the presence of the kinetic energy of the liquid in the OHP, 4) reduced pressure drop of the liquid by the vapor flow due to the same direction of liquid and vapor flow, 5) simple design and fabrication process, causing low cost to build, due to the absence of the wick structure, 6) less affected by the orientation of the pipes.

## 3.2. Fundamental Operating Principle

### *3.2.1. Mechanism Of Oscillation Motion*

The driving force for the oscillation motion of working fluid in OHP is basically the pressure difference in the evaporator and the condenser as heat is supplied to the evaporator and dissipated from the condenser. The liquid slugs of the working fluid are distributed randomly throughout the OHP before heat is supplied to the evaporator. When heat is supplied to the evaporator, the liquid slugs move to the condenser very slowly as the vapor volume is increased slightly. At a certain critical heat input, the liquid slugs start to oscillate and this point is called startup point. During the oscillation motion, the vapor plugs at the evaporator have higher pressure by evaporation and their temperatures increase as the heat is supplied, while the vapor plugs in the condenser have lower pressure due to condensation and their temperatures decrease. This causes pressure difference between the evaporator and condenser, and, in turn, causes liquid slugs to oscillate between the evaporator and condenser. The oscillation motion is maintained as long as enough pressure difference between the evaporator and the condenser is maintained.

### *3.2.2. Force Balance Analysis*

There have been trials to describe the motion of the liquid slugs in the OHP. Shafii et al. [33] developed an equation to predict the motion of *i*th liquid slug as shown in Figure 5 and the equation is expressed as

$$\frac{dm_{li}v_{li}}{dt} = \left(P_{vi} - P_{v(i+1)}\right)A - \pi dL_{li}\tau - (-1)^n m_{li}g\,, \tag{2}$$

where $n$ indicates the tube number. This equation takes into account the force by pressure difference of the two vapor plugs adjacent to the *i*th liquid slug, frictional force by shear stress, gravitational force acting on the liquid slug and inertial force.

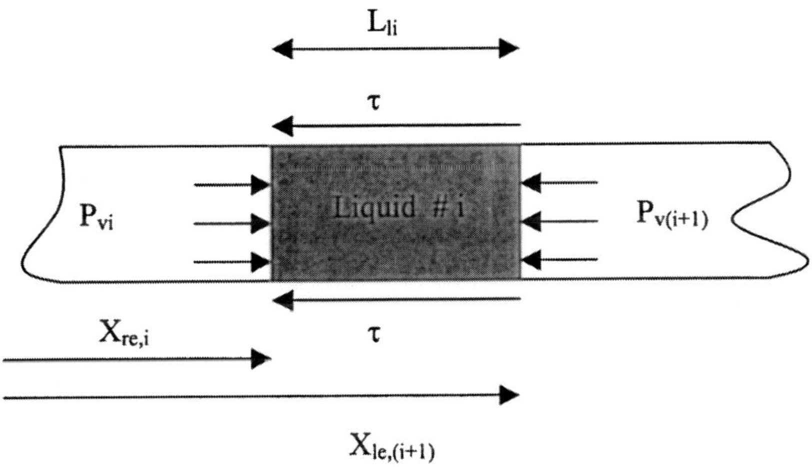


Figure 5. Control volume of the *i*th liquid slug in the channel [33].

Ma et al. [34] analyzed the motion of liquid slugs by considering the force balance. The vapor bubble compression and expansion were considered as a spring effect. They suggested that four different forces act on the liquid: driving force, force by frictional shear stress, force due to the deformation of compressible vapor bubble (vapor bubble spring effect), and inertial force. The driving force is caused by the pressure difference between the evaporator and the condenser, causing oscillation motion of working fluid, and is expressed as

$$F_d = \Delta pA = \left(\frac{Ah_{fg}\rho_{v,c}}{T_e}\right)\left(\frac{\Delta T_{max} - \Delta T_{min}}{2}\right)\left(1+\cos\left(\omega\tau\right)\right) \tag{3}$$

When the liquid flows through the capillary channel, the frictional force arises between the liquid/vapor and pipe wall, resulting in pressure drop. The total pressure drop due to the frictional shear stress is described as

$$\Delta p_f = \left[ (f_l \cdot \mathrm{Re}_l) \left( \frac{\mu_l L_l}{2D_h^2} \right) + (f_v \cdot \mathrm{Re}_v) \left( \frac{\mu_v L_v}{2D_h^2} \right) \right] \left( \frac{dx}{d\tau} \right) \tag{4}$$

The force due to the deformation of the compressible vapor bubbles are expressed as

$$\Delta p_v = \frac{\rho_v RT}{L_v} x \tag{5}$$

The inertial force is expressed as

$$F_i = m \frac{d^2 x}{d\tau^2} \tag{6}$$

Then, all these forces are combined into the one equation as force balance and this is described as

$$(L_l \rho_l + L_v \rho_v) A \frac{d^2 x}{d\tau^2} + \left[ (f_l \cdot \mathrm{Re}_l) \left( \frac{\mu_l L_l}{2D_h^2} \right) + (f_v \cdot \mathrm{Re}_v) \left( \frac{\mu_v L_v}{2D_h^2} \right) \right] A \left( \frac{dx}{d\tau} \right)$$
$$+ \frac{A \rho_v RT}{L_v} x = \left( \frac{A h_{fg} \rho_{v,c}}{T_e} \right) \left( \frac{\Delta T_{\max} - \Delta T_{\min}}{2} \right) \left( 1 + \cos(\omega \tau) \right) \tag{7}$$

### *3.2.3. Heat Transfer Analysis*

The heat transfers from the evaporator to the condenser by convection and thin film evaporation in the OHP. That is, the heat supplied to the outer wall of the evaporator is transferred to the inner wall by conduction. It is then carried by oscillating liquid slugs from the inner wall of the evaporator to the inner wall of the condenser. At this stage of transport, some proportion of heat is transferred by the convection of liquid slugs and vapor plugs (sensible heat), while the rest is transferred by evaporation at the evaporator and condensation at the condenser (latent heat). Zhang and Faghri [35, 36] reported that sensible heat is the main factor for heat transfer in the OHP, rather than latent heat. Once the heat reaches the inner wall of the condenser, it transfers to the outer wall by conduction.

Ma et al. [37] performed a heat transfer analysis theoretically at the evaporator and condenser separately. The total heat transfer coefficient at the evaporator was divided into two terms, microscopic and macroscopic heat transfer coefficient, and was expressed as [38]

$$h = h_{mic} + h_{mac},\tag{8}$$

where *mic* accounts for the nucleate boiling (microscopic) and *mac* is for the bulk convection (macroscopic). The microscopic heat transfer coefficient developed by Forster and Zuber [39] is expressed as

$$h_{mic} = 0.00122\left[\frac{k_l^{0.79} c_{pl}^{0.45} \rho_l^{0.49}}{\sigma^{0.5} \mu_l^{0.29} h_{lv}^{0.24} \rho_v^{0.24}}\right]\left[T_w - T_{sat}(P_l)\right]^{0.24} \times \left[P_{sat}(T_w) - P_l\right]^{0.75} S,\tag{9}$$

where S is a suppression factor and is expressed as

$$S = \left(1 + 2.56\times10^{-6}\,\mathrm{Re}_{tp}\,k_{tp}^{1.17}\right)^{-1},\tag{10}$$

The two-phase Reynolds number, $\mathrm{Re}_{tp}$, is defined by

$$\mathrm{Re}_{tp} = \mathrm{Re}_l\left[F\left(X_{tt}\right)\right]^{1.25},\tag{11}$$

where the liquid Reynolds number, $\mathrm{Re}_l$, is

$$\mathrm{Re}_l = \frac{G(1-x)D_i}{\mu_l},\tag{12}$$

and the Martinelli parameter, $X_{tt}$, is

$$X_{tt} = \left(\frac{1-x}{x}\right)^{0.9}\left(\frac{\rho_v}{\rho_l}\right)^{0.5}\left(\frac{\mu_l}{\mu_v}\right)^{0.1}\tag{13}$$

The function $F(X_{tt})$ in Equation (11) is decided by the Martinelli parameter as

$$F\left(X_{tt}\right) = 1 \quad \text{for } X_{tt}^{-1} \leq 0.1\tag{14}$$

$$F\left(X_{tt}\right) = 2.35\left(0.213 + \frac{1}{X_{tt}}\right)^{0.736} \quad \text{for } X_{tt}^{-1} > 0.1.\tag{15}$$

The quality, $x$, in the Equation (13) is defined as

$$x = \frac{\rho_v - \phi\rho_v}{\bar{\rho}},\tag{16}$$

where $\phi$ is the liquid filling ratio. The average density in Eq. (16) can be found as

$$\frac{1}{\bar{\rho}}=\frac{x}{\rho_v}+\frac{1-x}{\rho_l} \tag{17}$$

The macroscopic heat transfer coefficient in Eq. (8) is defined as

$$h_{mac}=F\left(X_{tt}\right)h_l, \tag{18}$$

where the liquid heat transfer coefficient, $h_l$, is calculated as

$$h_l=0.023\left(\frac{k_l}{D}\right)\mathrm{Re}_l^{0.8}\,\mathrm{Pr}_l^{0.4} \tag{19}$$

In the condenser, most heat transfer by condensation occurs in the thin film region because the thin film thickness is much thinner than the thickness of menisci of the liquid slugs. To calculate the heat transfer at the condenser, it is necessary to find the thin film thickness because, as stated, most heat is transferred through thin film. The pressure drop in the liquid film due to the viscous flow is

$$\frac{dp_l}{ds}=\frac{f\,\mathrm{Re}_\delta\,\mu_l D_0 q_c'' s}{2\delta^3 D_i \rho_l h_{fg}}, \tag{20}$$

where

$$\mathrm{Re}_\delta=\frac{\bar{U}_{l,c}\delta\rho_l}{\mu_l} \tag{21}$$

The total pressure drop along the half of the vapor plug can be found by integrating Eq. (20) from $s = 0$ to $L_{c,v}/2$, i.e.,

$$\Delta p_l=\int_0^{L_{c,v}/2}\left(\frac{f\,\mathrm{Re}_\delta\,\mu_l D_0 q_c'' s}{2\delta^3 D_i \rho_l h_{fg}}\right)sds \tag{22}$$

The filling ratio, $\phi$, is defined as

$$\varphi=\frac{V_l}{V} \tag{23}$$

and the idealized single vapor slug in the condenser is

$$L_{c,v} = L_c(1-\varphi) \tag{24}$$

The capillary pressure in the thin film in the condenser is found as

$$\frac{dp_c}{ds} = 2\sigma \frac{dK}{ds}, \tag{25}$$

and the total capillary pressure is calculated by integration from $K = 1/r_0$ to $K = 1/r_c$ and is expressed as

$$\Delta p_c = \sigma\left(\frac{2}{r_c} - \frac{1}{r_0}\right), \tag{26}$$

where $r_o$ is the meniscus radius of the liquid-vapor interface at the line of symmetry (s=0), and $r_c$ is the meniscus radius of the liquid-vapor interface at the liquid slug ( $s = L_{c,v}$ ), which can be calculated as

$$r_0 = \frac{1}{2}(D_i - 2\delta_0), \tag{27}$$

and

$$r_c = \frac{1}{2}\left(D_i - 2\delta_{0.5L_{c,v}}\right) \tag{28}$$

By substituting Eqs. (27) and (28) into Eq. (26), the total capillary pressure is expressed as

$$\Delta p_c = \sigma\left(\frac{2}{\left(\frac{1}{2}\left(D_i - 2\delta_{0.5L_{c,v}}\right)\right)} - \frac{1}{\left(\frac{1}{2}(D_i - 2\delta_0)\right)}\right) \tag{29}$$

Then, the condensation film thickness can be determined from the Eqs. (29) and (22) with given heat flux, $q_c''$.

### 3.3. OHP Channel Sizing

In order to form a train of vapor bubbles in an OHP, the liquid and vapor must not be able to stratify in the channel. The surface tension, $\sigma$, of the working fluid and the curvature of liquid-vapor interface, $K$, are responsible for determining the maximum allowable OHP diameter in an OHP. For a static bubble or droplet in one-g environment, the shape of a bubble or droplet depends on the gravitational force and the pressure difference across the liquid-vapor interface. When the meniscus radius of the curvature at the liquid-vapor interface becomes constant, the bubble displays a perfect round sphere. When the radius of the bubble becomes larger in one-g environment, the gravitational force is greater than the force due to the surface tension and results in the shape deformation of the vapor bubble from the perfect round sphere. It has been demonstrated that the key determining the bubble or droplet shape in one-g environment is the ratio of the gravitational to the surface tension, the Bond number, i.e.,

$$Bo = \frac{g(\rho_l - \rho_v)D^2}{\sigma} \text{(31)} \qquad (30)$$

where $g$ is the acceleration due to gravity, $\rho_l$ and $\rho v$ are the working fluid liquid and vapor densities, and $D$ is the bubble or droplet diameter. Rearranging Eq. (30) yields the bubble or droplet diameter of

$$D = \sqrt{Bo \frac{\sigma}{g(\rho_l - \rho_v)}} \qquad (31)$$

Equation (31) can be used to estimate the tube diameter of the oscillating heat pipe. Experimental investigations show that when the Bond number is ranging from 1.84 to 2.00 in the one-g environment [8].

## 4. NANOFLUIDS OHP FABRICATION PROCESSES

### 4.1. Designing of Oscillating Heat Pipe

One of the keys to design is to determine the channel diameter, and this can be calculated using Eq. (31) as mentioned before. Choosing the working fluid is also another important factor to determine the performance of the OHP. For selecting the working fluid, there are fluid characteristics to consider as follows: surface tension, thermal conductivity, latent heat, specific heat, and viscosity. The higher surface tension enables the larger diameter of the channel. However, it also causes higher pressure drop of the fluid in the channel, which is not desirable to achieve high performance, because higher heat input is required to overcome the pressure drop and maintain oscillation motion.

Latent heat is the factor to determine the evaporation rate of the fluid. The oscillation motion of the liquid is started and maintained by the vapor pressure difference between the evaporator and condenser. Thus, to maintain the oscillation motion, higher vapor pressure at the evaporator and the lower vapor pressure in the condenser are required. When the latent heat is low, the liquid in the evaporator is evaporated easily, in turn, leading to high vapor pressure to maintain the oscillation motion. The thermal conductivity directly affects the performance of the OHP because the heat transport is mainly due to the oscillation motion of the bulk fluid (sensible heat). Thus, the fluid that has higher thermal conductivity is preferred for the better performance.

The specific heat is important in determining the amount of heat transferred from the evaporator to the condenser by forced convection of the bulk liquid. Since forced convection of the bulk liquid (sensible heat) is more responsible than the evaporation (latent heat), to determine the performance of OHP, higher specific heat is preferred to achieve better performance.

The viscosity is to determine the pressure drop in the channel. When fluid flows through the channel, it encounters the shear stress causing a pressure drop. To overcome the pressure drop and maintain oscillation motion, more powerful pumping power is required by supplying higher heat input. For the designing of an OHP, it is important to make working fluid oscillate at a lower temperature difference between the evaporator and condenser. This means that the OHP needs to be designed to work at the lower heat input; otherwise it will stop working when the heat input does not meet the requirement to maintain the oscillation. Thus, low viscosity of the fluid is preferred to achieve higher performance.

Charging ratio is the volume ratio of working fluid to the total volume of the channel. Adequate working fluid needs to be charged because there might not be enough vapor bubbles to pump the liquid to oscillate when it is over charged. On the contrary, when it is less charged, there might not be enough convective liquids to carry the heat from the evaporator to the condenser or, all the liquid could be evaporated causing dry-out. There is no exact equation to determine the right charging ratio as far as the authors know. However, there are some experimental reports to determine the charging ratio that leads to the best performance. However, it is different for each case by case.

Number of turns and the size of the evaporator and condenser also need to be considered in the OHP design. Unfortunately, there is no equation or formula to decide those parameters.

## 4.2. Nanofluid Selections

The elevated temperatures within the evaporator cause the vapor to expand and liquid to evaporate or boil. In the condenser the vapor contracts and condenses, while it expands as liquid evaporates in the evaporator. This expansion and contraction creates pressure variations within the OHP. The dominant pressure difference is between the evaporator and the condenser; however, lesser pressure differences occur between channels due to the random arrangement of the liquid and vapor. These pressure differentials result in chaotic fluid motion within the OHP [3]. Because the vapor and liquid move together in the OHP, there is no counter flow limitation that is present in the conventional heat pipe. Also, the fluid motion automatically rewets the surfaces, which enhances the evaporation when the vapor bubbles move over the wetted regions. All of these characteristics help the OHP passively transfer high heat fluxes. Even though pressure change and phase change contribute to the OHP motion, the primary mechanism of heat transfer was shown to be sensible heat by Shafii et al. [4]. Therefore, the working fluid in the OHP is a significant factor for its operation.

There are a number of important properties to consider when picking a working fluid: low viscosity reduces the pressure drop; high surface tension increases the maximum channel diameter of the capillary tube; high specific heat of the liquid increases the amount of energy the liquid slugs can transport; high thermal conductivity of the liquid allows the heat to quickly transfer into and out of the liquid; high latent heat increases the amount of energy moved via phase change; low density of the liquid phase reduces the effect of gravity, and high $\partial p/\partial T$ causes increased pressure variations within the OHP. These considerations will increase the oscillation motion, and in turn, transfer high heat flux for the better performance. The selection of the latent heat depends on the power input.

Because the OHP cannot function below the freezing point or above the critical temperature of its working fluid, the selected base fluid for nanofluid must be within this range. The selection of nanoparticles for a given base fluid depends on the compatibility with the base fluid, and agglomerations of nanoparticles. When the base fluid is determined, the nanoparticles are to be added in the base fluid. Extensive investigations on the thermal properties of nanofluids have been conducted and demonstrated that the nanoparticles added in the base fluid can modify the thermal conductivity, convection heat transfer coefficient, specific heat, and viscosity.

## 4.3. Building, Charging and Testing OHP

In the fabrication process, preparation, charging the OHP with nanofluids and sealing it are inevitable requiring careful attention for successful testing. Charging and sealing methods and testing procedures could be different for each study. In this chapter, the methods and procedures discussed were developed by Ma et al. [9], which represents both OHP preparation and testing as well.

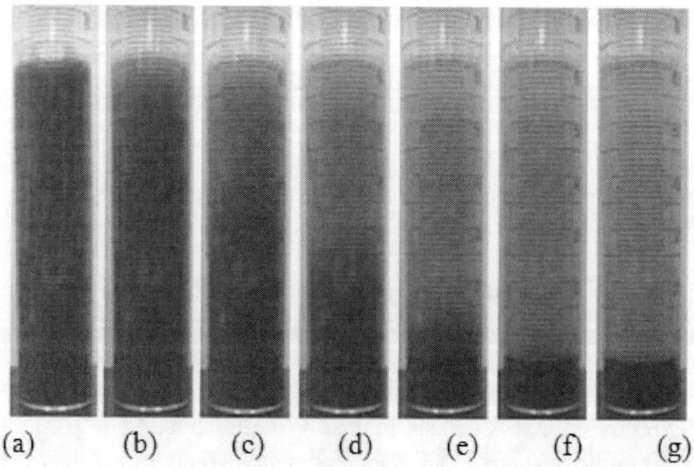


Figure 6. Settling down of nanoparticles at times of (a) 0 min, (b) 1 min, (c) 2 min, (d) 3 min, (e) 4 min, (f) 5 min and (g) 6 min [9].

For the preparation of the OHP, (Figure 9) the diameter of the channel needs to be considered first. The diameter can be approximately calculated using Bond number, Eq. (30) or its rearranged form Eq. (31). In the experiment of Ma et al. [9], the Bond number used was 1.84. Alloy 122 copper tubing with inner diameter of 1.65 mm and outer diameter of 3.18 mm was used. The number of turns is 12 and the size of the evaporator is 59.9 mm × 301.5 mm, and the one of the condenser is the same. The distance between the evaporator and condenser was 101.6 mm. For the preparation of the nanofluid, diamond nanoparticles with HPLC grade water were used. The nanoparticles were fabricated by 20 kW RF plasma with a high frequency of 13.56 MHz and their sizes are from 5 to 50 nm. The fabricated nanoparticles were added into HPLC grade water directly. While most of the nanoparticles are settled down as shown in Figure 6, some of them, with a size of less than 10 nm, can be suspended in the motionless water as shown in Figure 7, photographed by TEM. For the charging, the built OHP was placed on the scale and vacuumed. After the air was removed, the nanofluid was introduced into the OHP. Once the intended volume percentage was filled, the OHP was sealed. The filling ratio was 50%. Experimental setup is shown in Figure 8. A copper plate with dimensions 59.9 mm × 301.5 mm × 6.6 mm was used to reduce the thermal contact resistance between the heater and evaporator. The same method was used for the condenser. The OMEGATHERM "201" thermal paste was placed between the tube and the copper plate to reduce the thermal contact resistance. A strip heater was used to supply the heat to the evaporator. Two water-cooled aluminum blocks were attached to the condenser plate to dissipate the heat from the OHP. A Julabo F34 circulator was used to control the cooling water temperature and circulation to the condenser. Type T thermocouples were attached on the top surface of the OHP to monitor the temperature change. This will detect the oscillation motion of the working fluid. The entire OHP was wrapped by insulation material to ensure that the heat transfer occurred only at the evaporator and condenser.

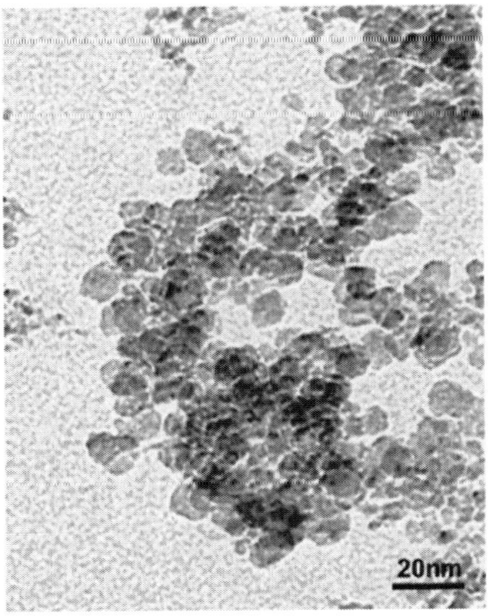


Figure 7. TEM image of diamond nanoparticles suspended in the motionless water [9].

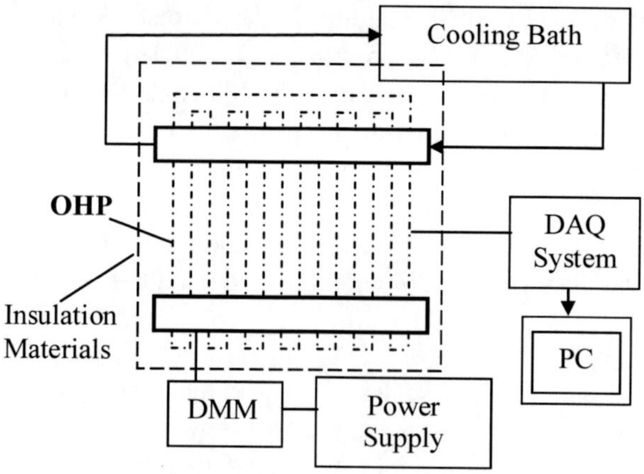


Figure 8. Experimental setup [9].

Before the testing, the OHP system was allowed to achieve equilibrium to reach the steady state. Then, the input power was supplied and increased in small increments. For each power input increment, the OHP was allowed to reach the steady-state. The temperature data for each successive power input increment and corresponding steady-state were collected. The thermal power and the temperature data were recorded simultaneously.

## 5. Parameters Affecting Heat Transfer Performance

The thermal characteristics of nanofluid are suited for the OHP. The improved thermal properties of the nanofluid aid the heat transfer in the OHP and the OHP's motion helps keep the nanoparticles suspended. Currently the experimental and theoretical understanding of the nanofluid OHP is rudimentary. Because it is new, most nanofluid OHP experiments are significantly different and only general trends can be determined from their combined results.

### 5.1. Operating Temperature Effect

Nanofluid was first used as the working fluid in an OHP in 2006 by Ma et al. [5], and more extensive tests of the same OHP and fluid were published later [9]. Ma et al. [5, 9] charged a 12 turns OHP with diamond nanofluid. The diamond nanoparticles, 20-50 nm in diameter, were added to pure water at a 1 vol% concentration. The bulk of the nanoparticles was found to settle in a matter of minutes when undisturbed. The remaining suspended nanoparticles were found to be less than 10 nm in diameter.

The nanofluid was charged into a closed loop, 12-turn OHP at a 50% filling ratio (Figure 9). The heat pipe was made out of 1.65 mm inner diameter copper tubing bent into a sinusoidal pattern and embedded into copper heat spreading plates for the evaporator and condenser. It was tested up to 336 W with condenser temperatures of 10 °C through 70 °C at 10 °C increments. It was vertically oriented so that the evaporator was gravitationally below the condenser. The results, Figure 10, show a significant improvement when using nanofluids compared to results where the base fluid of water served as the working fluid. The improvement is especially significant at low power; however, even at high power the improvement is measurable. It was also found that the thermal resistance decreased with increasing condenser temperature and power input with a minimum resistance of 0.03°C/W found at 336W and 70°C. Therefore, a nanofluid OHP exhibits improved thermal performance compared to a base fluid filled OHP.

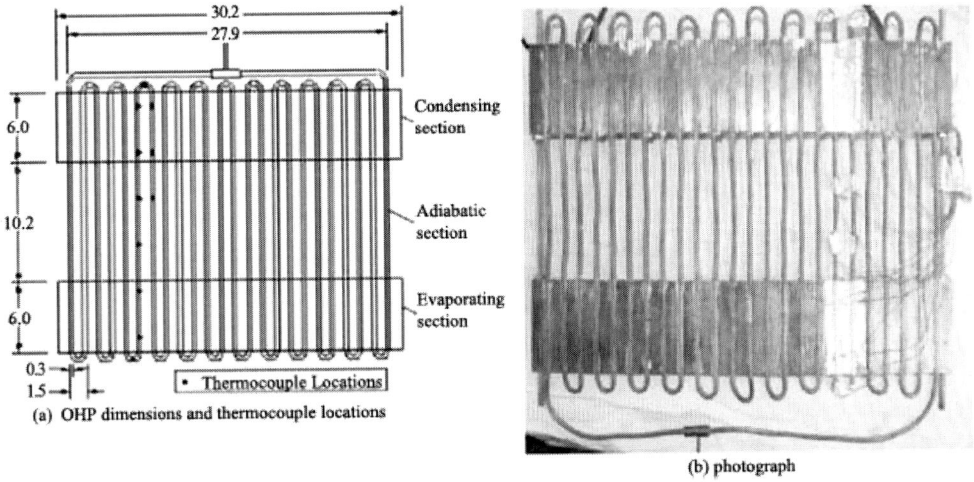


Figure 9. 12-turn nanofluid oscillating heat pipe (dimensions in cm) [5].

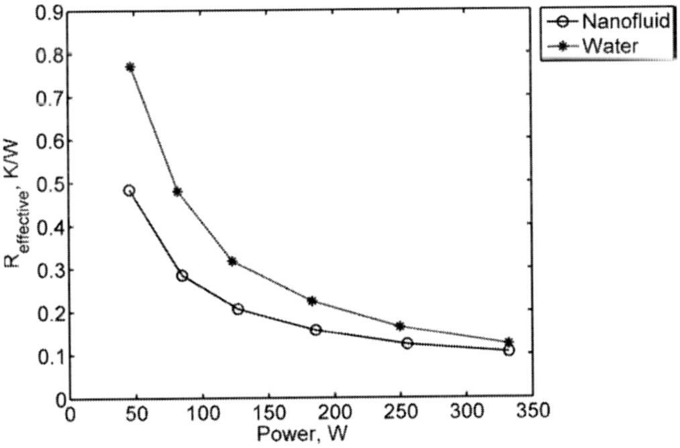


Figure 10. Nanofluid effect on the heat transport capability in a 12-turn OHP (The OHP was vertically oriented with bottom heating and circulator temperature of 20°C) [5].

## 5.2. Nanoparticle Effect on the Startup and Nucleation

For a conventional wicked heat pipe, when the nucleation occurs and bubbles form in the wicks, the boiling limitation has been reached; this should be avoided. For an OHP, the heat transport process is mainly due to the thermally excited oscillating motion. A number of investigators have performed visualization experiments using glass OHPs and demonstrated that the nucleation takes place in the evaporator section which can help the oscillating motion in an OHP [41-43]. More recently, Li et al. [44] added the nanoparticles in an OHP and demonstrated that the nanoparticles added in the OHP can significantly help the nucleation of bubbles and increase the nucleation site density.

Two factors affecting the boiling heat transfer on a surface are the nucleation of bubbles and nucleation site density. The initiate nucleation of the bubble on a surface depends on the cavity pore size and the superheat on the surface. If a vapor bubble to be initiated on a cavity has a meniscus radius of $R$, the superheat needed to grow the bubble can be found as

$$\Delta T = \frac{2\sigma T_{sat}}{h_{lv}\rho_v R} \tag{32}$$

where the meniscus radius $R$ depends on the cavity size. When the cavity size increases, the superheat needed to initialize the bubble decreases. Hsu [45] developed a semi-theoretical model to predict the effects of the cavity size and superheat on the nucleation of bubbles, bubble growth and release process. For a given thermal boundary layer thickness of $\delta_t$, the minimum meniscus radius, $r_{min}$, and the maximum meniscus radius, $r_{max}$, of the liquid-vapor interfaces of the cavities can be found as

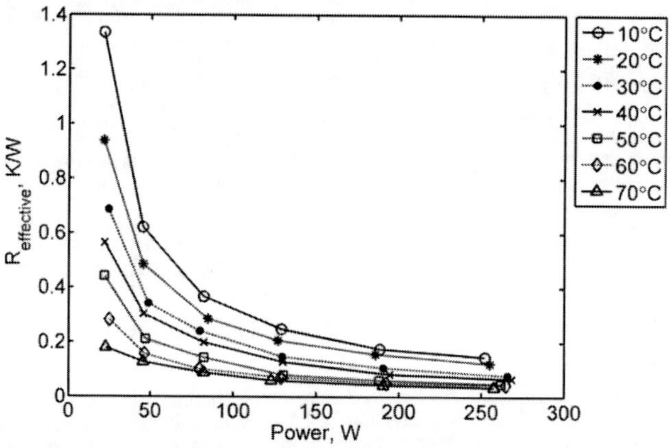


Figure 11. Thermal resistance at various heat loads and operating temperatures of the 12-turn nanofluid OHP oriented vertically with bottom heating [9].

$$\begin{Bmatrix} r_{c,\max} \\ r_{c,\min}\end{Bmatrix} = \frac{\delta_t}{4}\left[1-\frac{T_{sat}-T_\infty}{T_w-T_\infty}\begin{Bmatrix}+\\-\end{Bmatrix}\sqrt{\left(1-\frac{T_{sat}-T_\infty}{T_w-T_\infty}\right)^2-\frac{12.8\sigma T_{sat}}{\rho_v h_{lv}\delta_t\left(T_w-T_\infty\right)}}\right] \tag{33}$$

Equation (33) demonstrated that as the superheat, $T_w$-$T_\infty$, increases from zero, the calculated maximum meniscus radium, $r_{c,max}$, and the minimum meniscus radius, $r_{c,min}$, are imaginary. It means that the cavities are not active. When the superheat, $T_w$-$T_\infty$, increases, the discriminant,

$\left(1-\dfrac{T_{sat}-T_{\infty}}{T_w-T_{\infty}}\right)^2-\dfrac{12.8\sigma T_{sat}}{\rho_v h_{lv}\delta_t\left(T_w-T_{\infty}\right)}$, becomes zero and one solution for the meniscus radius can be obtained. When superheat, $T_w$-$T_\infty$, further increases, two real values will be obtained from Equation (33). It means that the cavities ranging from $r_{c,min}$ and $r_{c,max}$ will become active sites. In other words, if the surface does not have the cavities in the range between $r_{c,min}$ and $r_{c,max}$, the boiling does not take place for this given superheat. From Eq. (33), it can be found that when the superheat becomes larger, the range of active sites increases. It can be summarized that the model developed by Hsu [45] can result in two important features of the nucleation on a heated surface with cavities: 1) a certain minimum value of wall superheat is needed to activate the nucleation sites, and 2) when the wall superheat is above this minimum value, a finite range of cavities exist, and when the superheat increases, this range increases.

Qu and Wu [46] compared the performance of two different nanofluids in an OHP. They used $SiO_2$ with a diameter of 30 nm and $Al_2O_3$ with a diameter of 56 nm; both were spherical in shape and the base fluid was pure water. The nanoparticles were stabilized by adjusting the pH value. The $SiO_2$-water nanofluid was prepared in 0.1, 0.3, and 0.6 wt% concentrations. The $Al_2O_3$-water nanofluid was prepared in 0.1, 0.3, 0.6, 0.9, and 1.2 wt% concentrations. Two 6-turn closed loop OHPs were manufactured out of stainless steel capillary tubes with an inner diameter of 2 mm and an outer diameter of 3 mm. The lengths of evaporator, adiabatic, and condenser sections were 50 mm, 105 mm, and 70 mm, respectively. The OHPs were tested in the vertical orientation with the evaporator gravitationally below the condenser. The filling ratio for each nanofluid was 50%. One of the two OHPs was reserved for the $SiO_2$-water nanofluid and the other for the $Al_2O_3$-water nanofluid.

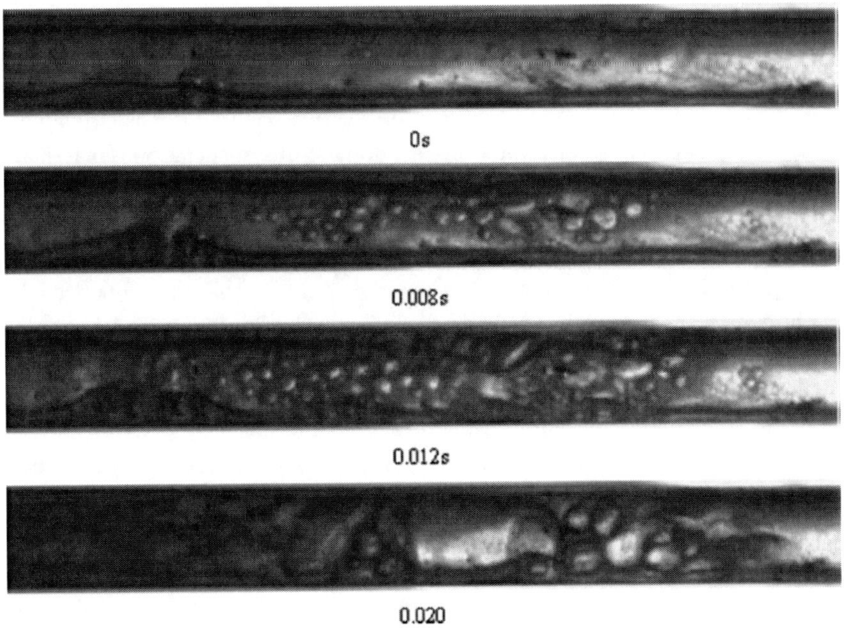


Figure 12. Strong boiling in the thick liquid film around a long vapor slug at Q=57 W [47].

The result showed that the thermal resistance for the $Al_2O_3$-water nanofluid was lower than that of pure water and had an optimal concentration of 0.9 wt%. The optimal thermal resistance was 0.057°C/W which was 25.7% better than pure water. This was similar to previous experiments [7]. However, the thermal performance of $SiO_2$-water nanofluid was worse than pure water and became worse with increasing concentration. The worst thermal performance occurred at the thermal resistance of 0.075°C/W, which was 23.7% worse than pure water. The suggested reason for the different thermal performance is that it is due to the different deposition pattern of the corresponding nanoparticles at the evaporator and condenser causing different surface conditions. The enhanced heat transfer when using $Al_2O_3$-water nanofluid is because the deposition of the nanoparticles, mostly at the evaporator, increases the surface nucleation site, while the decreased heat transfer by suing $SiO_2$-water nanofluid is because the deposition of the nanoparticles decreases the surface nucleation site and contact angle. Nanofluids surface effect will be discussed later in a separate section.

Li et al. conducted a visualization experiments of two-phase flows in an OHP charged with deionized (DI) water and a nanofluid (0.268% v/v) [47]. The OHP was made of quartz glass tube (with an inner diameter of 3.53 mm and an outer diameter of 5.38 mm) and coated with a transparent heating film in its evaporating section. The internal two phase flows at different heat loads were recorded by a charge-coupled device (CCD) camera. Experimental results demonstrated that when the nanoparticles were added in the OHP, the nucleation of bubbles can be easily started as shown in Figure 12. For a quartz glass tube, the surface is very smooth, i.e., the cavity size is very small. The superheat required to activate the nucleation is much larger. When the nanoparticles are added in the OHP as shown in Figure 12, the nanoparticles are deposited on the glass surface to form a number of nanoparticle clusters. Although the individual nanoparticle size is very small, the size of a nanoparticle cluster is much larger, which significantly decreases the superheat required to activate the nucleation of bubbles. When nanoparticles are added in an OHP fabricated with a very smooth wall surface like glass, the nanoparticles added in the OHP can help to increase the nucleation sites. The Li study [47] did not clearly mention why the enhanced heat transfer is achieved in their experiment. It might be expected that the some of deposited nanoparticles at the surface acted as crevice to form the boiling bubbles; this speculation has to be confirmed with further studies.

Ma et al. [9] investigated a copper OHP charged with nanofluids. Commercial alloy 122 copper tubing with an inside diameter of 1.65 mm and an outer diameter of 3.18 mm was used for the OHP. The inner surface of the copper tubing has no any enhanced cavities. In order to compare the heat transfer performance of the nanofluid OHP, the heat pipe charged with only HPLC grade water was tested. While the heat transport capability for the OHP charged with nanoparticles is much higher than that without nanoparticles, the startup power was significantly decreased. In addition to the higher thermal conductivity and higher heat transfer coefficient, the nanoparticles significantly help to increase the nucleation of the bubbles and site density. As shown when the input power was 80.0 W, the temperature difference from the evaporator to the condenser can be reduced from 40.9 °C for the one just charged with water to 24.3 °C for nanofluids. As the heat input increases, the oscillating motion becomes stronger and stronger. As a result, the temperature difference between the evaporator and the condenser would not increase as the power increased as shown in Figure 13. Due to the strong oscillating motion, the relative contribution of the nucleation site effect on the heat transport capability in an OHP goes down. Consequently, the thermal performance difference between the nanofluid OHP and that without nanoparticles becomes smaller as the power continuously increases.

The difference in thermal performance must be due to the nanofluid effect on the heat pipe. While the thermal conductivity, heat capacity, and dynamic viscosity of the nanofluid fluid affect the heat transport capability in an OHP, the nanoparticles added in the OHP modified the surface condition in addition to the modification of the nucleation of bubbles. Qu et al. [7] found that the nanofluid affected the surface of the OHP channel. Scanning electron microscope (SEM) images were taken of evaporator and condenser sections of the OHP tubing. It was found that both OHPs had nanoparticles deposited on the surface; however, $SiO_2$-water nanofluid had larger agglomerations in both the evaporator and condenser sections. In the $Al_2O_3$-water nanofluid, there were many nanoparticles deposited in the evaporator but negligible amounts of deposited nanoparticles in the condenser region.

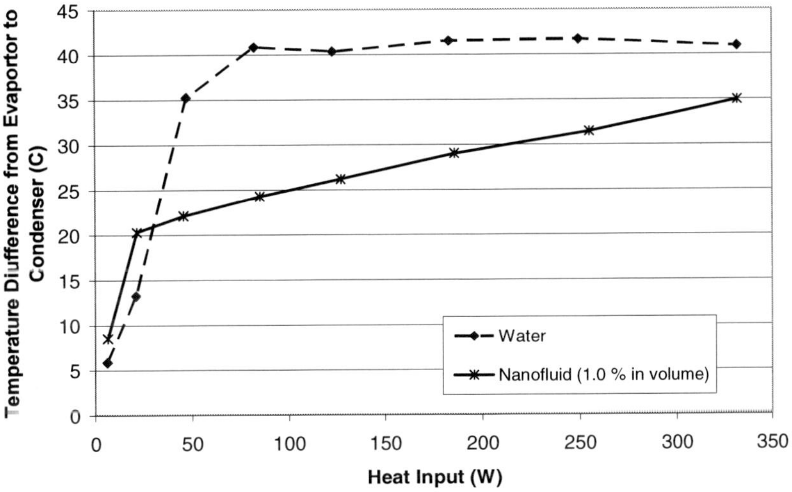


Figure 13. Effect of nanofluid on the heat transport capability in an oscillating heat pipe (vertical position, filled liquid ratio = 50%, nanoparticles: diamond, 20 – 50 nm, 1.0% in volume) [9].

Using boiling theory, Qu et al. [7] investigated the parameters relevant to the increase in OHP performance. The resistance of boiling is defined as

$$R_{evp} = \frac{1}{hA_e} \tag{34}$$

where $h$ is the boiling heat transfer coefficient and $A_e$ is the surface area of the evaporator. Substituting in Rohsenow's nucleate boiling equation gives,

$$R_{evp} = \frac{1}{2N_a D_b^2 A_e \sqrt{f} \sqrt{\pi k \rho c_p}} \tag{35}$$

where $N_a$ is the surface active nucleation site density, $D_b$ is the bubble release diameter, $f$ is the bubble release frequency, $k$ is the liquid thermal conductivity, $\rho$ is the liquid density, and $c_p$ is the liquid specific heat. Comparing the nanofluid properties to the base fluid properties shows that $k$, $\rho$ and $c_p$ are negligible. Also, the increase in surface area due to deposited nanoparticles is less than 2.1%. The nucleate evaporative resistance is,

$$R_{evp} \propto \frac{1}{N_a D_b^2 \sqrt{f}} \tag{36}$$

Therefore, the nucleate boiling resistance is only dependent on the number of active sites, the bubble diameter, and the release frequency. The active site density can be determined by,

$$N_a \propto N_c \left(1-\cos(\theta)\right) \tag{37}$$

where $N_c$ is the micro-cavity density and $\theta$ is the surface contact angle. Therefore, larger contact angles increase the density of active nucleation sites. The contact angle of the nanofluid was measured and found to be 65° for a clean surface, 29° for a $SiO_2$-water nanofluid deposited surface and 54° for an $Al_2O_3$-water nanofluid deposited surface. Based on this analysis, the $SiO_2$-water nanofluid would have the highest nucleate boiling resistance. This correlates to the $SiO_2$-water OHP results. However, the $Al_2O_3$-water OHP should also have worse results than pure water.

Using an atomic force microscope (AFM), the surface cavities and deposited nanoparticles were investigated for a clean surface and the nanofluid deposited surfaces. The clean surface was found to have a root-mean-square roughness of about 3 μm. The silica nanoparticle agglomerates were found to be an average 2-3μm in diameter. Therefore, these large agglomerates would fill the cavities and reduce the micro-cavity density on the surface. The nanoparticles also formed a thick layer on the surface adding to the thermal resistance of the wall. In the condenser the deposited nanoparticles would have increased wetting of the surface. This increased wetting would reduce condensation. Therefore, it was theorized that the poorer performance of the $SiO_2$-water nanofluid OHP could be attributed to the combination of the decreased number of active nucleation sites due to fewer cavities and reduced contact angle, the increased resistance of the nanoparticle layer, and increased condensation resistance. This would also explain why increasing concentrations of $SiO_2$ resulted in a worse performance.

As mentioned earlier, in the experimental report by Qu and Wu [46], even thou gh when using the $SiO_2$-water nanofluid, OHP performed worse than the base fluid, the $Al_2O_3$-water nanofluid OHP performed better. This can be attributed to the smaller agglomerations, larger contact angle, and minimal deposition in the condenser. It was theorized that the agglomerations did not completely fill the cavities, resulting in an increased number of nucleation sites. As the concentration increases, the agglomerations increase in size until they completely fill the cavities resulting in reduced thermal performance.

## 5.3. Effects of Nanoparticle Concentration and Filling Ratio

There are many independent parameters in the nanofluid OHP that should be analyzed for a complete understanding of the device. Lin et al. [6] analyzed several of these by varying the nanofluid concentration, filling ratio and heating power. They used silver nanofluid as the working fluid in a 5-turn close loop OHP built out of 2.45 mm inner diameter tubing (Figure *14*). The heat pipe was wound with heating wire in the evaporator region and cooled via 4.5m/s 30°C forced convection in the condenser. The OHP was tested at 20%, 40%, 60%, and 80% filling ratio with pure water, 100 ppm and 450 ppm silver nanofluid. The power was applied from 5 to 85 W at 10 W increments for each case. The same OHP was recharged for all tests.

It is expected that the best filling ratio would be 40 – 60 % because a lower filling ratio could result in dry-out, while higher filling ratio might hinder the oscillation of the fluid. The results show that, for the pure water test, the filling ratio of 20% above 35W was found to perform the worst, and, at 80%, the OHP will not start until 25W but performed well otherwise.

When comparing the concentrations, 100 ppm concentration silver nanofluid shows better performance than 450 ppm concentration silver nanofluid. One key reason to add nanoparticle to the heat transfer fluid is to utilize the high thermal conductivity of the solid particles. From this idea, it might be expected that higher concentration nanofluid has better heat transport. However, the experimental result from this study shows that the lower concentration nanofluid has better heat transport. This is because the higher concentration nanofluid has higher viscosity, causing less bubble formation. The vapor bubble has an important role for thermal performance because the driving force is due to the vapor pressure difference between the evaporator and condenser. Thus, there should be an adequate amount of bubbles to maintain the oscillation and to help convective heat transfer between the liquid slug and wall. The higher viscosity also might produce high shear stress, causing frictional force; in turn, leading to the pressure drop. Thus, even though the higher concentration nanofluid might be expected to have high heat transport theoretically, the lower concentration nanofluid shows a better thermal performance. Additionally, the best filling ratio was 60% with the 85 W heating power, and its thermal resistance was 0.092°C/W. Thus, it might be concluded that the filling ratio effect of nanofluids are the same with normal heat transfer fluids, and lower concentrations might show higher thermal performance. However, it does not necessarily mean that lower concentration is always better. At an extremely lower concentration, there is possibility that the thermal performance might be poor, which needs to be confirmed by further studies.

A study by Wannapakhe et al. [10] looked at the effect of evaporator length, orientation, and more closely at nanoparticle concentration. Silver nanoparticles with a diameter of less than 100 nm were combined with deionized water at mass concentrations of 0.25, 0.5, 0.75 and 1%w/v, and sonicated for five hours. The resulting nanofluid was shown to be stable for at least 48 hours. The nanofluid was charged into a 40-turn closed loop OHP with an inner diameter of 2mm. Two ball check valves were included in the OHP to enhance the flow pattern within the OHP (Figure 15).

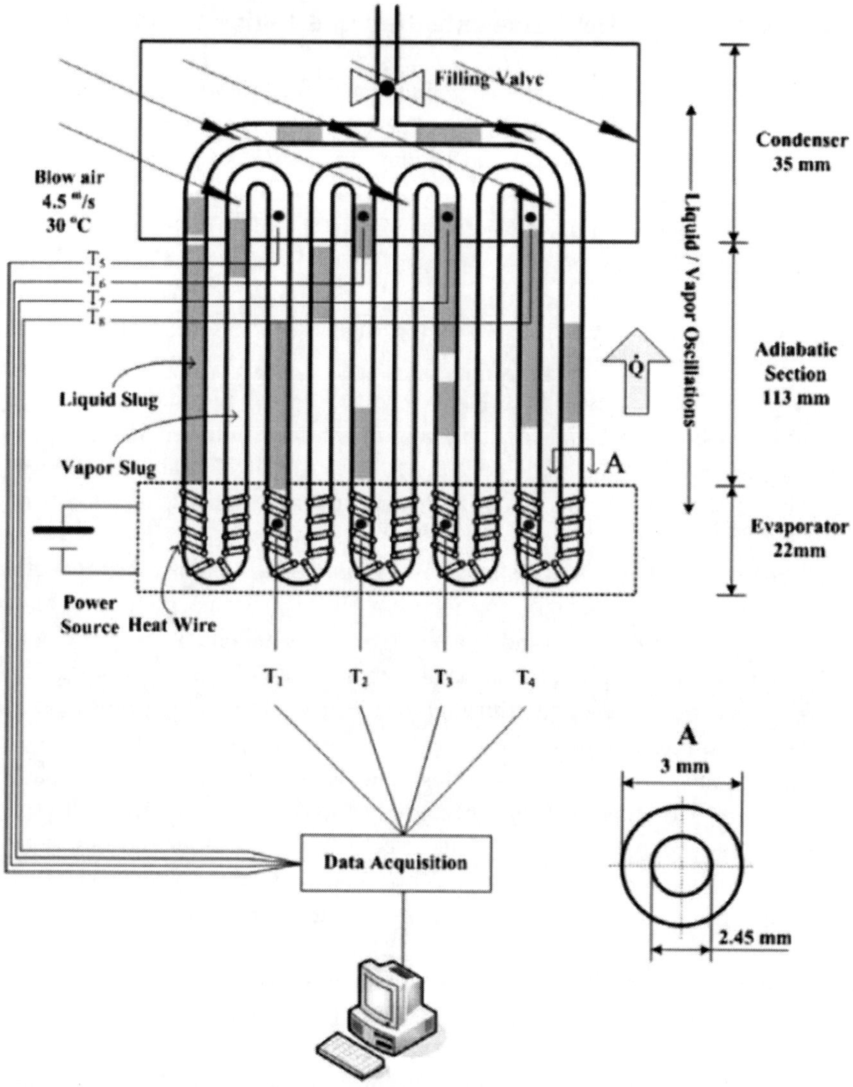


Figure 14. Schematic of the silver-nanofluid experiment [6].

The OHP was heated and cooled by two constant temperature baths flowing at 0.5 L/min. The evaporator temperature was set 40 °C, 50 °C, and 60 °C, respectively. The angle of the OHP was varied from 0, 20, 40, 60, 80, and 90° from the horizontal plane and the evaporator length ratio was 25, 50 and 75. The evaporator aspect ratio is,

$$\phi = \frac{L_e}{D_i} \tag{38}$$

where $L_e$ is the evaporator length, and $D_i$ is the tube inner diameter. The adiabatic length was 100 mm and the condenser was 150 mm.

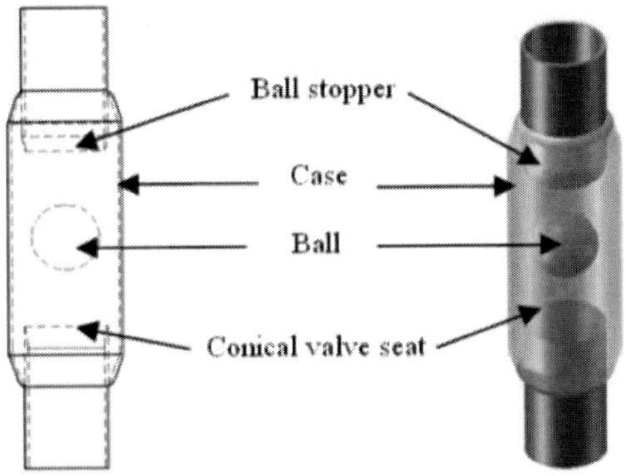


Figure 15. Check valve design [10].

It was found that the optimal parameters were a vertical orientation (90°) with 0.5%w/v silver nanoparticles, an evaporator aspect ratio of 25, and an evaporator temperature of 60°C. This case had a heat transfer rate of 13.19 kW/m$^2$. The vertical orientation was the best orientation, the performance improved as the evaporator aspect ratio decreased, and the performance improved as the evaporator temperature increased. The experiment also found that nanofluid routinely performed better than pure water at all orientations and aspect ratios. It also showed that the lower concentration generally has better performance. The optimal concentration was achieved with a 0.5%w/v silver nanofluid. It was suspected that the increased thermal performance of the higher concentration was outweighed by the increased viscosity, which confirms the result by Lin et al. [6]. However, sometimes, the concentration of 0.75 % shows better heat transport than that of the concentration of 0.25 %. It means that, sometimes, the heat transport is influenced by the thermal conductivity rather than the viscosity. Thus, this experimental result shows that, generally, lower concentration is preferred; however, there might some exceptional cases where a higher concentration is desirable for better performance. To confirm,, there needs to be more studies theoretically and experimentally.

## 5.4. Surface Effect of Nanofluids

An experiment by Qu et al. [7] looked at the surface effects of nanoparticles in OHPs. They used $Al_2O_3$-water nanofluid as their working fluid at weight percentages of 0.1%, 0.3%, 0.6%, 0.6%, 0.9% and 1.2%, respectively. The nanoparticles had a mean diameter of 56 nm and were stabilized using small amounts of hydrochloric acid to adjust the PH to 4.9 to prevent agglomeration and insure the solution remained stable for at least three days. The OHP had 6 turns with an inner diameter of 2 mm and an outer diameter of 3 mm. The lengths of evaporator, adiabatic and condenser section were 50, 105 mm, and 70 mm, respectively. The condenser was held at a constant temperature and the evaporator was heated up to nearly 140 W. The OHP was charged with pure water or nanofluids at filling ratios of 50%, 60%, and 70%, respectively. All tests were conducted in the vertical orientation with the evaporator gravitationally below the condenser. Qu et al. [7] further found that the thermal resistance of nanofluid was less than pure water and that a lower filling ratio was better. The optimal nanofluid concentration was 0.9 wt% with the thermal resistance increasing as with lower and higher concentrations.

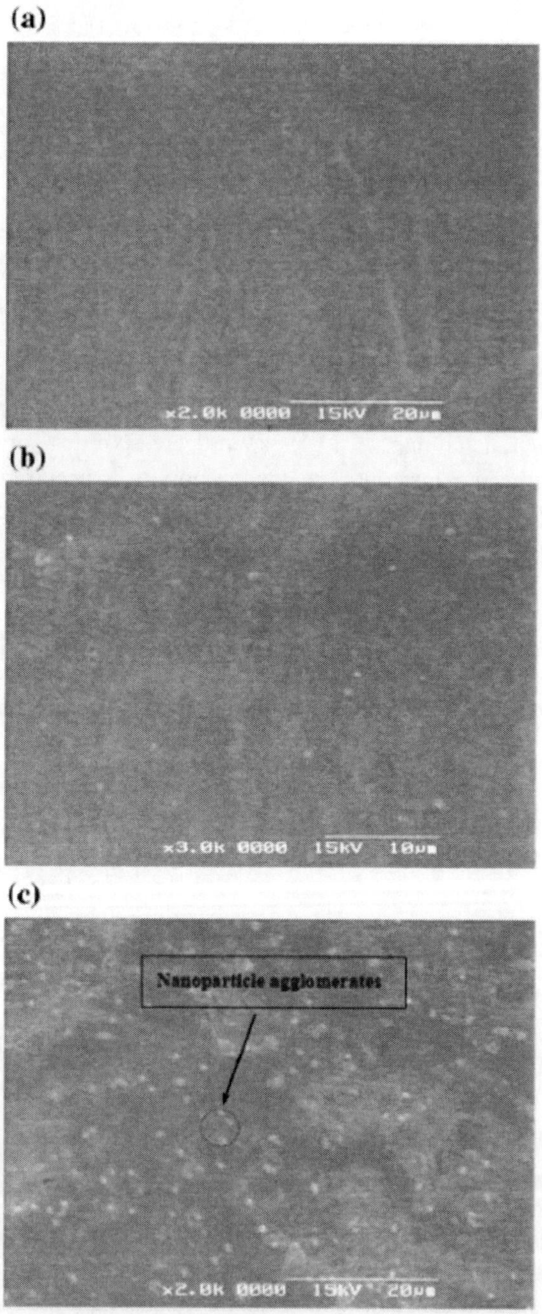


Figure 16. SEM images of (a) a clean substrate boiled in pure water, (b) a nanoparticle-deposited substrate at the condenser, and (c) a nanoparticle-deposited substrate at the evaporator [7].

They also investigated the effect of nanofluids on the channel surface of the OHP and its relation to boiling and evaporation. Tube samples were taken from the evaporator and condenser of a nanofluid OHP and one from a pure-water OHP. The channel surface was observed with a scanning electron microscope. They found that the pure fluid channel was clean; however, the nanoparticles deposited and agglomerated on the evaporator and condenser surface in the nanofluid OHP. The number of nanoparticle agglomerations on the surface in the condenser was negligible, but the evaporator had a significant number (see Figure 16).

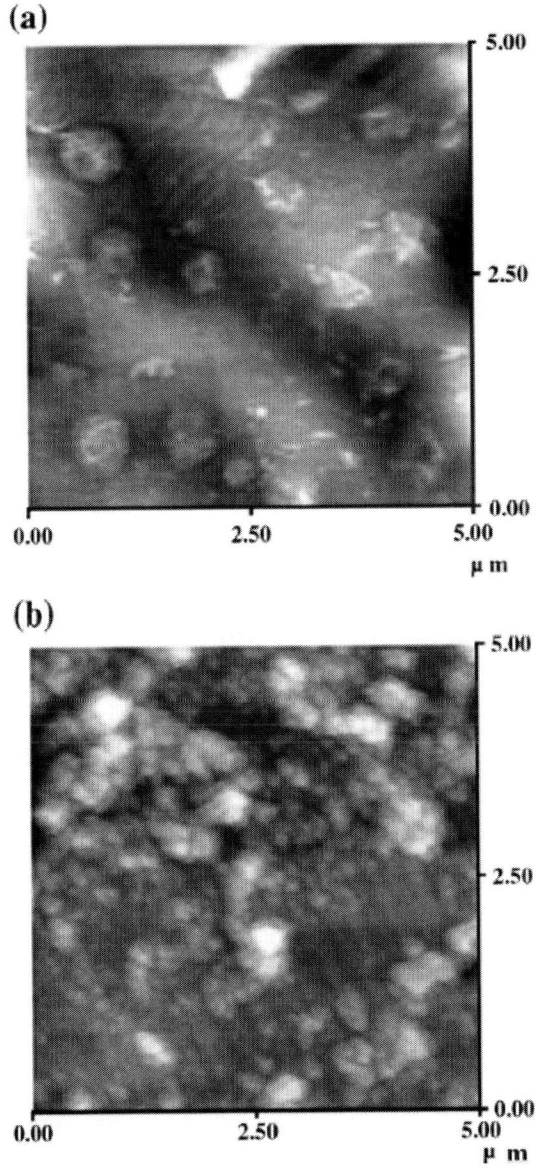


Figure 17. 2D atomic force microscope images of (a) a clean substrate boiled in pure water and (b) a nanoparticle-deposited substrate at the evaporator [7].

The nanoparticles on the surface of the evaporator will change the evaporative and boiling properties of the surface. Because nucleation sites are related to surface roughness, the increased surface roughness due to nanoparticles and their agglomerations would increase the bubble generation. The cavities on the clean surface were observed to be 2-3 μm in diameter (Figure 17). Those cavities in the nanofluid OHP would be filled with nanoparticles, thereby increasing the number of nucleation sites. The increase of nucleation sites and frequency improve the thermal performance of the nanofluid OHP. It was also suspected that increasing the concentration of nanoparticles beyond a certain value would fill in the nucleation sites too much, which is decreasing the nucleation sites, and therefore reduce the thermal performance.

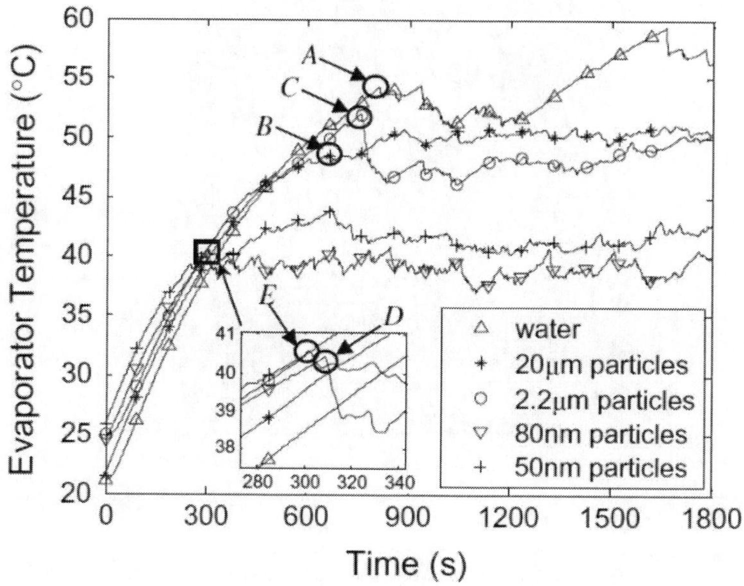


Figure 18. Particle size effect on the startup temperature [48].

## 5.5. Effects of Nanoparticle Size and Shape

An experiment by Ji et al. [48] investigated the particle size effect on the performance of the OHP. They used a 6-turn closed loop OHP with copper heat spreading plates in the evaporator and condenser. A copper tube with 1.65mm inner diameter and 3.18mm outer diameter was imbedded into the plates. The OHP had an evaporator length of 40 mm, adiabatic length of 51 mm, and condenser length of 64 mm. The OHP was charged with $Al_2O_3$-water nanofluid, with particle sizes of 20 μm, 2.2 μm, 80 nm and 50 nm, respectively. Each of these particles was mixed with the base fluid at a concentration of 0.5 wt%. All tests were conducted in the vertical orientation with a condenser operating temperature of 25°C. The heat pipe was tested up to 200 W. First, the optimal filling ratio was determined using pure water. It was found that the thermal performance was improved with decreasing filling ratio and the optimal filling ratio was found to be 50%. The start-up temperature was compared for the different nanofluids and pure fluid. Start-up temperature is defined as the minimum evaporator temperature at which the OHP will oscillate. Pure water had the highest start-up temperature of 54.2°C (Figure 18). For the nanofluid OHPs, as nanoparticle size decreased the start-up temperature decreased. They also found that as particle size decreased from 20 μm to 80 nm, thermal resistance decreased; however, the 50 nm nanofluid performed slightly worse than the 80 nm nanofluid. Therefore, the optimal nanofluid size was 80 nm. The lowest OHP thermal resistance was 0.113 °C/W with 80 nm nanoparticles at 25 °C and 200 W.

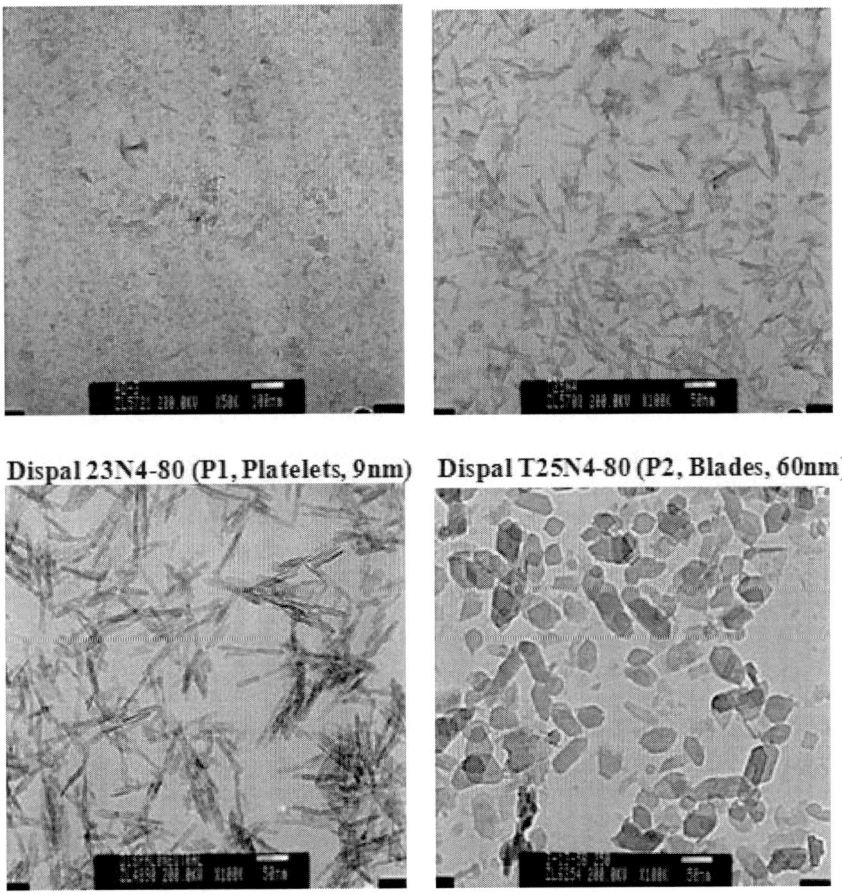


Figure 19. TME images of alumina nanoparticles [49].

Ji et al. [49] investigated the particle shape effect on the performance of the OHP. They used the same OHP as the one for the study of the nanoparticle size effect [48], i.e., a 6-turn closed loop OHP with copper heat spreading plates in the evaporator and condenser. A copper tube with 1.65mm inner diameter and 3.18mm outer diameter was imbedded into the plates. The OHP had an evaporator length of 40 mm, adiabatic length of 51 mm, and condenser length of 64 mm. In their study, the nanoparticles of boehmite alumina with different shapes (platelet, blade, cylinder and brick, Figure19) were used. Ethylene glycol (EG) 99+% (Fisher) and deionized water was mixed 50/50 by volume was used as the base fluid for all preparations. The particles were directly added into the base fluid at concentrations of 0.3 vol%, 1 vol%, 3 vol% and 5 vol%. Figures 20 illustrate the particle shape effect on the OHP heat transfer performance at the operating temperature of 20°C and 60°C respectively. In these figures, P1, P2, P3 and P4 stand for plateletlike, bladelike, cylinderlike and bricklike shape particles respectively, and V03, V1, V3 and V5 stand for the volume fraction of 0.3%, 1%, 3% and 5% respectively. So, the combination of P and V can stand for different nanofluids. BF means the working fluid is the base fluid without any particles. From Figure 20, it can be found that at the operating temperature of 20°C, the heat transport capability depends on the particle shape and volume fraction. When the input power is less than 100W, the OHP charged with P1 (volume fraction<3%), P2 (volume fraction<1%), P3 (volume fraction<3%) and P4 (volume fraction<1%), respectively can enhance the heat pipe performance. The heat transfer performance largely depends on the volume fraction. For the OHPs charged with P1, P2 and P4, respectively, the optimum volume fraction is about 0.3% while for the OHP charged with P3, the optimum fraction is about 1%. At a power input less than 100 W and a volume fraction of 0.3%, the OHP charged with P3 (cylinder) obtained the best heat transfer performance while the OHP charged with P4 (brick) showed the lowest among four types of particles. The sequence of heat transfer enhancement from the highest to lowest is: P3 (cylinder)>P2 (blade)>P1 (plate)>P4 (brick). However, when the input power is higher than 125 W, the OHP charged with P4 (brick) obtained the best heat transfer performance. The sequence of heat transfer enhancement from the highest to lowest becomes: P4 (brick)>P3 (cylinder)>P1 (plate)>P2 (blade). It can be concluded that the alumina nanoparticles added in the OHP can enhance the heat transfer performance of OHP significantly, but it depends on particle shape and volume fraction.

## CONCLUSION

In most cases it was found that nanofluids can substantially improve the heat transfer performance of the OHP and can reduce the startup power. The field is still very new and we are just starting to understand the mechanisms for this improvement. Currently most experiments are vastly different so it is difficult to draw direct conclusion. However the overall conclusion is that nanoparticles added in the base fluid can modify the heat transport capability and an optimal nanofluid concentration exists for the OHP. Below and above this concentration, the thermal resistance increases. There are two proposed theories about the cause of the increase in resistance at high nanoparticle concentrations: 1) the nanofluid viscosity increases reducing the fluid motion, and 2) the nucleation cavities fill with nanoparticles preventing nucleation and increasing thermal resistance due to deposited nanoparticles. It is also worth noting that the effect of nanoparticle deposition on thin-film evaporation has not been investigated and might play a significant role in the thermal performance of the nanofluid OHP. These aspects warrant further investigation.

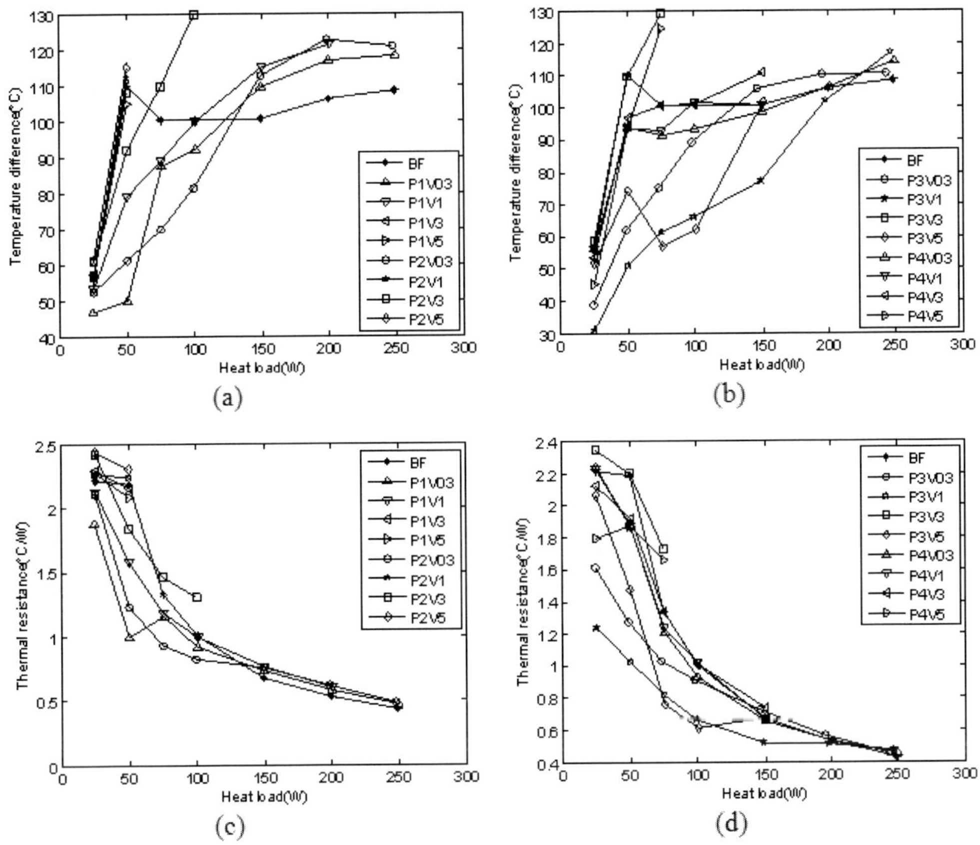


Figure 20. Particle shape effect on (a) temperature difference (P1, P2), (b) temperature difference (P3 and P4), (c) thermal resistance (P1 and P2), and (d) thermal resistance (P3 and P4) (operating temperature: 20℃, filling ratio: 50%, P1:platelet, P2:blade, P3:cylinder and P4:brick) [49].

## ACKNOWLEDGEMENTS

The work presented in this chapter is supported in part by the Office of Naval Research under grant number N000141110334 and US National Science Foundation under grant number CBET -1066917.

## REFERENCES

[1] Akachi H., "Structure of a Heat Pipe," U.*S. Patent #4,921,041.*

[2] Zhang Y., and Faghri A., 2008, "Advances and Unsolved Issues in Pulsating Heat Pipes," *Heat Transfer Engineering*, 29(1), 20–44.

[3] Song Y., and Xu J., 2009, "Chaotic behavior of pulsating heat pipes," *International Journal of Heat and Mass Transfer*, 52(13-14), 2932–2941.

[4] Shafii M. B., Faghri A., and Zhang Y., 2002, "Analysis of heat transfer in unlooped and looped pulsating heat pipes," *International Journal of Numerical Methods for Heat & Fluid Flow*, 12(5), 585–609.

[5] Ma H. B., Wilson C., Borgmeyer B., Park K., Yu Q., Choi S. U. S., and Tirumala M., 2006, "Effect of nanofluid on the heat transport capability in an oscillating heat pipe," *Applied Physics Letters*, 88(14), 143116.

[6] Lin Y.-H., Kang S.-W., and Chen H.-L., 2008, "Effect of silver nano-fluid on pulsating heat pipe thermal performance," *Applied Thermal Engineering*, 28(11-12), 1312–1317.

[7] Qu J., Wu H., and Cheng P., 2010, "Thermal performance of an oscillating heat pipe with Al2O3–water nanofluids," *International Communications in Heat and Mass Transfer*, 37(2), 111–115.

[8] Taft B., Williams A., and Drolen B., 2011, "Working Fluid Selection for Pulsating Heat Pipes," *42nd AIAA Thermophysics Conference,* Honolulu, Hawaii, USA.

[9] Ma H. B., Wilson C., Yu Q., Park K., Choi U. S., and Tirumala M., 2006, "An Experimental Investigation of Heat Transport Capability in a Nanofluid Oscillating Heat Pipe," *Journal of Heat Transfer*, 128(11), p. 1213-1216.

[10] Wannapakhe S., Rittidech S., Bubphachot B., and Watanabe O., 2010, "Heat transfer rate of a closed-loop oscillating heat pipe with check valves using silver nanofluid as working fluid," *Journal of Mechanical Science and Technology*, 23(6), 1576–1582.

[11] Qu J., and Wu H., 2011, "Thermal performance comparison of oscillating heat pipes with SiO2/water and Al2O3/water nanofluids," *International Journal of Thermal Sciences*, 50(10), 1954–1962.

[12] Choi S. U. S., 1995, "Enhancing Thermal Conductivity of Fluids with Nanoparticles," *Developments and Applications of Non-Newtonian Flows,* eds. D. A. Singer and H. P. Wang, American Society of Mechanical Engineers, New York, 99–105.

[13] Buongiorno, J., Venerus, D. C., Prabhat, N., McKrell, T., Townsend, J., Christianson, R., Tolmachev, Y. V., Keblinski, P., Hu, L., Alvarado, J. L., Bang, I. C., Bishnoi, S. W., Bonetti, M., Botz, F., Cecere, A., Chang, Y., Chen, G., Chen, H., Chung, S. J., Chyu, M. K., Das, S. K., Di Paola, R., Ding, Y., Dubois, F., Dzido, G., Eapen, J., Escher, W., Funfschilling, D., Galand, Q., Gao, J., Gharagozloo, P. E., Goodson, K. E., Gutierrez, J. G., Hong, H., Horton, M., Hwang, K. S., Iorio, C. S., Jang, S. P., Jarzebski, A. B., Jiang, Y., Jin, L., Kabelac, S., Kamath, A., Kedzierski, M. A., Kieng, L. G., Kim, C., Kim, J.-H., Kim, S., Lee, S. H., Leong, K. C., Manna, I., Michel, B., Ni, R., Patel, H. E., Philip, J., Poulikakos, D., Reynaud, C., Savino, R., Singh, P. K., Song, P., Sundararajan, T., Timofeeva, E., Tritcak, T., Turanov, A. N., Van Vaerenbergh, S., Wen, D., Witharana, S., Yang, C., Yeh, W.-H., Zhao, X.-Z., Zhou, S.-Q., 2009, "A Benchmark Study on the Thermal Conductivity of Nanofluids," *J. Appl. Phys,* 106, 094312.

[14] Kalteh, M., Abbassi, A., Saffar-Avval, M., Frijns, A., Darhuber, A., Harting, J., 2012, "Experimental and numerical investigation of nanofluid forced convection inside a wide microchannel heat sink," *Applied Thermal Engineering*, 36, 260-268.

[15] Wen D., Lin G., Vafaei S., and Zhang K., 2009, "Review of nanofluids for heat transfer applications," *Particuology*, 7(2), 141–150.

[16] Das S. K., Choi S. U. S., and Patel H. E., 2006, "Heat Transfer in Nanofluids—A Review," *Heat Transfer Engineering*, 27(10), 3–19.

[17] Wang X. Q., and Mujumdar A. S., 2008, "A Review on Nanofluids – Part I: Theoretical and Numerical Investigations," *Brazilian Journal of Chemical Engineering*, 25(4), 613 – 630.

[18] Keblinski P., Phillpot S. R., Choi S. U., and Eastman J. A., 2002, "Mechanisms of Heat Flow in Suspensions of Nano-Sized Particles (Nanofluids)," *International Journal of Heat and Mass Transfer*, 45, 855–863.

[19] Eastman J. A., Phillpot S. R., Choi S. U. S., and Keblinski P., 2004, "Thermal Transport in Nanofluids," *Annual Review of Materials Research*, 34(1), 219–246.

[20] Yu W., and Choi S. U. S., 2003, "The role of interfacial layers in the enhanced thermal conductivity of nanofluids: A renovated Maxwell model," *Journal of Nanoparticle Research*, 5, 167–171.

[21] Xue L., Keblinski P., Phillpot S., Choi S. U.-S., and Eastman J., 2004, "Effect of liquid layering at the liquid–solid interface on thermal transport," International Journal of Heat and Mass Transfer, 47(19-20), 4277–4284.

[22] Geiger G. H., and Poirier D. R., 1973, *Transport Phenomena in Metallurgy,* Addison-Wesley Readings, PA.

[23] Ashcroft N. W., and Mermin N. D., 1976, *Solid State Physics, Holt,* Rinehart and Winston, New York.

[24] Yu W., and Xie H., 2012, "A Review on Nanofluids: Preparation, Stability Mechanisms, and Applications," *Journal of Nanomaterials*, 2012, 1–17.

[25] Akoh H., Tsukasaki Y., Yatsuya S., and Tasaki A., 1978, "Magnetic properties of ferromagnetic ultrafine particles prepared by vacuum evaporation on running oil substrate," *Journal of Crystal Growth*, 45, 495–500.

[26] Eastman J. A., Choi U. S., Li S., Thompson L. J., and Lee S., 2011, "Enhanced Thermal Conductivity through the Development of Nanofluids," MRS Proceedings, 457.

[27] Lee S., Choi S. U.-S., Li S., and Eastman J. A., 1999, "Measuring Thermal Conductivity of Fluids Containing Oxide Nanoparticles," Journal of Heat Transfer, 121(2), 280-289.

[28] Wang X., Choi S. U. S., and Xu X., 1999, "Thermal Conductivity of Nanoparticle - Fluid Mixture," *Journal of Thermophysics and Heat Transfer*, 13(4), 474–480.

[29] Xuan Y., and Li Q., 2000, "Heat Transfer Enhancement of Nanofluids," *International Journal of Heat and Fluid Flow*, 21, 58–64.

[30] Eastman J. A., Choi S. U. S., Li S., Yu W., and Thompson L. J., 2001, "Anomalously Increased Effective Thermal Conductivities of Ethylene Glycol-Based Nanofluids Containing Copper Nanoparticles," *Applied Physics Letters*, 78(6), 718-720.

[31] Patel H. E., Das S. K., Sundararajan T., Sreekumaran Nair A., George B., and Pradeep T., 2003, "Thermal Conductivities of Naked and Monolayer Protected Metal Nanoparticle Based Nanofluids: Manifestation of Anomalous Enhancement and Chemical Effects," *Applied Physics Letters*, 83(14), 2931-2933.

[32] Grover G. M., Cotter T. P., and Erickson G. F., 1964, "Structures of Very High Thermal Conductance," *Journal of Applied Physics*, 35(6), 1990-1991.

[33] Shafii M. B., Faghri A., and Zhang Y., 2001, "Thermal Modeling of Unlooped and Looped Pulsating Heat Pipes," *Journal of Heat Transfer*, 123(6), 1159-1172.

[34] Ma H. B., Hanlon M. A., and Chen C. L., 2005, "An investigation of oscillating motions in a miniature pulsating heat pipe," *Microfluidics and Nanofluidics*, 2(2), 171–179.

[35] Zhang Y., and Faghri A., 2003, "Oscillatory Flow In Pulsating Heat Pipes With Arbitrary Numbers Of Turns," *Journal of Thermophysics and Heat Transfer*, 17, 340–347.

[36] Zhang Y., and Faghri A., 2002, "Heat transfer in a Pulsating Heat Pipe with Open End," *International Journal of Heat and Mass Transfer*, 45, 755–764.

[37] Ma H. B., Borgmeyer B., Cheng P., and Zhang Y., 2008, "Heat Transport Capability in an Oscillating Heat Pipe," ASME *Journal of Heat Transfer*, 130, 081501.

[38] Chen, I.C., 1963, "A Correlation for Boiling Heat Transfer to Saturated Fluids in Convective Flow," *ASME preprint 63-HT-34, Presented at 6th National Heat Transfer Conference,* Boston, MA.

[39] Forster, H.K., and Zuber, N., 1955, "Dynamics of Vapor Bubbles and Boiling Heat Transfer," *AIChE Journal*, 1, 531-535.

[40] Cai, Q., Chen, R., and Chen, C. L., An Investigation of Evaporation, Boiling and Heat Transport Performance in Pulsating Heat Pipe, *Proc. ASME International Mechanical Engineering Congress & Exposition,* New Orleans, Louisiana, USA, 2002.

[41] Nikkanen, K., Lu, C.G., and Kawaji M., Effects of Working Fluid, Fill Ratio and Orientation on Looped and Unlooped Pulsating Heat Pipes, Proc. of ASME Summer Heat Transfer Conference, San Francisco, CA, USA, 2005. Tong, B. Y., Wong, T. N., and Ooi, K. T., Closed-Loop Pulsating Heat Pipe, *Applied Thermal Engineering*, 21, 1845–1862, 2001.

[42] Katpradit T., Worngratanapaisarn T., Terdtoon P., Ritthidech S., Chareonsawan P., and Waowaew S., 2004, "Effect of aspect ratios and Bond number on internal flow patterns of closed end oscillating heat pipe at critical state," *Proc. 13th Int. Heat Pipe Conf.,* Shanghai, PR China.

[43] Khandekar, S., Schneider, M., Schafer, P., Kulenovic, R., and Groll, M., Thermofluiddynamic Study of Flat Plate Closed Loop Pulsating Heat Pipes, *Microscale Thermophysical Engineering*, 6(4), 303–318, 2002.

[44] Li Q.-M., Zou J., Yang Z., Duan Y.-Y., and Wang B.-X., 2011, "Visualization of Two-Phase Flows in Nanofluid Oscillating Heat Pipes," *Journal of Heat Transfer*, 133(5), p. 052901.

[45] Hsu, Y.Y., 1962, "On the Size Range of Active Nucleation Cavities on a Heating Surface," ASME *Journal of Heat Transfer*, 84, 207–216.

[46] Qu, J., and Wu, H.Y., Thermal Performance Comparison of Oscillating Heat Pipes with $SiO_2$/Water and $Al_2O_3$/Water Nanofluids, *Journal of Thermal Science*, 50(10), 1954-1962.

[47] Li, Q.M., Zou, J., Yang, Z., Duan, Y.Y., and Wang, B.X., "Visualization of Two-Phase Flows in Nanofluid Oscillating Heat Pipes," ASME *Journal of Heat Transfer*, 133(5), 052901.

[48] Ji, Y., Su, F., Wang, G., Ma, H.B., 2011, "Particle Size Effect on the Heat Transport Capability in an Oscillating Heat Pipe," *Experimental Thermal and Fluid Science*, Vol. 35, 724-727.

[49] Ji, Y., Wilson, C., Chen, H.-H., and Ma, H., 2011, "Particle Shape Effect on Heat Transfer Performance in an Oscillating Heat Pipe," *Nanoscale Res Lett*, 6(1), 296.

# INDEX

## A

acetic acid, 178
acetone, 263, 333
acid, 59, 141, 147, 155, 156, 161, 173, 183, 359
acrylate, 157
acrylic acid, 155, 192
additives, 2, 3, 4, 5, 7, 8, 9, 10, 28, 30, 31, 32, 41, 46, 47, 48, 49, 50, 57, 91, 142, 164, 205, 224
adhesion, 154
adhesive interaction, 286
adsorption, 143, 144, 145, 154, 155, 173, 177
agglomeration, 86, 117, 135, 195
aggregation, 58, 62, 65, 93, 98, 100, 104, 105, 106, 117, 120, 121, 163, 177, 268
Al2O3 particles, 111
aluminum, 58, 85, 114, 179, 205, 278, 306, 314, 315
ammonium, 141, 317
anisotropy, 157, 166, 167
Annealing, 85
antibody, 176
anticancer drug, 156, 159, 172
apex, 280, 282, 284, 290, 291, 293
aqueous solutions, 3, 315
aqueous suspension, 54
argon, 152
aspartic acid, 155
atmosphere, 44, 45, 140
atmospheric pressure, 225
atomic force microscope, 356, 361
atoms, 2, 65, 140, 141, 158, 337
Au nanoparticles, 314
Avogadro number, 127

## B

bacillus, 174
bacterium, 141
base fluid, 117, 327
behaviors, 41, 48, 104, 106, 112, 140, 142, 146, 246, 278
Beijing, 272
benefits, 4, 334
benzene, 99, 209, 217, 224
biocompatibility, 143, 173, 175, 177
biological activities, ix, 139, 172
biological fluids, 174
biological processes, 144
biomarkers, 175
biomedical applications, 157, 180
biomolecules, 143, 175
biosciences, 173
biosensors, 139, 140, 144, 152, 160, 172, 180
bioseparation, 153
biosphere, 140
biotechnology, 173, 180
birefringence, 146, 148
bismuth, 278, 282, 284, 285, 286, 287, 288, 289
block polymers, 155
blood, 140, 174, 175, 176
boiling, v, 118, 242, 243, 275, 299, 300, 305, 306, 312, 324, 325, 326, 327, 328, 329, 368
Boltzman constant, 121, 231, 234
Bond number, 346, 349, 368
bonding, 141, 143, 144, 147, 160, 175
bonds, 141, 144, 145, 146, 155, 163
bone marrow, 174
Boundary condition, 206, 207, 208, 209, 210, 211
breakdown, 311, 316
breast cancer, 177
breast carcinoma, 176
Brinkman model, 60
Brownian diffusion, ix, 31, 32, 80, 201, 219, 229, 231, 232, 233, 234, 235, 236, 335
Brownian motion, 31, 33, 41, 61, 62, 64, 65, 66, 70, 71, 72, 79, 80, 97, 98, 100, 102, 106, 107, 108,

112, 120, 121, 122, 133, 145, 228, 229, 230, 231, 234, 236, 237, 334, 335, 337
Brownian particle, 127, 133
Bubble, 275, 278, 289, 291, 292, 297, 324, 326, 327, 328
Bubble growth, 292
building blocks, 140, 147
burnout, 111, 311, 316
butyl methacrylate, 154

## C

calibration, 11, 16, 17, 18, 20, 26, 30, 45, 253
cancer, 155, 157, 159, 174, 175, 176, 180
capillary, 278, 318, 331, 332, 340, 342, 345, 348, 353
carbon nanotubes, 57, 58, 100, 220, 224, 236, 315, 338, 339
carboxyl, 173
carboxylic acid, 155
carboxylic groups, 286
carcinoembryonic antigen, 176
catastrophic failure, 111
cation, 139, 146
cavity, 21, 114
cellulose, 158
ceramic, 143, 160, 161, 225, 249, 264, 339
ceramic materials, 339
chain transfer, 154
charge density, 59, 145, 148, 150
chemical bonds, 145
chemical reactions, 15
chemical stability, 174
chemical vapor deposition, 4, 93
chemotherapeutic agent, 175
chemotherapy, 172, 175, 180
CHF, 111, 224, 242, 243, 326, 329
China, 53, 86, 199, 245, 272, 368
chitosan, 186
chlorine, 160
chromatography, ix, 139, 172, 173
circulation, 263, 349
clinical application, 175
clinical diagnosis, 173
clinical trials, 177
clustering, 3, 4, 7, 31, 41, 49, 58, 102, 103, 120, 229, 335, 337
cluster(s), 4, 97, 100, 102, 134, 163, 180, 337, 354
CMC, 61
coatings, 1, 50, 142, 180
cobalt, 169, 177
collisions, 64, 70, 80, 124, 166
colloid, 54, 84, 88, 114, 117, 185, 191, 192, 194, 195, 196, 197, 198, 324, 325, 326, 327
combined effect, 105, 167
combustion, 152
commercial, 4, 5, 46, 106, 140, 141, 163, 170, 178
common fluids, 1, 8, 11, 49, 338
complexity, 3, 32, 42, 49, 50, 56, 74
composites, 68, 153, 161, 162, 224
composition, 98, 99, 152
compression, 341
computer, 18, 178, 180
concentration, 61, 114, 181, 185, 206, 207, 208, 209, 210, 211, 240, 243, 279, 285, 327, 329, 357
condensation, 4, 5, 93, 341, 342, 344, 345, 356
conduction, 12, 21, 25, 27, 30, 31, 32, 33, 34, 37, 39, 40, 41, 53, 54, 66, 68, 87, 70, 99, 116, 120, 129, 135, 136, 224, 226, 235, 236, 333, 334, 335, 342
conductivity, 3, 4, 5, 8, 10, 11, 12, 13, 14, 15, 16, 18, 19, 20, 21, 22, 23, 24, 25, 26, 27, 28, 29, 30, 31, 32, 33, 34, 37, 38, 40, 41, 44, 46, 47, 49, 51, 54, 56, 57, 58, 60, 61, 64, 65, 66, 68, 74, 78, 81, 82, 102, 103, 106, 108, 111, 112, 114, 116, 119, 120, 121, 122, 123, 124, 125, 126, 127, 130, 133, 139, 215, 216, 217, 218, 220, 221, 222, 224, 225, 228, 229, 230, 232, 233, 236, 245, 246, 247, 248, 251, 337, 338, 339, 346, 347, 348, 354, 355, 356, 357, 359, 367
conductor, 179
conference, 85, 118
conjugation, 176
conservation, 67, 75, 76, 77, 79
construction, 13, 122, 249, 263
contaminated water, 153
contamination, 178
convection, 86, 89, 90, 114, 199, 201, 202, 205, 225, 226, 227, 241, 242, 271, 329, 333
convection model, 120
COOH, 143, 155, 157, 158, 176
cooling, 1, 4, 7, 10, 24, 42, 49, 52, 65, 155, 202, 203, 204, 205, 216, 219, 222, 225, 237, 245, 246, 247, 248, 249, 250, 252, 253, 255, 257, 259, 261, 263, 264, 265, 266, 268, 270, 299, 331, 332, 334, 349
copolymer(s), 154, 155, 156, 157, 158, 161, 177
copper, 5, 16, 17, 23, 24, 25, 33, 37, 39, 40, 44, 93, 97, 99, 104, 105, 111, 117, 144, 210, 217, 221, 224, 225, 236, 249, 263, 277, 306, 314, 315, 339, 349, 350, 354, 362, 363
core, 143, 182, 183, 186, 190, 194, 195, 197
correlation, 36, 45, 62, 67, 75, 76, 78, 82, 98, 119, 122, 124, 125, 126, 133, 150, 204, 214, 215, 218, 219, 223, 232, 233, 261, 262, 270
corrosion, 15, 23
cosmetics, 140

cost, 5, 13, 57, 82, 173, 177, 178, 247, 332, 338, 340
Critical heat flux, 308, 310, 312
critical state, 368
critical value, 108, 142, 171
criticism, 30, 49
Cross-section, 15, 264
CRP, 154, 155
crystal structure, 225
crystalline, 14, 335, 337
cyclodextrins, 157
Cylinder, 54
cytotoxicity, 154, 175

## D

damping, 179
Darcy friction factor, 45
data processing, 266
decomposition, 152, 159
deformation, 140, 341, 342, 346
degradation, 222
deposition, 6, 57, 111, 178, 224, 278, 298, 300, 301, 303, 305, 306, 311, 314, 322, 323, 331, 354, 356, 364
deposition rate, 6, 311
depth, 24, 151
derivatives, 142, 146, 155, 160, 172, 179
desorption, 173
destruction, 172
detachment, 289, 292, 299, 300, 306
detection, 139, 161, 172, 173, 176
deviation, 13, 19, 78, 110, 130, 154
diamonds, 178
dielectric constant, 64
dielectric permittivity, 150
differential scanning calorimeter, 110
diffusion, 31, 32, 50, 62, 65, 70, 80, 83, 88, 99, 124, 130, 136, 145, 159, 175, 181, 198, 201, 219, 229, 230, 231, 232, 233, 234, 235, 236, 335, 337
diffusiophoresis, ix, 31, 201, 233, 235
diffusivity, 12, 85, 119, 121, 122, 133, 229, 236, 251
dilute gas, 121
dimethacrylate, 160, 194
dipoles, 142, 149, 166
direct adsorption, 143
discs, 153, 156
diseases, 176
dispersion, 4, 5, 6, 46, 57, 58, 59, 60, 61, 93, 98, 99, 104, 105, 106, 124, 140, 141, 142, 144, 145, 148, 153, 157, 158, 163, 164, 229, 233, 236, 249, 338
distilled water, 18, 19, 20, 26, 27, 30, 32, 41, 45, 46, 47, 48, 98, 213, 217, 221, 250, 261
distribution, 4, 6, 7, 13, 24, 32, 33, 34, 35, 37, 40, 41, 44, 64, 99, 105, 145, 147, 148, 150, 154, 156, 158, 165, 167, 170, 171, 177, 202, 225, 232, 246, 276, 277, 278, 298, 300
DNA, 3, 51, 173, 174, 186, 196, 197
dopamine, 155, 176
dopants, 176
dosage, 142, 158, 163, 172
drainage, 32, 201, 233
drug delivery, 139, 144, 152, 158, 161, 172, 174, 175
drug release, 159, 175
drug toxicity, 177
drugs, 157, 159, 175, 176
dyes, 140
dynamic viscosity, 46, 47, 105, 120, 124, 127, 201, 355
dysprosium, 160

## E

Effective thermal conductivity, 40
Einstein equation, 214, 224
Einstein model, 66, 72, 104
electric field, 139, 162, 163, 164
electrical resistance, 25
electrolyte, 144
electromagnetic, 11, 139, 140, 141, 144, 146, 172, 176, 263
electromagnetic fields, 11
electromagnetic waves, 141
electron, 65, 88, 144, 298, 300, 333, 355, 361
electron microscopy, 298
Electron Paramagnetic Resonance, 182
electrophoresis, 173
embryonic stem cells, 174
emission, 65, 160
emulsion polymerization, 157, 158
emulsions, 4, 10, 140
encapsulation, 176
endothelial cells, 157
energy, ix, 2, 4, 50, 54, 59, 64, 65, 67, 75, 76, 77, 79, 85, 86, 95, 113, 114, 118, 127, 139, 141, 147, 149, 150, 163, 166, 168, 172, 176, 180, 184, 201, 222, 228, 229, 236, 245, 246, 255, 258, 260, 267, 326, 335, 337, 340, 348
energy transfer, ix, 2, 139, 172
engineering, 20, 82, 143, 144, 145, 147, 201, 301, 322
environment(s), 2, 50, 111, 174, 178, 221, 246, 346
enzyme, 173
epoxy groups, 155
equilibrium, 75, 76, 109, 110, 111, 112, 147, 235, 276, 278, 279, 281, 289, 292, 306, 307, 308, 350

equipotential surface, 177
erosion, 4, 56, 141, 334
erythrocytes, 173
ester, 155
ethanol, 61, 153, 278
ethylene, 2, 5, 6, 18, 19, 20, 56, 61, 95, 96, 97, 98, 99, 100, 102, 103, 104, 105, 143, 154, 160, 176, 177, 178, 183, 184, 190, 208, 220, 236, 245, 247, 249, 263, 264, 265, 266, 267, 268, 269, 270, 338, 339
ethylene glycol, 5, 6, 18, 19, 20, 56, 61, 92, 95, 96, 97, 98, 99, 100, 102, 103, 104, 105, 160, 177, 178, 183, 184, 208, 220, 236, 245, 247, 249, 263, 264, 265, 266, 267, 268, 269, 270, 339, 363
ethylene oxide, 143, 154, 190
evaporation, 5, 6, 7, 52, 157, 191, 276, 278, 300, 301, 314, 319, 325, 331, 338, 340, 341, 342, 346, 347, 361, 364, 367, 368
evidence, 64, 65, 99, 112, 278, 304, 306, 308
evolution, 140, 298, 299, 306
exchange rate, 229
exercise, 333
experimental condition, 130
exploitation, 180
exposure, 6, 100
external magnetic fields, 65
extraction, 173
extravasation, 175

## F

fabrication, 16, 85, 113, 117, 118, 151, 187, 196, 198, 200, 333, 338, 346
FEM, 33, 39, 40
ferric ion, 176
ferrite, 148, 149, 150, 177
ferroelectrics, 139, 172
Ferrofluids, 142, 146, 147, 148, 151, 162, 163, 172, 177, 178, 179, 181, 195, 196, 200
ferromagnetic, 142, 148, 367
fiber, 8, 264
film thickness, 165, 344, 345
films, 142, 147, 165, 177, 178
filtration, 173
flow, 41, 42, 43, 44, 46, 52, 53, 54, 85, 86, 87, 88, 89, 90, 116, 117, 118, 136, 197, 206, 207, 208, 209, 210, 211, 237, 239, 240, 241, 242, 243, 271, 272, 325, 326, 328, 329, 366, 367, 368
flow field, 232
Flow regime, 206, 207, 208, 209, 210, 211
fluctuations, 8, 61, 310
fluid, 3, 5, 6, 8, 10, 11, 13, 16, 17, 18, 19, 20, 21, 22, 24, 25, 27, 28, 30, 31, 32, 33, 34, 37, 38, 39, 41, 74, 79, 81, 93, 95, 96, 97, 98, 99, 100, 103, 104, , 122, 124, 125, 126, 127, 128, 129, 130, 133, 140, 141, 146, 147, 148, 150, 151, 162, 163, 164, 165, 166, 167, 168, 169, 170, 172, 173, 176, 178, 179, 211, 212, 213, 216, 217, 218, 219, 221, 222, 223, 224, 225, 235, 237, 268, 270, 277, 289, 299, 301, 306, 307, 318, 323, 346, 347, 348, 349, 350, 351, 353, 355, 356, 357, 359, 361, 362, 364, 366
fluorescence, 161, 172
folic acid, 176
force, 2, 33, 41, 59, 64, 65, 74, 76, 77, 79, 80, 81, 82, 145, 151, 162, 163, 164, 165, 179, 228, 235, 236, 275, 276, 277, 278, 279, 280, 283, 286, 288, 291, 292, 293, 294, 295, 297, 298, 300, 302, 303, 305, 311, 323, 338, 340, 341, 342, 346, 357
formula, 347
fouling, 318
fractal, 89, 116
fractal theory, 70
France, 51, 84, 117, 240, 241, 242
free energy, 127
free rotation, 169
freezing, 166, 167, 168, 348
friction, 8, 28, 41, 42, 45, 47, 48, 50, 64, 169, 206, 207, 208, 209, 210, 211, 237, 238, 252
Friction factor, 206, 207, 211

## G

gadolinium, 176
gastrointestinal tract, 176
gel, 152, 160
gene therapy, 152
genes, 174, 176
genomics, 180
geometry, 21, 31, 33, 39, 41, 56, 67, 202
glycerol, 61, 106, 107, 111
glycol, 2, 5, 19, 20, 61, 92, 97, 99, 100, 102, 103, 104, 105, 106, 143, 175, 176, 177, 178, 207, 220, 236, 264, 266, 268, 269, 339, 363
gold nanoparticles, 117, 278, 297
grain boundaries, 165
gravitational field, 140, 164
gravitational force, 79, 341, 346
gravitational pull, 120
gravity, 21, 22, 25, 32, 79, 163, 201, 202, 233, 235, 236, 237, 275, 277, 280, 282, 283, 288, 289, 323, 346, 348
growth, 145, 152, 153, 160, 161, 164, 275, 276, 278, 279, 292, 297, 298, 299, 306, 318, 352

## H

HaCaT cells, 174
Hamilton-Crosser model, 222
health care, 180
heat capacity, 12, 60, 61, 110, 130, 212, 230, 331, 333, 355
heat loss, 23, 24, 25, 27, 44, 225, 249, 264
heat removal, 201, 239, 246, 248, 300, 332
heat transfer, 1, 2, 8, 10, 12, 13, 19, 21, 23, 24, 27, 31, 32, 34, 37, 41, 42, 45, 47, 48, 49, 50, 51, 55, 56, 57, 58, 60, 61, 62, 63, 64, 65, 66, 67, 74, 78, 202, 203, 204, 205, 206, 207, 208, 209, 210, 211, 212, 213, 214, 215, 216, 217, 218, 219, 220, 221, 232, 233, 235, 236, 237, 239, 245, 246, 247, 248, 259, 260, 261, 276, 307, 308, 309, 310, 311, 312, 313, 314, 315, 316, 317, 318, 319, 320, 322, 323, 331, 332, 333, 334, 354, 355, 357, 359, 364, 366
heat transfer characteristics, 2, 3, 9, 10, 41, 42, 45, 50, 55, 62, 82, 205, 217, 225, 245, 247, 269
heat transfer coefficient, 257
heat transfer enhancement, 8, 32, 41, 48, 55, 56, 58, 60, 61, 62, 64, 65, 66, 79, 82, 98, 105, 216, 219, 224, 225, 228, 233, 245, 247, 248, 255, 261, 270, 364
heterogeneity, 32, 37
hexane, 153, 156, 167, 168
homogeneity, 288, 309
homogeneous model, 233
horizontal, 89, 325, 326, 327, 328
host, 66, 147, 148, 153, 168, 169, 171
hot spots, 318
hybrid, 2, 9, 10, 12, 50, 140, 141, 142, 144, 151, 157, 160, 161, 167, 177, 182, 189, 192, 193, 194, 198, 272
hybridization, 174
hydrazine, 158
hydrogels, 157
hydrolysis, 155
hydroxide, 141, 152, 177, 317
hydroxyethyl methacrylate, 158, 194
hydroxyl, 59, 148, 286
hyperthermia, ix, 139, 144, 155, 156, 159, 172, 176, 177, 180
hypothesis, 2, 283
hysteresis, 105, 112, 142, 176, 281, 309, 312
hysteresis loop, 142

## I

identity, 99
images, 144, 153, 156, 160, 161, 162, 175, 355, 360, 361, 363
immobilization, 172
immune system, 154
in vivo, 154, 161, 175, 189
independent variable, 67
India, 91, 139, 201, 225, 239
industry, 5, 10, 82, 179, 180, 201, 205, 246, 338
inertia, 31, 179, 201, 233, 234
inhomogeneity, 98
initial state, 162
initiation, 275, 276
insulation, 6, 11, 13, 14, 15, 24, 25, 227, 249, 263, 349
integration, 37, 345
integrin, 157
interface, 18, 102, 111, 140, 141, 143, 152, 229, 281, 335, 336, 337, 345, 346, 367
interfacial layer, 62, 116, 367
internalization, 175
intravenously, 174
ion-exchange, 152
ions, 144, 150, 157, 158, 177
iron, 111, 156, 158, 161, 176, 177
irradiation, 140, 143, 152
isolation, 172, 173, 174

## J

jet, 241, 245, 246, 249, 261, 263, 266, 271, 272, 273
jet impingement, 245, 249, 261, 271, 272
jet impingement cooling technique, 245, 246
Jordan, 181, 196, 199

## K

Kapitza resistance, 121
keratinocyte, 174
kerosene, 96, 165
kidney, 177
kinetics, 159

## L

Lagrangian approach, 78
laminar, 8, 9, 28, 41, 45, 50, 53, 63, 85, 86, 88, 89, 98, 115, 202, 203, 204, 205, 206, 208, 209, 212, 213, 215, 216, 219, 221, 222, 224, 227, 228, 229, 230, 234, 235, 236, 237, 238, 240, 242, 243, 272, 325
laplace equation, 281, 282, 289, 292
laser ablation, 93, 113

latent heat, 346
lead, 2, 13, 50, 67, 144, 151, 169, 171, 180, 201, 229, 232, 246, 286, 319
lecithin, 141
length, 54, 83
ligand, 143, 144, 155, 156, 173
light scattering, 158
linear dependence, 102
liposomes, 177
liquid, 64, 87, 88, 89, 116, 136, 169, 177, 180, 181, 200, 205, 239, 240, 271, 272, 278, 279, 324, 325, 326, 327, 328
liquid crystals, 140
liquid interfaces, 141, 335
liquid phase, 143, 164, 168, 278, 292, 309, 348
liquids, 13, 56, 58, 91, 92, 93, 98, 103, 113, 119, 122, 129, 141, 152, 177, 202, 277, 284, 292, 334, 335, 339, 347
liver, 175, 177
low temperatures, 168
lower lip, 21
lubricants, 10, 338
luminescence, 139, 172, 173
lymph node, 175
lysine, 157, 191

## M

mAb, 176
macromolecules, 8
magnet, 142, 178, 179, 228
magnetic field, 65, 95, 100, 101, 102, 103, 112, 141, 142, 146, 162, 163, 164, 169, 170, 171, 172, 173, 174, 175, 176, 178, 179, 209, 227
magnetic moment, 149, 150, 162, 163, 166, 167, 169, 170
magnetic particles, 103, 169, 228
magnetic properties, 147, 174, 179
magnetic relaxation, 170, 175
magnetic resonance, 139, 144, 161, 172, 175
magnetic resonance imaging, 139, 144, 172
magnetic structure, 178
magnetism, 142, 149
magnetization, 142, 149, 152, 162, 167, 168, 169, 171, 173
magnitude, 22, 57, 122, 129, 142, 148, 229, 236, 245, 246, 247, 303, 304, 320
Magnus effect, 32, 201, 233, 235, 237
fluid drainage, 201, 233
MALDI, 196
malignant tumors, 176
management, 1, 4, 10, 49, 112, 245, 246, 247, 332
manganese, 176
manipulation, 51, 162, 172, 173
manufacturing, 4, 47, 58, 140, 201, 322
mass, 2, 64, 67, 75, 76, 79, 96, 141, 153, 163, 179, 212, 214, 216, 221, 222, 224, 247, 252, 300, 304, 308, 309, 310, 311, 316, 322, 334, 357
Mass flow rate, 83
materials, 2, 3, 4, 49, 56, 57, 110, 119, 120, 140, 141, 144, 165, 175, 178, 180, 201, 248, 249, 277, 279, 280, 283, 284, 288, 333, 338
materials science, 140
matrix, 68, 174, 204
Maxwell, 31, 33, 36, 37, 39, 40, 41, 49, 53, 56, 66, 68, 69, 70, 71, 83, 88, 107, 116, 120, 121, 130, 134, 204, 239, 241, 333, 367
measurement(s), 4, 10, 11, 13, 14, 17, 18, 19, 20, 21, 25, 27, 28, 30, 45, 49, 60, 61, 66, 94, 95, 112, 114, 122, 123, 215, 220, 222, 224, 225, 232, 288
media, 33, 41, 143, 180
medical, 10, 146, 160, 198, 201
medicine, 140
membranes, 177
memory, 173, 178
MEMS, 89
mercury, 177
metal carbides, 338
metal complexes, 176
metal nanoparticles, 4, 58, 224
metal oxides, 100
metals, 5, 144, 152, 153, 160, 177, 178, 224, 338
methacrylic acid, 154
microchannel, 90, 209, 241, 300, 326
microconvection, 87, 135
microemulsion, 193
microgels, 144, 157, 158, 159, 161
micrometer, 56, 158, 164, 232
microparticles, 2
Microscale, 54, 135, 239, 327, 368
microscope, 278, 292, 300, 355, 361
microscopy, 304
microspheres, 158, 161, 162
microstructure(s), 32, 98, 99, 112, 142, 150, 152, 164
microwave heating, 143
migration, 65, 98, 175, 217, 228, 229, 230, 231, 232, 237
Mitoxantrone, 199
mixing, 5, 6, 64, 87, 95, 108, 109, 110, 112, 228, 229, 264, 334, 338
mixture, 75, 86, 89, 116, 117, 134, 241, 243, 311, 313, 314, 315, 316, 317, 318, 319, 320, 367
MMA, 157
models, 1, 8, 31, 33, 50, 55, 66, 72, 74, 78, 103, 104, 109, 110, 120, 121, 130, 132, 144, 180, 204, 223, 232, 233

moderate magnetic, 139
modifications, 66, 129, 205, 239, 298, 308
molar volume, 101
molecular biology, 173
molecular dynamics, 64, 82, 180, 204
molecular liquids, 336
molecular structure, 2
molecular weight, 28, 154
molecules, 3, 7, 8, 64, 65, 70, 80, 122, 124, 127, 128, 129, 143, 144, 145, 146, 150, 158, 163, 171, 172, 174, 176, 276, 277, 279, 286, 322
momentum, 50, 64, 67, 75, 76, 77, 78, 79, 83, 141, 236, 289, 334
monoclonal antibody, 176
morphology, 62, 141, 143, 224, 298, 310
MRI, 144, 152, 155, 157, 175, 176, 182, 183, 188, 191, 196, 198, 199, 200
multiphase, 242, 328, 329
multiwalled carbon nanotubes, 100

## N

N, 52, 54, 83, 84, 85, 86, 87, 88, 89, 90, 111, 114, 115, 117, 118, 134, 135, 136, 155, 157, 158, 161, 169, 174, 177, 182, 183, 184, 185, 186, 187, 188, 189, 190, 191, 192, 193, 194, 195, 196, 197, 198, 199, 200, 239, 240, 241, 242, 272, 273, 324, 325, 327,328, 329, 366, 367, 368
NaCl, 14
nano-bubbles, 1, 2, 49
nanocomposites, 1, 2, 49, 142, 173
nanocrystals, 176
nano-droplets, 1, 2, 49
nanofibers, 1, 2, 49, 93, 248
nanofluid, 28, 51, 52, 54, 82, 84, 85, 86, 87, 89, 90, 100, 101, 104, 107, 108, 109, 113, 114, 115, 116, 117, 118, 119, 125, 134, 135, 184, 221, 237, 240, 241, 242, 245, 250, 253, 272, 273, 275, 277, 280, 292, 299, 300, 305, 308, 313, 315, 316, 317, 324,325, 327, 331, 333, 338, 347, 350, 351, 366, 368
nanolayers, 64, 336
nanomaterials, 2, 50, 162, 224
nanometer(s), 2, 55, 120, 141, 202, 334
nanometer scale, 334
nano-mixtures, 1, 30, 49, 50
nanoparticle(s), 1, 2, 3, 4, 5, 6, 7, 8, 10, 28, 31, 41, 42, 46, 47, 48, 49, 50, 56, 57, 58, 59, 60, 61, 62, 64, 80, 81, 82, 102, 103, 104, 105, 106, 108, 110, 111, 112, 119, 120, 121, 139, 140, 181, 184, 202, 205, 213, 214, 215, 216, 217, 219, 220, 221, 222, 223, 224, 225, 227, 228, 229, 230, 233, 234, 235, 236, 251, 252, 253, 255, 257, 258, 259, 277, 278, 279, 280, 282, 283, 284, 285, 300, 301, 302, 303, 314, 315, 316, 317, 318, 319, 320, 322, 349, 350, 352, 353, 354, 355, 356, 357, 359, 361, 362, 363, 364, 367
nanorods, 1, 2, 49, 102
nanoscale, 54, 84, 86, 87, 88, 89, 116, 135, 136, 239, 242, 243, 273, 327, 329, 368
nanostructures, 1, 10, 50, 182, 188, 191, 196, 199
nanotechnology, 51, 52, 56, 88, 139, 182, 186, 194, 198, 199, 202, 225, 245, 324, 325, 334
nanotube, 41, 46, 48, 92, 94, 99, 100, 111, 217, 224
National Bureau of Standards, 136
Newtonian fluid, 3, 8, 106, 232
NH2, 156
nickel, 24, 312
nitrides, 338
nitrogen, 6, 7, 42, 43
novel materials, 140
nucleation, 153, 275, 276, 299, 300, 301, 302, 303, 305, 306, 307, 312, 313, 314, 315, 318, 319, 351, 352, 353, 354, 355, 356, 362, 364
nucleic acid, 173, 176
Nusselt numbers, 259

## O

OH, 286
oil, 2, 5, 6, 8, 32, 50, 56, 63, 92, 97, 99, 100, 101, 102, 106, 107, 117, 156, 157, 158, 163, 171, 338, 339, 367
oleic acid, 146, 152, 153, 156, 157
one-step method, 6, 49
operating costs, 56
operating range, 45
optical properties, 147, 148, 165
optimization, 2, 6, 8, 50, 51
ordinary differential equations, 281, 282, 290
oscillating Heat Pipe (OHP), x, 331, 332
oscillation, 10, 103, 277, 340, 341, 346, 347, 348, 349, 357
ovarian cancer, 155
oxidation, 5, 136, 152
oxide, 53, 58, 85, 86, 113, 117, 134, 183, 184, 186, 188, 189, 190, 191, 192, 194, 195, 198, 200, 201, 205, 239, 240, 242, 243, 326, 367
oxide nanoparticles, 93, 104, 224
oxygen, 263

## P

PAA, 155, 197

parallel, 11, 20, 21, 22, 25, 28, 30, 31, 37, 38, 39, 41, 49, 103, 120, 164, 165, 209, 227
particle, 40, 65, 83, 86, 87, 89, 96, 101, 107, 116, 117, 118, 134, 135, 136, 171, 177, 187, 196, 197, 229, 239, 240, 243, 267, 362, 365, 368
particle morphology, 64
Peclet number, 61, 213, 215, 217, 228, 229, 231, 232, 233
penalties, 108
penetrability, 172
peptide, 155
percolation, 31
perfusion, 175
peripheral blood, 174
pH, 4, 7, 10, 28, 46, 59, 64, 67, 85, 93, 96, 97, 99, 102, 104, 108, 112, 115, 120, 135, 148, 151, 155, 185, 186, 195, 205, 217, 353
phonons, 65, 141, 147, 337
photons, 141, 147
physical characteristics, 1, 2, 12, 31, 50
physical phenomena, 50, 74, 82
physical properties, 4, 7, 51, 56, 64, 75, 100, 162, 172, 214, 275, 276, 277, 279, 282, 283, 284, 292, 295, 299, 322, 323
physics, 133, 139, 146
Planck constant, 127
platinum, 11, 14, 15, 16, 17, 18, 122, 123, 124, 209, 227
PMMA, 155
polar, 142, 144, 146, 147, 148, 152, 163, 177
polarization, 13, 69
polyacrylamide, 28
polycondensation, 155
polydimethylsiloxane, 107
polyisoprene, 155
polymer, 2, 8, 9, 10, 12, 28, 29, 30, 50, 139, 140, 143, 144, 145, 146, 154, 155, 156, 157, 158, 159, 160, 161, 163, 164, 173, 199, 224, 315, 338
polymer composites, 160
polymer molecule, 10, 143, 144, 145, 163
polymer solutions, 2, 8, 30, 338
polymerization, 154, 155, 156, 157, 158, 160, 161
polymers, 2, 8, 12, 28, 50, 140, 145, 154, 156, 157
polypeptide, 155
polystyrene, 24, 156, 157
polyvinyl alcohol, 174
Pool boiling, 224, 306, 315
Potential, 54, 151, 193, 195, 196, 198, 199, 200
Prandtl number, 78, 120, 203, 213, 221, 224
precipitation, 4, 114, 145, 148, 152, 157, 173
premature judgments, vii, 1, 2, 51
preparation, 55, 56, 57, 58, 60, 64, 82, 93, 95, 98, 154, 161, 180, 224, 249, 258, 269, 270, 331, 338, 339, 348, 349
pressure, 52, 83, 87, 206, 207, 208, 209, 210, 211, 225, 226, 237, 238, 240, 241, 260, 261, 268, 310, 324, 325, 326, 327, 329
pressure gauge, 226
pressure gradient, 228, 235
probe, 147, 149, 157, 168, 225
process control, 7
progenitor cells, 174
propagation, 27, 45, 119, 122, 127, 128, 129, 130, 131, 133, 311, 316, 337
proportionality, 234
propylene, 97, 103, 104, 175
prostate carcinoma, 177
protein immobilization, 144
proteins, 173, 174, 176
pumping power, 209
pure water, 95, 222, 224, 238, 248, 253, 261, 270, 309, 315, 316, 317, 320, 331, 350, 353, 354, 356, 357, 359, 360, 361, 362
purification, 173, 174
PVP, 28, 143, 158, 315
pyrolysis, 4, 152

## Q

quanta, 65
quantification, 173, 201
quartz, 354

## R

radial distance, 12
radiation, 7, 31, 65, 88, 164, 192
radical polymerization, 154, 155
radicals, 158
radiotherapy, 172, 180
radius, 13, 14, 15, 16, 17, 34, 36, 80, 122, 214, 216, 277, 278, 280, 281, 282, 284, 286, 288, 289, 291, 292, 295, 296, 297, 298, 306, 311, 345, 346, 352
reaction medium, 145, 158, 173
recombinant proteins, 173
reduction, 7, 8, 45, 53, 113, 187, 295, 307
reflectivity, 177
regression analysis, 130
relaxation process, 170
relaxation rate, 175
reliability, 42, 60, 82, 105, 227, 246, 332
replication, 11, 49

repulsion, 105, 112, 145, 146, 148, 150, 151, 163, 286
research laboratories, 201
researchers, 4, 8, 50, 58, 60, 62, 66, 82, 93, 94, 109, 113, 133, 224, 232, 233, 247, 280, 331, 334, 339, 340
resistance, 17, 18, 23, 24, 25, 37, 44, 56, 64, 99, 121, 123, 124, 218, 235, 249, 277, 306, 314, 340, 349, 351, 352, 354, 355, 356, 360, 363, 364, 365
response, 19, 155, 156, 172, 175, 178, 179, 307
responsiveness, 159
restrictions, 177, 279
rheology, 66, 139, 142, 147, 153, 162, 163, 172
ring magnet, 179
RNA, 174
room temperature, 97, 108, 111, 153, 166
root-mean-square, 356
roughness, 245, 275, 276, 277, 278, 288, 299, 300, 302, 303, 304, 305, 306, 307, 308, 309, 310, 312, 313, 314, 315, 316, 317, 318, 319, 320, 323, 356, 362
routes, 139, 146
Royal Society, 134

## S

safety, 154, 174
salmonella, 174
saturation, 111, 142, 146
Scanning electron microscope (SEM), 355
science, 82, 140, 172, 177, 179, 301, 323
sedimentation, 56, 59, 67, 74, 98, 99, 104, 112, 120, 141, 163, 202, 315, 334
seed, 153, 161
Seed-mediated, 152
self-ordering, 278
semiconductor, 178, 224
sensitivity, 140, 174, 176, 301
sensors, 25, 43, 45, 140, 156
serum, 155, 173
shear, 8, 46, 47, 62, 65, 72, 73, 80, 81, 103, 104, 105, 106, 163, 164, 169, 171, 217, 219, 220, 229, 230, 232, 235, 289, 292, 338, 341, 342, 347, 357
Shell, 143, 182, 183, 186, 190, 194, 195, 197, 198
silica, 28, 29, 41, 46, 47, 48, 104, 160, 161, 177, 197, 236, 356
silicon, 99, 218, 282, 284, 285, 286, 287, 288, 289
silver, 97, 100, 104, 106, 144, 177, 339, 357, 358, 359, 366
simulation(s), 64, 82, 180, 204, 335
Singapore, 55
single test, 26
SiO2, 46, 63, 96, 100, 101, 114, 160, 186, 199, 307, 308, 312, 318, 319, 320, 331, 353, 354, 355, 356, 366, 368
smart materials, 140
sodium, 99, 146, 158, 209, 217, 224, 228, 269, 316
software, 11, 18, 25, 30, 44
sol-gel, 160
solid phase, 120, 168
solid state, 140, 168
solid surfaces, 64
solution, 39, 135, 136, 180, 185, 187, 230, 282, 294, 353, 359
Solvent(s), 10, 190, 191
solvent molecules, 142
species, 144, 146, 147, 153, 155, 157, 158, 169, 171, 172, 173, 175, 235
specific gravity, 108
specific heat, 12, 66, 74, 91, 108, 109, 110, 112, 121, 122, 130, 205, 214, 221, 222, 251, 346, 347, 348, 356
specific surface, 102, 152, 174, 334
specifications, 93, 301
spleen, 175
sputtering, 52
stability, 22, 28, 55, 56, 57, 58, 59, 60, 82, 93, 95, 96, 97, 98, 99, 102, 105, 112, 139, 142, 144, 148, 149, 150, 152, 155, 156, 163, 164, 174, 178, 202, 225, 232, 249, 268, 310, 334
stabilization, 3, 4, 5, 47, 49, 93, 99, 105, 142, 145, 177, 180, 185, 200
stable ferrofluids, 139, 148, 151
starch, 173, 176
Static model, 31
steel, 23, 25, 43, 44, 218, 225, 277, 278, 298, 300, 302, 303, 304, 309, 311, 312, 316, 353
stoichiometry, 152
storage, 5, 45, 57, 93, 159, 338
strategic position, 7
stress, 15, 72, 81, 89, 106, 140, 219, 232, 243, 289, 292, 341, 342, 347, 357
strong interaction, 64
structure, 9, 10, 32, 47, 61, 99, 102, 108, 112, 141, 142, 143, 144, 145, 146, 153, 154, 157, 160, 161, 163, 164, 165, 167, 171, 174, 202, 205, 217, 335, 336, 340
styrene, 155, 157, 158, 161, 162, 174, 194
substitution, 35, 144
substrate, 46, 161, 275, 276, 277, 278, 279, 280, 281, 283, 284, 285, 286, 287, 288, 298, 299, 300, 301, 302, 303, 304, 305, 306, 307, 308, 311, 312, 313, 314, 315, 316, 317, 320, 323, 360, 361, 367
superparamagnetic, 142, 143, 160, 167, 177

surface, 59, 83, 110, 118, 139, 146, 153, 155, 181, 182, 183, 184, 185, 188, 189, 190, 191, 192, 195, 199, 200, 240, 242, 243, 263, 266, 268, 270, 271, 277, 303, 304, 309, 324, 325, 326, 327, 328, 359, 368
surface area, 7, 64, 80, 83, 120, 143, 161, 175, 246, 334, 355, 356
surface chemistry, 59, 143
surface energy, 10, 58
surface layer, 141, 142, 143, 146, 147, 148, 150, 155, 163
surface modification, 143, 154, 310
Surface roughness, 303, 304, 309
surface structure, 143
surface tension, 108, 110, 111, 113, 276, 277, 278, 279, 280, 281, 282, 284, 286, 287, 288, 290, 291, 292, 293, 294, 295, 297, 298, 299, 305, 309, 311, 314, 316, 319, 321, 323, 333, 346, 348
surface treatment, 307
surfactant(s), 1, 2, 3, 4, 7, 8, 49, 59, 67, 93, 99, 103, 110, 111, 120, 141, 142, 146, 147, 148, 152, 153, 154, 157, 158, 163, 170, 220, 224, 270, 316, 317, 338
susceptibility, 142, 168, 169, 175
suspension(s), 2, 5, 10, 11, 20, 30, 32, 41, 46, 49, 55, 56, 59, 65, 86, 88, 89, 93, 96, 99, 102, 104, 106, 108, 111, 120, 202, 204, 220, 239, 245, 250, 255, 259, 269
synthesis, 54, 85, 86, 87, 113, 114, 115, 139, 151, 152, 160, 180, 181, 182, 183, 184, 186, 187, 188, 189, 190, 192, 193, 194, 195, 198, 199, 200, 338

## T

tantalum, 14, 15
Taylor series, 72
techniques, 4, 10, 28, 56, 58, 60, 93, 94, 95, 105, 173, 179, 245, 246, 248, 338
technological developments, 180
technology, 2, 93, 140, 176, 179, 322, 333
tellurium, 24
TEM, 153, 156, 160, 349
temperature, 3, 4, 6, 7, 10, 11, 12, 13, 14, 17, 18, 19, 20, 22, 23, 24, 25, 26, 27, 31, 32, 34, 37, 40, 41, 43, 44, 45, 59, 60, 62, 64, 65, 67, 70, 71, 80, 95, 108, 112, 114, 118, 119, 120, 121, 122, 123, 124, 125, 126, 127, 129, 130, 133, 143, 149, 150, 152, 155, 156, 159, 164, 166, 168, 169, 173, 176, 178, 203, 205, 224, 226, 227, 229, 231, 232, 234, 236, 237, 246, 247, 249, 252, 259, 260, 263, 264, 266, 270, 300, 348, 349, 350, 351, 354, 358, 359, 362, 364, 365
Temperature dependence, 4, 125, 128, 169
tensile strength, 17
tension(s), 11, 16, 17, 110, 111, 112, 113, 276, 277, 278, 279, 286, 287, 292, 293, 295, 297, 299, 346
TEOS, 160
test data, 44
testing, 5, 19, 25, 30, 47, 49, 331, 348, 350
tetrahydrofuran, 161
theoretical approaches, 127
therapeutic agents, 175
therapeutics, 144
therapy, 139, 172, 175, 176, 177
thermal analysis, 12
thermal conductivity, 10, 12, 19, 29, 41, 58, 60, 83, 92, 95, 96, 97, 98, 100, 121, 251, 265, 333
thermal conductivy enhancement, viii
thermal decomposition, 156
thermal energy, 64, 149, 167, 170, 172, 332
thermal properties, 245, 248, 313, 348, 350
thermal resistance, 23, 37, 64, 94, 218, 306, 314, 331, 338, 340, 351, 352, 354, 356, 357, 359, 363, 364, 365
thermal stability, 142
thermodynamic method, 145
thermodynamics, 135, 326
thermophoresis, ix, 31, 32, 64, 100, 201, 219, 228, 229, 233, 234, 235, 236, 237
tissue, 172, 175
titanate, 102, 105, 208, 220
titania, 102, 208, 220, 236
titanium, 144, 178, 225
toluene, 158, 339
topology, 140, 142, 176
toxic effect, 175
toxicity, 177
transfer performance, 48, 56, 60, 61, 78, 82, 201, 245, 246, 247, 248, 267, 364
transition metal, 144
transition temperature, 168
transmission electron microscopy, 98
transport, 31, 32, 50, 64, 65, 66, 95, 96, 102, 110, 172, 174, 176, 202, 205, 228, 235, 236, 245, 247, 255, 258, 260, 269, 313, 331, 332, 333, 334, 335, 337, 340, 342, 347, 348, 351, 354, 355, 357, 359, 364, 366, 367
transportation, 4, 5, 57, 93, 338, 340
treatment, 8, 58, 59, 60, 62, 64, 67, 99, 153, 176, 225, 312
Triple line, 277
Tubes, 52, 86, 90, 118, 327
tumor, 157, 161, 172, 175, 176, 177
turbulence, 8, 233, 245, 246, 247, 255, 259
turbulent, 47, 48, 53, 63, 85, 86, 87, 89, 208, 210, 211, 240, 241, 243, 271

turbulent flows, 28, 50, 207, 238
two step method, 57, 58, 93
Two-phase flow, 82

## U

United Kingdom (UK), 83, 194, 226, 275
ultrasound, 44
umbilical cord, 174
urethane, 155
United States (USA), 83, 185, 188, 200, 242, 243, 248, 366, 368
user-interface, 45
USSR, 200

## V

vacuum, 5, 6, 159, 178, 367
valve, 45, 226, 249, 359
vapor, 5, 57, 93, 111, 276, 299, 300, 306, 314, 318, 319, 331, 332, 340, 341, 342, 344, 345, 346, 347, 352, 353, 357
vapor bubble, 300
velocity, 3, 8, 31, 44, 65, 71, 75, 76, 79, 80, 81, 100, 119, 120, 121, 122, 124, 125, 127, 128, 129, 130, 133, 202, 218, 221, 223, 224, 229, 230, 232, 233, 234, 235, 246, 252, 255, 256, 257, 260, 266, 267, 297
vibration, 4, 65, 124, 127, 337
vinylidene fluoride, 143, 184
viruses, 174
viscosity, 3, 4, 8, 9, 28, 29, 30, 46, 60, 61, 65, 66, 72, 74, 75, 78, 80, 81, 91, 99, 100, 103, 104, 105, 106, 107, 108, 112, 114, 117, 120, 126, 139, 142, 146, 162, 163, 164, 166, 169, 170, 171, 179, 202, 204, 205, 214, 215, 216, 217, 218, 220, 221, 224, 228, 229, 230, 232, 236, 237, 251, 252, 255, 260, 266, 269, 276, 292, 299, 309, 314, 316, 319, 321, 323, 333, 346, 347, 348, 357, 359, 364
vision, 177, 303
visualization, 352, 354

## W

wall temperature, 44, 45, 98, 209, 213, 218, 222, 225, 227, 229
waste, 82, 140, 246
water, 2, 5, 8, 10, 19, 20, 24, 28, 33, 37, 39, 40, 45, 46, 47, 48, 56, 61, 63, 65, 72, 73, 74, 93, 94, 95, 96, 97, 98, 99, 100, 101, 102, 103, 104, 105, 106, 226, 227, 234, 235, 236, 237, 238, 240, 245, 247, 248, 249, 250, 251, 252, 253, 254, 255, 256, 257, 258, 259, 260, 261, 263, 264, 265, 266, 267, 268, 269, 270, 277, 278, 282, 284, 285, 286, 292, 308, 319, 320, 331, 333, 338, 339, 340, 349, 351, 353, 354, 355, 356, 357, 359, 361, 362, 364, 366
weak interaction, 64
wettability, 113, 275, 276, 277, 278, 282, 283, 284, 287, 288, 295, 300, 302, 303, 304, 305, 306, 307, 308, 309, 312, 314, 316, 317, 318, 319, 320, 322, 323
wetting, 64, 275, 276, 277, 278, 279, 286, 289, 297, 299, 300, 302, 306, 323, 331, 333, 337, 356
Wheatstone bridge, 17, 18
wires, 11, 15, 16, 17, 23, 250, 307, 312
working conditions, 205

## X

X-Ray, 87
X-ray diffraction, 98

## Y

yield, 2, 4, 5, 50, 56, 106, 160, 203
Young-Laplace equation, 280, 281, 282, 284, 289, 290, 292, 293, 294

## Z

zinc, 178, 309, 317
ZnO, 61, 85, 96, 98, 101, 106, 107, 111, 117, 118, 183, 304, 313, 316